ECOLOGY AND CONSERVATION OF THE SAN PEDRO RIVER

ECOLOGY AND CONSERVATION OF THE SAN PEDRO RIVER

Edited by JULIET C. STROMBERG AND BARBARA TELLMAN

Foreword by W. James Shuttleworth

The University of Arizona Press
Tucson

The University of Arizona Press

www.uapress.arizona.edu

Library of Congress Cataloging-in-Publication Data

Ecology and conservation of the San Pedro River / edited by Juliet C. Stromberg and Barbara Tellman.
p. cm.
Includes bibliographical references and index.
ISBN 978-0-8165-2752-6 (cloth : alk. paper)
1. Riparian ecology—San Pedro River Valley (Mexico and Ariz.)
2. Geomorphology—San Pedro River Valley (Mexico and Ariz.)
3. Hydrology—San Pedro River Valley (Mexico and Ariz.)
4. Ecosystem management—San Pedro River Valley (Mexico and Ariz.)
I. Stromberg, Juliet C. II. Tellman, Barbara.
QH102.E282 2009
577.68—dc22 2008052971

Publication of this book is made possible in part by the proceeds of a permanent endowment created with the assistance of a Challenge Grant from the National Endowment for the Humanities, a federal agency.

Manufactured in the United States of America on acid-free, archival-quality paper containing a minimum of 30% post-consumer waste and processed chlorine free.

14 13 12 11 10 09 6 5 4 3 2 1

Contents

Foreword

Arid and semi-arid lands comprise one-quarter of the land surface in the United States and one-third of the land surface on Earth and include the most rapidly growing counties and U.S. states. Population pressures lead to increasing water demand, potentially leaving water resource managers reliant on nonsustainable pumping of groundwater supplies and resulting in overdraft of regional aquifers. Climate change and variability and changing vegetative cover and land use changes complicate management of water resources. As populations grow, competing uses and increased/changing water demands challenge water managers to meet demand while sustaining biodiversity and maintaining healthy riparian ecosystems.

The mission of the National Science Foundation Science and Technology Center for Sustainability of Semi-arid Hydrology and Riparian Areas (SAHRA) is to promote sustainable management of water resources by conducting water resources–related science, outreach, and education in the context of critical water management issues of semi-arid and arid regions. SAHRA carries out coordinated research relevant to improving water resource management in semi-arid catchments, focusing on research questions that can be addressed only by the sustained efforts of scientists of several disciplines working together. To maximize the benefit of our work, we coordinate our research with that of other researchers and with the requirements of the local water professionals and consumers who are our stakeholders. SAHRA selected the upper San Pedro River basin as a focus area when the center was first established and is building on research generated by past programs. This volume reviews the collective progress made by investigators in the basin, including recent efforts by SAHRA scientists and partner researchers.

The upper San Pedro River basin sits at the ecotone between the Sierra Madre Mountains to the south, the Rocky Mountains to the north, the Sonoran Desert to the west, and the Chihuahuan Desert to the east. The basin is one of the most ecologically diverse areas in the Western Hemisphere and contains numerous different biotic communities and supports several endangered plant and animal species. Groundwater largely sustains the system throughout its length during dry seasons. The San Pedro River is one of the last free-flowing perennial streams in the American Southwest and serves as an international flyway for more than 400 species of birds, and 60 km of riverine territory north of the U.S.-Mexico border is designated as a national conservation area. Because of continued population growth in the basin, there is great concern over the long-term viability of the San Pedro riparian system as an ecological resource. In addition, any decline in water availability will exacerbate the increasing competition for water resources between productive sectors, such as agriculture and industry, and domestic consumption.

Water resource management in the upper San Pedro River basin is complicated by the fact that it lies across the U.S.-Mexico border, with different laws governing water management and allocation in the two portions of the basin. However, from the perspective of SAHRA, the interaction between science and key stakeholders is enhanced by the growth and mobilization of stakeholder organizations, such as the Upper San Pedro Partnership in the U.S. portion of the basin, whose membership includes both water-related professionals and research scientists from SAHRA and elsewhere. This volume includes the results of research studies which address not only the complex scientific challenges associated with better understanding the issues regarding water quantity and quality and ecological diversity but also the intricate legal and multifaceted water management issues within the basin.

In editing this volume, Drs. Stromberg and Tellman have provided a comprehensive review of the contextual background and research activity in the San Pedro River basin. The contributions they have selected are written not only to meet the highest standards of professional scientific reporting but also to describe the relevance and applicability of their academic content. The 10 chapters in the first two sections provide detailed information on the extent and nature of the biological diversity within the basin and on current understanding of the interface between biological and physical systems and of associated interface processes. The eight chapters in sections 3 and 4 provide a wide-ranging description of the physical environment of the San Pedro River basin and are necessary reading for anyone seeking to understand the hydrology and geomorphology of this basin. These 18 chapters together provide an up-to-date description of the riparian ecosystem in the San Pedro River basin that is unique in terms of its completeness and depth of understanding. These chapters provide understanding that is relevant not only within the basin, but understanding that also has broader regional context.

Section 5 is somewhat unusual in a volume reviewing scientific progress in that it describes "science in action." Component chapters in this section describe relevant legal and policy issues and their interface with scientific understanding; activities related to mitigation, conservation, and restoration; and a prognosis of the potential for sustaining the basin's riparian ecosystem. The San Pedro River basin in general, and the upper San Pedro River basin in particular, is not just a location where separate experimental studies are being conducted but also provides a setting for scientific investigation in a broader context: an experiment in the process and mechanics of using science in support of societal needs, in which the feasibility of ecosystem preservation in a semi-arid basin can be fully explored, taking into consideration the many conflicting demands for water resources. SAHRA seeks to motivate and establish this broader multidisciplinary approach involving not only the physical and biological disciplines but also the social science disciplines to promote integrated basin management. The San Pedro River basin is a worthy prototype of the application of this approach, and this volume is an excellent prototype for the synthesis of the resulting understanding.

Professor W. James Shuttleworth
Director, NSF Science and Technology Center for Sustainability of Semi-arid Hydrology and Riparian Areas (SAHRA)

Acknowledgments

Many individuals provided reviews of chapter drafts: Edie Allen, Douglas Andersen, Greg Auble, Victor Baker, Rebecca Brown, Bala Chaudhary, Gail Cordy, Elizabeth Corley, Clifford Dahm, Joe Feller, Deborah Finch, Erica Fleishman, Jim Fogg, Julia Fonseca, Mark Fredlake, Edward Glenn, Michael Gooseff, Christine Hass, James Hogan, Andrew Keller, Jeffrey Kelly, William Kepner, Lin Lawson, James Leenhouts, Michael McClain, Steven McLaughlin, David Merritt, Mark Nelson, Waite Osterkamp, Steven Overby, Duncan Patten, Don Pool, Tom Sheridan, Andrew Smith, Dale Turner, Peter Unmack, Ruth Valencia, H. Maurice Valett, and Margaret White. The book as a whole benefited from three anonymous reviewers. For crafting figures, assisting with illustrations, or providing images, we thank Philip Amorosi, John Callahan, William Kepner, James Leenhouts, Lainie Levick, Robert Luce, Kerrith McKay, Alexander Smith, Dale Turner, Carl Unkrich, Robert Webb, Mara Weisenberger, David Wood, and their organizations, which include the Arizona Department of Environmental Quality, The Nature Conservancy, U.S. Department of Agriculture–Agricultural Research Service, U.S. Environmental Protection Agency, and U.S. Geological Survey; special thanks go to Mike Buffington. Many people, including Kathleen Blair, Bill Childress, Troy Corman, Mark Fredlake, David Hall, Robin Llewellyn, Peter

Mayne, Shawn Sartorius, Martin Schlaepfer, Cecil Schwalbe, Jeff Simms, and Eric Wallace provided valuable information to individual chapter authors. We thank all of these people. Also, we are grateful to the National Science Foundation Science and Technology Center for Sustainability of Semi-arid Hydrology and Riparian Areas (SAHRA) for assisting with publication costs and to Allyson Carter of The University of Arizona Press for her editorial guidance and support.

JULIET C. STROMBERG
AND BARBARA TELLMAN

Introduction

Why Another Book?

Southwest rivers and their historical changes have been described by many authors (Rea 1983, Brown 1994, Logan 2002, Baker et al. 2004, Webb et al. 2007). There are books specific to the San Pedro River, including a popular guide for hikers (R. Hanson 2001) and a scholarly study on alternate scenarios for the future of the area (Steinitz et al. 2003). In addition, there are myriad book chapters, journal articles, and conference proceedings on San Pedro River hydrology, ecology, hydro-politics, and science-policy connections (e.g., Glennon 2002, Browning-Aiken et al. 2006, Webb and Leake 2006). These contributions are all valuable. We saw a need, however, for a multi-disciplinary treatise on scientific pattern and process, historical and ongoing change, and conservation efforts for this particular river.

As is typical of desert rivers, fluctuating conditions are the norm on the San Pedro. However, some long-term stream gages along the San Pedro River show declines in streamflow in recent decades, raising concerns about long-term sustainability of the aquatic and riparian ecosystems. Many land use changes are ongoing: urbanization and associated water development are expanding in certain reaches of the river (Steinitz et al. 2003), while in others irrigated agriculture is being retired. Extensive areas along the river have been designated as federal or private riparian preserves, but streamflow is affected not only by local actions but also by actions throughout the watershed. Stakeholders along the river are striving to meet the challenge of maintaining or restoring streamflows for riparian and aquatic ecosystems while also sustaining thriving human communities. Scientists are actively studying the linkages between abiotic and biotic elements within the riparian ecosystem and the watershed, with an emphasis on interrelationships between stream hydrology and riparian biota.

As one of the last undammed (mostly) perennial rivers in the desert Southwest, the tale of the San Pedro is timely. As a regional reference river, it can reveal ecological lessons that inform efforts to restore physical processes and biotic communities to more heavily regulated rivers in the Southwest. Additionally, the condition of its riparian ecosystems may be the canary in the coal mine with respect to sustainable water use; time will tell if thriving riparian communities and expanding urban-agricultural centers can co-exist and how transitions from agricultural to residential and conservation-based landscapes will influence river condition. The extensive scientific research that has been conducted on the San Pedro, in concert with the active stakeholders who are implementing science-driven management, seemed a story begging to be told.

Our goal was to describe the ecological patterns and processes of the San Pedro River and to explore the ongoing science-driven efforts to sustain and restore its riparian ecosystems. We were motivated by several factors. Given our respective professional backgrounds as ecologist and environmental historian, we want river management efforts to be solidly grounded in scientific knowledge and believe that a sense of a river's history is critical for anticipating its future. Another underlying motive—to aid efforts to sustain or restore the river's flows and its riparian biota into the future—reflects our personal backgrounds and biases.

We tapped into the expertise of the biologists, ecologists, geomorphologists, historians, hydrologists, lawyers, and social and political scientists who have studied the San Pedro River in recent years to provide a broad perspective. We envision this edited volume as a resource for college students in interdisciplinary courses in riparian ecology, eco-hydrology, ecosystem management, and environmental sustainability, as well as for stakeholders interested in river ecology and conservation. We also hope it will foster more communication and understanding among scientific disciplines.

What Are Riparian Ecosystems?

Riparia riparia, the bank swallow, owes its name to its predilection for roosting along river banks. The term "riparian" can refer to the narrow strip of land immediately adjacent to a river channel but also is used more broadly to encompass all of the zones, including banks, floodplains, and terraces, that are influenced by surface and/or subsurface river flows or by lake water (National Research Council 2002, Naiman et al. 2005). Wetlands are nested within those portions of riparian zones that have a year-round, ample supply of water.

Riparian ecosystems also can be envisioned as interfaces between aquatic and terrestrial ecosystems (Gregory et al. 1991, Naiman and Décamps 1997). As such, they are shaped by terrestrial processes, such as rainfall, and by aquatic processes, such as streamflows. Often, a spatial gradient exists in the degree of reliance of the biota on river flows, with plants and animals becoming increasingly dependent on rainfall versus surface water or groundwater with increasing distance from the river channel. The zone of influence of the river can extend for miles in the case of animals that have large home ranges and periodically use the river's resources. Although riparian ecosystems occupy a very small portion of the land area in dryland regions, they are hotspots of productivity and diversity and exert great influence on biota throughout the watershed.

"Nothing Is Permanent Except Change" (Heraclitus)

Dryland rivers have some of the most variable flow regimes in the world, as wet periods alternate with dry periods, river channels widen and contract, water levels peak and recede, and vegetation waxes and wanes. Dryland rivers also receive and transport large amounts of sediment from the watershed, producing an ever-changing geomorphic template. Although physical processes are paramount in shaping dryland riparian ecosystems, the biota also exert influence. Plants transpire water, insects pollinate plants, and birds feed on insects. Beavers build dams, and people build larger ones.

Over the ages, people have harnessed rivers to provide water, food, and minerals, with the extent of these manipulations increasing with the rise of engineering technology (Josephson 2002). This has allowed vast cities to develop in desert regions seemingly adverse to human life, but the associated water and land uses have modified these freshwater ecosystems (Tellman et al. 1997, Patten 1998, Stromberg et al. 2004, Webb and Leake 2006). In the desert Southwest, humans have shaped rivers such as the San Pedro first as hunters and gatherers, then as small-scale irrigating farmers and miners, often heavily dependent on fuelwood, and finally as industrialized farmers, miners, and city dwellers. People have knowingly or unwittingly influenced the riparian ecosystem by altering resource levels (water and nutrients) and disturbance regimes (flood and fire patterns) and by re-shuffling biotic communities (e.g., introductions of herbivores, extirpation of predators). Concurrently, climate fluctuations produce ongoing change in the river's form and function. However, the very unpredictability of streamflows in dry regions, over time, has produced ecosystems with high resilience. Despite having undergone extensive change, the San Pedro River today sustains productive and diverse biotic communities.

The San Pedro River in Context

In the greater scheme of things, and even within the American Southwest, the San Pedro is not a very large river. Years ago when the river was visited by the San Pedro Expert Study Team, a group appointed by the tri-national Commission for Environmental Cooperation to assess riverine policy strategies, a team member from the Pacific Northwest exclaimed, "This is it?!" Indeed, during the dry season one can easily walk or wade across the channel in most places. Although small, this bi-national dryland river, which spans portions of the Chihuahua and Sonoran Deserts, has high scientific importance and conservation value. And many watershed groups are looking to the San Pedro as a model for river-protection efforts.

The San Pedro River has been recognized by many conservation organizations as having natural heritage values of global significance (USBLM 1990, The Nature Conservancy 1991, CEC 1999a, 1999b, American Bird Conservancy 2003). The river supports long stretches of Sonoran cottonwood/willow riparian forest, a rare woodland type (Noss et al. 1995), as well as several rare ciénegas (a type of wetland), sacaton grasslands, and mesquite bosques. It serves as a migratory corridor of international importance for neotropical birds, bats, and insect pollinators (Krueper 1996a). The American Bird Conservancy named the San Pedro as a Globally Important Bird Area because it provides important stopover and nesting habitat for neotropical migrants (birds that nest in temperate zones and overwinter in the tropic); it also provides nesting habitat for many year-round residents. Terrestrial insects have not been exhaustively catalogued, but about 100 species of butterflies and 250 species of bees have been recorded along the San Pedro River, some only recently described. The San Pedro's communities of mammals and herpetofauna are diverse, with about 60 species each. It has a known flora of more than 750 species (with some areas yet to be botanized) and harbors 2 endangered plant species in its watershed. In part because of this high diversity, The Nature Conservancy deemed it one of the "Last Great Places" in the Northern Hemisphere.

These biotic communities are sustained by the river's flow regime. The San Pedro retains a more natural flow regime than many desert rivers, most of which have multiple large dams (Poff et al. 1997, Graf 1999). Some reaches are influenced by small diversion dams, but plans to build a flow-regulating dam in the Charleston Hills area never materialized. Its streamflows thus reflect climatic and hydrologic processes occurring throughout the watershed. Seasonal rains produce pulses of surface runoff and river flooding, and

Fig. 0.1. The San Pedro River near Charleston, April 24, 1950. Photo by the Miller Expedition during their research on fish of the Southwest (see R. R. Miller 1961).

groundwater inflowing from the regional aquifer contributes to dry-season streamflows. These inputs, interacting with local geology, create a spatially intermittent river wherein perennially flowing reaches alternate with reaches with more intermittent flow and deeper water tables. Limited portions flow through narrow canyons, but over most of its length the San Pedro roams broadly within wide, alluvial floodplains.

History of Scientific Inquiry

There is a long history of scientific inquiry and data collection in the San Pedro basin (figs. 0.1,0.2). U.S. and Mexican boundary surveys during the mid-1800s collected fishes in the San Pedro that were new to science, and botanists in the late nineteenth century collected plant specimens from the upper San Pedro River. Mearns (1907) reported on mammals of the boundary, and a Smithsonian Institution entymologist conducted studies in San Pedro tributaries. Many ornithologists described the birds, including Swarth (1905) in the early and Brandt (1951) in the mid-twentieth century. Photographs of the border and of southern Arizona taken prior to 1900 have been repeated through time (Humphrey 1987, Turner et al. 2003). Physical science studies, too, have a long history; riparian lands were surveyed in the late 1800s, and some stream gages have been in existence for over 100 years. Hydrology studies intensified during the mid-twentieth century, conducted, in part, through the USDA Walnut Gulch Experimental Watershed, the University of Arizona, and the U.S. Geological Survey. In the past 20 years, the pace of scientific inquiry,

Fig. 0.2. Line drawing of the regal horned lizard (*Phrynosoma solare*) by Spencer Baird (originally referred to as *Phrynosoma regale*). Plate XXVIII in *Reptiles of the Boundary* (Emory 1859).

especially in the Sierra Vista subbasin, has accelerated rapidly. Because of the long history of scientific inquiry, a critical baseline exists in many areas.

The San Pedro's Environmental Setting: Overview

The San Pedro River lies within the Basin and Range Physiographic Province (Hunt 1967, Dohrenwend 1987). The province is typified by bedrock mountains bounded by the faults along which they rose and surrounded by the sediments that eroded from them through time. Within the province is an area called the Sky Islands, which encompasses the mountain ranges that arise from the valley floor in northern Sonora and southern Arizona and New Mexico, resembling islands in an ocean.

The San Pedro River itself extends for about 225 km (fig. 0.3), with perennial flow occurring over about 40 percent of its length (fig. 0.4). It arises in northern Sonora, Mexico, formed by the confluence of the Rio Nutrias (aka Rio Fresnos) and other tributaries. It then flows north to northwesterly through Cochise County in southeastern Arizona and north through Pima and Pinal counties to the Gila River, a major drainage that flows into the lower Colorado River. The San Pedro has a long and narrow hydrologic basin, with most tributaries draining short, relatively steep catchments oriented perpendicular to the main valley axis. The river flows through several deep, sediment-filled basins bounded by mountain ranges including the Huachuca, Mule, Winchester, Dragoon, Rincon, Galiuro, Santa Catalina, and Tortilla Mountains. The San Pedro's watershed is about 12,000 km^2.

The mountain ranges in the San Pedro watershed support many biotic communities, including Rocky Mountain conifer forests and Madrean evergreen oak-pine woodlands (Brown 1994). Many of the narrow canyons that drain from the mountains to the valley floor support Interior riparian deciduous forests, typified by a diverse mix of tree and shrub species, while

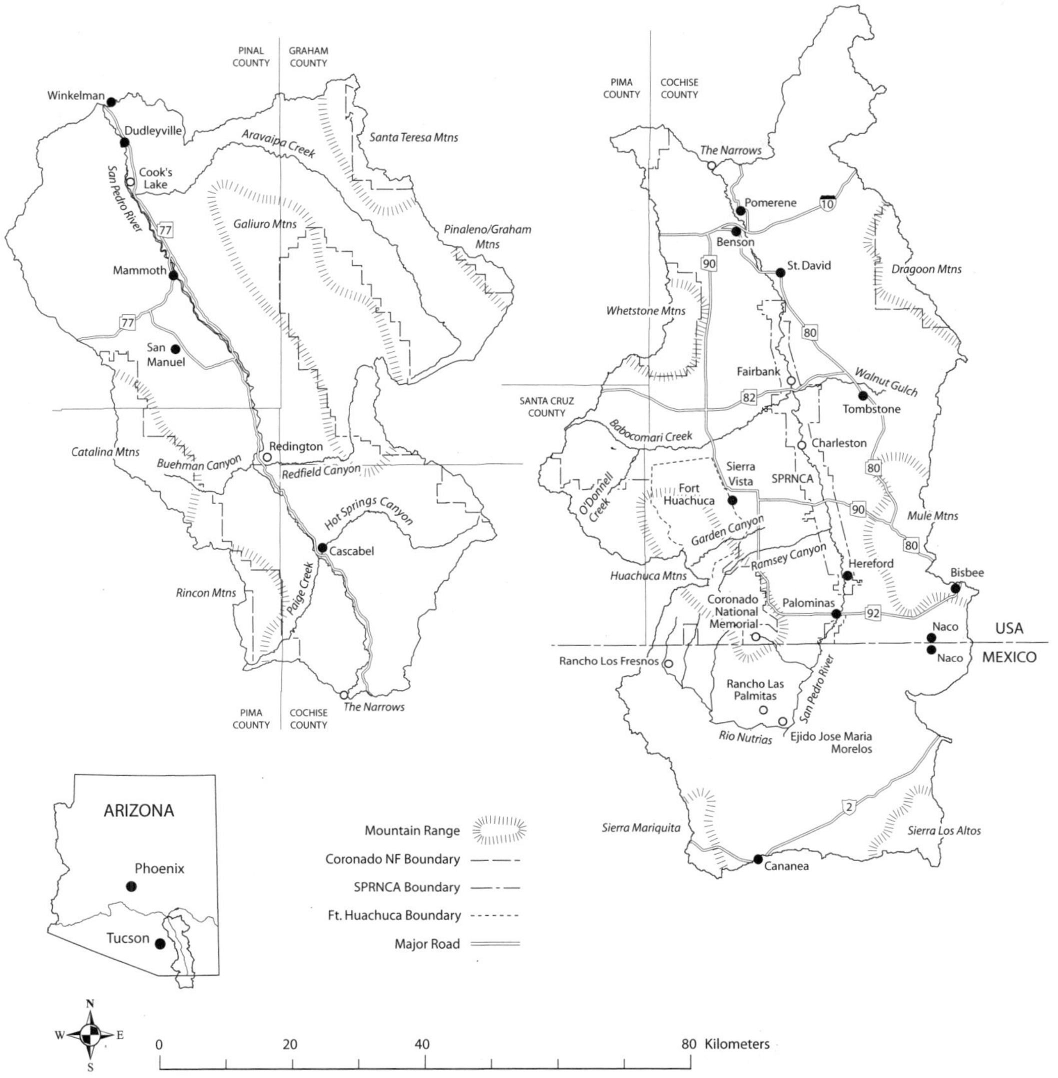

Fig. 0.3. Line drawing of the San Pedro watershed, showing major towns and tributaries and locations of other notable features including the longest conservation preserve (San Pedro Riparian National Conservation Area). Illustration credit: Carl Unkrich, USDA-ARS Southwest Watershed Research Center.

the many ephemeral drainages (washes, arroyos) distributed throughout the valley support xeroriparian communities (Szaro 1989). The basin floor is vegetated largely by semidesert grasslands and Chihuahuan desert scrub (upper reaches) or Sonoran desert scrub (lower reaches).

The San Pedro basin is divided into the upper and lower portions at the Narrows, a geologic constriction about 16 km north of Benson. The upper basin itself is divided into three subbasins: the Sonora subbasin, the Sierra Vista subbasin (U.S.-Mexico border to Fairbanks), and the Benson subbasin (Fairbanks to the Narrows). The lower San Pedro River basin begins at the Narrows and ends at its confluence with the Gila River near Winkelman, Arizona, a river distance of about 130 km. Valley-floor elevations range from about 1,460 m in the Mexican headwaters to 1,000 m at the Narrows and 595 m near the confluence with the Gila River.

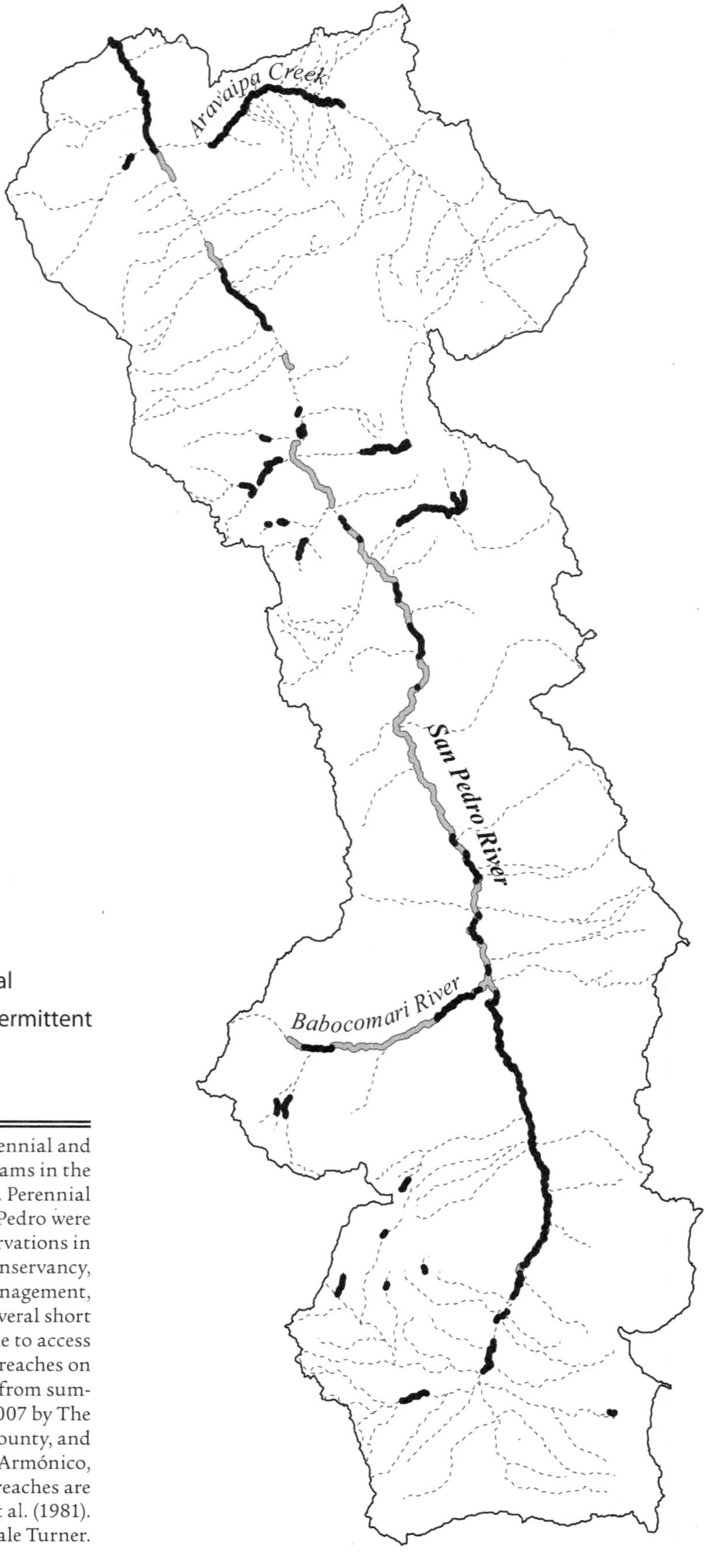

Fig. 0.4. Location of perennial and formerly perennial streams in the San Pedro River watershed. Perennial reaches on the San Pedro were determined from field observations in June 2007, by The Nature Conservancy, U.S. Bureau of Land Management, and many volunteers; several short reaches were not surveyed due to access restrictions. Perennial reaches on tributaries were determined from summer observations in 2000–2007 by The Nature Conservancy, Pima County, and Biodiversidad y Desarrollo Armónico, A.C. Formerly perennial reaches are redrawn from D. Brown et al. (1981). Map prepared by Dale Turner.

The climate is hot and dry, with annual rates of potential evapotranspiration far exceeding precipitation. The period from April through June constitutes the beginning of the growing season for many perennial plant species but is typically very dry. During July through September (monsoon season), locally heavy convective precipitation occurs, derived from moisture from the Gulf of California, the eastern Pacific, and the Gulf of Mexico (Sheppard et al. 2002). October and November are typically relatively dry and cool, but large precipitation events associated with dissipating tropical storms occasionally occur during late summer and fall. Finally, a cool rainy season occurs during winter and early spring, when more widespread, longer-duration rains fall, derived from moisture in the eastern Pacific.

Climatic conditions along the river vary along a north-to-south gradient, with a more Sonoran climate (50 percent summer, 50 percent winter precipitation) at the northern end of the basin and a more Chihuahuan climate (65 percent summer, 35 percent winter precipitation) at the southern end. Temperatures vary by elevation with average maximum daily temperatures of about 25°C along the upper river and 29°C at the Gila confluence.

Structure of the Volume

This book is divided into five sections, with heaviest weight given to the science components. Two sections introduce the biological communities of the river, with chapters ranging from mycorrhizae to mammals. We emphasize the riparian biota but do include one chapter on aquatic organisms (fishes). Two sections describe the physical conditions and processes, notably stream hydrology and geomorphology, that underpin the biota. The final section looks at conservation and management challenges and programs in three sections of the San Pedro River from its headwaters in Mexico to its confluence with the Gila River. The book touches on many types of natural and cultural influences on riparian ecosystems, from climate to livestock grazing and urbanization, although a central theme is streamflow regimes. Several chapters delve into the river's history, and a few speculate about its future.

Notes on Species Nomenclature

Generally throughout the book, we use lowercase when referring to common names of species, but in some taxon-specific chapters, we follow the traditions of the respective discipline and use uppercase. For some taxonomic groups, including plants and arthropods, we list the scientific and common name at the first mention of a species and use common names thereafter. For taxonomic groups with standardized common names, such as birds, herpetofauna, and mammals, we usually list only the common name in the text and provide scientific names in tables. For some taxa, such as many fungi, bacteria, and insects, we list only scientific names because common names do not exist.

Final Thought

> Writing a book is an adventure. To begin with, it is a toy and an amusement. Then it becomes a mistress, then it becomes a master, then it becomes a tyrant. The last phase is that just as you are about to be reconciled with your servitude, you kill the monster and fling him about to the public.
>
> WINSTON CHURCHILL

Please enjoy our flingings.

Vegetation of the Riparian Ecosystem

JULIET C. STROMBERG,
SHARON J. LITE,
MARK D. DIXON,
AND RONALD L. TILLER

Riparian Vegetation

PATTERN AND PROCESS

ONE

Introduction

A riparian corridor can be viewed as a mosaic of patches, each supporting a vegetation type reflective of particular hydro-geomorphic characteristics (Hupp and Osterkamp 1996). The vegetation types vary with distance from the stream edge and with distance downstream, partly in response to changes in resource availability and disturbance (fig. 1.1). The relative abundance of the vegetation types also changes through time, in response to changes in flood and fire patterns, water availability, climate, and animal (including human) activity. Over the past century on the San Pedro, riparian grasslands and marshlands have declined, while riparian shrublands and forests have increased (table 1.1). Complex patterns of change in the cottonwood forests have been set in motion by past channel entrenchment. Ongoing human actions and climate variability continue to drive vegetation change.

This chapter addresses the main riparian vegetation types of the San Pedro River, recognizing that these are artificial groupings that are stable in neither time nor space. We discuss associations of the wettest geomorphic surfaces, including ciénega wetlands, which presently (but not historically) occur mainly in areas beyond the active floodplain, and riverine marshes, which line the perennial river channel. We also discuss the pioneer forests

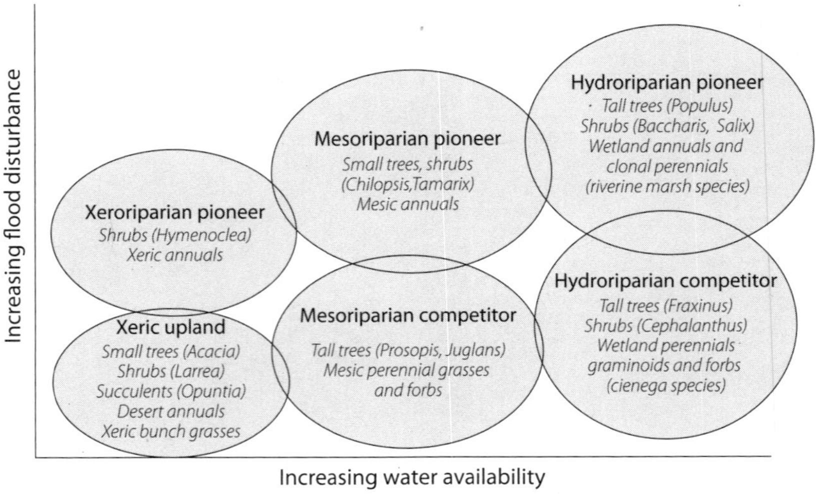

Fig. 1.1. Vegetation types along desert rivers such as the San Pedro are arrayed along gradients of water availability and flood intensity and frequency.

and shrublands of the floodplains and the sacaton grasslands and mesquite woodlands that occupy the river terraces. After providing overviews of each vegetation type, we discuss the ecological influences of physical processes (e.g., water flows, flood and fire disturbance) and biotic interactions (e.g., herbivory and competition) as a basis for understanding present distribution, historical patterns, and future changes in vegetation patterns.

Overview of Vegetation Types and Their Spatio-Temporal Patterns

CIÉNEGA WETLANDS

A ciénega is a mid-elevation (1,000–2,000 m), warm-temperate wetland of the Southwest that occurs along small, low-energy rivers. Ciénegas have low frequency of scouring floods and typically are sustained by groundwater inflow (Hendrickson and Minckley 1984). The soils are permanently saturated and rich in organic matter (see chap. 14). Herbaceous wetland plants are abundant, and often grow with wetland trees and shrubs including Goodding's willow (*Salix gooddingii*), velvet ash (*Fraxinus velutina*), and common buttonbush (*Cephalanthus occidentalis*).

Wide expanses of ciénegas once bordered the San Pedro and other desert rivers, but today they are much reduced in size and occurrence (Davis 1982, Hendrickson and Minckley 1984). Much of the ciénega vegetation along the San Pedro was destroyed near the turn of the nineteenth century during a period of river entrenchment when high-energy floods deepened and widened channels and led to abrupt drops in water table levels (see chap. 12). Ciénegas in the San Pedro watershed today persist in areas where soils are saturated but rarely scoured, such as along low-energy headwater streams or on spring-fed surfaces above the zone of intense flooding.

Extensive ciénegas occur in the headwater of the San Pedro in Sonora,

TABLE 1.1. Primary riparian vegetation types along the San Pedro River and patterns of change in abundance over time and space.

	General trend since ca. 1850s	General trends with distance downstream
Ciénega wetlands	Decrease	Decrease
Sacaton grasslands	Decrease	Decrease
Riverine marshlands	Increase?	No pattern
Cottonwood/willow forests	Increase	Shift to younger forests
Saltcedar shrublands	Increase	Increase
Mesquite forests	Increase	Increase
Xeroriparian shrublands and open areas	Increase	Increase

Mexico (see chap. 23), and three occur along the Babocomari River, a tributary in the upper basin. St. David Ciénega, at 30 ha, is the largest San Pedro ciénega on the U.S. side of the border. It lies about 400 m west and 7 m above the river channel and is sustained by artesian springs. Along the lower San Pedro River there are two major wetland complexes—Bingham Ciénega has about 55 acres of emergent wetland, forests, and open water (Fonseca 1998), and Cook's Lake wetland, at the San Pedro floodplain periphery, has swamplands of western buttonbush and Goodding's willow. Palynological data indicate that these and other woody species were less common in the San Pedro wetlands during past eras when fire was more common (Davis et al. 2002). Fires, set by local peoples and ignited by lightning in adjacent grasslands (McPherson 1995), caused mortality of fire-sensitive riparian trees and shrubs.

RIVERINE MARSHLANDS

Riverine marshes are wetlands that develop along actively flooded, perennial streams (Hendrickson and Minckley 1984). They are most common along high-order streams below about 1,000 m in elevation, reflecting the high-intensity floods in such areas, and occur intermittently over the length of the San Pedro. They are vegetated mainly by graminoids (grasses and grass-like plants) in the Juncaceae and Cyperaceae families. These marsh plants line the banks of the perennial low-flow channel and also form pockets in oxbows, scour pools, and beaver ponds.

Riverine marshes are transitory communities that fluctuate with flood and drought cycles. Following periodic large fall or winter floods, streamflows can be elevated for several months, causing wetland plants such as Torrey's rush (*Juncus torreyi*) to increase in abundance along the San Pedro channel (Bagstad et al. 2005). Marshes also can temporarily expand after erosive floods widen channels and reduce the effective elevation of the new channel bed above the water table. Along the Hassayampa River in central Arizona, for example, wetland plants became abundant in a channel widened by a large winter flood; two years later, these wetlands were buried by a large depositional flood, revealing their transient nature in these ecosystems (Stromberg et al. 1997).

COTTONWOOD/WILLOW FORESTS

With canopy heights of up to 30 m, Fremont cottonwood (*Populus fremontii*)/Goodding's willow gallery forests form a visually prominent element of the San Pedro riparian corridor. In spring and summer, these broadleaf, winter-deciduous trees create a ribbon of green that stands out against the desert background. Other co-occurring woody species are narrowleaf willow (*Salix exigua*), a clonal shrub, and seepwillow (*Baccharis salicifolia*) and Emory's baccharis (*Baccharis emoryi*), both evergreen shrubs in the sunflower family (Szaro 1989).

The abundance of cottonwood/willow forests on southwestern rivers has changed in complex ways over the past century and a half (Stromberg, Beauchamp et al. 2007). The forests have declined on heavily modified drainages due to dewatering and damming (Rea 1983, Fenner et al. 1985, Howe and Knopf 1991, Snyder and Miller 1992, Busch and Smith 1995), but have increased on some others including the upper San Pedro (Webb et al. 2007). Although locally abundant along the San Pedro prior to the twentieth century, the trees did not form the continuous gallery forest that extends along much of the river today. Prior to 1900, the San Pedro River had a lower-energy flood regime because flood pulses and sediment inputs were attenuated by dense grasslands in the uplands and by dense growth of riparian grasses and marsh plants in the wide floodplain. Thus, there were fewer opportunities for establishment of disturbance-dependent tree species such as Fremont cottonwood and Goodding's willow. Frequent fire, particularly in the upper basin, in conjunction with very high water tables and soil anoxia, also contributed to the lower abundance of these trees.

The age structure of a population provides a window into its past, and that of the cottonwood/willow forests along the San Pedro River reflects the legacy of historic geomorphic changes interacting with the ever-changing flood regime (see chap. 16). Most of the cottonwoods present at the beginning of the twentieth century were destroyed during the large floods that were the proximate triggers of historic channel entrenchment (Turner et al. 2003; and see chap. 12). The few surviving centenarians are on pre-entrenchment surfaces (some associated with old homesteads) and are very large, many with trunks over 2 m in diameter.

As the San Pedro floodplain began to widen following the period of inci-

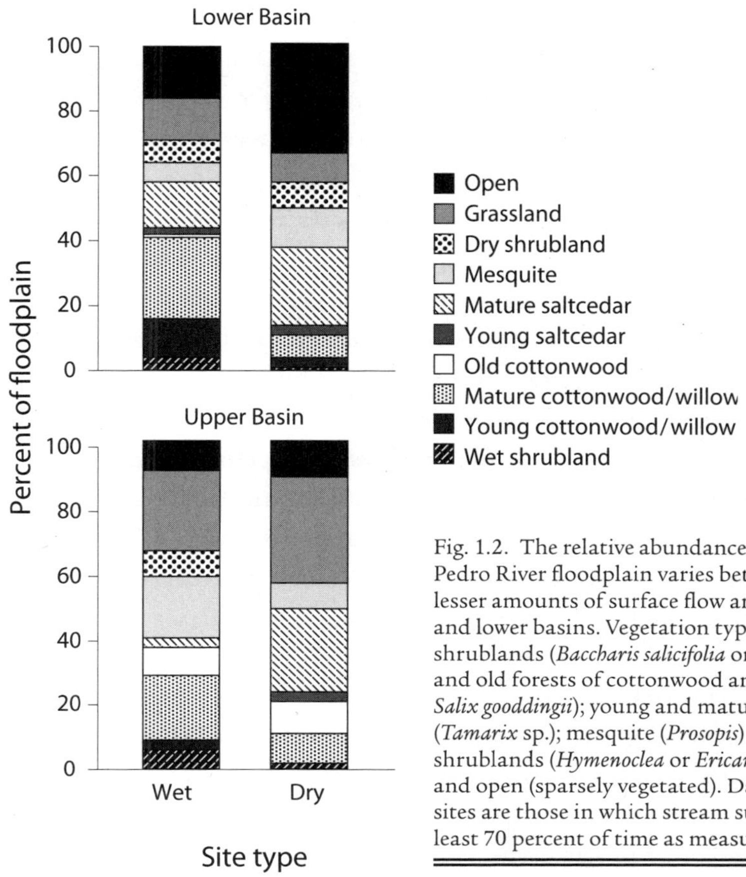

Fig. 1.2. The relative abundance of vegetation types in the San Pedro River floodplain varies between sites with greater and lesser amounts of surface flow and between sites in the upper and lower basins. Vegetation types are wet (hydroriparian) shrublands (*Baccharis salicifolia* or *Salix exigua*); young, mature, and old forests of cottonwood and willow (*Populus fremontii* and *Salix gooddingii*); young and mature shrublands of saltcedar (*Tamarix* sp.); mesquite (*Prosopis*) patches; dry (xeroriparian) shrublands (*Hymenoclea* or *Ericameria*); grasslands/forblands, and open (sparsely vegetated). Data are based on 18 sites. (Wet sites are those in which stream surface flow was present at least 70 percent of time as measured over a two-year period.)

sion, the stage was set for replacement of marshlands by pioneer tree and shrub species adapted to high-energy flood conditions and to periodically dry surface soils. Cottonwoods and willows established on the widening floodplain during the 1920s to early 1940s, a period of favorable flood conditions (Stromberg 1998a). Describing the San Pedro River in the 1940s, Brandt (1951:231) said, "Along the broad course of the river large old cottonwoods rise in leafy luxuriance. They seldom line more than one side of the stream at a time, because the bed of the river is ever restless." Conditions for establishment again became favorable in the 1960s and subsequent decades, owing to increased frequency and size of fall/winter floods and declines in summer flooding (see chap. 16). Today, the river is more geomorphically stable, and much of the low-flow channel in the upper basin is lined not just on one side but on both sides by trees that established in the 1970s or 1980s. While many of the older cohorts occupy wide swaths of the floodplain, the younger cohorts form only narrow bands.

Cottonwood/willow age structures differ between the lower and upper basins because flood intensities and flood signatures change over the length of the river (Stromberg 1998a; and see chap. 16). At sites with high flood stream power, the frequencies of plant scour, sediment erosion, and channel movement increase, thereby increasing rates of tree establishment and mortality. Floodplains in the lower San Pedro are dominated by young pioneer trees (fig. 1.2) because of the very large size of some recent floods in this part of the river. None of the cottonwood trees that have been aged in the lower basin pre-date the large, scouring flood of September 1926.

Although floods are the primary disturbance that shapes the San Pedro floodplain forests, fire also plays a role. Reduced grassland cover, an abundance of livestock, and active fire suppression during much of the twentieth century contributed to the reduced fire frequency that allowed for high survivorship of cottonwood trees. Recent changes in land management over parts of the river, notably the removal of livestock, have again allowed for seasonal proliferation of dry, grassy fuel loads and set the stage for the return of fire as an ecological process. Concurrently, increased human presence has increased fire ignitions. Among the effects of the increased frequency of riparian fire are reduction in cottonwood forest cover and increase in densities of young willow trees (Rychener 2006; Stromberg et al. forthcoming).

SALTCEDAR SHRUBLANDS

Saltcedar or tamarisk (*Tamarix chinensis*, *T. ramosissima*, or hybrids), like other members of its genus, is a stress-tolerant pioneer species. Along river courses in Eurasia and northern Africa, *Tamarix* have co-occurred with *Populus*, *Salix*, and other riparian genera for millennia (Kurschner 2004). In North America, saltcedar is a relative newcomer, having been introduced for soil erosion control and landscaping purposes. It occupies a position in the fluvial landscape intermediate to that of cottonwood and mesquite (Bagstad et al. 2006). Similar to cottonwood, saltcedar produces many small wind- and water-dispersed seeds and requires bare, moist mineral soils for seedling establishment, but unlike cottonwood, saltcedar is tolerant of many stresses (Glenn and Nagler 2005).

Saltcedar has increased on many rivers of the American West over the past century (Friedman et al. 2005), becoming particularly abundant in reaches affected by damming, diversion, groundwater decline, salinization, and livestock grazing (Stromberg, Lite et al. 2007). Saltcedar was first reported along the Gila River in 1916 (Robinson 1965) and was abundant there by the 1940s and 1950s (Turner 1974). The oldest documented saltcedar on the San Pedro dates back to the 1950s (Stromberg 1998a). Today, saltcedar is most abundant along the river in its drier, lower reaches, where it grows as an understory species in the cottonwood/willow forests and also forms discrete shrubland patches.

The relative abundance of saltcedar along the San Pedro has varied through time in response to flood and drought cycles and changing land use. For example, in the lower basin, saltcedar were more abundant than cottonwood/willow during the 1950s through 1970s, a period of drought, extensive groundwater pumping, and widespread livestock grazing (Stromberg 1998a). By the 1990s, a wetter period, cottonwood/willow patches had become dominant at some sites. In parts of the upper basin, in contrast, saltcedar has increased in abundance in recent decades, perhaps relating to changing streamflow conditions (Leenhouts et al. 2006).

XERORIPARIAN SHRUBLANDS

Also abundant in the San Pedro floodplain today are patches sparsely vegetated by small (<2 m tall) xeroriparian pioneer shrubs including burrobrush (*Hymenoclea monogyra*), rubber rabbitbrush (*Ericameria nauseosa*), and yerba de pasmo (*Baccharis pteronioides*). Burrobrush's drought tolerance is evidenced by small, needle-like leaves, and its disturbance tolerance is evidenced by its ability to propagate vegetatively from stem or root fragments. These xeroriparian shrublands are most abundant in the lower basin, and they likely increased in abundance following historic channel entrenchment.

Soils in the xeric shrublands are sandy and dry and have little organic matter (see chap. 14). Following rains or floods, herbaceous plants with affinity for coarse-textured soils, such as needle grama (*Bouteloua aristidoides*) and sand dropseed (*Sporobolus cryptandrus*), become abundant in the shrub interspaces (Stromberg et al. 1996).

SACATON GRASSLANDS

Riparian grasslands vegetated by big sacaton (*Sporobolus wrightii*), alkali sacaton (*S. airoides*), vine mesquite (*Panicum obtusum*), and tobosa grass (*Pleuraphis* mutica) occupied millions of acres of low-gradient streams and alluvial fans in the semi-desert grassland biome of the southwestern United States and northern Mexico in the nineteenth century (Humphrey 1958). These grasslands slowed the velocity of flood runoff and thus reduced erosion rates, trapped sediment that enhanced soil fertility, and enhanced the infiltration of water that contributes to the formation of shallow groundwater tables and perennial streams (Thornber 1910, Humphrey 1958). In addition to providing habitat for wildlife (see chap. 6), these grasslands provided forage and hay for livestock (Bock and Bock 1978, 1979, Bahre 1987).

Today, these riparian grasslands occupy a small percentage of their original area (Cox et al. 1983). Big sacaton, a large, warm-season bunchgrass associated with fine-textured soils (see chap. 14), often is the dominant species. Many sites in the San Pedro watershed once vegetated by sacaton grasslands now support mesquite forests or sacaton-mesquite savannahs (Fonseca 1998). This shift was caused by a complicated interplay of climatic and anthropogenic influences in the late 1800s and early 1900s, including extreme flooding following extended drought, channel incision or arroyo formation, overgrazing by livestock, and alteration of the fire regime (Griffiths et al. 1915, Bryan 1928a, Humphrey 1958, Bahre 1991). Further, given channel entrenchment, reductions in overbank flooding, and the advent of industrialized farming, thousands of acres of sacaton terrace grassland became better suited for agricultural production and were converted to cropland by the mid-1900s.

Sacaton grows in the San Pedro floodplain, but the grasslands are most extensive on terraces. The terrace grasslands may be remnants that have persisted since channel incision. Repeat photography provides evidence of some recent establishment of sacaton on the terraces (Turner et al. 2003), but only at very low rates. Examination of stratigraphy at cut banks of an arroyo shows sacaton plants surviving through several episodes of fire and silt sedimentation (Flory 1937), suggesting great longevity for individual sacaton plants. Although the maximum age of sacaton is not known, tussock-forming grasses generally have very long life spans (Harper 1977).

Although sacaton historically was abundant over the entire San Pedro River, today it shows a pattern of decline with distance downstream. On pre-entrenchment terraces with deep water tables and no overbank floods, precipitation is sacaton's primary water source. Along the lower reaches of the river, temperatures are hotter and evaporative stresses greater, and rainfall is not sufficient to sustain sacaton on terraces. Even in the wetter upper basin, the sacaton grasslands on terraces, particularly where water tables are deep, are less productive than those on floodplains (Tiller 2004).

MESQUITE FORESTS

A high degree of phenotypic plasticity (Givnish 2002) allows mesquite to grow in a wide range of hydrogeomorphic settings including river floodplains, river terraces, ephemeral washes, alluvial fans, and desert uplands. Where groundwater is within its rooting depth, mesquite trees can reach canopy heights of 18 m and stem diameters of 1.3 m and form dense bosques (forests or woodlands; Swarth 1905, Brandt 1951). In contrast, mesquite that survive strictly on rainfall grow to only 4 or 5 m tall, have a shrubby growth form, and occur at low density (Stromberg et al. 1993, Martinez and Lopez-Portillo 2003). With increasing distance from the stream channel, groundwater deepens, and water availability declines; mesquite uses proportionately more water from shallow soil layers and undergoes declines in height and canopy cover (Stromberg et al. 1992).

Patterns of change in mesquite abundance along southwestern rivers are complex. Mesquite trees historically were abundant along the San Pedro River only in locally entrenched reaches (see chap. 12). Today, most of the river is entrenched, and mesquite is considerably more common. River entrenchment (and dropping water tables) reduced habitat suitability for sacaton grass, while favoring the more deeply rooted mesquite (Bryan 1928b). Mesquite also have become denser in upland parts of the San Pedro watershed, including areas once vegetated by desert grasslands. Introduction of livestock and horses, reduced fire frequency, changes in rodent populations, increased carbon dioxide levels, coarsened soil textures, and climate change all may have contributed to the increased abundance of mesquite in both upland and riparian zones (Humphrey 1958, Bahre and Shelton 1993, Fredrickson et al. 1998, Van Auken 2000; and see chap. 6).

On the other hand, many factors have contributed to regional loss of mesquite forests. Conversion of riparian lands to agricultural use, wood cutting, and extreme groundwater depletion from aquifer pumping and stream diversion all have led to declines in mesquite acreage (Brown 1994). In the Tres Alamos to Redington reach of the lower San Pedro River, for example, about half of the mesquite forests were cleared for irrigated agriculture (Zimmerman 1969).

Although mesquite is abundant along the San Pedro, it undergoes a shift in structure from short shrublands to tall, dense forests with distance downstream (Leenhouts et al. 2006). This river-wide pattern may relate to one or more factors that change with distance downstream including frequency and intensity of frost (Glinski and Brown 1982), summer vs. winter precipitation patterns, fire patterns, and past locations of wood cutting.

OLD-FIELD VEGETATION

The many abandoned agricultural fields that occupy San Pedro River terraces (see chap. 11) are undergoing old-field succession (Oleksyszyn 2001). A variety of annual plants (e.g., carelessweed, *Amaranthus palmeri*) and perennial herbaceous plants (e.g., field bindweed, *Convolvulus arvensis*) are common on the fields that pre-cultivation were predominantly inhabited by mesquite and sacaton vegetation types. Mesquite is actively re-vegetating the fields, particularly in the lower basin where one study showed two to eight years elapsing between field abandonment and mesquite establishment (Oleksyszyn 2001). There is limited establishment of sacaton, however, on the old fields (Richter et al. 2002).

Influence of Water Quantity and Quality on Riparian Vegetation

WETLANDS

Vegetation patterns in dryland riparian zones are strongly shaped by water availability. Water quality, particularly salinity, also plays a key role. Within ciénegas, for example, zones of vegetation are discernible that reflect gradients of water quantity and quality (Yatskievych and Jenkins 1981, Cross 1991). In St. David Ciénega, California loosestrife (*Lythrum californicum*), cutleaf water

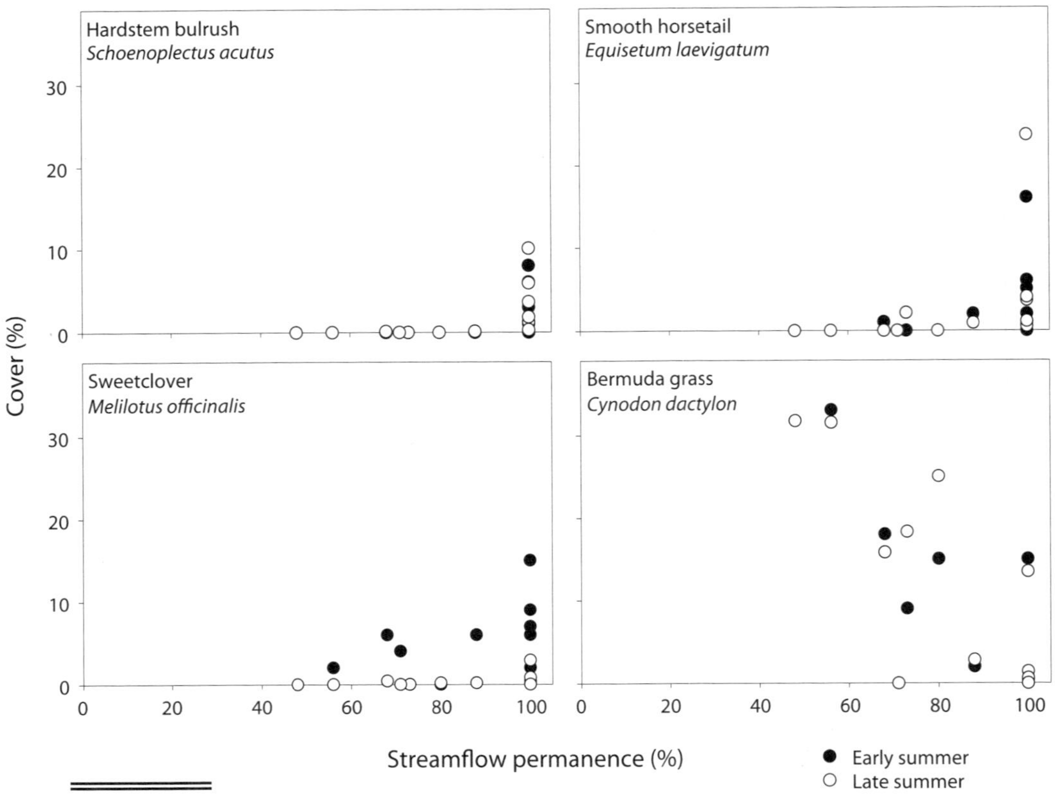

Fig. 1.3. Mean cover values for four herbaceous species in the San Pedro River streamside zone, during pre-monsoon and monsoon periods, in relation to site streamflow permanence during 2002. Each data point represents a study site in the upper basin.

parsnip (*Berula erecta*), and other plants that tolerate anoxic soils form dense stands in the central areas where water can be ankle-deep (Makings 2006). Clustered field sedge (*Carex praegracilis*) and beaked spikerush (*Eleocharis rostellata*) are encountered as one moves outward to areas with saturated soils but no standing water. Halophytes, including alkali marsh aster (*Almutaster pauciflorus*), alkali sacaton (*Sporobolus airoides*), and alkali muhly (*Muhlenbergia asperifolia*), grow in the drier and more saline soils at the perimeter of the ciénegas.

Riverine marsh plants also are hygrophytes that require a permanent water source. They thrive where influent groundwater maintains wet soils and/or standing water throughout the year, and are associated with perennial surface flow (Stromberg et al. 2005). Hardstem bulrush (*Schoenoplectus acutus*) and spikerush (*Eleocharis montevidensis*) are among the marsh species common in perennial reaches of the San Pedro (fig. 1.3). As surface flows become intermittent, these wetland plants are replaced along the active channel by more drought-tolerant and more deeply rooted species including bermuda grass (*Cynodon dactylon*).

The composition of the wetland plants along the channel differs from that in ciénegas owing to differences in flood frequency and water chemistry between the lotic (flowing water) and lentic (standing water) environments

(Cross 1991). For example, the streambanks of the upper San Pedro River are vegetated largely by a different species of bulrush (*S. acutus*) than occurs in St. David Ciénega (*S. americanus*), perhaps reflecting greater salinity tolerance of the latter species. Availability of macronutrients also can influence wetland plant communities, but the effects on streamside plants of the elevated levels of nitrogen observed in agricultural reaches of the river (see chap. 17) remain to be investigated.

PIONEER FORESTS AND SHRUBLANDS

Fremont cottonwood and Gooding's willow are phreatophytes, meaning they extract water from aquifers or the capillary fringe above the water table (Meinzer 1927). Further, both are obligate phreatophytes, meaning that groundwater is essential to their survival and forms a primary component of their water supply (S. D. Smith et al. 1998). Fremont cottonwood and Goodding's willow both have moderately deep tap roots but also develop whorls of adventitious roots from buried stems that may allow for uptake of nutrient-rich flood water (Leffler and Evans 1999; and see chap. 2). Seepwillow also appears to be an obligate phreatophyte (Gary 1963, Schade et al. 2001, Rood et al. 2003).

Cottonwoods and willows are intolerant of drought (Pockman and Sperry 2000, Amlin and Rood 2002). Because fine roots are concentrated just above the water table, the plants are sensitive to fluctuation in water table depth, particularly on sandy soils with a narrow capillary fringe. Even if floods create moist seed beds, establishment can be precluded if water levels decline at rates that exceed downward root growth of seedlings (Segelquist et al. 1993, Mahoney and Rood 1998, Shafroth et al. 1998) or if water tables decline throughout the summer to levels 1 or 2 m below the establishment surface (depending on soil texture, weather, and species) (Kalischuk et al. 2001, Amlin and Rood 2002; but see Cooper et al. 1999). Water table decline during the summer dry season also can influence productivity and survivorship of older plants. Seasonal declines of as little as 1 m can kill saplings (Shafroth et al. 2000) as well as adults (M. Scott et al. 1999, 2000, Horton et al. 2001b). Rate, extent, and duration of the water table decline all influence plant growth and survivorship.

Abundance and age-class diversity of Fremont cottonwood and Goodding's willow along the San Pedro River are linked to groundwater availability. Although individual cottonwood trees grow where depth to groundwater can be 4 to 6 m or more (Zimmerman 1969, Stromberg et al. 1996), dense, multiple-aged stands develop only where groundwater averages less than 3 m deep (dry-season value) and fluctuates by less than 1 m annually (Lite and Stromberg 2005). Cottonwood/willow abundance along the San Pedro River also is related to streamflows, with forests becoming sparser as flows

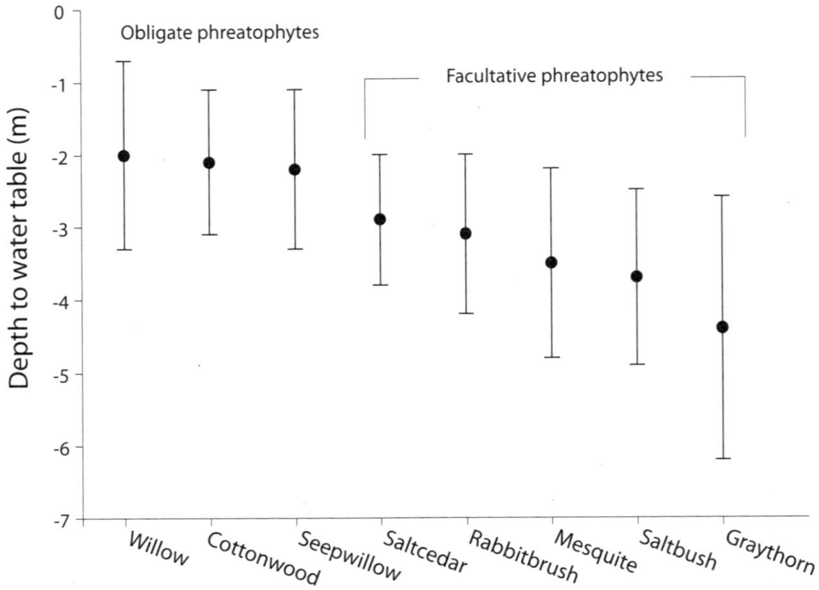

Fig. 1.4. Trees and shrubs in the San Pedro riparian zone are arrayed along gradients of depth to groundwater. Values shown are means plus and minus one standard deviation for maximum depth to groundwater during water year 2002 at sites occupied by several common tree and shrub species. Data are from the upper basin. Species are Goodding's willow (*Salix gooddingii*), Fremont cottonwood (*Populus fremontii*), seepwillow baccharis (*Baccharis salicifolia*), saltcedar (*Tamarix* sp.), rubber rabbitbrush (*Ericameria nauseosa*), velvet mesquite (*Prosopis velutina*), fourwing saltbush (*Atriplex canescens*), and graythorn (*Ziziphus obtusifolia*).

become more intermittent. This correlation likely exists because streamflow permanence is correlated with depth to the water table and degree of variation in water table level. However, perennial surface flows may also directly benefit the trees by increasing humidity (Horton et al. 2001a).

Hydrology and geomorphology are key drivers of riparian vegetation dynamics, and as these physical factors change, so does the biota. Whereas cottonwoods are most abundant along the San Pedro River where depth to the water table is less than about 3 m (fig. 1.4), saltcedar remain abundant where the water table is at depths of 7 m (Lite and Stromberg 2005). In contrast to the pattern for cottonwoods and willows, saltcedar increases in abundance as the water table deepens and fluctuates more, and as streamflows become more intermittent. The drier reaches are more prevalent in the lower basin, and so too are the saltcedar shrublands (fig. 1.2). Saltcedar thrives at sites with deep and fluctuating water tables because of a suite of morphological and physiological adaptations (Horton and Clark 2001, Glenn and Nagler 2005). For example, studies suggest capacity for rooting depths of 10 m or more in the *Tamarix* genus (Zimmerman 1969, Gries et al. 2003). Additionally, saltcedar is a facultative phreatophyte, using groundwater when available but also readily using water from unsaturated soil layers (see chap. 2).

Growth and survivorship of cottonwood, willow, and saltcedar also are influenced by soil chemistry. Nitrogen, in particular, can be a limiting resource in floodplains (see chap. 18). High levels of this nutrient (from agricultural runoff or proximity to mesquite trees) can increase shoot and root growth and also increase ability of the trees to tolerate environmental stress (Marler et al. 2001, Adair and Binkley 2002, Pataki et al. 2005).

Salt levels also influence riparian forest growth. Saltcedar is halophytic, while cottonwoods and willows are salt-intolerant glycophytes (Glenn et al. 1998). With increasing salinity, germination rate of saltcedar increases while that of Fremont cottonwood decreases (Siegel and Brock 1990, Shafroth et al. 1995). Under saline conditions, saltcedar maintains high water use and thus higher competitive ability (Vandersande et al. 2001). Ability to exude salts from leaf glands is one trait that allows saltcedar to thrive in salty environments. However, although the distribution of saltcedar along some rivers is linked to high salinity levels, soil salinities along the San Pedro are low because of a lack of source salts in conjunction with frequent flushing floods (see chap. 14).

TERRACE GRASSLANDS AND FORESTS

Few grass species are as deeply rooted as big sacaton. At arroyo cut-banks, we have observed roots of this grass to be dense in upper soil horizons and to extend to a depth of 3.8 m (fig. 1.5). This rooting pattern enables sacaton to use different water sources by season (see chap. 2). During the pre-monsoon season, sacaton uses water from deep soil layers, allowing it to maintain some productivity at a time when most grasses are dormant; a water table depth of 3 to 4 m appears to be a threshold level beyond which sacaton vigor and productivity decline (Tiller 2004). With the onset of the monsoon, sacaton uses shallow soil water derived from rainfall, and biomass production peaks (Cox 1985). For plants growing in the floodplain, soil moisture from overbank floods also provides a seasonal water source.

Overstory and understory vegetation are often de-coupled in riparian ecosystems, with each vegetation layer responding to a different suite of environmental controls. This de-coupling is particularly evident on river terraces vegetated by mesquite. Mesquite trees have wide-spreading lateral roots that allow for uptake of rain and floodwater but are most well known for having very deep taproots, with 53.5 m recorded for *Prosopis juliflora* (Phillips 1963, Stone and Kalisz 1991). Whereas mesquite can tap into the stream aquifer, which often is 5 or 10 m below the terrace surface, most of the understory plants (with the notable exception of sacaton) appear to be shallow-rooted responders to seasonal rains (see chap. 2). Among the perennial forbs and grasses that become abundant in mesquite understory and interspaces following late summer rains are toothleaf goldeneye (*Viguiera dentata*), bristlegrass

Fig. 1.5 Big sacaton (*Sporobolus wrightii*) grasslands on a San Pedro River terrace (right; 1.5a) and along an incised tributary (opposite; 1.5b). Photo credit: Ron Tiller.

(*Setaria leucopila*), and Mexican passionflower vine (*Passiflora mexicana*) (fig. 1.6). Common shrubs include graythorn (*Ziziphus obtusifolia*), pale desert-thorn (*Lycium pallidum*), and littleleaf sumac (*Rhus microphylla*). Quantitative data on rooting depth and water sources, however, are not available for most of these understory plants (Jackson et al. 1996).

Rooting depth and water sources also are poorly known for burrobrush and most other xeroriparian shrubs. Rubber rabbitbrush, however, is known to have roots that extend 4 m deep and to be phreatophytic in some settings (Groeneveld and Crowley 1988, Donovan and Ehleringer 1992, Toft 1995). Burrobrush and other xeroriparian shrubs likely are sustained by floodwater

when growing in the San Pedro floodplain and in the many ephemeral washes that drain into it.

Edaphic Factors

By influencing processes such as water movement and nutrient retention, the physical structure of soil influences vegetation development and primary productivity (Marks 1950, Stromberg et al. 1996, Merigliano 2005). Sacaton seedlings, for example, are larger, and seeds can emerge from deeper soil depths when growing in the fine-textured soils that characterize San Pedro terraces vs. the coarse-textured soils that characterize its floodplains (unpublished data of authors). Seedlings of mesquite show faster root growth in sandy loam soils than in clay soils, perhaps enabling greater access to deep water tables (Brock 1986). Thus, shifts towards coarser soil textures may be one factor favoring mesquite over sacaton in portions of the San Pedro River (see chap. 14). Sediment texture also influences competitive outcomes of cottonwood vs. saltcedar seedlings (Sher and Marshall 2003). Given the important role of soil texture, there is a need for greater study of the linkages between watershed land cover changes, sediment inputs to streams, and riparian plant community composition.

Floods

In riparian corridors, flood disturbance events create a shifting mosaic of vegetation patches over time (Pickett and White 1985). As floods of high velocity

Fig. 1.6. Velvet mesquite (*Prosopis velutina*) forest with dense monsoon-season herbaceous understory. Photo credit: Juliet Stromberg.

and shear stress exceed thresholds for vegetation removal, certain vegetation patches are destroyed, followed by development of new patches that have a different species composition or age structure (M. Scott et al. 1996). Large floods, of several days' duration, can scour portions of the floodplain and deposit thick lenses of sediment elsewhere (Stromberg et al. 1997). Such actions can, for example, cause mature cottonwood forests to be replaced by xeric shrublands, marshlands, or young cottonwood forests.

Several traits of a flood pulse, including magnitude, duration, timing, and rate of change (Poff et al. 1997), influence vegetation dynamics and the post-flood establishment process. Flood magnitude and duration are important, as they drive geomorphic change and create seed beds. Flood timing is important as it relates to the availability of viable seeds. Rate of change is important as it relates to seedlings' ability to maintain contact with moist soils.

PIONEER FORESTS AND SHRUBLANDS

Fremont cottonwood and Goodding's willow, as well as narrowleaf willow, seepwillow baccharis, and saltcedar, have traits that adapt them for life in frequently disturbed environments—rapid growth rate, young age of reproductive maturity, and high investment in reproduction (Karrenberg et al. 2002). Members of the *Populus* genera are among the fastest-growing tree species in North America, with trunk diameter increments of 3 to 4 cm per year and height increases of 1 m per year common for young trees. Cottonwoods often reach sexual maturity within about 10 years (Farley et al. 1994), and saltcedar does the same at about 2 or 3 years of age.

Cottonwood, willow, and seepwillow are dioecious, with male and female flowers produced on separate individuals. Female plants of these species annually produce large numbers of small, wind- and water-dispersed seeds (Bessey 1904, Siegel and Brock 1990). Saltcedar also produces large numbers of seeds annually, of a size somewhat smaller than those of cottonwood (R. Marler et al. 2001). Given their moderately small seed size, seedlings of all these species are not strong competitors and require bare, moist mineral soil for establishment. In human-modified habitats, suitable establishment conditions are created by actions such as excavation of irrigation canals. Along river courses, suitable conditions are created by floods.

The short-lived seeds of Fremont cottonwood and Goodding's willow are dispersed during spring (Fenner et al. 1984, Young and Clements 2003), and seedlings thus tend to establish in years with wet winters (Stromberg et al. 1991, Shafroth et al. 1998) (fig. 1.7). The large winter (or late fall) floods erode and re-deposit sediments, creating a patchwork of potential seed beds, and the floodwaters then recede and expose the requisite substrates during the period of seed dispersal. Fremont cottonwood initiates dispersal a bit earlier than Goodding's willow, and this phenological difference allows the species to partition their environments: compared to cottonwoods, the later-dispersing willows establish on surfaces that are slightly lower (by one or more decimeters) and closer (one or more meters) to the stream channel, reflecting the streamward contraction of the zone of moist soil exposed by receding floodwaters. Seepwillow and saltcedar establish later in the sequence (Stromberg 1997), as both have later and more protracted seed dispersal periods. However, seedlings of saltcedar often overlap spatially with the earlier-

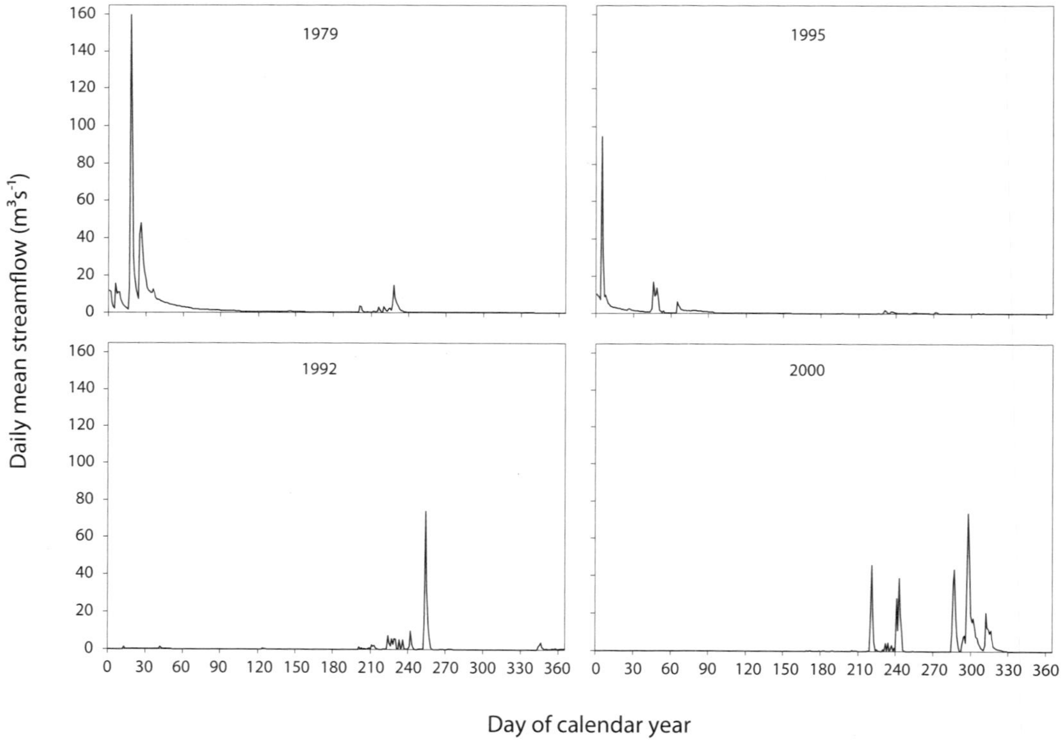

Fig. 1.7. Stream hydrograph for the San Pedro River–Redington gaging station (USGS #09472000/50) depicting daily mean streamflow during four separate calendar years. The hydrograph of most years (such as 1992, bottom left) is dominated by summer monsoonal floods. Fremont cottonwood (*Populus fremontii*), a spring-seeding tree species, establishes in years with large winter floods and elevated spring flows, such as occurred in 1979 and 1995 (top panel).

seeding cottonwoods and willows, because of wider tolerance range for soil moisture.

Requirements for seedling establishment (i.e., a species' regeneration niche) are more restrictive than are those for adult survivorship (Grubb 1977), and for Fremont cottonwood and Goodding's willow, the hydrologic conditions allowing for seedling establishment along the San Pedro River have been variable through time. Each cohort of trees that establishes occupies a linear band or patch, and over time, fluvial processes create complex spatial patterns of forest patches in the floodplain. Owing to channel meandering, avulsion, and other fluvial processes (see chap. 13), older cohorts are farther from the active channel on higher surfaces where scouring forces, inundation depths, and durations are all lessened (Stromberg et al. 1997).

Trees that grow fast often die young. Analysis of Fremont cottonwood treerings from streams in Arizona reveals maximum ages of about 65 (lower San Pedro), 85 (Hassayampa), 105 (upper San Pedro), and 130 years (Sonoita Creek) (unpublished data of authors), with the youngest ages on the streams with the most intense flood regimes. Most other studies of Fremont cotton-

Fig. 1.8. Fremont cottonwood (*Populus fremontii*) forest on the upper San Pedro River floodplain. Photo credit: Elizabeth Makings.

wood also report maximum age less than 130 years (Irvine and West 1979, Brotherson et al. 1983, Howe and Knopf 1991, Hinchman and Birkeland 1995).

After being damaged by floods, many riparian trees and shrubs grow new stems. Seepwillow, for example, propogates vegetatively after being prostrated by floods (Stromberg et al. 1993), and Goodding's willow, Fremont cottonwood, and saltcedar produce new stems from branch fragments (Everitt 1968). In narrow canyons, such as along the short confined sections of the San Pedro, seedling establishment can be restricted by flood scour, and vegetative propagation can be the primary recruitment mode (Everitt 1995, Karrenberg et al. 2002).

FOREST SUCCESSION

Over time, the surfaces occupied by cottonwood and willow trees often become elevated due to accretion of sediments around the tree bases and to incision of the channel (see chap. 13). As the floodplain aggrades, flood inundation frequency declines, surface soils become drier, depth to groundwater increases, and soils increase in organic matter, silt, clay, and litter cover (see chap. 14). Because of the elevated surfaces and the dense canopy, seedlings of cottonwood and willow do not establish in their own understory (fig. 1.8).

Fig. 1.9. Within the San Pedro River floodplain, big sacaton (*Sporobolus wrightii*) increases in cover with age of the Fremont cottonwood (*Populus fremontii*) forest. Data are based on measurements taken in the San Pedro Riparian National Conservation Area.

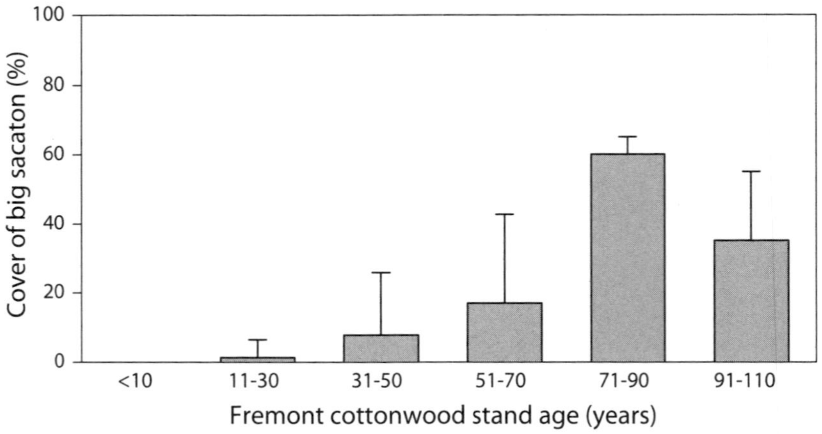

Forest composition shifts towards species adapted to drier and more stable sites, including velvet mesquite (*P. velutina*), net-leaf hackberry (*Celtis reticulata*), and Texas mulberry (*Morus microphylla*). Velvet mesquite has a lifespan of over 300 years, and as the shorter-lived cottonwood and willow trees die, it can become the dominant tree species in the forest patch.

Mesquite, like many later successional species (Drezner et al. 2001), produce large, animal-dispersed seeds. Although quite large (30 mg; Mooney et al. 1977), velvet mesquite seeds are produced in abundance, with up to 140,000 seeds per tree per season (Glendening and Paulsen 1955). Flowers are produced in early to mid-summer and again in late summer with up to four crops per year possible (Mooney et al. 1977). Owing to prolific flower and seed production, mesquite woodlands sustain a wide variety of animal species (see chap. 7). As animals such as coyotes and cows feed on the mesquite pods, they disperse the seeds to locations far from the parent plant. Passage through the digestive tracts increases germination rates by scarifying the hard seed coat and by killing the bruchid beetle larvae which feed on the developing seeds (Mooney et al. 1977).

Mesquite seedlings are intolerant of saturated soils and intense flood scour. Thus, within the floodplain of the San Pedro River, mesquite establishes on aggraded surfaces, sometimes in the open and sometimes in cottonwood understory. Large seed size allows seedlings of mesquite and other tree species to tolerate shade and drought and to emerge though dense litter cover (Stromberg and Patten 1990b, Leishman and Westoby 1994), and in fact, the seeds germinate at higher rates if covered by soil or litter (Mooney et al. 1977). Seed germination is stimulated by the combination of high summer temperatures (20 to 40° C) and moist soils, such as produced by late summer rains or floods (Siegel and Brock 1990). During their first year, velvet mesquite seedlings develop deep and extensive root systems, enabling access to groundwater at depths of 2 to 3 m (Brock 1986).

Sacaton establishes with some regularity on high surfaces of San Pedro floodplains. Its abundance in the floodplain increases with forest age, to become the dominant ground cover in the interspaces and understory of old cottonwood stands (fig. 1.9). Although the seedlings arise from tiny seeds, they can emerge from below 3 cm of sediment, such as might be deposited in a summer flood (Tiller 2004). Sacaton seeds have greatest germination at high (24 to 38° C) and fluctuating temperatures, conditions that are typical of canopy light gaps.

Fire

Riparian tree and shrub species differ in ability to survive fire and to regenerate after fire. Survivorship and regeneration rate also vary with plant age, moisture status, and fire intensity. Very hot fires along the San Pedro have completely combusted some cottonwood trees, in some cases leaving only a hole in the ground. Overall, though, fire causes less mortality of cottonwoods and willows on the upper San Pedro than on dam-regulated rivers in the Southwest where litter and dead stems accumulate (Busch 1995, Ellis 2001, Stromberg et al. forthcoming). This is probably because San Pedro floods routinely remove woody debris and maintain a more open structure within the floodplain, and perhaps because a high water table keeps the vegetation well hydrated.

Goodding's willow, saltcedar, and velvet ash all resprout vigorously following fire, as long as their basal meristems have been spared. Fremont cottonwoods do not resprout abundantly, although rates vary depending on tree age, fire intensity, and, perhaps, ecotype (Busch 1995, Ellis 2001). Owing to between-species differences in mortality resprout rates, fire can lead to dominance shifts from cottonwood towards willow and, in drier reaches, towards saltcedar. In wet reaches of the San Pedro, however, fire has caused declines in abundance of the saltcedar growing in cottonwood understory. In addition to causing shifts in species dominance, fire along the San Pedro River shifts tree-age structure from older to younger stands and converts contiguous gallery forests into more fragmented patches (Paxton et al. 1996, Rychener 2006).

Presumably, fire historically was a regular occurrence in sacaton grasslands (Davis 1982) and allowed the grasses to maintain competitive dominance over woody plants such as mesquite and cottonwood. Sacaton, where well watered, is resilient to fire, with biomass recovering rapidly after winter or summer burns (Bock and Bock 1978, Bock and Bock 1986). Fire typically kills young mesquite, particularly if it occurs with high frequency; older trees, however, either withstand fire with little biomass loss or vigorously resprout from buds at the base of the tree to produce a short, multi-stemmed growth form (Reynolds and Bohning 1956, Busch 1995).

Biotic Interactions

HERBIVORY

Although abiotic factors are key factors shaping dryland riparian vegetation patterns, biotic interactions also are important. Mycorrhizal fungi, for example, influence plant productivity and survivorship (see chap. 4). Beaver, as ecosystem engineers, contribute to expansion of marsh vegetation through their dam-building and tree-felling activities (see chap. 6). The many herbivores in riparian ecosystems, ranging from insects to large mammals, have widely varying effects, ranging from minor leaf loss to whole-plant mortality.

Cottonwood trees have relatively few defenses against herbivores, allocating most resources to growth and reproduction and relying on a survival strategy of rapid regrowth of consumed plant parts. Bagworms, for example, can be abundant along the San Pedro River, and after the worms defoliate the cottonwoods, the trees produce another leaf flush. This regrowth strategy is effective for mature plants but not necessarily for seedlings.

In contrast to cottonwood, saltcedar is well defended against some herbivores. Where cottonwood, willow, and saltcedar seedlings co-mingle, cattle will selectively feed on the former, leading to dominance shifts towards unpalatable saltcedar (Stromberg 1997). Seepwillow is another plant that increases in relative abundance with livestock grazing. The flavinoids and other chemical compounds in its resinous, odorous leaves deter foraging by many animals. In similar fashion, livestock grazing can favor the structurally well-defended (i.e., spiny) mesquite over less-well-defended grass species. Sacaton, however, has high silica content and is less palatable than some other perennial grasses (Cox and Morton 1986).

Large herbivores such as cattle affect vegetation by trampling as well as browsing (Auble and Scott 1998), and livestock removal can lead to a pulse of recruitment of pioneer tree species and increased structural complexity of riparian vegetation (M. Scott et al. 2003). In some settings, however, removal of cattle from riparian zones and uplands may ultimately lead to reduced abundance of woody pioneer vegetation. Cover of streamside herbaceous vegetation has increased substantially in areas of the San Pedro River where livestock are excluded (Krueper et al. 2003) (figs. 1.10a, 1.10b), and this dense growth can restrict opportunities for establishment of pioneer trees by stabilizing river bank sediments, inducing local channel deepening, and reducing formation of open surfaces during channel migration (see chap. 13).

INTERACTIONS BETWEEN PLANTS

Plants can modify the abundance of other plant species through facilitation and competition (Levine 2000b). Mesquite, for example, may facilitate high productivity of understory vegetation through their symbiotic relation-

Fig. 1.10. Vegetation became more abundant along the low-flow channel of the San Pedro Riparian National Conservation Area after livestock were excluded. These repeat photos show a site along the upper San Pedro River in 1987 (top; 1.10a) and in 1991 (bottom; 1.10b), four years after cattle exclusion. Photos courtesy of U.S. Bureau of Land Management.

ship with nitrogen-fixing bacteria which convert atmospheric nitrogen to a form usable by plants. As the mesquite leaf litter decomposes, soils become enriched in nitrogen, particularly on geomorphic surfaces that are above the zone of frequent inundation (see chap. 14). Shallow-rooted herbs and shrubs also may benefit from the hydraulic redistribution of water from deep to shallow soil layers that occurs in areas occupied by mesquite (see chap. 2).

Plant species differ in their ability to compete for limiting resources, and shifts in hydrologic regimes can bring about changes in plant species abundance by altering competitive hierarchies. Of note, the shifting abundance of saltcedar vs. cottonwood/willow along spatial and temporal moisture gradients is a result of differing competitive abilities coupled with inter-specific differences in environmental tolerance ranges. Where conditions are suitable for cottonwood seedlings to establish, the cottonwoods competitively reduce growth rates of saltcedar seedlings (Sher et al. 2000, Sher et al. 2002, Sher and Marshall 2003); however, where soils are dry or salty, or where livestock are abundant, saltcedar "wins" the competition.

Summary

The San Pedro riparian corridor supports many vegetation types, the component species of which are influenced in varying ways by water availability, water quality, flood and fire disturbance, climate, soil traits, and various biotic interactions including herbivory and competition. Riverine marshes and ciénegas occupy the wettest fluvial surfaces, with the riverine marshes being transitory depending on flood and drought cycles, and ciénegas persisting mainly as remnants on elevated surfaces with groundwater inflow and minimal flood disturbance. Fremont cottonwood/Goodding's willow forests predominate on floodplains in wetter reaches of the river and give way to deeper-rooted saltcedar shrublands where water tables are deeper and streamflows more intermittent. Pioneer xeroriparian shrublands of burrobrush and rabbitbrush also are common on coarse flood deposits and in dry reaches of the river. Cottonwoods, willows, and saltcedar all establish after floods, and young individuals of these woody pioneer species are more abundant in downstream reaches where floods are more intense. Velvet mesquite forests and sacaton grasslands are best developed on the pre-entrenchment river terraces, where groundwater can be several meters below the surface. Both species are facultative phreatophytes, but mesquite is the more deeply rooted of the two. Mesquite predominates in the hotter and drier lower basin, while sacaton is abundant only in the upper basin. To some degree, the spatial differences in vegetation type abundance from the upper to lower river parallel the temporal changes that have occurred over the past century and a half (table 1.1). The riparian vegetation will continue to change in response to ongoing forest succession and to changes in riparian water tables, flood patterns, fire regimes, beaver abundance, and livestock grazing, as driven by climatic fluctuations, human land and water use, and legacies of past events including historic river entrenchment.

DAVID G. WILLIAMS
AND RUSSELL L. SCOTT

Vegetation-Hydrology Interactions

DYNAMICS OF RIPARIAN PLANT WATER USE

TWO

Introduction

Riparian vegetation intercepts surface and subsurface water flowing from drainage basins and forms a functionally important interface between terrestrial and aquatic ecosystems. The influence of riparian vegetation on hydrological processes (Malanson 1993, Tabacchi et al. 2000) and, conversely, the impact of hydrological processes on riparian vegetation (Rood and Mahoney 1990, Auble et al. 1994, Johnson 2000; and see chap. 1) have been the focus of considerable scientific investigation. Through such investigations, ecologists and hydrologists have formed productive, collaborative relationships and together have generated broad conceptual understanding of hydrological factors controlling riparian ecosystem structure and function and associated feedbacks with stream hydrology and geomorphology (Vannote et al. 1980, Minshall et al. 1985, Junk et al. 1989, Gregory et al. 1991, Stromberg et al. 1993).

Riparian vegetation influences hydrological processes through effects on runoff and through control of uptake, storage, and return of water to the atmosphere (Tabacchi et al. 2000). Interactions between riparian vegetation and hydrological processes are especially strong in arid and semi-arid basins where the quantity and spatio-temporal distribution of water clearly determine the composition and structure of vegetation and associated biogeochemical processes. The stream catchment is commonly used as the fundamental unit for hydrological study, and it is within this unit that ecologists and hydrologists make measurements at a compatible scale.

This chapter describes the current state of knowledge of vegetation-hydrology interactions in riparian ecosystems of arid and semi-arid basins, focusing on studies conducted along the upper San Pedro River. We highlight some key concepts related to the influence of riparian plants on stream hydrology, describe patterns of water use by dominant vegetation communities along the San Pedro River, and evaluate some of the principal methodologies used to make these determinations. The upper portion of this river near the international border with Mexico has been the focus of numerous ecohydrological studies (D. Williams et al. 1998, Goodrich, Chehbouni et al. 2000, Goodrich, Scott et al. 2000, Schaeffer et al. 2000, R. Scott et al. 2000a, 2004, R. Scott, Goodrich et al. 2003, R. Scott, Goodrich et al. 2006, Snyder and Williams 2000, 2003, Yepez et al. 2003, Hultine et al. 2004). The principal goal of this research is to understand the impact of climate variability, land-cover variation and change, and hydrogeology on the water balance of riparian vegetation. These studies have focused particularly on evapotranspiration by riparian vegetation with special emphasis on groundwater use by velvet mesquite (*Prosopis velutina*) and sacaton (*Sporobolus* spp.) bunchgrass on terraces and Fremont cottonwood (*Populus fremontii*) on floodplains. These three, together with Goodding's willow (*Salix gooddingii*), and the shrubs seepwillow (*Baccharis salicifolia*) and saltcedar (*Tamarix* sp.), are the dominant vascular plants along primary and secondary channels, floodplains, and terraces of the river (see chap. 1).

The hydrogeomorphic, geologic, and climatic setting of the upper San Pedro River is complex (see chaps. 13 and 15). Variation in the type and depth of underlying bedrock, texture and composition of alluvial materials, and channel morphology produces substantial spatial heterogeneity in depth to groundwater, streamflow, and unsaturated zone hydraulic properties. Layered over this complex spatial geomorphology and hydrology is riparian vegetation characterized by heterogeneous patterns and agents of disturbance. Precipitation, an important driver of ecohydrologic patterns and processes, is seasonally and interannually variable in this semi-arid environment. The complexity of riparian vegetation structure, hydrogeomorphic setting, and climate in the upper San Pedro River basin creates significant challenges and opportunities for research aimed at understanding processes linking vegetation and hydrology and scaling these processes to the catchment level (Goodrich, Scott et al. 2000, R. Scott, Goodrich et al. 2006).

The need for better understanding and modeling of vegetation-hydrology interactions in riparian zones of arid and semi-arid basins is critical because

(1) a high level of regional biological diversity in dry landscapes occurs within riparian ecosystems, (2) riparian vegetation plays an important role in nutrient biogeochemistry affecting water quality (Peterjohn and Correll 1984; and see chap. 18), (3) riparian environments are valuable to humans for recreation and aesthetic appeal, and (4) abstraction of groundwater or diversion of streamflow for other human uses requires quantitative and objective evaluation of potential impacts on vegetation (Rood and Mahoney 1990, Hughes and Rood 2003).

Vegetation-Hydrology Interactions in Riparian Ecosystems

The role of riparian plants in stream hydrologic processes is best illustrated using the concept of a hydrologic balance (eq. 1).

$$Q_{in} + G_{net} + P - Q_{out} - ET - \Delta S = 0 \qquad \text{[eq. 1]}$$

Here, Q_{in} and Q_{out} are streamflow into and out of the defined reach, G_{net} is net input of groundwater, P is precipitation occurring within the defined reach, ET is evapotranspiration from plants, soil and open water, and ΔS is the change in water storage in the alluvial floodplain deposits. Riparian plants can physically influence the hydraulic properties of the streambed and floodplain surface, thereby affecting transmission and storage of water in alluvium (Tabacchi et al. 2000). Further, the structural and physiological characteristics of dominant riparian plants control patterns, sources, and rates of ET at the landscape scale (Sala et al. 1996).

THROUGHFALL, STEMFLOW, AND INFILTRATION

Riparian plants influence the capture and cycling of precipitation. That fraction of precipitation falling on vegetation which does not quickly evaporate passes through the canopy as throughfall or is routed via stemflow to the plant base, where it can infiltrate the soil (Crockford and Richardson 2000). Living and dead plant roots and coarse debris at the ground surface create and alter infiltration flow paths, facilitating rapid transmission of direct precipitation, throughfall, and stemflow to deeper soil layers.

HYDRAULIC REDISTRIBUTION

Roots create macro-pores for infiltration and percolation and also modify water transport processes in riparian floodplains through "hydraulic redistribution" (Dawson and Pate 1996, S. Burgess et al. 2001, Hultine, Cable et al. 2003, Hultine, Williams et al. 2003, Hultine et al. 2004). When living root systems span soil layers with different moisture levels, water is passively redistributed along gradients from wet to dry soil layers through uptake and leakage by fine roots (Caldwell et al. 1998). Hydraulic redistribution, known as "hydraulic lift," when soil water is transported upward in the soil profile,

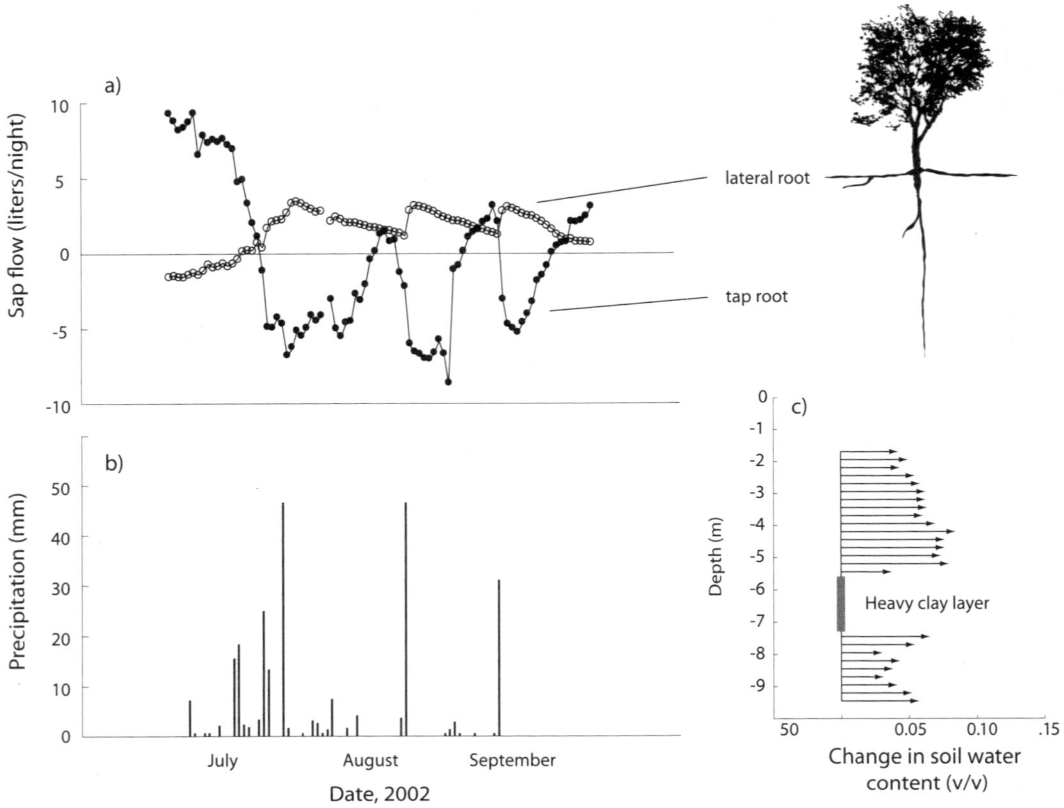

Fig. 2.1. (a) Total nocturnal sap flow of a tap root and lateral root of a large mesquite tree (*Prosopis velutina*) on a pre-entrenchment terrace along the San Pedro River in southeastern Arizona. Negative values indicate flow from tree to soil, and positive values represent flow from soil to tree. (b) Daily precipitation totals at this field site during the 2002 summer growing season. (c) Change in soil water content measured by cross-borehole ground penetrating radar (GPR) at the same field site in deep unsaturated soil layers (below layers directly wetted by precipitation and above the capillary fringe of the water table) from June 17, 2002, before the onset of the summer monsoon, to September 25 of the same year at the end of the monsoon. From Hultine et al. (2004). Used with permission.

typically occurs at night when root water potentials (the relative potential energy of water) rise as transpiration-induced tension on plant xylem water is reduced. The rise in energy of root water compared to water in the surrounding soil allows water to move passively from inside the root to the soil, in the opposite direction from daytime patterns. The wetted soil stimulates microbial activity, facilitating mineralization of important plant nutrients (Caldwell et al. 1998). Although the ecological significance of hydraulic redistribution may, to a large degree, be associated with effects on nutrient availability, the amount of water redistributed by plant root systems is not trivial (D. Smith et al. 1999, Hultine et al. 2004, Ryel et al. 2004).

It is still questionable whether hydraulic redistribution substantially alters landscape-level hydrologic balances. Hultine et al. (2004) revealed that sustained periods of downward hydraulic redistribution by velvet mesquite on upper river terraces of the San Pedro during the summer monsoon corresponded to progressive accumulation of water deep in unsaturated alluvial substrates (fig. 2.1). If this accumulation at depth was strictly a function of hydraulic redistribution of groundwater and rainfall by mesquite roots, as the data suggest, and not rapid transmission of rainfall through soil macro-pores, then widespread conversion of sacaton grassland to mesquite-dominated communities may profoundly alter the relationships between precipitation, moisture storage in shallow and deep soil layers, and ET in riparian floodplain environments. Increased moisture storage in deep unsaturated layers resulting from hydraulic redistribution by mesquite would enhance the available water for this deep-rooted shrub during dry periods. Hydraulic redistribution by mesquite root systems takes place also during winter months when mesquite canopies are dormant and leafless; thus, mesquite presence potentially affects the spatial and temporal dynamics of both winter and summer precipitation storage (ΔS) and use within these ecosystems (Hultine et al. 2004).

EVAPOTRANSPIRATION OVERVIEW

Evapotranspiration is a dominant, but poorly characterized, component of the riparian corridor water balance. ET includes water taken up and transpired by plants to the atmosphere and direct evaporation of water from the stream, soil, and living and dead plant surfaces. Evaporation of precipitation intercepted by plant canopies is included within this term and can account for as much as 30 percent of the total ET, depending on microclimatic conditions, plant species composition, and rainfall intensity and duration (Shuttleworth 1988, Crockford and Richardson 2000). Evaporation from soil surfaces and interception constitutes a substantial fraction of ET release in upland ecosystems of arid and semi-arid environments; however, in riparian ecosystems of drylands, transpiration is by far the most important component of ET (Yepez et al. 2003).

Riparian plants control transpiration rates by regulating stomatal pores on leaf surfaces. Excessively high rates of transpiration are avoided through closure of stomatal apertures to prevent tension-induced breakage (cavitation) of the continuous hydraulic connection that exists from soil, through roots and up to leaf surfaces (Sperry 1995, Sperry et al. 1998). The embolisms (i.e., the air-filled and dysfunctional xylem conduits) formed as a result of cavitation reduce the water-conducting capacity of the xylem, potentially exacerbating physiological stress in the plant. Indeed, tension-induced cavitation is likely the cause of branch dieback and tree mortality in riparian cottonwoods subjected to water table decline (Tyree et al. 1994). Riparian plant species such as mesquite, saltcedar, willow, and cottonwood differ in their vulnerability to tension-induced xylem cavitation (Pockman and Sperry 2000) and regulate stomata to different degrees when subjected to soil or atmospheric drought (S. D. Smith et al. 1991, Horton et al. 2001a). Such species-specific stomatal behaviors and drought-stress vulnerabilities influence species' distributions and vegetation-hydrology interactions at a large scale.

In arid and semi-arid environments, ET of riparian vegetation greatly exceeds precipitation due to access to groundwater by deep-rooted plants (R. Scott et al. 2004, R. Scott, Goodrich et al. 2006). Potential ET in the San Pedro basin along the river corridor is ~1.8 m per year (R. Scott, Goodrich et al. 2006), and when the water table is shallow, actual ET in riparian vegetation approaches this maximum potential ET set by the available energy driving the vaporization of water (Gazal et al. 2006). Actual ET generally is less than potential ET due to stomatal limitation (Gazal et al. 2006, R. Scott, Goodrich et al. 2006), leaf phenology patterns of cold-deciduous species, and disequilibria between vegetation successional development and water availability (Gregory et al. 1991). Under conditions of high atmospheric evaporative demand or low soil-water availability, riparian vegetation transpires at rates well below potential ET (R. Scott, Watts et al.2003, Gazal et al. 2006). As a result, simple models relying strictly on potential ET do not adequately predict uptake and release of water to the atmosphere by riparian vegetation.

Water Sources of Dominant Plant Species

The key vegetation process affecting riparian zone hydrology is uptake of water from the phreatic zone (groundwater) by deep-rooted plants followed by release of this water to the atmosphere. Unfortunately, groundwater uptake is one of the more difficult processes to characterize. Obligate phreatophytes (e.g., cottonwood and willow) are plant species that do not survive without access to groundwater. Facultative phreatophytes (e.g., mesquite and saltcedar) are plants that take up groundwater if it is available but are capable of using alternative water sources (e.g., rainfall) and can survive without access to groundwater.

Because of their structural dominance in riparian ecosystems of the American Southwest, the physiological ecology and water relations of willow, cottonwood, mesquite, and saltcedar have been intensively studied (S. D. Smith et al. 1991, Busch et al. 1992, Busch and Smith 1995, Sala et al. 1996, Cleverly et al. 1997, Devitt et al. 1997, Glenn et al. 1998, Snyder and Williams 2000, 2003, J. Horton et al. 2001a, Nagler et al. 2003, Hultine et al. 2004, Glenn and Nagler 2005, Gazal et al. 2006). Stable isotope tracer studies, which take advantage of variation in the abundance of deuterium (^2H) and ^{18}O in different source waters (Ehleringer et al. 2000), confirm that transpiration by willow and cottonwood originates predominantly from the phreatic

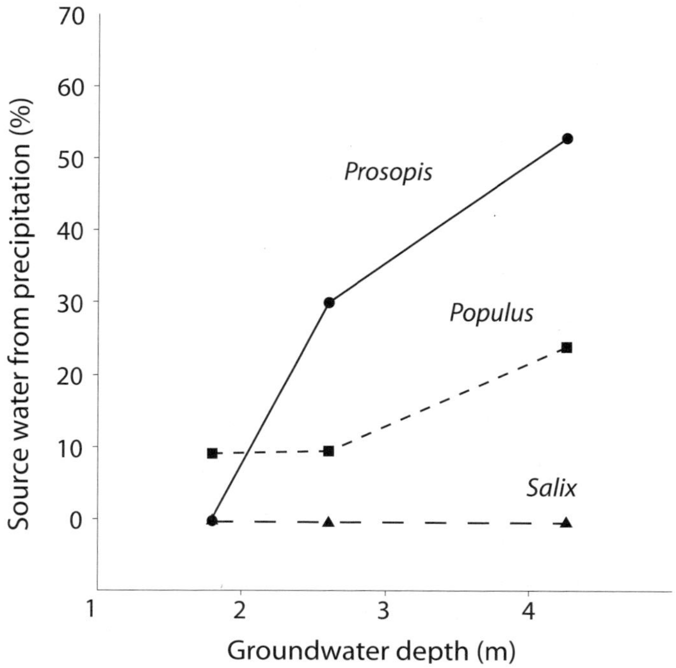

Fig. 2.2. Percentage of total plant transpiration in populations of mesquite (*Prosopis velutina*), cottonwood (*Populus fremontii*), and willow (*Salix gooddingii*) obtained from recent rainfall inputs at three sites along the San Pedro River in southeastern Arizona. The sites represented a gradient in depth to groundwater and reveal the heterogeneous nature of vegetation-water interactions in riparian ecosystems. From K. Snyder and Williams (2000). Used with permission.

zone (Busch et al. 1992, Snyder and Williams 2000). However, where water table depths for cottonwood exceeded 4 m, approaching the depth threshold for survival (see chap. 1), cottonwood obtained a significant proportion (>20 percent) of transpiration water from unsaturated soil layers wetted by rainfall (Snyder and Williams 2000) (fig. 2.2).

The facultative phreatophytes mesquite and saltcedar switch water sources more readily than cottonwood and willow, depending on depth to groundwater and availability of water in unsaturated alluvial soils (Busch et al. 1992, Snyder and Williams 2000, 2003). Across a gradient of water table depth from 1.8 to 4.2 m, Snyder and Williams (2000) observed shifts in the contribution of groundwater to mesquite transpiration from 100 to less than 50 percent (fig. 2.2). Saltcedar is similar to mesquite in its capacity to tolerate drought stress and its flexibility to use moisture either from groundwater or unsaturated soil layers (Busch and Smith 1995). Not fully resolved is whether uptake of rainfall by these riparian species replaces or adds to water taken from groundwater. This last point is important for understanding the influence of riparian phreatophytes on groundwater resources.

Unraveling the complex controls on actual ET and identifying sources of plant transpiration (groundwater, surface water, soil water, etc.) in the riparian floodplain are the two critical pieces of information needed to understand how riparian vegetation influences the groundwater hydrology of stream catchments. Identifying and partitioning plant water sources is fairly straightforward using stable isotope tracer methods (Ehleringer and

Dawson 1992, Ehleringer et al. 2000); however, measuring and predicting ET and its source across a complex landscape is somewhat more challenging.

Quantification of Evapotranspiration in Riparian Ecosystems

An accounting of ET across the riparian landscape requires fairly sophisticated measurement and modeling approaches. The techniques we review here have been employed in research on the upper San Pedro River. These include water budget, in situ measurement, modeling, and remote sensing approaches.

WATER BUDGET APPROACHES

Researchers use various forms of the water balance equation (eq. 1) to quantify riparian ET (Freethey 1982, Corell et al. 1996, Goodrich, Scott et al. 2000). To do this, ET is determined as the residual of the balance between inflow and outflow terms over a defined river reach and over a specified time interval. Because of the difficulty of estimating many of the equation's components over a river reach, a number of simplifying assumptions are often made to reduce the number of terms. For example, along the upper San Pedro River, researchers have used long-term streamflow data to compute an estimate of ET from groundwater by comparing the river baseflows (groundwater-derived streamflow) during winter months, when transpiration by the riparian vegetation is minimal, to those of summer, when the vegetation is active (e.g., Corell et al. 1996). On a smaller reach and for a shorter period, Goodrich, Scott et al. (2000) synthesized independent estimates for each term in eq. 1 for a 90-day, dry-season period and found that the water balance was within 5 percent of total inflows. However, such a detailed water balance is not feasible over larger reaches and longer time intervals due to the dynamics of precipitation, net groundwater inputs, and storage terms.

DIRECT MEASUREMENTS

Direct measurements of ET are necessary for validating and calibrating indirect measurement and for providing data input for modeling approaches (see next section). Direct approaches to measure riparian vegetation ET include sap flow and micrometeorological approaches.

Sap flow techniques. Sap flow techniques are used widely in riparian vegetation research because of their remote field portability, low cost, and capacity for high temporal and spatial resolution targeted to individual plants (Schaeffer et al. 2000, Nagler et al. 2003, Hultine et al. 2004, Gazal et al. 2006, R. Scott, Goodrich et al. 2006). This technique has advantages over other direct measurement techniques because it does not require a large, homogeneous patch of vegetation for application. It is well suited for strips of streamside vegetation (Schaeffer et al. 2000) where micrometeorological approaches present difficult challenges (Hipps et al. 1998).

The approach measures (1) the velocity of a heat pulse introduced to the flowing sap (Barret et al. 1995), and (2) the dissipation of heat applied continuously from heaters inserted into the wood (Granier 1987) or completely enveloping a branch (Baker and van Bavel 1987). When calibrated properly for variation in wood properties and damage from probe insertion, sap flow techniques provide highly reliable and accurate data. The difficulty with this approach lies in scaling the very localized point measurements on a tree to the stand and landscape levels. Compounded errors associated with spatial scaling can be significant (+/- ca. 10 to 20 percent), but not insurmountable (Schaeffer et al. 2000).

Micrometeorological techniques. Micrometeorological techniques, including the Bowen ratio and eddy covariance methods, monitor patch-scale ET in riparian areas (Devitt et al. 1997, Cleverly et al. 2002, Dahm et al. 2002, R. Scott et al. 2000a, 2004, R. Scott, Goodrich et al. 2006). These techniques involve measuring above-canopy meteorological conditions over a relatively homogeneous and extensive (>~200–500 m) patch of land (fig. 2.3), which limits their applicability in many riparian settings (Baldocchi et al. 1988, Schmid 1997). The advantages of micrometeorological data are their high temporal (often <1 hour) resolution, near-continuous sampling (data sets can be collected over months to years), and patch-scale spatial resolution, which is readily amenable to scaling up spatially. The disadvantages are its high cost ($15,000–$45,000) and low portability.

The Bowen ratio technique is simpler and less costly than eddy covariance measurement but requires complete measurements of the surface energy balance to determine ET. Over the last several years, the eddy covariance technique has become the predominant micrometeorological technique for measuring ecosystem ET. While the instruments are expensive and the data analysis sophisticated, it has the advantages of measuring ET directly and not requiring assumptions about energy balance. As a bonus, many eddy covariance setups can also estimate carbon dioxide exchange between the atmosphere and the land surface to understand the relationship between ecosystem productivity and water use (Baldocchi et al. 2001, R. Scott et al. 2004, R. Scott, Huxman et al. 2006).

The accuracy of micrometeorological techniques for estimating ET depends on many site-specific factors and is generally difficult to ascertain. This is especially true for riparian ET measurements because one cannot rely on a simple surface water balance ($ET = P - \Delta S$) as a check since the contributions from the groundwater source are unknown. For eddy covariance measurements, a standard check is to evaluate how the surface energy

Fig. 2.3. (top; 2.3a): A 14-m tall micrometeorological tower measures evapotranspiration from a riparian mesquite woodland along the San Pedro River. (bottom; 2.3b): Micrometeorological instrumentation to measure evapotranspiration from a floodplain sacaton grassland along the San Pedro River. Photo credits: Russell Scott, USDA-ARS.

balance is closed. Along the San Pedro River, the eddy covariance sites have not demonstrated complete closure (R. Scott, Goodrich et al. 2006). Yet, this is a common occurrence for most sites around the world, and there has been no consensus as to whether this indicates the possibility that ET is underestimated using this approach (Wilson et al. 2002).

MODELING APPROACHES

There is a huge array of models that range from simple empirical relationships to very detailed ecohydrology land-surface models that estimate ET for specific vegetation types (Shuttleworth 1991). The simplest are based on derived empirical relationships between more easily measured atmospheric measurements—like air temperature, relative humidity, wind speed, and solar radiation—and ET measurements. However, it is often problematic to apply empirical relationships developed for a specific site and vegetation type elsewhere, and these models should be proven locally to validate their application. For example, the Arizona Department of Water Resources (ADWR 1991) modeled annual ET for several vegetation types over a large reach of the San Pedro River using a map of riparian vegetation and an empirical model (Blaney and Criddle 1950) driven by mean monthly air temperatures and the monthly percentage of daytime hours of the year (see R. Scott et al. 2000a for a summary of these model-based estimates). The vegetation-specific ET parameters for this model were derived from a study on the lower Gila River Valley in Arizona (Gatewood et al. 1950), and the results of this study were not validated due to the lack of data on the San Pedro. Numerous measurements of annual ET made later (table 2.1) are considerably less (500 mm yr^{-1} less for some vegetation types) and reveal the danger of using unsubstantiated model results.

The more complex models are physically based and try to represent the key physical mechanisms that control the ET process, yet applying these models is difficult due to the greater number of vegetation-specific parameter values required. Thus, it is often necessary to perform model calibration (again, requiring direct measurements for comparison) to determine the often unknown, or even unmeasurable, parameters to ensure the model's accuracy. For example, Goodrich, Scott et al. (2000) used sap flow estimates of cottonwood/willow ET (Schaeffer et al. 2000) to calibrate the stomatal resistance term in the well-known Penman-Monteith model (Monteith 1965). The calibrated model was then used with hourly air temperature, relative humidity, wind speed, and solar radiation to determine annual cottonwood/willow ET (fig. 2.4).

REMOTE SENSING

Remote sensing has the promise of making measurements of land-surface variables (e.g., surface temperature, solar radiation, soil moisture, absorbed photosynthetically active radiation, vegetation greenness) relevant to the ET process over large areas and at regular intervals (from sub-hourly to weeks). Researchers have developed many ways to utilize remote sensing measurements to drive ET models, ranging from empirical to physically based models. An example of an empirical approach is a regression relationship between

TABLE 2.1. Estimates of total growing season ET and ET from groundwater along the San Pedro River.

Cover type	Streamflow	Geomorphic surface	Vegetation notes	Date	Depth to GW [m]	ET [mm]	ET_{gw} [mm]	Canopy cover [%]	ET'_{gw} [mm]
Cottonwood/ Willow	Perennial	Floodplain		1997	0.5–2	755	755	100	755
Cottonwood	Perennial	Floodplain	Mature stand	2003	1.6	966	966	100	966
	Intermittent	Floodplain	Mature stand	2003	3.3	484	410	100	410
Mesquite	Intermittent	Floodplain	Woodland	2001–2003	10	638–694	488–510	74	532–689
	Perennial	Floodplain	Dense shrubland	2003	6.5	565	380	55	691
	Perennial	Floodplain	Sparse shrubland	1997	9.5	330	157	32	491
Sacaton	Perennial	Floodplain		1997	>3.5	272	0	NA	0
		Floodplain		2003	2.5	554	374	65	575
Seepwillow	Perennial	Floodplain	Cottonwood understory	2003	1–1.5	819	NA	100	NA
Open water	Perennial	Channel	Under mixed cottonwood canopy	2003	NA	1156	1156	NA	1156

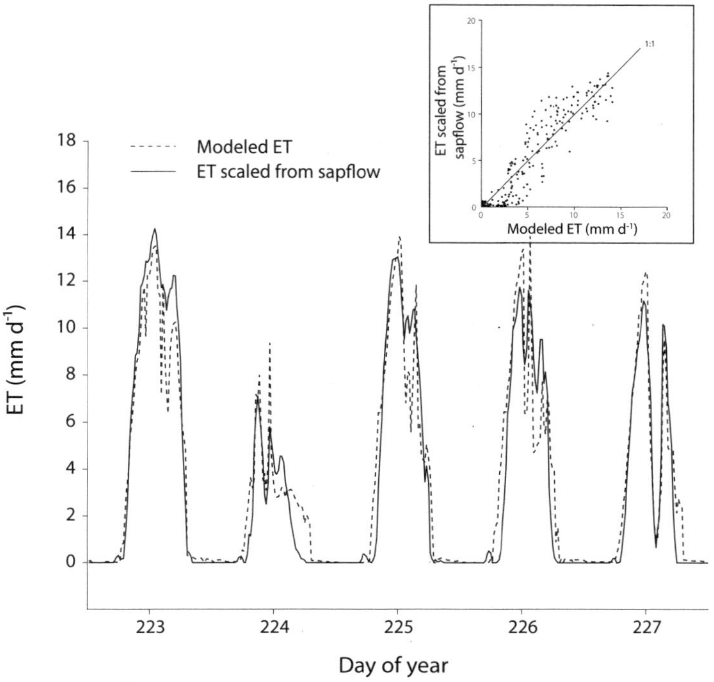

Fig. 2.4. Comparison of calibrated Penman-Monteith model for cottonwood/willow evapotranspiration versus scaled sap flow estimates for a five-day period during the monsoon in 1997. The study site was along a perennial section of the upper San Pedro River with a dense, but structurally heterogeneous, canopy of riparian trees. From Goodrich, Scott et al. (2000). Used with permission.

a vegetation index and ET, whereas a more physically based approach would use remote sensing to estimate light absorption or surface temperature as inputs in models that specifically represents the surface energy balance (Bastiaanssen et al. 1998, Davis and Roberts 2000, Kustas et al. 2003, Nagler, Cleverly et al. 2005).

With the increased availability of remote sensing products from programs like NASA's Earth Observing System, there have been many recent attempts to use remote sensing to assess transpiration from riparian plant canopies. One approach is to determine ET indirectly from remotely sensed canopy temperature and in situ air temperature (Qi et al. 1998, Nagler et al. 2003). This approach is based on the cooling effect of transpiration on leaf surfaces and the remote measurement of canopy temperature (T_c) relative to surrounding air (T_a) (Jackson et al. 1981).

Another empirical approach is to use relationships between satellite-derived vegetation indices and ET. For example, Nagler, Cleverly et al. (2005) found strong correlations between MODIS-derived indices (Huete et al. 2002) and riparian ET along the Rio Grande and developed an empirical model using these indices and local air temperature to skillfully predict ET. This regression relationship has been updated (Nagler, Scott et al. 2005) with additional measurements of different cover types along the San Pedro and lower Colorado Rivers. They found that one relationship could be used to predict ET from a range of riparian cover types from different Western rivers, and they used this relationship to predict ET for entire reaches.

In contrast to these empirical approaches, Bastiaanssen et al. (1998) developed a more physically based model that uses radiances recorded by satellite-based sensors, plus ordinary meteorological data, to solve the energy balance at the earth's surface. This model is currently being applied and tested over many different riparian vegetation types (http://www.sebal.us). One advantage of this more physical approach is that the model does not require that the vegetation be well watered, which might be the reason why the vegetation index approach of Nagler, Scott et al. (2005) has worked in its application so far but might not work for riparian vegetation along ephemeral flowing rivers that experience more seasonal water stress. One possible setback to the use of remote sensing for riparian ET is that the 0.25–1 km scale of many of the satellite measurements like surface temperature or some vegetation indices can be greater than the scale of homogeneous riparian vegetation patches of interest.

Patterns of Evapotranspiration in the Upper San Pedro Riparian Ecosystem

Many of the in situ estimates of riparian ET along the upper San Pedro River come from the Semi-arid Land-Surface-Atmosphere (SALSA) Program (Goodrich, Chehbouni et al. 2000), the National Science Foundation's Sustainability of Semi-arid Hydrology and Riparian Areas (SAHRA) center, and the Upper San Pedro Partnership (USPP) Water Needs Study (Leenhouts et al. 2006). For these studies, the amount of ET derived from groundwater was quantified so that the total groundwater use by phreatophytes in the basin, an important component of the basin's groundwater budget, could be determined. ET rates, summarized in table 2.1 and shown diagramatically in figure 2.5, were quantified for various vegetation types in a range of hydrogeomorphic settings (Schaeffer et al. 2000, Goodrich, Scott et al. 2000, R. Scott et al. 2000a, 2004, R. Scott, Goodrich et al. 2006). In addition to seasonal ET, estimates were made for the amount of ET derived from groundwater (ET_{gw}, table 2.1) using a mass balance approach for the micrometeorological studies (R. Scott, Goodrich et al. 2006) or by isotopic identification of source water (Snyder and Williams 2000). Finally, ET_{gw}, normally given in units of water volume per land surface area per time, was divided by the fractional phreatophyte cover in order to provide an estimate of total riparian groundwater use for 100 percent land cover (ET'_{gw}) so that a total ET for a reach could be determined.

INTER-ANNUAL VARIABILITY

For vegetation types with multi-year ET estimates (the cottonwoods along a perennial reach and the mesquite woodland), there was considerable inter-

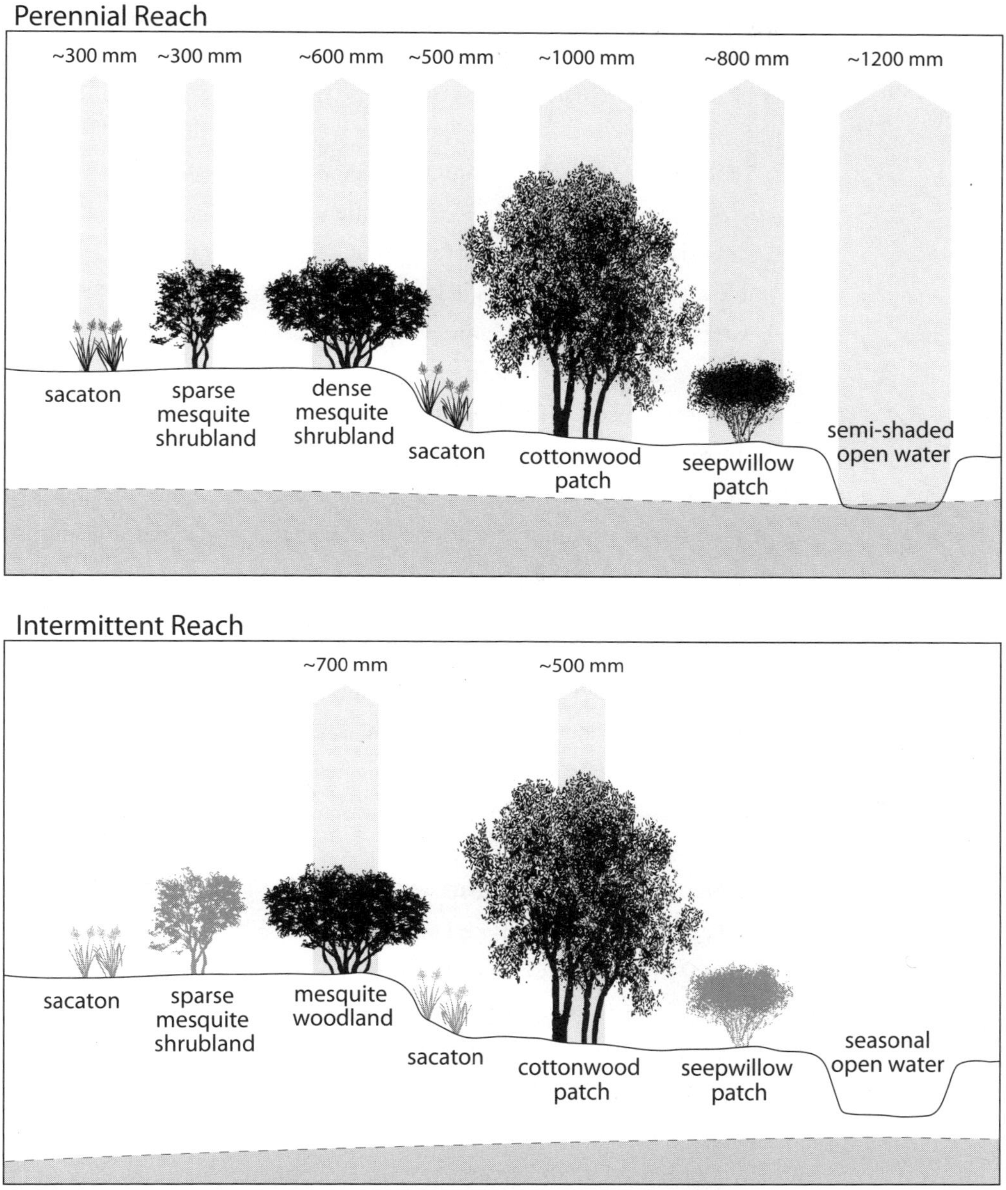

Fig. 2.5. Stylized representation of growing season evapotranspiration (ET) rates for riparian vegetation types (and open water) on different fluvial surfaces (floodplains and terraces) and in perennial and intermittent reaches of the San Pedro River riparian corridor (values are based on measurements given in table 2.1). ET values are not available for vegetation patches shown in gray. Illustration credit: Mike Buffington.

annual variability (table 2.1). We speculate that most of this variability is controlled by climate, namely, spring and fall temperatures (frost events) and precipitation. The spring and fall temperatures determine the length of the transpirational growing season and are quite variable from year to year (R. Scott, Goodrich et al. 2006). The amount of precipitation within the growing season also impacts total ET due to use of shallow soil water by plants such as mesquite. Wintertime precipitation also appears to play a role in the functioning of mesquite later in the year, likely through storage of water by hydraulic redistribution (Hultine et al. 2004, R. Scott et al. 2004).

VARIATION IN ET WITHIN AND AMONG PLANT COMMUNITIES

In general, the highest rates of ET along the upper San Pedro River are for open water surfaces followed by cottonwood, seepwillow, mesquite, and sacaton (table 2.1). For cottonwoods, ET and ET_{gw} were higher along a perennial reach with shallow groundwater but markedly reduced along an intermittent reach with a greater depth to, and fluctuation in, groundwater. ET was highly variable among mesquite communities, but ET_{gw} was less variable, indicating that the groundwater use by mesquite trees was similar across sites with different depths to groundwater (R. Scott, Goodrich et al. 2006). Sacaton had high ET and used groundwater at one site but had low ET and did not use groundwater at another site with a greater depth to groundwater; a depth to groundwater of 3 m is a likely cutoff that determines whether this grass does or does not use substantial quantities of groundwater. Finally, seepwillow ET was large, despite the reduced atmospheric demand present in the cottonwood understory occupied by these shrubs. Apparently, the cottonwood canopy structure allows sufficient (radiative and/or advected) energy to reach the understory shrubs, allowing them to transpire at high rates.

Evapotranspiration in Stands of Saltcedar (*Tamarix* spp.)

Reported estimates of saltcedar ET vary widely. Variable environmental conditions and differing structural characteristics observed among saltcedar stands may explain these differences. Evapotranspiration in any riparian vegetation type is spatially and temporally variable due to differences in vegetation density, groundwater depth, seasonal plant phenology, and available energy. Collectively, studies reporting ET in saltcedar reveal magnitudes and levels of temporal and spatial variability comparable to those reported for stands of native riparian woody vegetation (Robinson 1970, Weeks et al. 1987, Sala et al. 1996, Devitt et al. 1998, Cleverly et al. 2002, Dahm et al. 2002). On a per leaf area or plant basis, saltcedar has transpiration rates similar to native riparian taxa (Busch and Smith 1995, Sala et al. 1996, Cleverly et al. 2002, Dahm et al. 2002). Where saltcedar occurs at very high densities and has a high leaf area index, its ET may be greater than for other riparian vegetation (Sala et al. 1996, Nagler et al. 2004).

However, several recent studies employing eddy covariance methods show that saltcedar ET is not greater than riparian stands dominated by native woody taxa (Weeks et al. 1987, Cleverly et al. 2002, Dahm et al. 2002). Total annual ET of saltcedar stands ranged from 74 cm in an unflooded stand to 122 cm in a flooded stand (Cleverly et al. 2002). For comparison, total growing season ET was 48 to 97 cm in unflooded stands of cottonwood and 33 to 69 cm in unflooded stands of mesquite along the San Pedro River (table 2.1). Clearly, saltcedar is not an extraordinary water user in all situations. Indeed, leaf area index, a measure of the evaporative surface area of leaves per unit of ground area, which is highly correlated to actual ET in riparian vegetation (Sala et al. 1996, Schaeffer et al. 2000, Cleverly et al. 2002), was, on average, lower in stands of saltcedar than in stands of cottonwood and willow along the lower Colorado River (Nagler et al. 2004).

Scaling Evapotranspiration up to the Catchment Level

Riparian ET at the stream reach or catchment level can be a major component of the surface water and groundwater budgets of watersheds in arid and semi-arid environments, and ET at these scales is often the most relevant for watershed stakeholders. Often the surface water and groundwater budgets are determined independently, so it is important to identify the riparian ET source water so that ET can be partitioned into surface water (ET_{sw}) and groundwater (ET_{gw}) sources. In the Sierra Vista subbasin of the upper San Pedro River basin, the focus has been on determining groundwater use by riparian vegetation because perennial flow in the river, the condition of the riparian vegetation, and people living in the basin all depend on water from this aquifer source.

Prior to the SALSA Program (Goodrich, Chehbouni et al. 2000), numerous estimates of upper San Pedro River riparian corridor groundwater usage were derived indirectly using conceptual and numerical models (e.g., Freethey 1982, Vionnet and Maddock 1992, Corell et al. 1996). More recently, Goodrich, Chehbouni et al. (2000) and R. Scott, Goodrich et al. (2006) have quantified riparian corridor groundwater use from direct measurements of ET (see Quantification of Evapotranspiration in Riparian Ecosystems, p.44). These estimates of ET_{gw} (table 2.1) for each vegetation type were multiplied by the area within the riparian corridor covered by that vegetation and then totaled to determine riparian groundwater use. A major advantage to this approach is that it is based on actual measurements rather than on model outputs, which have considerable uncertainty at large scales. A major disadvantage stems from the implicit assumption that ET of a particular vegetation type can be accurately represented from measurements at a few locations. Site-to-site heterogeneity in vegetation age, density, and environmental conditions is the rule along most rivers, and the effects of this heterogeneity on ET are often unknown. Instrumentation networks need to be expanded to sample the range of conditions occupied by each particular vegetation type, but the costs and feasibility of such sampling can be prohibitive.

Table 2.2 presents a detailed riparian groundwater-use budget for the San Pedro River from the international border to the Tombstone gage during

TABLE 2.2. Riparian groundwater use in 2003 along the main stem of the San Pedro River from the international border to the Tombstone gaging station (from R. Scott, Goodrich et al. 2006).

Cover type	Minimum amount[d] [ha]	Maximum amount[d] [ha]	2003 ET_{gw} [mm yr^{-1}]	Min. GW Use [1000 m^3 yr^{-1}]	Max. GW Use [1000 m^3 yr^{-1}]
Mesquite	723	973	689	4983	6706
Cottonwood/Willow (perennial flow)	253	253	966	2444	2444
Cottonwood/Willow (intermittent flow)	118	118	410	484	484
Sacaton (<3 m to groundwater)	113	167	575	650	961
Open water	43	43	1156	497	497
Saltcedar[a]	1	3	689	7	21
Total				9065	11112
Total from Corell et al. (1996)[b]					8758
Total from Goodrich, Scott et al. (2000)[c]					8130

[a]Saltcedar *ET* was not measured and assumed the same as mesquite.

[b]Using baseflow information from the Palominas, Charleston, and Tombstone gages and subtracting their estimate of 600 ac-ft/yr for the Babocomari River.

[c]From the international border to the Tombstone Gage.

[d]Vegetation amounts vary due to the polygon-based map, which did not specify exact percentages of vegetation cover for some of the sacaton and mesquite polygons.

2003 (R. Scott, Goodrich et al. 2006). The median estimate of 2003 groundwater withdrawal by riparian vegetation along the San Pedro and within the Sierra Vista subwatershed is 10,089,000 m^3 yr^{-1} (table 2.2), and the natural groundwater recharge estimated to have occurred in 2002 within this basin is 22,203,000 m^3 yr^{-1} (ADWR 2005a). This reveals the importance of riparian ET in the water budget.

The estimates of riparian groundwater use from R. Scott, Goodrich et al. (2006) and Goodrich, Scott et al. (2000), the two most recent studies that used a similar approach, differ by nearly 3,000,000 m^3 yr^{-1} or 27 percent of the larger amount. The difference between these two studies can be explained by revisions and improvements made by R. Scott, Goodrich et al. (2006) to the approach of Goodrich, Scott et al. (2000) including (1) revised estimates of water use of different cover types, based on improved measurement methodology and sampling of a greater number of sites and a greater range of hydrologic conditions, and (2) use of a more detailed vegetation map. The difference in vegetation areas between the two vegetation maps is alarming. This type of scaling depends on an accurate vegetation map, and differences between the two maps highlight the difficulty in vegetation classification.

Uncertainty in the estimates of R. Scott, Goodrich et al. (2006) is difficult to quantify and arises from measurement errors, the application of site-specific measurements across the reach, and uncertainty in vegetation amounts. Given these complications, uncertainty will always be a part of these and future reach-scale estimates even while we hope that their accuracy improves with each successive iteration.

Conclusions and Challenges for Future Research

Due to spatial and temporal heterogeneity in water availability and a high-frequency pulse-disturbance flood regime, the structure and hydrologic function of riparian plant communities in arid and semi-arid environments are extremely dynamic. Understanding the effects and feedbacks associated with spatio-temporal heterogeneity and scaling these responses to the catchment level represent great challenges to researchers focused on ecohydrological interactions in riparian ecosystems. Riparian vegetation influences the hydrologic cycle directly through controls on interception, soil water redistribution, and evapotranspiration, and indirectly through effects on stream geomorphology and accumulation of coarse woody debris. Evapotranspiration is particularly difficult to quantify and model in riparian ecosystems. However, the combined application of sap flow, eddy covariance, isotope, and remote sensing techniques has greatly reduced the uncertainty in estimates of ET and clarified our understanding of vegetation-groundwater relationships.

We still have limited understanding of how riparian vegetation water use responds to discrete pulses of rainfall, flood events, and rapid groundwater fluctuations. A more complete integration of plant ecophysiological properties and vegetation dynamics within a hydrological context will lead to models of vegetation-hydrology interactions that include pulsed resource and disturbance processes. Such models, when calibrated with landscape characteristics and vegetation parameterizations, offer researchers a valuable scientific tool for synthesizing information on key processes and identifying limitations to conceptual understanding. For stakeholders who wish to negotiate the most prudent policies for managing riparian vegetation and groundwater resources, such models could form the basis for identifying risks and potential outcomes of management activities.

An accurate accounting of riparian ET is required to properly apportion water for both human and environmental needs, especially along rivers in dryland regions like the southwestern United States. Accordingly, efforts to estimate riparian ET similar to those used for the San Pedro have been, or are being, employed in the middle Rio Grande and lower Colorado basins (Brower 2004, USBOR 2002). A common approach of using state-of-the-art direct measurements of ET over representative vegetation types to calibrate and improve models driven with local meteorological and/or remote sensing measurements has emerged in these efforts. As these efforts progress, a more general understanding of the riparian ET process and how to quantify this process will emerge.

MARK D. DIXON,
JULIET C. STROMBERG,
JEFF T. PRICE,
HECTOR GALBRAITH,
ALEXANDER K. FREMIER,
AND ERIC W. LARSEN

Potential Effects of Climate Change on the Upper San Pedro Riparian Ecosystem

THREE

Climate Change

Over the last two centuries, exponential growth of the human population, coupled with technological advancement, has transformed the surface of the planet. This transformation has been powered by the burning of fossil fuels, returning carbon that had been captured in photosynthesis hundreds of millions of years ago back to the atmosphere. Levels of atmospheric CO_2, a key greenhouse gas, have risen over 35 percent since pre-industrial times and approximately 20 percent since the 1950s (IPCC 2007). Increases in levels of methane, another greenhouse gas, have been even greater (an approximately 150 percent increase since 1750) (IPCC 2007).

Greenhouse gases function as a kind of global insulation, trapping re-radiated heat from the earth's surface, warming the lower atmosphere, and moderating Earth's climate. In tandem with the increases in greenhouse gases, global mean temperatures have risen an estimated 0.74°C (0.56–0.92°C) over the last century. Coincident with recent warming have been widespread glacial retreat, polar ice thinning, sea level rise, and precipitation increase in the Northern Hemisphere (IPCC 2007). Biota have undergone changes in phenology of reproduction and migration and have expanded their ranges poleward and upward in elevation, suggesting significant ecological responses to recent climate change (Walther et al. 2002, Parmesan and Yohe 2003, Root et al. 2003).

Models of the global climate system project significant changes in temperature and precipitation patterns over the next 50 to 100 years in response to the rising CO_2 levels. Although much uncertainty remains, average model projections across a range of fossil fuel emissions scenarios suggest increases in global average temperature of 1.8–4.0°C over the next 100 years (IPCC 2007). The climate changes are projected to modify global hydrologic processes, with higher rates of both precipitation and evaporation. Greater year-to-year variability in rainfall and runoff may also result, with more extreme climatic and hydrologic events (Easterling et al. 2000, Arnell 2003, IPCC 2007). On balance, the world is projected to become a warmer and wetter place, but effects will vary considerably by region (Ragab and Prudhomme 2002, Arnell 2003, Manabe et al. 2004, IPCC 2007). In general, regions in the high latitudes are expected to become wetter, while subtropical regions are expected to become drier. Most of the current General Circulation Models (GCMs) project a warmer and drier American Southwest in decades to come (IPCC 2007, Seager et al. 2007), although warmer sea surface temperatures could also spur stronger and more-frequent El Niño–associated winter storms (Garfin and Lenart 2007). Recent satellite observations, however, suggest that current GCMs may substantially underestimate precipitation increases accompanying recent and future global warming (Wentz et al. 2007)

CLIMATE CHANGE AND RIVERS

Climate change will modify runoff and streamflow in river basins. Effects on rivers are likely to be complex because the delivery of water to the channel is influenced by the interaction of precipitation, temperature, land use, and vegetation (Pfister et al. 2004). Precipitation and temperature interact to influence the ratio of runoff/infiltration vs. evaporation and thus affect total annual river discharge. River flows are further influenced by basin vegetation and land use, which may also change in response to climate, and affect ratios of runoff to infiltration and evapotranspiration.

Effects of climate change on rivers will differ by region. River basins at high latitudes are projected to have increased winter precipitation, with earlier snowmelt and an increased proportion of precipitation as rain rather than snow, and thus greater but more variable spring peak flows (Arnell and Liu 2001, Arora and Boer 2001, Nijssen et al. 2001, Arnell 2003). On snowmelt-dominated mid-latitude rivers, including many in the western United States, projected climate changes would reinforce the recent trajectory towards earlier snowmelt and earlier and less-predictable spring peak flows (Gleick and Chalecki 1999, Lettenmaier et al. 1999, Dettinger et al. 2004, Stewart et al. 2004, Gibson et al. 2005). Many arid or semi-arid river basins may exhibit declines in runoff due to increased evaporative demand from increased temperatures, leading to streamflow and flood size reduction (Nijssen et al. 2001, Ragab and Prudhomme 2002, Arnell 2003, Manabe et al. 2004), although increases in storm intensity (e.g., in tropical cyclones; IPCC 2007) could lead to periodic, larger-magnitude peak flood events.

Climate change could impact human use of water resources and exacerbate the effects of flow regulation and water extraction on riparian and aquatic ecosystems. On the Colorado River, warming temperatures and modest decreases in precipitation could lead to declines in reservoir storage and significant shortfalls in meeting human water demands by the mid-twenty-first century (Christensen et al. 2004). On other impounded rivers, high human demand for water coupled with climatically driven declines in runoff could decrease flow releases downstream of the dam, further reducing the frequency of floodplain inundation and severing ecological connections between the historic floodplain and the aquatic ecosystem (Gibson et al. 2005). On the other hand, potential increases in storm size could increase the frequency of flood releases, important for sustaining many ecological functions such as providing spawning habitat for fish, stimulating reproduction of riparian plants, recharging soil moisture and aquifers, and stimulating decomposition and movement of woody debris.

CLIMATE CHANGE PROJECTIONS FOR THE SOUTHWEST

In the Southwest, the naturally high variation in temperature and precipitation across time and space (Andrade and Sellers 1988, Sheppard et al. 2002), the importance of long-term climatic cycles (like the Pacific Decadal Oscillation and El Niño–Southern Oscillation), and the relatively short history of climate records make it difficult to separate the signal of climate change from natural variation. However, climate records suggest that average annual temperatures have increased 1.1–1.7°C over the last century in the Southwest (Colorado, Utah, Nevada, New Mexico, Arizona, and southeastern California [SRAG 2000]). Changes in precipitation patterns have been more variable, with increases in some areas (e.g., central Arizona) and decreases in others (e.g., southern Arizona).

Climate models used in the U.S. National Assessment of the Potential Consequences of Climate Variability and Change project warming through the next century in the Southwest (SRAG 2000) but disagree about the direction, timing, and magnitude of precipitation changes (table 3.1). The Hadley and Canadian Climate Centre global climate models project warming of mean seasonal temperatures of 1.5–4°C by 2060 and 2–7°C by 2090 in the region, with the greatest warming in the winter (table 3.1). Both models also project a potential doubling of annual precipitation by 2090, with most of the increase in the winter. A regional model developed by the National Center

TABLE 3.1. Means of temperature and precipitation projections for the Southwest Region from the Hadley 2, Canadian, and NCAR High Resolution Climate Models (see http://www.ispe.arizona.edu/research/swassess/pdf/complete.pdf).

Year	2030		2060			2090	
	Hadley Centre Model 2	Canadian Climate Centre Model	Hadley Centre Model 2	Canadian Climate Centre Model	NCAR Regional Climate Model	Hadley Centre Model 2	Canadian Climate Centre Model
Mean temperature change (°C)							
Winter	+2.5	+3.0	+2.0	+4.0	+4.0	+4.0	+7.0
Spring	+1.5	+2.0	+1.5	+3.0	+4.0	+2.0	+6.0
Summer	+1.5	+2.0	+2.5	+3.0	+5.0	+3.0	+5.0
Fall	+1.5	+1.5	+3.0	+4.0	+4.0	+3.0	+5.0
Mean precipitation change (mm/day)							
Winter	+1.0	+1.5	+1.5	+1.5	-1.0	+5.0	+4.5
Spring	+0.5	+0.3	+0.3	+0.5	+0.3	+2.0	+1.0
Summer	+0.3	0.0	0.0	-0.3	-0.3	0.0	0.0
Fall	0.0	+0.5	-0.3	+1.0	0.0	+3.0	+1.0

for Atmospheric Research (NCAR) also projects significant seasonal warming (4–5°C) by 2060, with the greatest warming in the summer but projects a decline in annual and winter precipitation (table 3.1). None of the three models projects strong changes in summer precipitation; current climate models do not adequately represent convective precipitation events, such as those dominating the North American Monsoon (Grimm et al. 1997). Recent high-resolution model runs across the continental United States, however, suggest that the Southwest could experience significant reductions in summer precipitation, along with increases in the frequency and duration of extreme high temperatures (Diffenbaugh et al. 2005). Most of the GCMs used in the recent Intergovernmental Panel on Climate Change (IPCC) assessment also suggest that the southwestern United States will become drier in coming decades (IPCC 2007). If so, the recent multi-year drought in the Southwest may become the norm, with perpetual drought conditions of an intensity rivaling the megadroughts of the past (Seager et al. 2007).

Any climate change scenario will interact with and overlay existing climate patterns driven by large-scale atmosphere-ocean connections. Historical, empirical patterns suggest that El Niño conditions could become more frequent under a warmer climate (Hunt 1999, Timmermann et al. 1999, Wang and Schimel 2003, Garfin and Lenart 2007), potentially bringing greater frontal precipitation to the Southwest during winter. Alternatively, winter and annual precipitation could be lower over the next 2–3 decades if the Pacific Decadal Oscillation (PDO) transitions into its negative (cool) phase (Mantua et al. 1997, Hamlet and Lettenmaier 1999, McCabe et al. 2004). Negative phases of the PDO tend to be associated with reductions in the intensity and frequency of El Niño events. The early-1950s drought coincided with the joint occurrence of cool conditions in the Pacific (negative PDO) and warm conditions in the north Atlantic (McCabe et al. 2004), and the negative phase of the PDO predominated during the 1950s–mid-1970s period, during which fall and winter floods were infrequent (see chap. 16).

SOUTHWESTERN U.S. RIVERS AND CLIMATE CHANGE

Dynamics of aquatic and riparian ecosystems in the southwestern United States are sensitive to climatic change, given the strong linkage between riparian zone pattern and processes and precipitation-driven fluvial geomorphic and hydrologic processes (Dahm and Molles 1992, Grimm and Fisher 1992; and see chaps. 1 and 13). However, projecting the influence of climate change on rivers and riparian ecosystems in the Southwest is complicated by several factors. First, although there is general agreement that the region will warm, there is high uncertainty about the magnitude, seasonality, and even the direction of precipitation change. Second, some key climatic drivers for the region, such as the North American Monsoon (Sheppard et al. 2002), orographic effects, and multi-year atmospheric-oceanic teleconnections like the El Niño–Southern Oscillation (ENSO) and PDO, are poorly represented in existing climate models (Grimm et al. 1997). Third, heterogeneous terrain, differences in elevation, and gradients in rainfall seasonality, magnitude, and important rainfall-generating storm systems mean that responses to climate change will not be uniform across the region (Grimm and Fisher 1992). In addition, the high species diversity of riparian ecosystems and the unique sets of environmental tolerance ranges displayed by individual species make it difficult to project all of the possible permutations of response to the direct (temperature, precipitation) and indirect (streamflow rates, flood regimes, sediment dynamics, fire regimes) effects of changing climates. Finally, ongoing human alterations of rivers, riparian zones, and watersheds will interact in complex ways with the effects of climatic change (Grimm et al. 1997, Tellman et al. 1997, Patten 1998, Christensen et al. 2004, Gibson et al. 2005).

The Upper San Pedro as a Case Study

The upper San Pedro is a well-studied river, but there has been little consideration of the potential effects of climate change on this system (but see Skirvin et al. 2000). Here, we use a modeling approach to explore the sensitivity of the San Pedro riparian ecosystem, particularly the composition and dynamics of riparian vegetation, to a range of climate change scenarios, bounded by the results of global and regional climate models used in the Southwest Regional Assessment (table 3.1). Our projections are based on the direct effects of climate on plant growth and its indirect effects on surface water dynamics (flood magnitude and timing) and disturbance (fire, floods). Because our models have not yet been rigorously tested against field data, our conclusions should be considered preliminary.

CLIMATE SCENARIOS AND MODEL OVERVIEW

We simulated long-term vegetation change under five climatic scenarios (table 3.2) for eight study transects that span a gradient of hydrologic conditions within the San Pedro Riparian National Conservation Area (table 3.3). We refer to these scenarios as no change, warm, warm wet, warm very wet, and warm dry. Although some of the scenarios may seem extreme (e.g., doubling of winter precipitation in 100 years), they fall within the range of projections for the region (table 3.1). Weather inputs to the model were developed from historic (1951–2002) daily temperature and precipitation data from Tombstone and other local weather stations, adjusted for transient changes expected under each scenario for 2003–2102. Only "winter" (October 1–May 31) precipitation levels were modified from historic conditions, as none of the scenarios projected significant regional changes in summer monsoonal rains.

TABLE 3.2. Summary of climate change scenarios developed for modeling vegetation change on the San Pedro River.

Climate scenarios for vegetation simulation	Temperature change from historic 1951–2002 data, by 2102	Precipitation change from historic 1951–2002 data, by 2102
1. No Change	None	None
2. Warm	+4°C daily maximum, +6°C daily minimum	None
3. Warm Dry	+4°C daily maximum, +6°C daily minimum	50% decline in Oct.–May daily precipitation
4. Warm Wet	+4°C daily maximum, +6°C daily minimum	50% increase in Oct.–May daily precipitation
5. Warm Very Wet	+4°C daily maximum, +6°C daily minimum	100% increase in Oct.–May daily precipitation

TABLE 3.3. Characteristics of study sites modeled in simulation runs.

Site name	Flow permanence	Flow permanence class	Percent cw/willow coverage	Percent saltcedar coverage
Palominas 3	61%	Wet intermittent	26%	0%
Palominas UA	58%	Wet intermittent	81%	10%
Kolbe	100%	Perennial	77%	0%
Hereford	95%	Perennial	34%	0%
Moson	100%	Perennial	100%	0%
Contention	48%	Dry intermittent	3%	0%
Summers	69%	Wet intermittent	75%	0%
St. David	33%	Dry intermittent	0%	46%

Notes: Sites are listed from upstream to downstream. Permanence of streamflow is the percent of months in 2002–2003 with observable surface flow (Stromberg et al. 2006). Coverage of cottonwood/willow and saltcedar patches is expressed as the percent of floodplain occupied by each.

We developed a model using STELLA II® dynamic simulation modeling software (ISEE Systems, Lebanon, NH) to project floodplain vegetation change under these five scenarios. The model simulates the potential effects of different climate scenarios on the plot-level survival and growth of 11 dominant species or functional groups of riparian plants along the San Pedro. In addition to the direct effects of temperature and precipitation on plant growth, vegetation will change in response to many climate-linked factors. Those that we modeled were changes in streamflow, using a watershed runoff model (SWAT) (Arnold et al. 1994); rates of channel change, using a channel migration model (MEANDER) (Larsen and Greco 2002, Larsen et al. 2006); and the frequency and intensity of wildfire (Rothermel 1983). A more detailed description of modeling methods will be published in a forthcoming paper.

The channel migration model was useful for representing the effects of climate on geomorphic processes that drive riparian patch dynamics: the loss of old patches to bank erosion and the creation of new patches for recruitment by cottonwood and other pioneer species. One limitation of the geomorphic model, however, was that it could be used only for representing lateral migration of the channel, based on the existing channel geometry and lateral shifts observed under past flow events. Hence, potential changes in channel width (e.g., widening) or depth (e.g., downcutting), which were important drivers of change on the San Pedro in the early decades of the twentieth century, could not be represented. Because of this, our modeling assumed that precipitation and flow thresholds for severe channel incision and floodplain "unraveling" (J. Smith 2004) would not be exceeded in our simulations. These thresholds are poorly known but have apparently not been exceeded on the upper San Pedro over the last 50 years (see chap. 12).

Some other potentially important drivers of vegetation change (i.e., changes in the incidence of severe frosts, effects of CO_2 concentrations on plant growth) were not included in the model. Effects of other human actions, including groundwater pumping or future land management, also were not examined. These human actions may interact with climate change to either partially counteract or amplify its effects.

MODEL RESULTS

Streamflow and stream geomorphology changes. Based on the SWAT model, flood flow projections differ markedly among the five scenarios (fig. 3.1). While warming alone has only a small effect on flood size, changes in winter precipitation have a strong influence on flood size, seasonality, and frequency. Relative to the no-change scenario, increases in winter precipitation under the warm wet and warm very wet scenarios result in large increases in the frequency and magnitude of winter-season floods. There is a threefold increase in flood size given a 50 percent increase in winter precipitation (warm wet) and a five- to sixfold increase with a doubling of winter precipitation (warm very wet). Occurrence of the annual flood peak in the winter becomes the norm under these scenarios. Under the warm dry scenario, in contrast, a progressive decline in winter precipitation essentially eliminates winter floods after about 2050.

Differences in flood magnitudes among the scenarios result in sharply different rates of simulated channel movement. There is very little cumulative channel migration under the warm dry scenario and substantial movement under the warm wet and warm very wet scenarios over the 100-year simulation (fig. 3.2). These differences in channel dynamics among scenarios have important implications for modeled rates of recruitment by pioneer species, such as cottonwood.

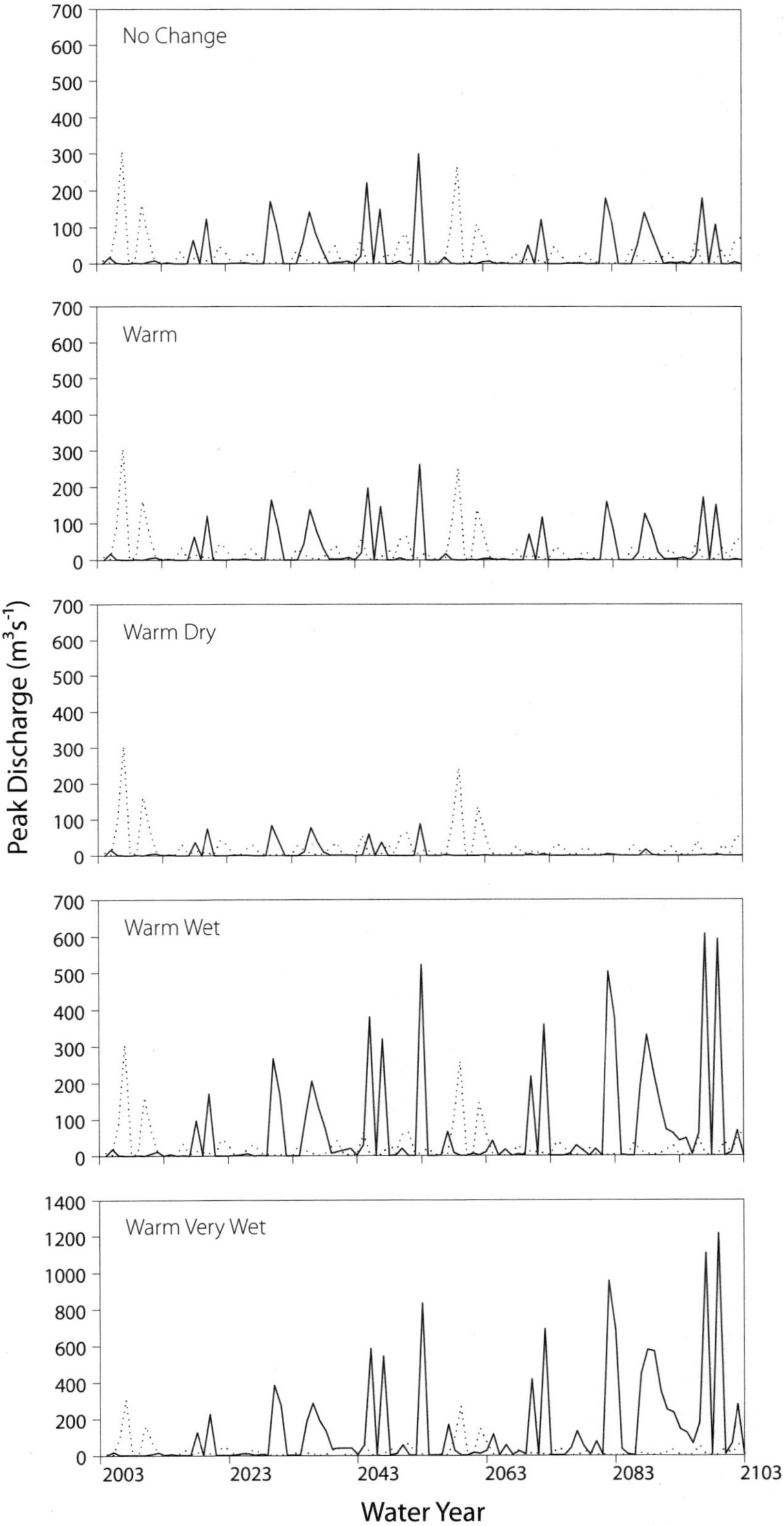

Fig. 3.1. Seasonal flow peaks projected by SWAT modeling under each climate scenario. Solid lines represent "winter" (October–May) floods, dotted lines represent "summer" floods (June–September). Note the difference in the scale of the y-axis between the warm very wet scenario (bottom graph) and the other scenarios.

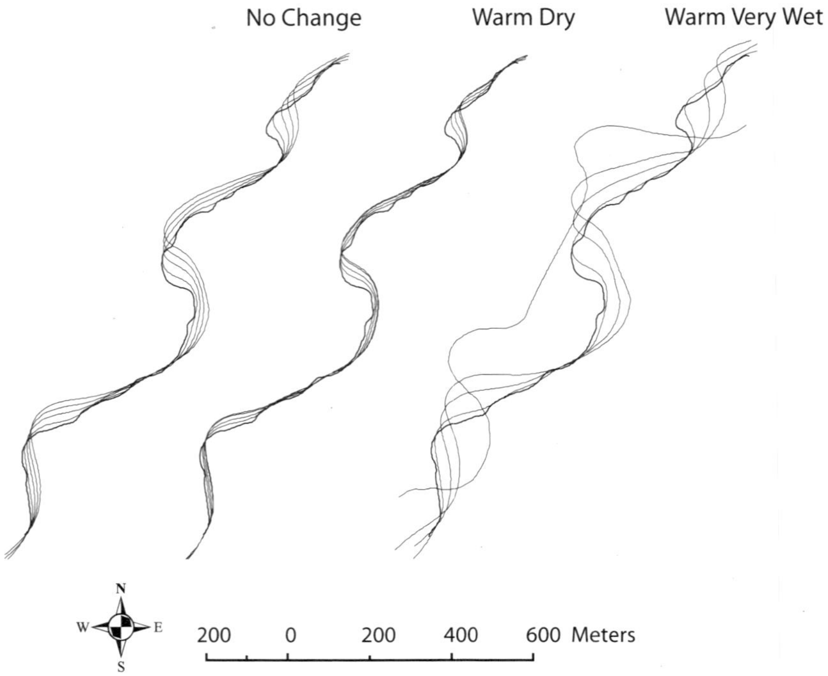

Fig. 3.2. Projected channel migration at the Palominas study site for 2003–2103 under three climate scenarios. Channel traces are shown at 25-year intervals. The darkest trace represents the 2003 channel location.

Cottonwood/willow forest changes. Significant changes in coverage of the various vegetation types over the next 100 years are projected under all scenarios (fig. 3.3). Averaged across study sites, the strongest patterns are a decrease (28 to 80 percent) in the total width of cottonwood/willow-dominated patches (*Populus fremontii/Salix gooddingii*) and a dramatic increase in coverage by patches dominated by mesquite (*Prosopis velutina*) under all climate scenarios. Sacaton (*Sporobolus wrightii*) grasslands increased under some scenarios.

Model projections across scenarios suggest that, independent of climate change, the area of cottonwood/willow forest on the upper San Pedro is already on a trajectory of decline. Under a continuation of historic conditions (no climate change) or with warming alone, our model projects a two-thirds decline in cottonwood/willow coverage from 2003 to 2103. Total declines in cottonwood/willow patch coverage are least under the warm very wet climate scenario (decline of 28 percent) and greatest under the warm dry scenario (decline of 80 percent). Declines in basin groundwater levels from groundwater pumping could further exacerbate cottonwood/willow declines.

Fremont cottonwood and Goodding's willow are relatively short-lived pioneer species that require open, moist mineral substrates for seedling recruitment and rarely regenerate within established stands. Generation of new patches requires periodic flood disturbance by channel migration, channel widening, or overbank deposition; these processes create seedbeds for recruitment of young trees (M. Scott et al. 1996; and see chap. 1). Hence, the present-day spatial and age distribution of floodplain forest is a function of past channel processes that created conditions suitable for the establishment of those

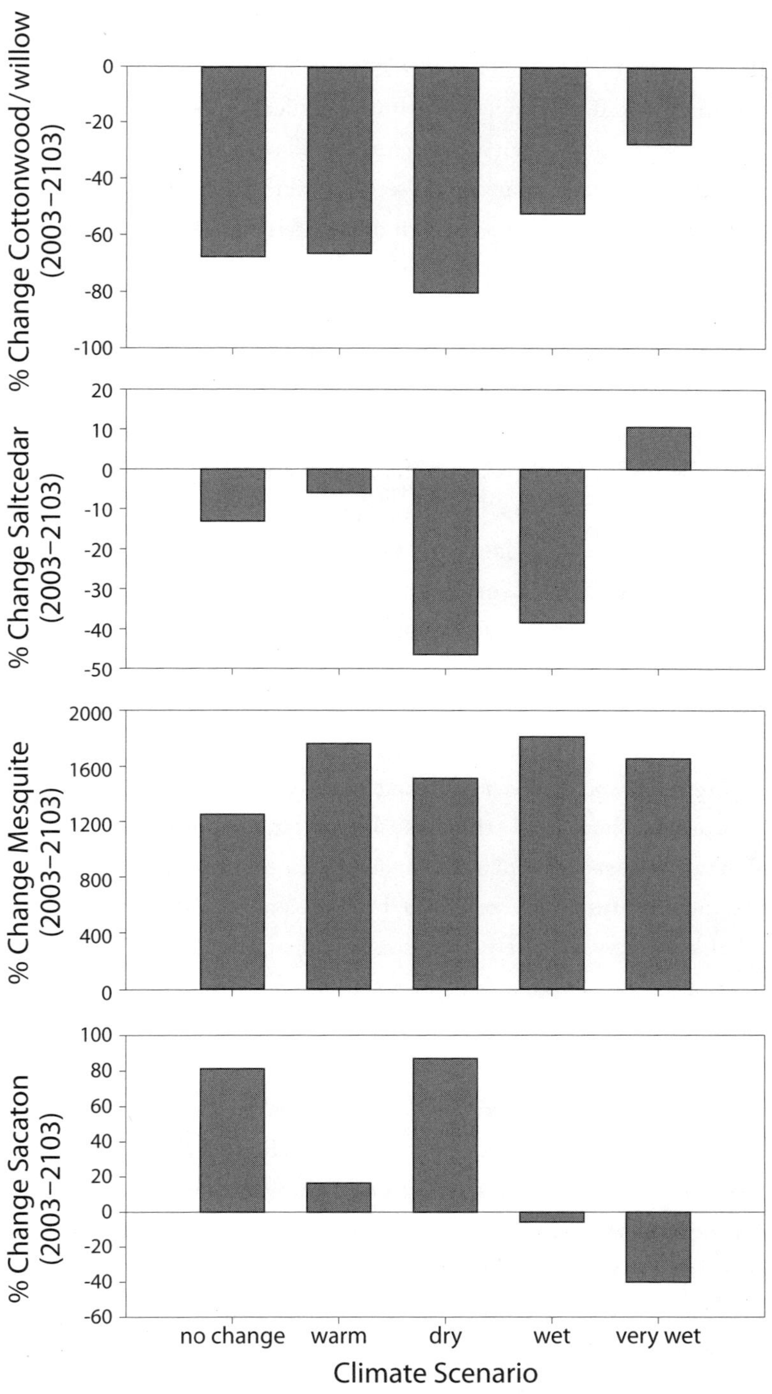

Fig. 3.3. Simulated percent changes from 2003 to 2103 in coverage of major vegetation patch types across all eight study sites.

stands. The current widespread extent of cottonwood/willow stands on the upper San Pedro is a legacy of the several decades of channel narrowing and vegetation recruitment that followed the extreme channel widening of the late nineteenth and early twentieth century and is not in equilibrium with the currently low rates of channel dynamics and small areas of seedling recruitment (Stromberg 1998a). Hence, the future coverage of pioneer trees and shrubs in the floodplain will be a balance between the rates of senescence and erosion of old patches established under past channel dynamics and the rates of new patch formation and seedling establishment under future channel dynamics. Cottonwood/willow patch area declined in the model because projected rates of future channel migration and seedling recruitment were insufficient (Johnson 1992, Lytle and Merritt 2004) to counterbalance the loss of established riparian forests by senescence and erosion. The situation may be quite different on the lower San Pedro, where channels remain highly dynamic.

Our model assumes that channel dynamics will be expressed only through lateral channel migration rather than through catastrophic widening or incision. If this assumption is valid, then only very high rates of channel migration would be sufficient to maintain the current area of cottonwood/willow forest over the next 100 years. Barring major channel change, cottonwood/willow patches in the floodplain will senesce, leaving only a narrow band of pioneer trees lining the banks of the future San Pedro. The area of these near-channel patches will vary as a function of the rate of lateral channel migration and/or periodic small-scale widening and subsequent re-narrowing.

Although the possibility remains that a large rainfall event, particularly one associated with a dissipating tropical storm (chap. 16), could overwhelm system thresholds and cause widespread channel change, vegetation changes in recent decades have made this less likely. First, flashiness of watershed runoff has decreased as upland range conditions improved during the mid-twentieth century (Hereford 1993, but see Kepner et al. 2004). Second, streambank stability has likely increased with historic riparian forest development and increased vegetation cover from recent cattle removal (Krueper et al. 2003).

Saltcedar abundance changes. Saltcedar (*Tamarix ramosissima*) coverage decreases under most model scenarios with the strongest declines in the warm dry and warm wet climates and a slight increase under the warm very wet (fig. 3.3). As with cottonwood/willow, saltcedar declines are linked with projected channel migration rates that were too low to balance out the loss of old patches from senescence and erosion. We assumed that saltcedar, like cottonwood, could not recolonize within old stands but would be replaced by later successional species. To our knowledge, however, there have been no studies that document long-term successional dynamics or replacement patterns within old-growth saltcedar stands.

The relative success of cottonwood/willow versus saltcedar in new recruitment patches depended on site hydrology: cottonwood/willow dominated at wet (perennial and wet intermittent) sites, while only saltcedar was able to recruit at the drier sites (fig. 3.4). Late spring groundwater depths greater than 1.5 m below the channel bed at the dry intermittent sites (table 3.3) exceed

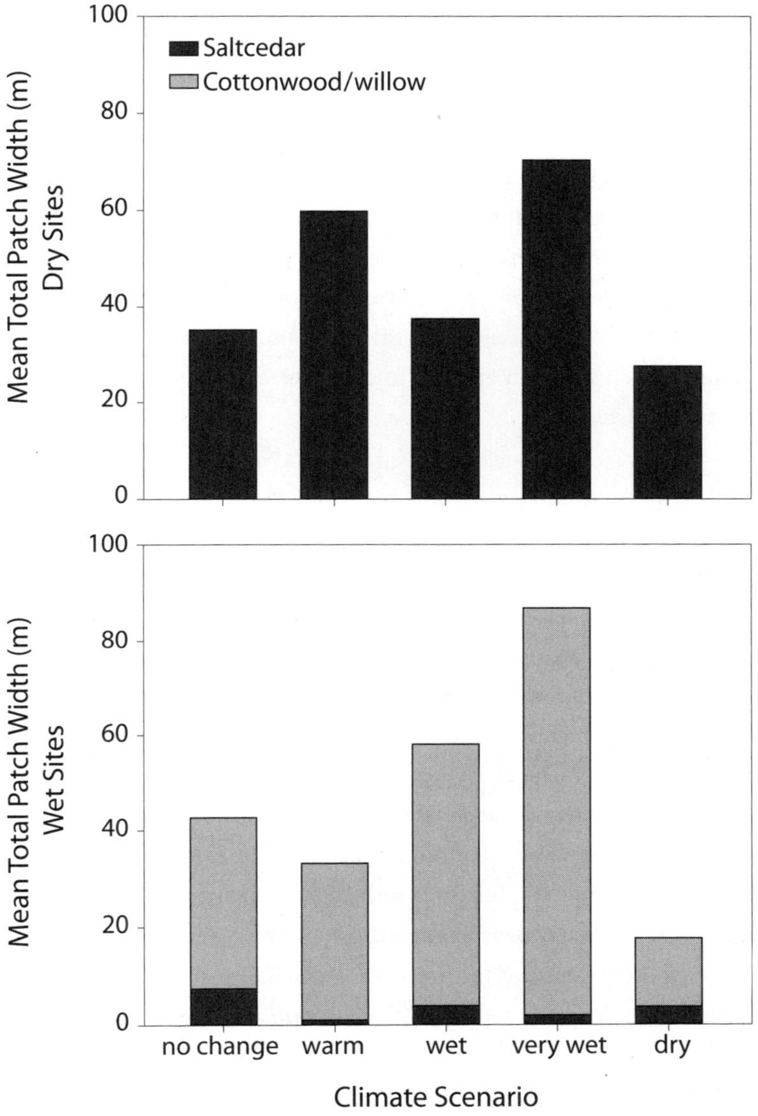

Fig. 3.4. Mean total width of new saltcedar (*Tamarix*) and cottonwood/willow (*Populus/Salix*) recruitment patches by 2103, by site, for the two dry intermittent sites (top) and the six wet intermittent and perennial sites (bottom).

mortality thresholds for cottonwood and willow seedlings and saplings but may permit recruitment of saltcedar (see chap. 1). (Additional studies are needed to confirm the validity of these modeled thresholds.) Recruitment dynamics on new patches formed by channel migration thus reinforce the original dominance of saltcedar on the driest sites (or in one case resulted in a shift from cottonwood to saltcedar) and dominance by cottonwood at the wet intermittent and perennial sites (table 3.3).

Assessing the effects of climate on groundwater dynamics was beyond the scope of our model. A recent modeling study, however, suggests that regional climate change, in tandem with continued groundwater pumping, will significantly reduce groundwater levels, streamflow, and riparian evapotranspiration on the upper San Pedro by 2100 (Serrat-Capdevila et al. 2007). Notably, these declines occur under both drier and modestly wetter climate scenarios, as estimated recharge is insufficient to offset the effects of ongoing groundwater pumping. The strong effects of site hydrology on relative cover of cottonwood/willow vs. saltcedar suggest that declines in alluvial groundwater levels, whether due to climate, anthropogenic pumping in the basin (chap. 15), or a combination of these factors, would increase the abundance of saltcedar relative to cottonwood. Saltcedar's higher drought tolerance and higher temperature optimum for photosynthesis (Horton et al. 2001b) would likely favor its spread under warmer and/or drier future climatic conditions (Zavaleta and Royval 2001), but only if flood frequency and magnitude remain high enough to create sufficient opportunities for seedling recruitment.

Dynamics of mesquite and sacaton. In contrast to cottonwood, willow, and saltcedar, mesquite does not require open sediment bars for recruitment. It can colonize open and forested surfaces on both floodplain and terrace surfaces and is long-lived (up to several hundred years) (chap. 1). In the model runs, the increase of mesquite patches across climate scenarios is partly a function of succession of aging cottonwood/willow stands to mesquite, and partly a function of mesquite expansion in sacaton grasslands and other patch types.

Mesquite increases the most under warm and wet conditions, while sacaton grassland increases the most under no climate change, or a hotter, drier (less winter precipitation) climate (fig. 3.3). The increased success of mesquite under warmer conditions is related to its treatment in the model as a subtropical species (Brown 1994) with a high temperature optimum for growth and photosynthesis (Hanson 1982, Mauchamp et al. 1994). Hence, warmer conditions lead to a longer growing season and higher growth rates for mesquite, with sacaton not responding as positively to the warmer conditions. The increased success of mesquite under wetter winter conditions may be related to interactions between deep soil moisture storage and competitive outcomes of trees vs. grasses. Deep recharge of soil moisture from winter precipitation and floods may favor growth and recruitment of deeply rooted woody shrubs and trees more than the more shallowly rooted grasses, which benefit the most from summer precipitation (Burgess 1995, Weltzin and McPherson 1997, Gao and Reynolds 2003, McAuliffe 2003). Sacaton, a deeply rooted grass and facultative user of groundwater (see chap. 2), may be more versatile in its use of different moisture sources than other grasses but presumably still is more dependent on summer precipitation and shallow soil moisture than is mesquite (R. Scott et al. 2000a). Our projections of mesquite expansion with wetter winters are consistent with observed increases in upland mesquite in the San Pedro basin (Kepner et al. 2002) and in other areas in the region (Brown et al. 1997) during the period of higher winter precipitation from the late 1970s to 1990s.

The future ratio of mesic grassland to mesquite woodland also will be influenced by fire regime. Fire exerts considerable influence on the relative coverage of trees and grass in semi-arid savannas and woodlands (McPherson 1995, Scholes and Archer 1997). Although relationships between climate and fire are complex (Swetnam and Betancourt 1998), climate warming and increased climatic extremes would likely increase the probability of intense fires in the Southwest (Brown et al. 2004). Intense fires (fireline intensity >500 kW/m^2) in the San Pedro floodplain would be more frequent under warm dry conditions and warm wet conditions than under conditions of no climate change (unpublished data of authors). As a vigorously resprouting species, mesquite is well adapted to occasional fires, although very frequent or high-intensity fires may cause high mortality (Glendening and Paulsen 1955, McPherson 1995) and conversion to grassland. Sacaton is well adapted to recurrent fire, although the combination of drought stress and fire can be lethal (Cox 1988).

Although not included in the model, increasing CO_2 levels could have a strong direct effect on plant growth rates and an indirect effect on species composition in the San Pedro riparian zone. Increases in CO_2 could increase productivity differentially among species and functional types, favoring species utilizing the C3 photosynthetic pathway (e.g., shrubs like mesquite) more than those utilizing the C4 pathway (e.g., warm-season grasses like sacaton). Historic CO_2 increases have been proposed as one possible contributing mechanism for the historic spread of shrubs in desert grasslands throughout the Southwest and the increases in shrub cover throughout semi-arid and arid grasslands and savannahs throughout the world (Idso 1992, but see Archer et al. 1995).

IMPLICATIONS FOR RIPARIAN ECOSYSTEM FUNCTION

Changes in the composition and patch configuration of riparian vegetation would have significant effects on riparian ecosystem function in the upper San Pedro floodplain. As cottonwood/willow forest and woodland decline in total area, connectivity, and width and are replaced by mesquite woodland or sacaton grassland, transpiration of groundwater by riparian vegetation will decline (see chap. 2). Changes in riparian patch composition and landscape configuration within the riparian corridor will also affect animal diversity. Reductions in the area and connectivity of cottonwood/willow patches would reduce abundance of species that rely heavily on these multi-level gallery forests, including Yellow Warbler, Common Yellowthroat, Cassin's Kingbird, Vermilion Flycatcher, and the threatened Western Yellow-billed Cuckoo (Galbraith et al. 2004; and see chap. 8). On the other hand, expansion of mesquite forests would increase habitat for birds (e.g., Lucy's Warbler) that have high fidelity to this habitat type, while shifts from riparian forest to grassland could result in increased population sizes of grassland bird species, like Botteri's Sparrow, as well as of many butterfly species (see chap. 7).

Summary

Regardless of the direction and magnitude of changes in future climate, our analysis suggests that several successional changes in the riparian vegetation of the upper San Pedro floodplain are likely. First, barring a catastrophic flood event that reshapes the floodplain, the area of cottonwood/willow forest will shrink, as larger, older stands senesce and narrower, new stands form with lateral channel migration. Barring substantial groundwater change, cottonwood/willow woodland and forest will remain important patch types on sites with shallow groundwater but will occur in narrower bands on the channel banks. Older stands of cottonwood/willow will succeed to other patch types, particularly mesquite and grassland. Recent trajectories of woody plant dynamics in the Southwest and in semi-arid and arid regions throughout the world are consistent with the idea that mesquite coverage will increase at the expense of grassland or other herbaceous communities. Increases in CO_2 levels, temperatures, and perhaps precipitation would likely accentuate this process. However, our analyses suggest that grassland may also increase, particularly if climate changes do not include significant winter precipitation increases.

All in all, future vegetation changes, whether related to climate, direct human influences, or simply to ongoing successional processes, will change the nature of the upper San Pedro. Consideration of these trajectories and their potential effects, including the potential interactions between climate change and basin-level groundwater depletion, will increase our ability to make informed decisions regarding management and conservation of the upper San Pedro ecosystem. A warmer, drier future climate could exacerbate current water use conflicts between people and accelerate the decline of cottonwood/willow gallery forests, while a wetter one could partially ameliorate the impacts of human water use and counteract current successional trajectories toward cottonwood-forest decline. If basin-wide groundwater pumping continues at high rates into the future, however, losses of cottonwood/willow forests under either drier or wetter scenarios may be greater than our projections.

JEAN C. STUTZ,
VANESSA B. BEAUCHAMP,
JEFFERY JOHNSON,
LINDA J. KENNEDY,
BRANTLEE SPAKES RICHTER,
AND KATHRYN M. JACOBSON

Mycorrhizal Ecology

FOUR

Introduction

Mycorrhizae are symbiotic associations between certain types of soil fungi and roots of plants. These associations are widespread in natural ecosystems with approximately 80 percent of terrestrial plant families colonized by fungal symbionts (Trappe 1987). Mycorrhizal fungi can benefit plants by aiding in the uptake of soil nutrients, including phosphorus and nitrogen, improving soil structure, providing protection from pathogens, and increasing stress tolerance, particularly to drought (S. E. Smith and Read 1997). Some mycorrhizal fungi play a role in nutrient cycling, especially in the mobilization of nitrogen from organic substrates (Read and Perez-Moreno 2003, Leake et al. 2004). Plants benefit fungi by providing photosynthetically derived carbohydrates for growth and reproduction. Not all plants respond equivalently to mycorrhizal fungi (Johnson et al. 1997, Jones and Smith 2004); thus, these fungi can alter competitive relationships between plant species and affect plant species composition and diversity (Bever et al. 2001). In this chapter we review the structure and functioning of mycorrhizae in riparian ecosystems with a focus on those in the deserts of the southwestern United States.

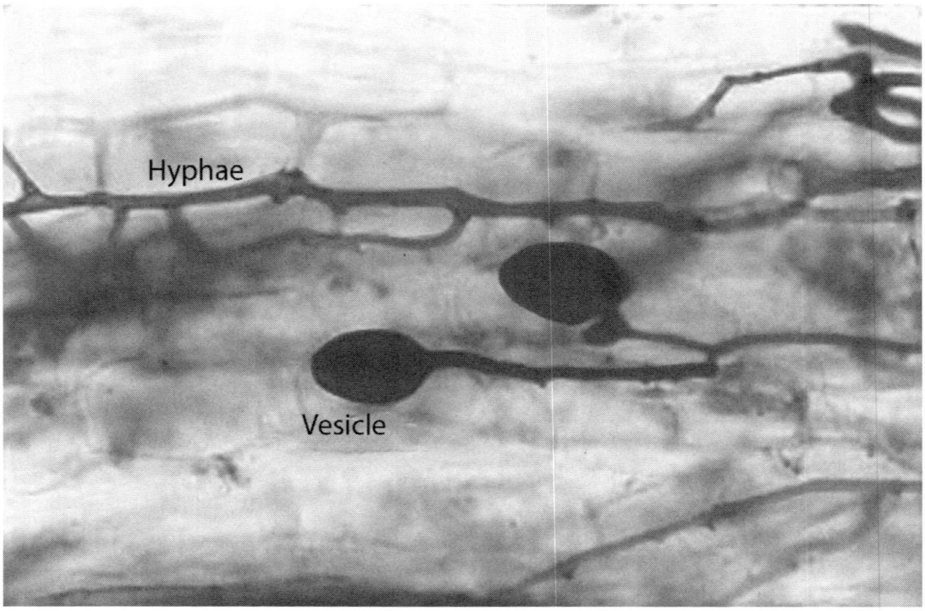

Fig. 4.1a (above) Big sacaton (*Sporobolus wrightii*) roots colonized by AM fungi; internal hyphae and swollen vesicle tip are noted. Fig. 4.1b (opposite) Black ectomycorrhizal root tips of cottonwood (*Populus*) surrounded by external hyphae, as viewed using a stereomicroscope. Fungi are stained with acid fuschin (red dye) and viewed using a light microscope. Photo credits: Jeff Johnson and Jean Stutz.

Types of Mycorrhizal Associations

The two most common types of mycorrhizal associations in riparian systems are arbuscular mycorrhizae (AM) (sometimes referred to as vesicular-arbuscular mycorrhizae [VAM] mycorrhizae) and ectomycorrhizae (EM). A third class of fungal-root associates called dark septate endophytes (DSE) may also be functionally important in desert riparian ecosystems.

AM FUNGI

AM fungi are a type of endomycorrhizae (roots are colonized internally by fungi) involving fungi in the phylum Glomeromycota (Schussler et al. 2001). AM fungi are believed to be an ancient group that played a role in the colonization of land by plants (Redecker et al. 2000), but fewer than 200 species of AM fungi have been described (Morton et al. 1995). These fungi form associations with many herbaceous and woody plants (Bever et al. 2001) and are the predominant type of mycorrhizal associations in grasslands (Hartnett and Wilson 2002), deserts (Stutz et al. 2000), and tropical forests (Janos 1983).

AM fungi colonize plant roots by forming internal structures called arbuscules between the cell wall and plasma membrane of root cortex cells. Arbuscules are intricately branched, maximizing contact between plants and fungi, and function in the exchange of mineral nutrients and photosynthates. The morphology of colonized roots is similar to that of uncolonized roots, and the fungi can be seen only if roots are appropriately stained and viewed using a microscope (fig. 4.1a).

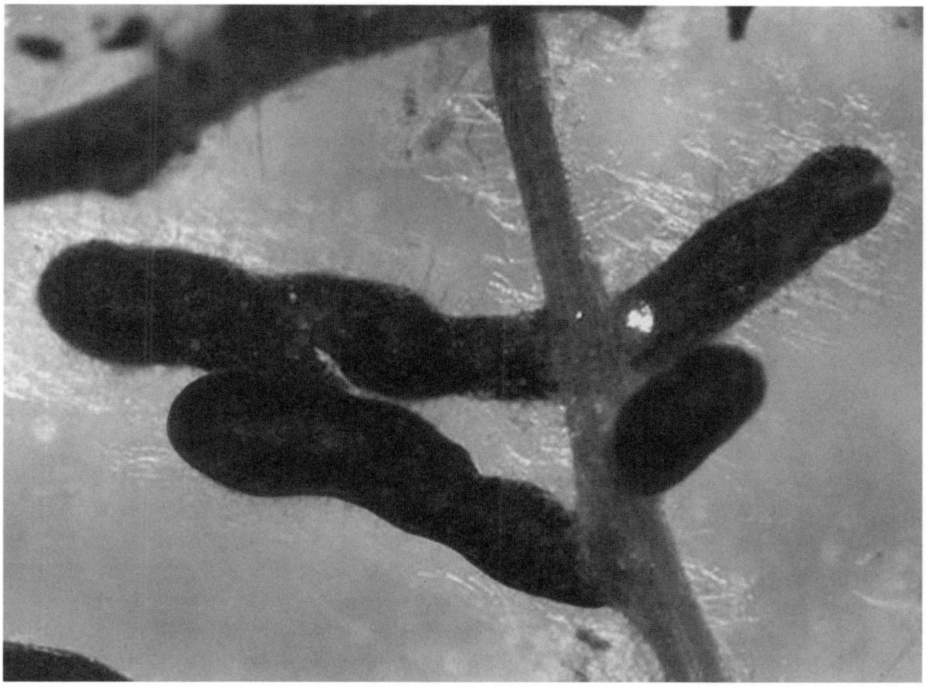

EM FUNGI

EM fungi are found in diverse orders within the phyla Basidiomycota and Ascomycota and most can form sexual fruiting bodies (sporocarps), such as mushrooms, in areas surrounding colonized plants (S. E. Smith and Read 1997). EM fungi form associations with woody plants and are the predominant mycorrhizal association in most of the world's boreal and temperate forests on acid soils (Dahlberg 2001, Horton and Bruns 2001) where nutrients are typically tied up in organic matter. This type of association is called ectomycorrhizae because colonized roots have a dense fungal sheath or mantle that covers the outside (ecto-) of colonized root tips. The external appearance of the root tip is altered, so ectomycorrhizae can be readily observed without staining and often without microscopy (fig. 4.1b). In addition, a network of fungal cells called a Hartig net grows between cells in the root epidermis or cortex and functions in the exchange of mineral nutrients and photosynthates.

Most plants form associations with either AM or EM fungi, but a few plant genera form tripartite associations (sensu Lodge 2000) in which both types of mycorrhizae are found within the root system of a single host species. Some common woody genera in temperate and boreal riparian habitats that form associations with both AM and EM fungi are *Alnus* (alder), *Populus* (cottonwood), *Salix* (willow), and *Quercus* (oak) species (Allen 1991, Lodge 2000).

DSE FUNGI

Dark septate endophytes (DSE) are an understudied group of microfungi primarily in the phylum Ascomycota. Fungi in this group can be cultured to form microspores, with the most well-known example being *Phialocephala fortinii* (Addy et al. 2005). DSE have been observed in over 600 plant species, ranging from herbs to trees (Jumpponen and Trappe 1998, Jumpponen 2001). DSE form dark-pigmented microsclerotia (inflated cells that may function in fungal survival) on the surface of roots. Thin-walled fungal cells (hyphae) with septa (cross-walls) form between root epidermis, cortex, and sometimes in vascular tissue (Barrow and Aaltonen 2001, Jumpponen 2001, Barrow 2003). The ecological function of DSE is unclear, but they are common in arid and arctic habitats (Addy et al. 2005) and may play a role in root protection and plant growth stimulation (Barrow and Aaltonen 2001, Jumpponen 2001, Barrow and Osuna 2002).

Mycorrhizal Relations of Dryland Riparian Plants

Cottonwood and willow species are the dominant pioneer trees in Sonoran Desert floodplains. Trees in these genera form tripartate associations with EM and AM fungi in riparian ecosystems throughout the world (Lodge 1989, Lodge and Wentworth 1990, Sasaki et al. 2001). Regionally, EM and AM fungi are associated with several cottonwood species, including *Populus deltoides* ssp. *wislizenii* (synonym *P. fremontii*), *P. acuminata*, and *P. angustifolia* on various rivers in New Mexico (Jacobson 2004). On the San Pedro River, roots of *P. fremontii* are colonized by EM and AM fungi as well as by DSE fungi (fig. 4.2).

Unlike cottonwood and willow, roots of *Prosopis* species (mesquite), a common tree on desert river floodplains and terraces, are colonized only by AM fungi (Stutz et al. 2000) (fig. 4.3). The shrub *Tamarix* sp. (tamarisk or saltcedar) occasionally forms associations with AM fungi but is essentially considered non-mycorrhizal (Beauchamp 2004, Titus et al. 2002), although it does form extensive associations with DSE (Beauchamp et al. 2005). The association with DSE is consistent with its ability to grow on fairly dry and saline sites. There has been no investigation of the mycorrhizal status of other common riparian shrub species.

Many perennial forbs and grasses in desert riparian floodplains and terraces are colonized by AM fungi (Kennedy et al. 2002, Beauchamp 2004). Species such as *Typha* (cattails) and *Juncus* (rush) that are prevalent in wetland portions of riparian corridors were once assumed to be non-mycorrhizal because of anoxic saturated soil conditions but now are known to be colonized by AM fungi (Rickerl et al. 1994, Stenlund and Charvat 1994, Miller and Bever 1999). The AM fungi have been observed in aerenchymatous tissue of the wetland plants (Brown and Bledsoe 1996).

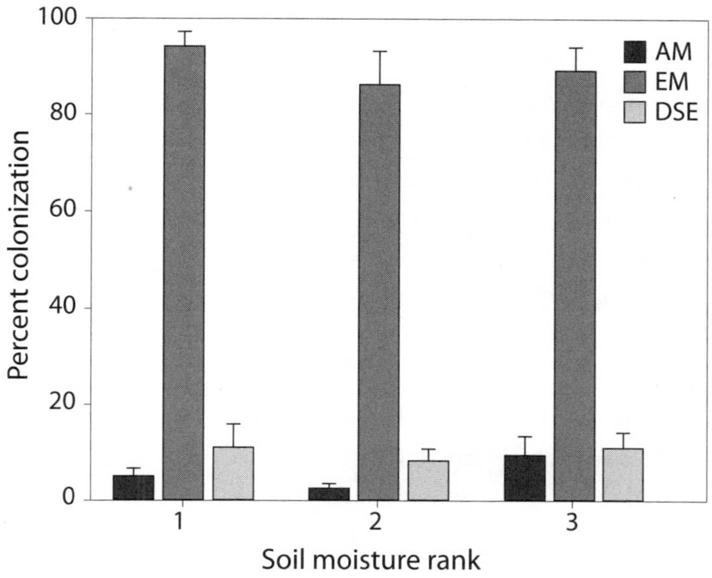

Fig. 4.2. Mean colonization of Fremont cottonwood (*Populus fremontii*) roots by arbuscular mycorrhizae (AM), ectomycorrhizae (EM) and dark septate endophyte (DSE) fungi at eight sites on the San Pedro River that were grouped by site hydrology (1=wettest and 3=driest). Bars indicate standard error of the mean.

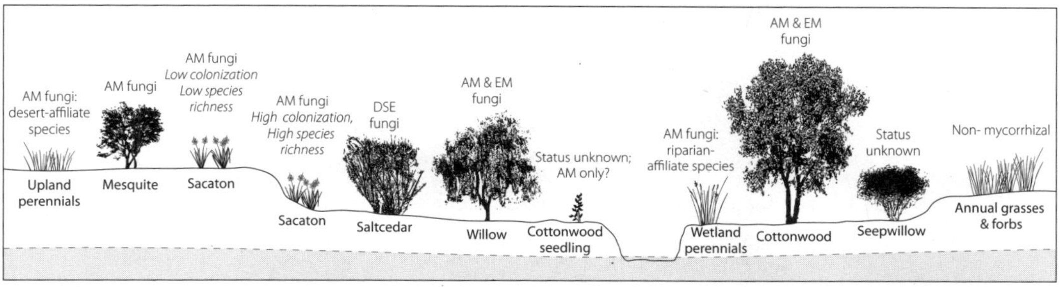

Fig. 4.3. Schematic diagram of mycorrhizal relationships for dominant plant types along the San Pedro River channel, floodplain, and terrace. Illustration credit: Mike Buffington.

Annual plant species comprise a significant portion of the flora in desert riparian areas, being seasonally abundant after rains and floods (see chap. 5). Annual plants tend to be non-mycorrhizal or not highly dependent on mycorrhizal fungi (Allen 1991, Jones and Smith 2004).

Temporal and Spatial Patterns

SEASONAL VARIATION

Abundance of root colonization of AM mycorrhizal fungi appears to be tightly coupled with seasonal growth patterns of the host plant. In a study of *Sporobolus wrightii* (big sacaton) growing in riparian areas of the San Pedro River, AM fungal colonization was highest in spring and early summer, when plants shift from quiescence to active vegetative growth, and declined in the fall and winter, when plants flower or are dormant (Kennedy et al. 2002). For wetland plants in Ohio, AM fungal colonization similarly was greatest in spring during new root and shoot growth, with rates lower in the summer (Bohrer et al. 2004).

Research has not been conducted in riparian areas on seasonal patterns of tree root colonization by AM and EM fungi, but a study of *Salix repens* (creeping willow) in sand dune ecosystems in Europe found seasonal differences, with highest levels of AM colonization in spring and lowest levels in fall (van der Heijden and Vosatka 1999). Levels of EM colonization, in contrast, remained relatively constant throughout the year. Seasonal effects, as opposed to site differences, could account for the lower rates of AM fungal colonization of cottonwood roots on the San Pedro River (fig. 4.2; samples collected in fall) compared to those on the upper Gila and Mimbres Rivers (28 to 60 percent reported by Jacobson [2004] for samples collected in mid-summer).

ONTOGENIC PATTERNS

In some tree species, fungal colonization patterns vary with life stage. Root colonization of *Quercus agrifolia* (coastal live oak) shifted from predominately AM in 1-year-old seedlings to predominately EM in 11-year-old saplings (Egerton-Warburton and Allen 2001). Similarly, in a microcosm experiment in which *Eucalyptus* seedlings were inoculated with AM and EM fungi, the AM fungi were largely replaced by EM fungi after several months of growth (Chen et al. 2000). In contrast, Sasaki et al. (2001) found that 2-month-old *S. gracilistyla* (rose-gold pussy willow) growing in a channel bar in Japan were colonized only by EM fungi, while most 1-year-old plants were colonized by both EM and AM fungi. They attribute the low levels of AM colonization in the seedlings to low propagule densities due to river flooding, and observed propagule densities to increase with time since flooding. Based on these studies, we anticipate that AM fungi would predominate in cottonwood and willow seedlings growing on channel bars (unless inoculum is limiting), while EM fungi would predominate in roots of trees growing in older stands farther from the active channel.

SUCCESSIONAL PATTERNS

In riparian corridors such as the San Pedro, the surfaces occupied by cottonwood and willow trees change over time; they can increase in elevation above the water table, decrease in inundation frequency, develop finer-textured soils with higher organic content, and accumulate a dense leaf-litter layer over successional time (see chap. 14). These edaphic changes likely impact mycorrhizal colonization patterns, because in many ecosystems there is a shift in predominance from AM to EM fungi as the principal source of nitrogen shifts from mobile mineral ions to organic forms (Read and Perez-Moreno 2003). EM fungi can produce enzymes that degrade complex structural polymers in organic matter, freeing nitrogen and phosphorus for transfer and use by host plants (Langly and Hungate 2003, Simard et al. 2002), and can access the recalcitrant organic compounds in litter (Langley and Hungate 2003). In *Populus tremuloides* (quaking aspen), AM fungi were common in roots in deeper mineral soil, while EM fungi dominated in roots in the shallow organic soil (Neville et al. 2002). These results suggest that EM fungi should be more prevalent than AM fungi in roots of cottonwood and willow trees growing in soils with high organic matter content and litter accumulation, but this topic has not yet been studied.

SPATIAL PATTERNS

Soil moisture and inundation frequency vary laterally across riparian zones and longitudinally between river reaches and may influence abundance of AM and EM fungi. Lodge (1989) found differences in AM and EM colonization of *P. deltoides* (eastern cottonwood) across a mesic riparian floodplain in the eastern United States that were likely related to soil moisture. AM fungal colonization was greatest in trees nearest and farthest from the river, at both extremes of the floodplain soil-moisture gradient. EM fungal colonization predominated in trees at intermediate distances from the river where soil moisture levels were intermediate.

Spatial position was related to the type of associations formed in roots of Fremont cottonwood on the upper Gila and Mimbres Rivers in New Mexico (Jacobson 2004). Levels of EM fungal colonization were highest at sites with frequent flooding (1–2 years) and a shallow water table, perhaps due to the higher soil-moisture levels in the upper rhizosphere, while trees at drier sites with infrequent flooding (>10 years) had higher levels of AM fungal colonization. On the San Pedro River, colonization levels in roots of Fremont cottonwood did not vary greatly among sites along a gradient of streamflow permanence and groundwater depth (fig. 4.2); EM fungal colonization levels were high at all sites (with a slight increase at wetter sites), while levels of AM and DSE fungal colonization were consistently low at all sites (with a slight decrease at drier sites) (Johnson 2005).

In deserts, AM fungal activity is positively correlated with soil moisture, while the reverse is true in wetland environments. Patterns of AM fungal colonization in Southwest riparian zones resemble those in desert environments. Along the Verde River in Arizona, AM fungal root colonization of forbs and grasses increased with proximity to the active channel, with values greatest in the lower, wetter fluvial surfaces (Beauchamp et al. 2006, 2007). Kennedy et al. (2002) found similar results in a study of big sacaton along the San Pedro River. Levels of root colonization by AM fungi were highest for plants growing in the floodplain at sites with perennial flow and were lowest in plants growing on drier terraces at sites with intermittent flow. In contrast, for wetland plants such as cattails and rushes, fungal colonization levels are higher at dryer sites than at wetter and more anaerobic sites (Rickerl et al. 1994, Miller and Bever 1999).

Functional Roles of Mycorrhizal Fungi

PLANT GROWTH

Plant species can be classified as obligate, facultative, or non-mycorrhizal depending on their growth response to inoculation (i.e., introduction into the plant tissue) and colonization by mycorrhizal fungi (Jones and Smith 2004). Little is known about the response of riparian tree species of the desert Southwest to inoculation, but positive growth responses occur for related species of cottonwood (Baum et al. 2002), willow (Jones et al. 1990, van der Heijden 2001), and mesquite (Bhatia et al. 1998). Although these plant species appear to be facultatively dependent on mycorrhizal fungi, the relationship between mycorrhizal fungi and plants has been shown to range from mutualistic to parasitic depending on factors such as plant and fungal genotype, developmental stage of the plant, competition between plants, and edaphic factors (Johnson et al. 1997, Jones and Smith 2004). In addition to influencing plant productivity and biomass, mycorrhizal fungi can alter fecundity and seedling survivorship, with these responses also dependent on environmental and genotypic factors (Jones and Smith 2004).

The functional role of DSE is ambiguous, with reports of beneficial, neutral, and parasitic plant responses to colonization by *P. fortinii* (Jumpponen 2001, Addy et al. 2005). Inoculation with *P. fortinii* increased foliar phosphorus concentrations in a few experiments (Jumponnen 2001). Cellulolytic and proteolytic enzyme activity has been associated with DSE fungi, but it is unclear if DSE fungi actively decompose organic matter in soil and transfer nutrients to host plants (Addy et al. 2005). The extent of external hyphae of DSE fungi in soil is not known, though dark septate hyphae are prevalent in soils from arctic and boreal regions (Jumpponen and Trappe 1998). Much research is needed on the functional aspects of DSE, as well as of AM and EM associations, in desert riparian ecosystems.

NUTRIENT UPTAKE

Mycorrhizal fungi can increase plant growth by enhancing nutrient uptake. In addition to producing hyphae inside plant roots, mycorrhizal fungi form external hyphae (the long, branching filaments that make up the body of fungi) that proliferate in soil beyond plant roots. External hyphae comprise a sizeable portion of fungal biomass in soil (Miller et al. 1995, Simard et al. 2002) and are important in nutrient uptake of many elements including phosphorus, nitrogen, copper, iron, and zinc (S. E. Smith and Read 1997, Govindarajula et al. 2005). Several mechanisms have been proposed for this, including the ability of hyphae to access very small soil pores, increases in the total surface area for absorption, alterations in the chemistry of the mycorrhizosphere, and production of enzymes that obtain phosphorus and nitrogen from organic sources (Simard et al. 2002, Leake et al. 2004). The mycorrhizal partnership may play a significant role in allowing riparian ecosystems to function as nutrient filters (see chap. 18).

DROUGHT RESISTANCE

Conferring drought resistance may be another functional role of mycorrhizal fungi important to desert riparian plants. While the mechanisms remain unclear, external hyphae of mycorrhizal fungi can impact host water balance in drying soils by contributing to root water absorption and changing soil water relations (Augé 2001). AM fungal communities collected from a river floodplain in southern Arizona had a positive effect on plant drought resistance in comparison to a *Glomus intraradices* isolate sold as commercial inoculum (Augé et al. 2003). The riparian fungi produced more of the external hyphal biomass that likely influence water relations of the host plant.

SOIL STABILITY

External hyphae of mycorrhizal fungi modify soil structure by helping to bind soil particles and form stable aggregates (Tisdall et al. 1997, Jastrow et al. 1998, Bogeat-Triboulot et al. 2004). Proteins, including glomalin, produced by AM fungi (Wright et al. 1996) persist in soil and also contribute to soil aggregation (Rillig et al. 2002, Harner et al. 2004).

RIPARIAN-AQUATIC INTERACTIONS

Mycorrhizal fungi in riparian zones may indirectly influence productivity of the stream ecosystem. Proteins from AM fungi were detected in river foam in the Gila, Jemez, and Rio Grande in New Mexico (Harner et al. 2004). Laboratory experiments demonstrate that these proteins can be washed from soil and through turbulent mixing become part of foam, where they may serve as a nutrient source for aquatic organisms (Harner et al. 2004).

PLANT COMMUNITY STRUCTURE

While impacts of mycorrhizal associations on plant nutrition and growth have been long recognized, only recently have experiments shown that mycorrhizal fungi can influence plant community structure by altering competitive relationships (West 1996, Hartnett and Wilson 2002). For example, Beauchamp et al. (2005) did not find a positive or negative growth response of Fremont cottonwood seedlings to inoculation with AM and EM fungi collected from a river floodplain but did find that inoculation altered the competitive interaction between cottonwood and saltcedar, with cottonwood seedlings having more of the combined biomass in microcosms that were inoculated than in those that were not inoculated. Mycorrhizal fungi also may play a role in successional changes and the success of other immigrant (exotic) plant species (Janos 1980, Gange et al. 1993, Marler et al. 1999).

PRIMARY PRODUCTIVITY

Although mycorrhizae can form between many combinations of plant and fungal partners, fungal species are not functionally equivalent. Plant responses can vary with different fungal partners (van der Heijden, Boller et al. 1998, van der Heijden and Kuyper 2001), and the composition of the AM fungal community can regulate ecosystem processes such as primary productivity and relationships between plant species richness and biomass (van der Heijden, Klironomos et al. 1998, Klironomos et al. 2000). Fewer studies have been conducted on EM fungal plant interactions, although increases in EM fungal species richness can increase plant productivity (Baxter and Dighton 2001).

Fungal Community Structure in Desert Riparian Ecosystems

AM FUNGI

AM fungi do not form above-ground fruiting bodies, and the species are difficult to distinguish based on vegetative chacteristics. This complicates our ability to characterize AM fungal communities. Species identification is based on morphological characteristics of spores that form in soil (Bever et al. 2001), and AM fungal communities have been studied by identifying and counting spores isolated from field soils (Morton et al. 1995). However, some cryptic AM fungal species do not readily sporulate in field soils (Bever et al. 2001), and in arid soils AM fungal spores are present in low concentrations (Stutz and Morton 1996), reducing the effectiveness of this method at capturing the full community. To overcome these problems, trap cultures have been used (Stutz and Morton 1996). In this technique, plants are grown in a greenhouse for several months in pots containing diluted field soils. This treatment induces the fungi to sporulate, enabling their detection and identification. Additionally, molecular identification methods are increasingly being used to identify AM fungal species (Redecker et al. 2003), and these methods may revolutionize studies of AM fungal communities. However, their use is hampered by several obstacles, including the polymorphic, multigenomic nature of AM fungi, the lack of a single primer to amplify AM fungal DNA, and the difficulty in matching sequence groups to morphospecies (Redecker et al. 2003).

Desert riparian AM fungal communities are fairly diverse. Using trap cultures, Kennedy et al. (2002) detected 15 species of AM fungi associated with big sacaton at four sites on the San Pedro River. Seventeen species were detected at one site when more intensive sampling was utilized (Stutz and Morton, unpublished data). In both studies, about a quarter of the species had not been previously described (Kennedy et al. 1999). There was a trend toward greater species richness in the floodplain sites than in the no-longer-inundated terraces with relict sacaton grasslands (Kennedy et al. 2002).

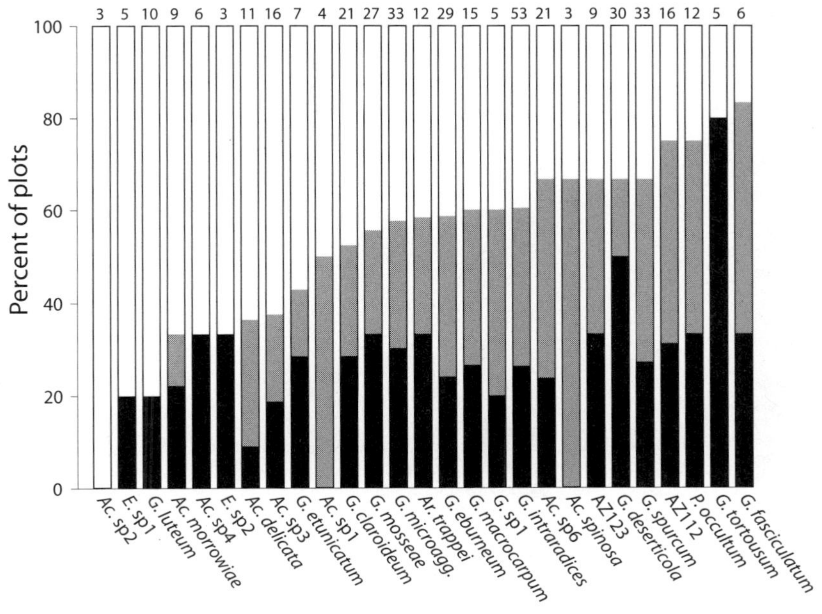

Fig. 4.4. Hydrologic affinity for AM fungal species from cottonwood/willow (*Populus/Salix*) stands on the Verde River. Each column shows the frequency of detection for each species in dry (black), intermediate (gray), and wet (white) plots. The numbers along the top are the total number of detections for each species.

Also using trap cultures, Beauchamp (2004) detected 30 AM fungal species associated with cottonwood/willow forests on the Verde River, 9 of which were new. Species richness of AM fungi in these desert riparian areas was higher than in the Sonoran desert (Stutz et al. 2000), but lower than in mesic regions. Bever et al. (2001), for example, using intensive sequential sampling over a long period and a variety of culture techniques, detected 37 species in a 1 ha old field in North Carolina.

Many environmental factors, including soil moisture, may structure the AM fungal community within riparian zones. In the Verde River floodplain, AM fungal species did not have a strict affinity for wet or dry sites, but some species occurred more frequently in wet sites near the active channel, while others were more frequent in dry areas (Beauchamp et al. 2006)(fig. 4.4). All of the dry-site species are common in the surrounding Sonoran desert (Stutz and Morton 1996, Stutz et al. 2000), but species with 60 percent or more of their occurrences in wet sites have not been detected in the desert. This group of riparian-affiliate AM fungi species included five undescribed species. Some of these riparian-affiliate species found along the perennial Verde River have also been detected from perennial reaches of the San Pedro River (Kennedy et al. 2002). These results are similar to those of S. P. Miller and Bever (1999), who found zonation of AM fungal species in a wetland in North Carolina with different species dominating at wet, intermediate, and dry locations.

EM FUNGI

Identification of EM fungi is based on characteristics of sporocarps, the morphology of ectomycorrhizal root tips, and molecular techniques (Horton and

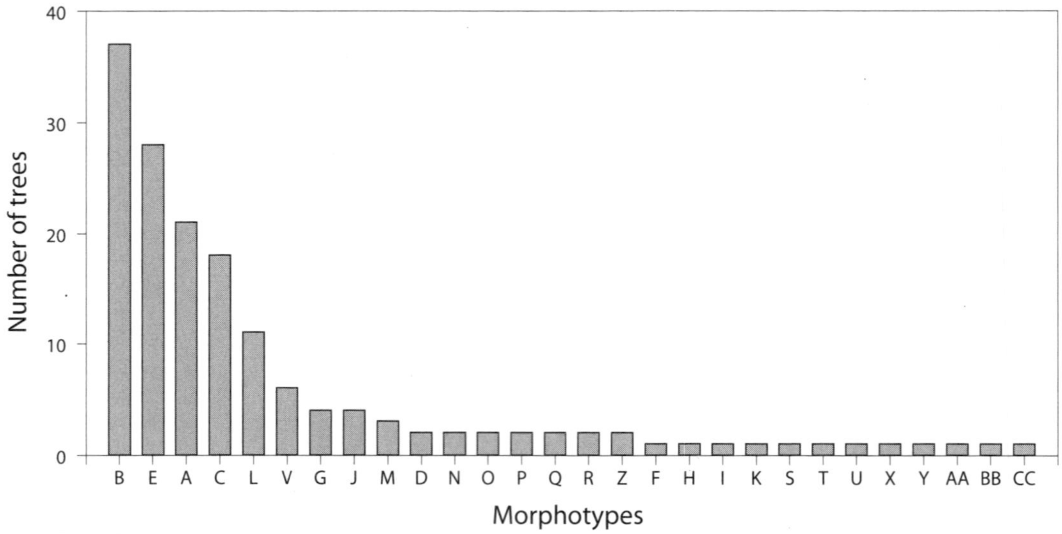

Fig. 4.5. Frequency distribution of EM fungal morphotypes associated with roots of Fremont cottonwood (*Populus fremontii*) trees (n=40) at eight sites on the lower San Pedro River. Based on J. D. Johnson (2005).

Bruns 2001). Morphotypes (aka morphospecies) are identified using morphological and structural descriptions of ectomycorrhizal root tips (Agerer 1991). Molecular techniques use polymerase chain reaction (PCR) combined with restriction fragment length polymorphism (RFLP) or DNA sequence analysis (Dahlberg 2001, T. Horton and Bruns 2001). Molecular techniques offer certain advantages over morphotyping because fungi associated with specific root types can be identified (Horton and Bruns 2001), and these techniques have become the dominant method to identify EM species.

Morphotyping was used to document the dominant types of EM root tips on Fremont cottonwood roots along the lower San Pedro River and to determine if there were community differences across spatial and temporal gradients within the floodplain (Johnson 2005). In this study, 28 morphotypes were detected (fig. 4.5) with a mean of 9 per site. Three morphotypes (A, B, and E) were detected at all sites and were associated with over half the trees sampled. Typical of general community patterns of commonness-rarity, almost half of the morphotypes detected were found in only one tree sample.

Dispersal of Mycorrhizal Fungi

AM and EM fungi disperse separately from their host plants, and seedling roots initiate associations with fungi at establishment. Newly emerged seedlings can be colonized by the fungi in existing hyphal networks if soil is undisturbed (Simard et al. 2002, Leake et al. 2004). AM fungi can be dispersed locally through the underground spread of hyphal networks and through activities of soil invertebrates, and over longer distances by wind erosion of soil (Warner et al. 1987) and by mammals (Allen 1987).

Wind dissemination is important in the dispersal of spores and hyphae of AM fungi in arid and semi-arid regions (Allen et al. 1989). Wind dispersal of propagules in dust from the surrounding desert into riparian areas may be an important mechanism for the establishment of AM fungal species that are common in both the desert and riparian areas. EM fungi spores also are dispersed by wind, after fungal sporocarps (fruiting bodies) form on the surface of soil near colonized plants. Although spore dispersal by wind is aided by mechanisms for forcible discharge, low wind velocities under forest canopies may limit dispersal distance of fungal spores in densely forested areas (Allen 1991).

Mammals are important in the dispersal of some EM fungi, especially those hypogeous species that form truffle and truffle-like sporocarps underground (Johnson 1996). These fungi rely on mammals, such as squirrels and other rodents, which dig up and ingest fruiting bodies, passing the spores through their gut and later depositing them in feces.

Studies of water dissemination of mycorrhizal fungi are limited (Koske and Gemma 1990) and have not been conducted in riparian systems. Some riparian mycorrhizal species likely disperse by water during flood events, with spores deposited on newly scoured surfaces on which seedlings of pioneer plant species establish. Additionally, fungal spores may be dispersed into riparian corridors from the uplands via sheet flow.

Little is known about the dispersal of DSE fungi, although fragments of the dark microsclerotia and asexual spores are possible propagules (Jumpponen and Trappe 1998). Vertical transmission (wherein the fungus is transmitted through the seed) may be important for DSE fungi because these fungi infect plants systemically and are carried in seeds (Barrow et al. 1997).

Environmental Changes

RIVER ENTRENCHMENT

The historic entrenchment of the San Pedro River and associated restructuring of the riparian ecosystem (see chap. 12) likely influenced the mycorrhizal communities. EM fungi may have been sparse at times and places where the marshlands and grasslands prevailed and may have increased in tandem with the colonization of the post-incision surfaces by cottonwood, willow, and other trees.

AGRICULTURE

Direct studies of livestock grazing effects on mycorrhizal networks in the San Pedro or other desert riparian ecosystems have not been undertaken. Overgrazing can alter abundance and composition of riparian plant species and accelerate soil and bank erosion (Belsky et al. 1999); this can destroy fragile hyphal networks (Gobl 1996), resulting in lower root colonization rates and reduced nutrient and water uptake for host plants. In some mesic grasslands, in contrast, grazing can increase AM root colonization and external hyphae density (Eom et al. 2001).

Many acres of river terrace along the San Pedro and other desert rivers have been converted to agricultural fields. Agricultural practices such as plowing, fertilization, and pesticides can reduce root colonization, species richness, and propagule levels of AM fungi (Kurle and Pfleger 1994). In Ohio, agricultural activities had greater impact on EM fungi than on AM fungi, probably because common crops do not form associations with EM fungi (Boerner et al. 1996). In this same area, old fields dominated by non-mycorrhizal plants had reduced levels of AM and EM fungi up to 25–30 years after abandonment (Boerner et al. 1996). In contrast, AM fungal propagules in 10- to 40-year-old abandoned agricultural fields along the San Pedro and other nearby rivers were equal to or higher than levels in adjacent riparian grasslands (Richter et al. 2002).

Efforts have been undertaken to restore perennial grasses to some of the abandoned agricultural fields along the San Pedro River. These efforts may benefit from inoculation of plants with mycorrhizal fungi, if prior land use has caused declines in fungal propagule levels or altered fungal species composition and function (Renker et al. 2004). In a riparian restoration trial on an abandoned field in southern Arizona that was dominated by non-mycorrhizal, ruderal plant species, inoculation of greenhouse-grown big sacaton transplants with AM fungi significantly increased the plants' growth and survival (Richter and Stutz 2002). This growth enhancement occurred despite the fact that AM fungal propagules were present in the soils at the site to colonize the transplants.

In general, experimentation with mycorrhizal inoculation in riparian restoration projects is in its infancy. In addition to increasing plant growth and survival, mycorrhizal inoculation may increase the competitiveness of perennial species against non-mycorrhizal annuals and increase the drought and pathogen resistance of new plantings (Allen 1991). The effective use of mycorrhizal inoculation in riparian restoration will require knowledge of mycorrhizal fungal community composition, the mycorrhizal dependency of the plant species targeted for restoration, the effects of anthropogenic alterations on mycorrhizal fungal communities, and the tolerances of various mycorrhizal fungal species to site factors such as aridity and salinity.

STREAMFLOW CHANGES

Soil moisture appears important in determining the relative dominance of AM versus EM fungi in cottonwood and willow (Lodge 1989, Jacobson 2004) and in influencing colonization levels of both. Thus, as changes in surface water and groundwater flow rates alter soil moisture in floodplain and channel sediments, abundance patterns of AM and EM fungi may be altered. Additionally, root colonization in arid plants has been shown to shift from AM to DSE with increasing aridity (Barrow and Aaltonen 2001, Barrow 2003), and if similar patterns occur in riparian zones, DSE fungi may be more prevalent than EM and AM fungi in dry riparian reaches. Dry reaches favor saltcedar (see chap. 1), a species that has DSE but only low levels of AM fungi and no EM fungi. However, studies on the San Pedro River indicate that propagule densities of mycorrhizal fungi differ little between areas dominated by saltcedar and those where cottonwood prevail (Johnson 2005).

Summary

The high numbers of AM and EM mycorrhizal fungi species in desert riparian ecosystems, coupled with the prevalence of mycorrhizal host plants, suggest that mycorrhizae are important to riparian ecosystem structure and function. Cottonwood and willow, the dominant pioneer trees in southwestern riparian areas, form tripartate relationships with AM and EM fungi. Mesquite, a common riparian tree, and big sacaton, a locally common riparian grass, are colonized by AM fungi, as are many other perennial grasses and forbs. Many of the AM fungi found in the riparian corridor also occur in the desert, but others (including several newly described species) are riparian-zone obligates. Overall, more species of AM fungi have been detected in riparian areas of the desert Southwest than in the Sonoran Desert upland.

Mycorrhizal fungi play a functional role in riparian ecosystems by increasing soil stability, nutrient uptake, and drought tolerance of riparian plants. Mycorrhizal fungal communities of the pioneer trees likely change over successional time, with AM fungi initially colonizing seedlings in flood-scoured seedbeds and EM fungi becoming more predominate in older stands, perhaps linked with the ability of the EM fungi to access nutrients in the soil organic matter that accumulates in the older forest stands.

Research on riparian ecosystems of the San Pedro and other rivers in Arizona and New Mexico demonstrates that riparian mycorrhizal fungi are sensitive to factors, including soil moisture and flood regimes, and that fungal communities show complex spatial and temporal patterns in riparian zones. Saltcedar, a riparian shrub that has become common in many desert rivers, including those that have become dry and saline, is colonized by a type of fungi known as DSE. Given its association with stressful environments, DSE fungi might be more prevalent than EM and AM fungi in riparian reaches with the driest conditions. However, mycorrhizal communities in the drier, saltcedar-dominated reaches of the San Pedro are similar to those of the wetter, cottonwood/willow-dominated reaches.

Alterations of the composition of the riparian mycorrhizal fungal community, such as from agricultural land conversion or severe stream dewatering, have the potential to affect plant health, competitive ability, and survival. Changes to the physical system can also drive changes in plant communities that in turn could alter fungal communities. Advances in the understanding of the structure and composition of mycorrhizal communities in desert riparian ecosystems should lead to studies on the effects of mycorrhizal communities on primary plant productivity, plant community interactions, and successional dynamics.

JULIET C. STROMBERG,
KENNETH J. BAGSTAD,
AND ELIZABETH MAKINGS

Floristic Diversity

FIVE

Introduction

Biological diversity exists at many levels of organization—from genes to populations, species, functional types, and landscapes—and is important in maintaining ecosystem functions and services of value to people (Ward and Tockner 2001, Duffy et al. 2007). Riparian zones are noted for having high levels of species diversity and landscape heterogeneity, and for contributing to regional diversity (Naiman et al. 1993, Nilsson and Svedmark 2002, Sabo et al. 2005). Species of both wetland and upland affinity intermingle in arid-region riparian corridors, and the combination of permanent water sources, seasonal rains, and flood disturbance creates a distinctive, and species-rich, plant community amidst the desert uplands.

In this chapter we focus on plant species diversity in the San Pedro riparian corridor. We address the influence on diversity of larger-scale geographic settings as well as of local factors such as disturbance regime and resource levels (Sarr et al. 2005). Recognizing that heterogeneity in flowing-water ecosystems vary over four dimensions (Ward 1989), we describe diversity patterns: (1) through time, (2) vertically, from soil depths through upper canopy, (3) laterally, across perpendicular gradients from the channel to uplands, and (4) longitudinally, over the length of the river. For the temporal component, we address seasonal and annual patterns as well as longer-term floristic changes, including extirpations and recent immigrations.

How Diverse Is the San Pedro Riparian Flora?

To date, over 750 vascular plant species have been identified within the San Pedro riparian corridor and bordering uplands. The most intensively surveyed area is the San Pedro Riparian National Conservation Area (SPRNCA) in the upper basin, where there are 625 documented taxa from 614 species, 358 genera, and 92 families (Makings 2006). The most species-rich plant families are Asteraceae (composites or sunflowers), Poaceae (grasses), and Fabaceae (legumes). Studies of the lower basin (unpublished data of authors) and Mexican reaches (G. Solis, unpublished data) have yielded additional species, but thorough floristic inventories remain to be conducted in these parts of the river. These species totals for the San Pedro are similar to some rivers elsewhere (Tabacchi and Planty-Tabacchi 2005) and higher and lower than for other rivers in the American Southwest (Wolden et al. 1994, Patrick 2000). However, most rivers in the region have yet to be systematically inventoried (McLaughlin 2004).

What Factors Influence Diversity?

FLORISTIC PROVINCES

Southeastern Arizona is located at the intersection of several floristic provinces, including the Sonoran, Chihuahuan, Apachian, Great Basin, and Southern Rocky Mountain (McLaughlin 1989, Bowers and McLaughlin 1996). The spatial overlap of these provinces contributes to high species diversity in the San Pedro basin. The basin's flora is dominated by Chihuahuan floristic elements such as viscid acacia (*Acacia neovernicosa*), whitethorn acacia (*A. constricta*), and tar bush (*Flourensia cernua*) and by Sonoran elements including alkali goldenbush (*Isocoma acradenia*) and desert broom (*Baccharis sarothroides*). Species representative of other floristic provinces include fineleaf hymenopappus (*Hymenopappus filifolius* var. *pauciflorus*) of Great Basin affiliation. The basin also harbors many widespread species, some of circumglobal distribution. (Note: floristic provinces differ somewhat from biogeographic provinces [see Brown 1994; and see chap. 6], the latter being based on distribution patterns of multiple taxa).

REGIONAL HETEROGENEITY AND GEOGRAPHIC ISOLATION

Also contributing to the high floristic diversity of the San Pedro basin is complex topography. The Sky Islands of southeastern Arizona rise to 1500 m over the desert floor, providing habitat for desert as well as montane species. Because of the geographical isolation of the mountain tops, some of the taxa have speciated (McLaughlin 1995). Arizona phacelia (*Phacelia arizonica*)

TABLE 5.1. Percentages of plants in the San Pedro River floodplain classified by life span and growth form, and percentage distribution by water requirements within the life span/growth form categories.

		Hydro-riparian[1]	Meso-riparian[2]	Xero-riparian	Total
Annuals and biennials	58%	18%	14%	68%	100%
Herbaceous perennials	29%	25%	15%	60%	100%
Woody plants	13%	23%	19%	58%	100%
Total	100%				

Note: Data are based on sampling of 18 sites in the lower and upper basin over a two-year period (unpublished data of authors).

[1]Species classified as obligate wetland or facultative wetland.

[2]Species classified as facultative or facultative upland, according to USDA Plants Database (http://plants.usda.gov).

and Chihuahuan brickellbush (*Brickellia floribunda*) are among the Sky Island endemics in the SPRNCA flora.

A wide range of geologic substrates and soil types are present in the basin, further increasing diversity. Rare and disjunct plant species grow in areas of azonal soils (i.e., soils that differ from the prevailing type), such as volcanic and granitic outcrops in the Charleston Hills area of the upper basin. A new species, *Eriogonum terrenatum* (Reveal), was recently collected in the SPRNCA from areas with a type of soil derived from Tertiary-period lacustrine (lake bed) deposits (Anderson 1996, Makings 2006). Further surveys may yet yield more undescribed species.

LANDSCAPE CONNECTIVITY

The diversity of a riparian zone echoes that of its watershed. Although many of the plant species along the San Pedro are riparian-zone obligates, many more are facultative riparian species that also grow in upland habitats. About two-thirds of the species encountered in the San Pedro floodplain are facultative riparian species (Bagstad et al. 2005) (table 5.1), similar to percentages reported for riparian zones elsewhere (Mouw and Alaback 2003). Some of these, such as plains lovegrass (*Eragrostis intermedia*) and Emory oak (*Quercus emoryii*), are perennial species a few of which extend their range downward in the cooler and wetter riparian corridors. Many others, such as Albert's creeping zinnia (*Sanvitalia abertii*) and western tansy mustard (*Descurainia pinnata*), are desert annuals that are seasonally abundant in the riparian zone and uplands alike. The abundance of these upland species in the San Pedro riparian corridor reflects the high degree of connectivity between the river and its watershed: desert grasslands and desert scrub abut the riparian corridor, while myriad tributaries, each with a unique assemblage of species

(J. Zimmerman et al. 1999, R. Richter and Stromberg 2005), provide corridors for species movement from mountains to river.

RESOURCE AVAILABILITY

High water availability in dryland riparian zones is one key factor contributing to their high species diversity (Tabacchi et al. 1996, Rundel and Sturmer 1998, Ali et al. 2000). About one-third of the species in the San Pedro riparian corridor are riparian obligates with high water needs (table 5.1). These include hydroriparian species such as Goodding's willow (*Salix gooddingii*) and Torrey's rush (*Juncus torreyi*), associated with very wet environments, and mesoriparian species such as Arizona walnut (*Juglans major*) and western ragweed (*Ambrosia psilostachya*), which require moisture beyond that provided by local rainfall.

LOCAL DISTURBANCE AND HETEROGENEITY

Floods contribute to high floristic diversity through many mechanisms. One mechanism is the reduction in competitive exclusion. Diversity levels typically increase as resources increase, but where resources are very abundant and disturbance minimal, a few competitively superior species can dominate a site (Pollock et al. 1998). Floods, by scouring vegetation, maintain plant cover below the levels at which competitive exclusion limits diversity (Ward et al. 2002, Stromberg 2007). For example, monsoonal floods routinely create new sediment bars in the San Pedro River, allowing for replacement of competitive perennial species such as bermuda grass (*Cynodon dactylon*) by short-lived, disturbance-adapted, ruderal species (sensu Grime 1973, 1977) including clammyweed (*Polanisia dodecandra*) (fig. 5.1a).

Floods also influence diversity patterns at larger spatial scales (Nilsson et al. 1989, Brown 1997, Nichols et al. 1998, Huston 1999). Powerful floods shape riparian landscapes, creating complex topographies that sustain complex biotic communities across the wide (up to 400 m) San Pedro floodplains. Floodwaters erode channels and deposit sediments on floodplains, producing gradients of inundation frequency, scouring forces, soil moisture, and depth to groundwater (Hupp and Osterkamp 1996). The vegetation that develops further modifies the physical environment. Understory light levels are patchy, and soil textures and nutrient pools are heterogenous (see chaps. 14 and 18). This environmental complexity results in high beta diversity (species turnover across gradients; Whittaker 1972; and see chap. 6) within the riparian corridor (Lite et al. 2005, Stromberg 2007) and allows for the co-existence of species with colonizing and competitive strategies (Connell 1978).

In addition to leaving a legacy of spatial heterogeneity, floods also increase temporal heterogeneity. Because floods occur at different times of the year,

Fig. 5.1. (top; 5.1a): Clammyweed (*Polanisia dodecandra*), an annual plant that grows on San Pedro River sand bars; (bottom; fig. 5.1b): Tripleleaf morning glory (*Ipomoea ternifolia*), an annual plant that becomes abundant on river floodplains and terraces following summer rains and floods. Photo credits: Elizabeth Makings.

and with different intensities, they create opportunities for establishment of species with varying phenological patterns and varying patterns of resource use and allocation. The periodic fires that occur in the San Pedro riparian corridor also influence diversity. Fire, like floods, increases diversity in the short term by allowing a variety of annual grasses and forbs to establish on sites previously dominated by perennial herbs and trees (Bock and Bock 1978, Stromberg et al. forthcoming).

Seasonal and Annual Patterns of Diversity

Some of the plants along the San Pedro River are sustained by permanent water sources such as shallow groundwater or perennial streamflow, while others germinate or grow in response to seasonal pulses of rain or floodwater. The seasonal rhythm of the vegetation thus reflects the bi-modal rain and flood pattern of the Chihuahuan and Sonoran Deserts regions (fig. 5.2). The number of plant species actively growing in the San Pedro riparian corridor is high in spring, following the winter rains, low during the early-summer dry season, and high again after the late-summer rains and floods. During one representative year, the number of herbaceous plants growing in the San Pedro floodplain doubled over the course of the summer, increasing from dry June to wet September (Bagstad et al. 2005).

Species numbers can vary sharply between years depending on the particular flood and rain patterns. During the growing season following one large fall San Pedro River flood, species diversity showed a 37 percent increase over prior-year values, owing largely to establishment of annual species (Bagstad et al. 2005). Another group that showed strong post-flood increases was hydroriparian species. This group increased because floods enhance streamflows and water levels in the stream aquifer. The extent of such increase varies between years with the magnitude, frequency, and duration of flooding.

Longer-term Changes: Immigrant and Extirpated Species

The biota of any region undergoes continuous change, although the rate of change itself may vary. Ecosystems are open (Odum 1971), and new seeds or spores are sporadically brought in from afar by various physical and biological dispersal processes. The millions of neotropical migrant birds flying north and south along the San Pedro riparian corridor bear plant seeds and spores on their feathers and feet, while mammals moving between the river and uplands transport them on fur and in feces. The thousands of people who use the river as a migration corridor from Mexico to the United States unknowingly provide transport for plant hitchhikers. Roads and railways border and cross the river and provide for movement of plant seeds from many directions. Overland runoff and tributary flows disperse seeds into the riparian corridor from farm fields and urban landscapes throughout

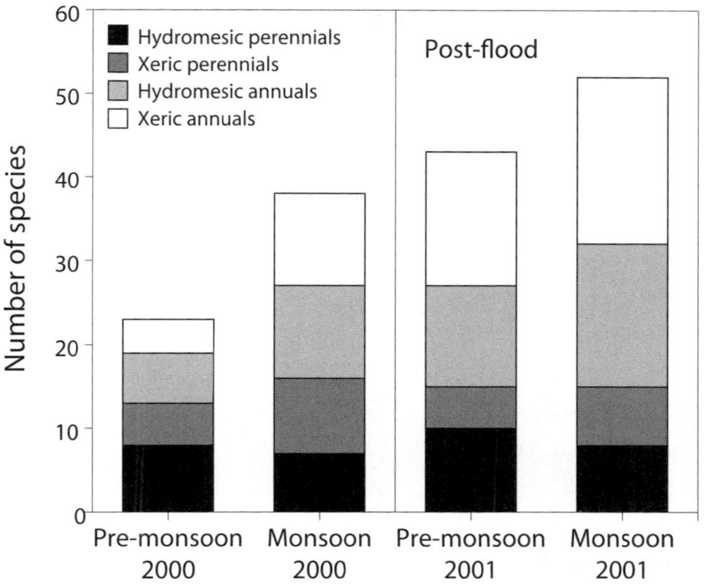

Fig. 5.2. Mean herbaceous species richness in floodplain belt transects (1 m wide by approximately 300 m long) during pre-monsoon (May/June) and monsoon (Aug./Sept.) seasons of 2000 and 2001, based on 18 sites along the San Pedro River. A large flood occurred in October 2000. Plants are classified by moisture affinity (hydromesic, xeric) and by life span (annual, perennial). Modified from Bagstad et al. (2005). Used with permission.

the watershed, while winds transport spores and seeds. In addition to these unplanned movements, several plant species, including Lehmann lovegrass (*Eragrostis lehmanniana*), bermuda grass, and Johnson grass (*Sorghum halepense*), have been intentionally introduced to the riparian corridor or surrounding uplands as forage plants for domestic livestock or to facilitate recovery of overgrazed lands (Tellman 2002).

The environmental setting into which the species immigrate also is ever-changing (see chap. 11). Over past centuries there have been changes in surface water and groundwater availability, intensity and frequency of flood and fire disturbance, and abundance of herbivore populations, including cattle and pocket gophers. Irrigated agriculture has occurred along the river for 2,000 years, and small urban centers and mining areas have come and gone. Climate and atmospheric CO_2 levels and nutrient inputs have changed, and continue to change, in response to global-scale phenomena (Dukes and Mooney 1999). Some plant species thrive under the new conditions, while others decline.

The long history of agriculture in the San Pedro River basin has played an important role in transforming the flora. Irrigated agriculture brings with it many crop weeds, many of which thrive in the mesic, disturbed habitats of riparian zones (Parker 1972). Livestock grazing contributes to the abundance of well-defended species—such as red brome (*Bromus rubens*) with its sharp awns, and sweet clover (*Melilotus officinalis*) and Johnson grass with their respective toxins coumarin and cyanide—while selecting against species that did not evolve with intense grazing pressure (Mack 1981, Whitson and Burrill 1992).

Riparian plant communities are largely structured by immigration-

TABLE 5.2. Percentage of vascular plant species in the San Pedro River floodplain that are considered introduced to North America, by category.

	Native	Exotic
Annuals and biennials		
Hydro- and mesoriparian	74%	26%
Xeroriparian	86%	14%
Herbaceous perennials		
Hydro- and mesoriparian	85%	15%
Xeroriparian	93%	7%
Trees, shrubs, and woody vines		
Hydro- and mesoriparian	94%	6%
Xeroriparian	100%	0%

Source: unpublished data of authors.

driven processes rather than by competition, allowing for rapid integration of new species (R. Brown and Peet 2003). Periodic floods and fires create fluxes in resource availability and thus provide frequent opportunities for species replacement (M. Davis et al. 2000). Thus, it is no surprise that immigrant (exotic) species form a large component of the flora of most rivers. Of the vouchered vascular plant taxa within the SPRNCA, 11 percent are considered to be introduced to the United States (Makings 2006); values of 20 to 30 percent are common for rivers around the world (Wolden et al. 1994, Dixon and Johnson 1999, Hood and Naiman 2000).

Many of the exotic species along the San Pedro are annuals (as are many of the natives), several with wetland affinity (table 5.2). Many Eurasian annuals have found a suitable climatic home in the Sonoran Desert (Mack 1996, Tabacchi et al. 1998), exemplified by London rocket (*Sisymbrium irio*), which is common on San Pedro terraces after winter rains in the nitrogen-rich understory of mesquite (*Prosopis*) forests. Warm-season perennial grass species of African origin, including bermuda grass, Johnson grass, and Lehmann lovegrass, are common in summer, as are a few warm-season annual species from Central or South America including tall morning glory (*Ipomaea purpurea*) (fig. 5.1b). White mulberry (*Morus alba*), a species often found by old homesteads, is one of the few woody exotics. Saltcedar (*Tamarix chinensis*, *T. ramosissima* or hybrids; Gaskin and Schaal 2002), a species classified as a small tree or large shrub, is another.

Sites along the San Pedro River that have high numbers of native species also have high numbers of exotic species (fig. 5.3). Other studies across spatial gradients also show that diversity of native and exotic species increases in tandem, as measured at large sampling scales: both groups respond in

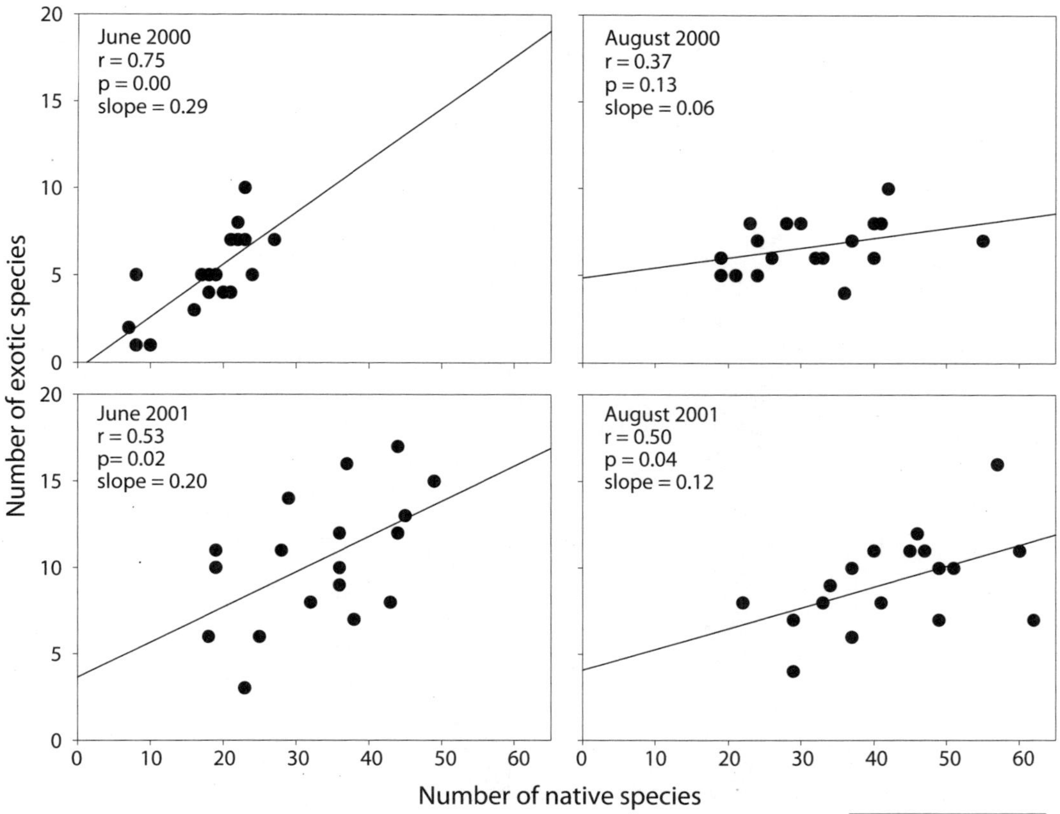

Fig. 5.3. Scatterplots and regression lines relating exotic plant species richness to native plant species richness in San Pedro River belt floodplain transects (see fig. 5.2) during four sampling times. Each data point represents a site (unpublished data of authors).

kind to ecosystem controls, such as spatial heterogeneity, on diversity levels (Stohlgren et al. 1999, Levine 2000a, Tabacchi and Planty-Tabacchi 2005).

RECENT IMMIGRANTS

A recent floristic study of the SPRNCA encountered five species new to Arizona, indicating ongoing floristic change (Makings 2006). Four of the five do not represent large range extensions from their previously known distributions. The fifth, wild marigold (*Tagetes minuta*), is originally from Argentina and recently was collected from California and southeastern Atlantic states, making the San Pedro population highly disjunct.

LOCAL EXTINCTIONS

It is challenging to determine which plant species, if any, have disappeared from the San Pedro. Environmental conditions in the San Pedro riparian zone are quite different than they were a century ago, prior to major channel incision. It is likely that falling water tables, intensive grazing pressure, and loss of ciénega habitat led to some extirpations; however, the San Pedro River has been only moderately studied by botanists over the past century. Nonetheless, there are a few historical records of species apparently no longer present along the upper San Pedro River. For example, rooted spikerush

Fig. 5.4. (top; 5.4a): Landscape view of St. David Ciénega wetland along the upper San Pedro River. (bottom; 5.4b): Yerba mansa (*Anemopsis californica*) and other wetland plants growing in a San Pedro ciénega. Photo credits: Elizabeth Makings.

(*Eleocharis radicans*), whorled marshpennywort (*Hydrocotyle verticillata*), and floating marshpennywort (*H. ranunculoides*) are in historical collections of the University of Arizona herbarium from "San Pedro Valley, Cochise County." Some of these were collected in 1908 by Leslie Goodding, and some in 1894 by J. W. Toumey.

The *Hydrocotyl* and *Eleocharis* species are wetland plants and may be testaments to the historically marshy character of the San Pedro. Some of the species that persist today in the San Pedro River ciénegas, including Arizona eryngo (*Eryngium sparganophyllum*), are regionally uncommon, reflecting the present rarity of the habitat (figs. 5.4a, 5.4b). The changing abundance of Huachuca water umbel (*Lilaeopsis schaffneriana* spp. *recurva*), a federally endangered species, also may be indicative of changes in hydrologic conditions. This semi-aquatic, rhizomatous perennial grows along wet reaches of the upper San Pedro, in areas that have a permanent water source and intermediate levels of flood disturbance. Without some flood disturbance, the tiny umbels are outcompeted by larger plants, but moderate flooding allows the plants to increase in abundance. Very intense flood scour, however, can cause local population extirpation.

Spatial Patterns of Diversity

VERTICAL PATTERNS: TREE CANOPY TO SOILS

High-canopy species. There are comparatively few woody species in the San Pedro riparian corridor: the 23 species of trees and 43 species of shrubs in the SPRNCA comprise only 11 percent of the flora. Two hemi-parasitic species of mistletoe occur in the forest canopy—Colorado desert mistletoe (*Phoradendron macrophyllum*) on cottonwood, willow, and ash (*Fraxinus*), and mesquite mistletoe (*P. californicum*) on legume trees—ever-so-slightly increasing canopy-layer diversity. Moving downward through the vegetation, diversity levels decrease from the tree canopy layer to the shrub (mid-stratum) layer and ground cover layers. Diversity is highest on the ground floor, reflecting the abundance of grasses and forbs (table 5.1).

Soil seed banks. There is more diversity in the riparian corridor than meets the eye, with many plant species at any given time existing only as dormant seeds in the soil. These seed banks are diverse along southwestern rivers (Boudell and Stromberg 2008). Wetland forbs and graminoids, such as seep monkeyflower (*Mimulus guttatus*) and rush (*Juncus* spp.), are common in the seed banks; seeds of sacaton (*Sporobolus*) and other mesic and xeric grasses and forbs also are abundant (Knipe and Springfield 1972). The small seeds are distributed broadly across the floodplain, sometimes buried deeply in the sediments; this long-term seed storage serves as a form of insurance against floods and drought by increasing the chances that a seed will be present to germinate wherever suitable conditions are created.

LATERAL PATTERNS: STREAM EDGE TO UPLANDS

Soils along the edge of the low-flow stream channel are wet and frequently disturbed, while those at the riparian perimeter are drier and more stable. Given these gradients of resource availability and disturbance, plant species composition and diversity change with distance from and above the river channel (Tabbachi et al. 1996, Lyon and Sagers 1998). On the San Pedro, we have observed these spatial diversity patterns to shift seasonally and annually (Lite et al. 2005, Stromberg 2007). During the summer dry season, herbaceous species diversity is greatest along the stream channel, but during the late-summer wet season, diversity is high on sites farther from and higher above the channel; in such seasons, abundant rain allows plants to establish even on high surfaces, while floods scour them from low surfaces. These spatial patterns are driven by herbaceous species, which respond rapidly to rain and flood events.

Riparian zones are often assumed to have more plant species per unit area than adjacent uplands, but surprisingly little information is available on this topic for dryland rivers. Some studies document higher plant species diversity in the riparian zone (Stohlgren et al. 1998), while others show higher diversity in the uplands (Hancock et al. 1996, Schade et al. 2003). In the San Pedro lower basin, upland vs. riparian diversity patterns varied by season (Stromberg 2007), but for the year as a whole, the floodplain did indeed have more species (fig. 5.5) and more functional types.

LONGITUDINAL PATTERNS: UPSTREAM TO DOWNSTREAM

Downriver in dryland basins such as the San Pedro, there are complex gradients of surface water and groundwater availability, flood intensity, fire patterns, stream geomorphology, rainfall amounts and seasonality, and temperature. Perennially flowing reaches are most common in upper portions of watersheds, with flows becoming intermittent as streams emerge from the mountains and lose water into the deep alluvium of inter-mountain basins. Flood intensities (total stream power) vary with distance downstream as watershed size, gradient, and drainage density change, and also vary with individual storm tracks (see chaps. 13 and 16). Upper reaches have higher rainfall and cooler temperatures. All of these factors interact to influence the local riparian floras (Baker 1990).

Surface water and groundwater. Where conditions are wetter along the San Pedro River, there are more tree and shrub species. In the floodplain, both woody species richness and a diversity index (Shannon's H', which integrates rich-

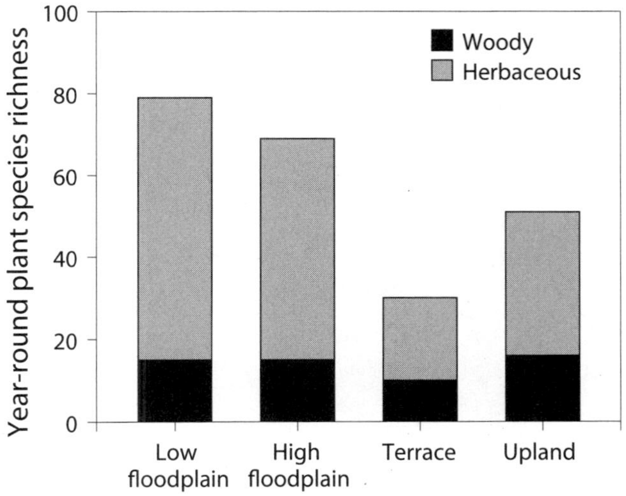

Fig. 5.5. Number of species in floodplain and terrace zones of the lower San Pedro River and in adjacent Sonoran Desert uplands, based on sampling during three seasons (March, May, September) of one year (1995). Herbaceous species were sampled in 20, 1 m^2 plots per zone; woody species were sampled in 20, 100 m^2 plots per zone. Based on Stromberg (2007). Used with permission.

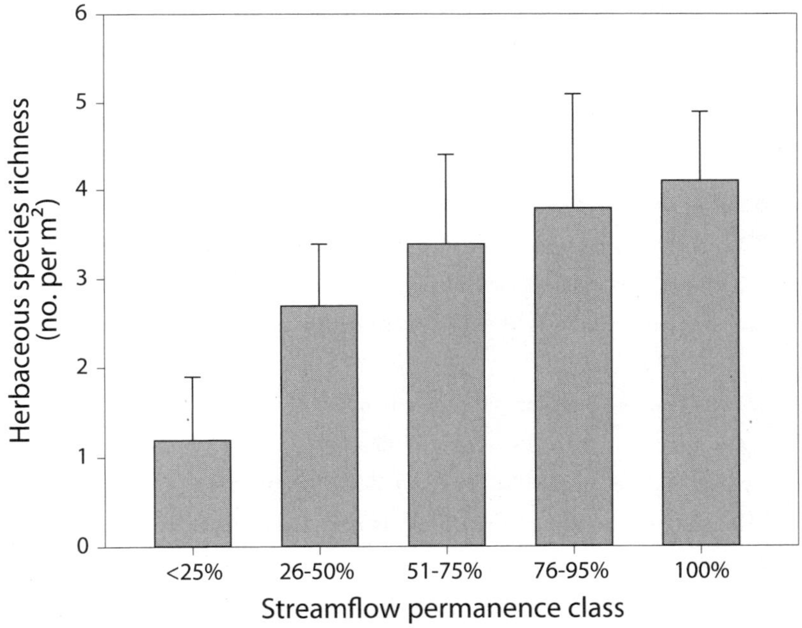

Fig. 5.6. Richness of herbaceous species within the low-flow channel zone of the San Pedro River during the pre-monsoon summer season in relation to streamflow permanence at the site. Values are based on data from multiple sites in multiple years. See Stromberg et al. (2005) for detailed methodology.

ness and evenness) increase among sites as flow permanence increases (Lite et al. 2005). Herbaceous species richness along channels and in the floodplain increased among sites with streamflow permanence and groundwater availability, but only when measured during the summer dry season (Stromberg et al. 2005, Lite et al. 2005) (fig. 5.6).

Flood intensity. For herbaceous species along the San Pedro River, sites with intermediate levels of flooding appear to have the greatest diversity. Richness, as measured after major flooding, varied with river-wide gradients of total stream power in a fashion consistent with the intermediate disturbance hypothesis (Connell 1978): it increased among sites as stream power increased and then

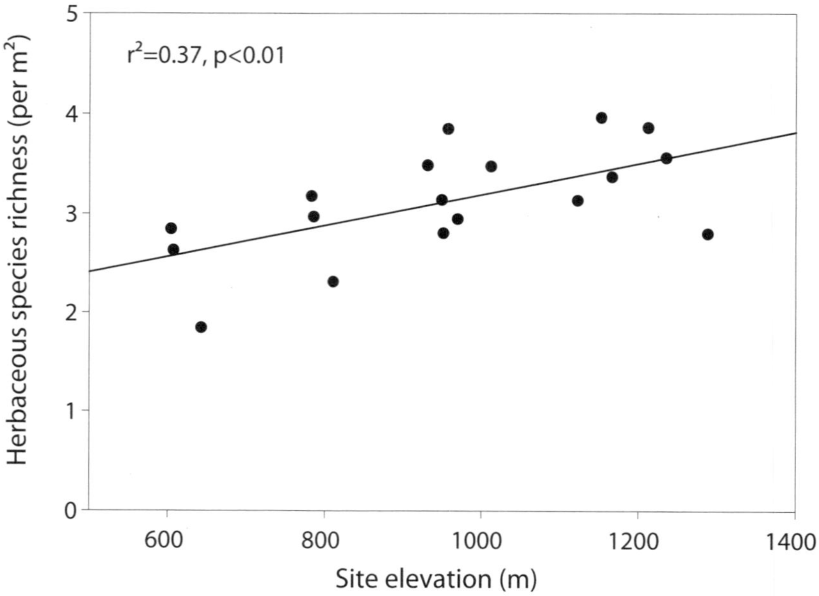

Fig. 5.7. Mean numbers of herbaceous species per 1 m^2 plot in the San Pedro floodplain. Values are averaged across four sampling seasons. From Lite et al. (2005). Used with permission.

declined at sites with greatest stream power (Lite et al. 2005). However, for woody species, this was not the case. Woody species diversity and flood power are positively correlated on some rivers (Bendix 1997), but woody species numbers did not change along a San Pedro gradient of flood stream power (Lite et al. 2005). Woody pioneer species were more abundant at sites with high flood intensity, but this trend was offset by decreases in later successional species.

Elevation and climate. Sites in the upper reaches of the San Pedro River have slightly greater herbaceous species richness and cover than do those in the lower basin (Lite et al. 2005) (fig. 5.7). This longitudinal pattern reflects rainfall and temperature gradients but is not fixed; rather, the pattern varies depending on weather conditions. During the monsoon season, rain-dependent annuals increased in diversity with distance upstream (paralleling trends for summer rains to increase upstream), but in early summer, the annuals declined in diversity with distance upstream (paralleling patterns for winter rains and temperatures to increase in the downstream, northerly direction) (Bagstad et al. 2005).

Differences among Vegetation Types

Each of the vegetation types that occur in the San Pedro riparian zone occupies a slightly different physical setting, and each supports different numbers and types of plant species (fig. 5.8). While some understory species are widely distributed among riparian vegetation types, others are more restricted (Merigliano 2005). Vegetation types that occur near the stream, such as seepwillow shrublands or young cottonwood/willow forests, tend to have wet soils and intermediate species diversity (Bagstad et al. 2006). Older cottonwood

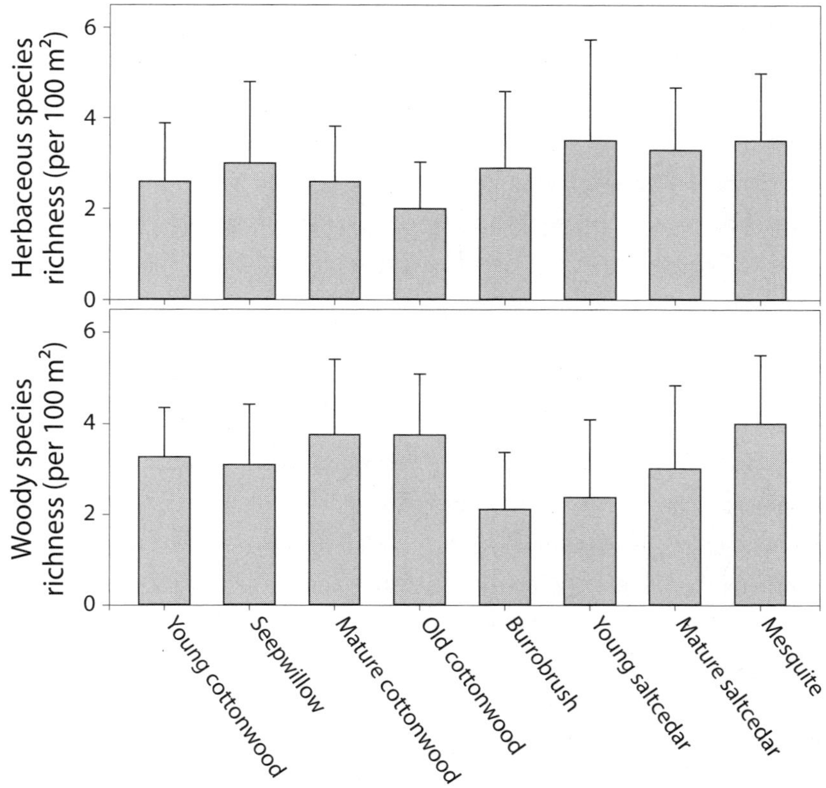

Fig. 5.8. Number of woody and herbaceous plant species sampled in various vegetation types in the San Pedro riparian zone. See Bagstad et al. (2006) for detailed methodology.

forests have relatively low diversity of herbaceous plants (but high diversity of woody plants), perhaps because establishment of small-seeded species is thwarted by dense shade, dense leaf litter, or dry surface soils (Nilsson et al. 1999, Xiong et al. 2003).

IMMIGRANT SPECIES AND DIVERSITY: THE SALTCEDAR CASE

Several review articles (e.g., Walker and Smith 1997, Di Tomaso 1998) and a few empirical studies report that saltcedar stands have little biological diversity. On the San Pedro River, however, a different story emerges: plant species diversity in saltcedar stands is similar to that in other vegetation types (Stromberg 1998b, Bagstad et al. 2006) (fig. 5.8). Why do patterns differ? It appears to be because environmental conditions along the frequently flooded San Pedro differ from those on other studied rivers, which have tended to be impounded and flow regulated. Low diversity in saltcedar patches on dammed rivers may be a consequence, in part, of river-level changes in the physical processes that regulate diversity, and the presence of saltcedar may simply be a corollary.

Summary

The San Pedro riparian corridor has high plant diversity, with over 750 vascular plant species. Species are abundant because water resources (groundwater, surface water, and growing-season rainfall) are ample and because an active flood regime creates high temporal and spatial heterogeneity. Additionally, the river is located within a floristically diverse region. While many plant species are riparian-zone specialists with high water needs, many others are facultative riparian species that also occur in upland areas of the watershed. Most of the species along the river are herbaceous, and many of these are annuals. The soil seed bank represents a hidden source of diversity, with many species maintaining themselves as seeds in between periodic floods and droughts.

Complex longitudinal gradients allow for a wide range of biological outcomes over the length of the San Pedro River. Reaches with perennial flow are more prevalent in the upper San Pedro, and these wet areas have greater woody species diversity and channel-side herbaceous species diversity than do drier areas. Reaches in the higher elevations also receive more rainfall, contributing to higher values for herbaceous species diversity. Perpendicular to the stream, there is a high degree of turnover of species across complex hydro-gradients, creating high beta diversity. Herbaceous species diversity patterns vary with distance from and above the river channel, although these patterns shift depending on seasonal differences in flood and rain patterns.

Species richness in the riparian zone varies widely between seasons and years. Richness shows pulse-type increases during the summer rain and flood season and can show more sustained increases (of several months' duration) after larger floods. Over past centuries, there has been considerable influx of species to the San Pedro River corridor and some local extirpations of wetland plants. One introduced woody species that has become common along the river is saltcedar. However, counter to some expectations, floristic diversity in saltcedar patches is similar to that in other riparian vegetation types.

SECTION 2

Animals of the Riparian Ecosystem

CANDAN U. SOYKAN,
L. ARRIANA BRAND,
AND JOHN L. SABO

Causes and Consequences of Mammal Species Richness

SIX

Overview

The San Pedro watershed is an internationally recognized "hotspot" for mammals, hosting one of the richest assemblages of mammal species in the United States (Simpson 1964, Hall 1981, Duncan 1988). Estimates of species richness range from 61 to 87 mammals, about half of the highest estimate in the world (139 mammals in Cocha Cashu/Pakitza, a Neotropical lowland rainforest site; Voss and Emmons 1996). These estimates are particularly high for a semi-arid habitat at this latitude (Boeer and Schmidly 1977). In this chapter, we consider the importance of geography, environmental gradients, and non-equilibrium processes in explaining the extensive mammalian diversity of the San Pedro watershed.

In the sections that follow, we review definitions of richness and diversity, provide some background on the mammal fauna of the San Pedro, describe broad ecological theories relating abiotic and biotic factors to species richness, and apply these to the San Pedro watershed with mammals as a case study. We close with a discussion of how mammals may (in turn) affect the habitats in which they live and a brief section on the impacts of human activity on the mammals of the San Pedro watershed.

Definitions of Species Richness and Diversity

According to Krebs (1999), species diversity is the total number of species weighted by their relative abundance. It includes two components: species richness, which refers to the total number of species in a community, and species evenness, which refers to the relative number of individuals of each species in a community. Thus, a community with 100 species is more species rich than one with 50 species. Likewise, a community in which the species are equally abundant is more even than one dominated by one or a few species. Although other definitions exist, in this chapter we adhere to Krebs's definitions of diversity, richness, and evenness.

Whittaker (1972) partitioned total species richness in a region, called gamma richness, into two components: local (alpha) richness and turnover of species between localities (beta richness), where Gamma Richness = Alpha Richness × Beta Richness. How one defines a locality depends on the question being asked and the taxon under consideration. Since localities can vary in size, scale plays an important role in any discussion of species richness (Sarr et al. 2005). In this chapter we treat each habitat type as a locality and use beta richness to refer to differences in species composition among habitats. Scale also affects calculations of species evenness. Clumped and/or random distributions of organisms provide different estimates of species evenness depending on the scale of measurement and the index being used (Wilson et al. 1999). For example, if a few sample quadrats happened to fall on clumps of a patchily distributed species, one would overestimate its abundance in the community.

Researchers have developed different diversity metrics to suit their needs. Commonly used metrics include the Simpson's Diversity Index and the Shannon-Wiener Diversity Function (Krebs 1999). Similarly, many different measures of evenness have been proposed, with no single index standing out as the best one. For simplicity, we will focus our discussion on species richness.

Mammals of the San Pedro

The number of mammal species that occur in the San Pedro watershed—estimated using indirect observations (e.g., sign and scat), captures, or field observations (Woolsey 1987, Duncan 1988, Hass 2001), as well as overlapping range maps (Cockrum 1960, Hall 1981, Davis 1982, Hoffmeister 1986)—may be as high as 87 species (see tables 6.1 and 6.2). Excluding extirpated species, and using only observations and occurrence records but not range maps, a more conservative estimate is 61 species (Woolsey 1987, Duncan 1988, Hass 2001) (table 6.1). By way of comparison, Boeer and Schmidly (1977) listed only 30 mammal species on the lower Rio Grande at Big Bend National Park, while 55 species are confirmed in the Bill Williams River National Wildlife

Refuge (K. Blair, personal communication). Estimates for mammal species richness in the nearby Verde River watershed range as high as 92 (but this includes humans, extirpated species, and subspecies; L. Stevens, personal communication). The 61 mammal species observed in the San Pedro watershed include 1 didelphid marsupial (order Didelphimorphia), 1 insectivore (order Insectivora), 29 rodents (order Rodentia), 3 lagomorphs (rabbits and hares, order Lagomorpha), 8 bats (order Chiroptera), 3 artiodactyls (even-toed ungulates, order Artiodactyla), and 16 carnivores (order Carnivora) (Woolsey 1987, Duncan 1988, Hass 2001) (table 6.1).

Among the species that have been regionally or locally extirpated are three top predators—the grizzly bear, jaguar, and gray wolf (Ceballos 1985, Duncan 1988, Hass 2001). However, recent jaguar sightings in southern and southeastern Arizona suggest that source populations are nearby and could potentially recolonize the San Pedro watershed. Likewise, gray wolves have been reintroduced in eastern Arizona, not too far from the San Pedro watershed. The long-tailed weasel has not been detected in the area since the 1950s (Hoffmeister and Goodpastor 1954, Woolsey 1987), and black-tailed prairie dogs were last sighted on the U.S. side of the border in 1938 (Hoffmeister 1986), though they still occupy the San Pedro watershed south of the Mexican border (Castillo-Gamez et al. 2005). Beavers were extirpated from the upper San Pedro River by the mid-1980s (Ceballos 1985, Duncan 1988). A small beaver population has recolonized reaches near the San Pedro's confluence with the Gila River, and several beaver were recently reintroduced to suitable habitats along the upper parts of the river. Muskrats have not been recorded along the San Pedro River since the 1890s (Allen 1895), but could possibly survive if reintroduced (Duncan 1988). River otters have not been detected for a long time, but were once widely distributed across the Southwest and likely occupied parts of the San Pedro watershed (Woolsey 1987).

Theory of Species Richness

Although no single explanation suffices, ecologists have found relationships between species richness and a host of environmental variables including latitude, climate, biological productivity, stress, disturbance, habitat heterogeneity, habitat complexity, and ecosystem size and isolation (MacArthur 1972, Whittaker 1975, Schluter and Ricklefs 1993, Brown 1995, Rosenzweig 1995). More specifically, species richness tends to increase with decreasing latitude (Latham and Ricklefs 1993), increasing area (Hamilton et al. 1963), decreasing isolation (Simberloff and Wilson 1969), increasing temperature and precipitation (Richerson and Lum 1980), increasing habitat heterogeneity (Simpson 1964) and habitat complexity (MacArthur and MacArthur 1961), and intermediate levels of disturbance and environmental stress (Connell 1978, Menge and Sutherland 1987). Thus, both geographic position and

TABLE 6.1. Mammal species of the San Pedro River.

Introduced species names are marked with an asterisk (*); extirpated species names are shaded gray.

Family	Common name	Scientific name
Didelphidae	Virginia opossum	*Didelphis virginiana*
Soricidae	Desert shrew	*Notiosorex crawfordi*
Phyllostomatidae	California leaf-nosed bat	*Macrotus californicus*
Vespertilionidae	Pallid bat	*Antrozous pallidus*
	Big brown bat	*Eptesicus fuscus*
	Red bat	*Lasiurus borealis*
	Cave myotis	*Myotis velifer*
	Townsend's big-eared bat	*Plecotus townsendi*
Molossidae	Western bonneted bat	*Eumops perotis*
	Mexican free-tailed bat	*Tadarida brasiliensis*
Leporidae	Antelope jack rabbit	*Lepus alleni*
	Black-tailed jack rabbit	*Lepus californicus*
	Desert cottontail	*Sylvilagus audubonii*
Sciuridae	Harris's antelope squirrel	*Ammospermophilus harrisii*
	Black-tailed prairie dog	*Cynomys ludovicianus*
	Spotted ground squirrel	*Spermophilus spilosoma*
	Round-tailed ground squirrel	*Spermophilus tereticaudus*
	Rock squirrel	*Spermophilus variegatus*
Geomyidae	Botta's pocket gopher	*Thomomys bottae*
Heteromyidae	Bailey's pocket mouse	*Chaetodipus baileyi*
	Hispid pocket mouse	*Chaetodipus hispidus*
	Rock pocket mouse	*Chaetodipus intermedius*
	Desert pocket mouse	*Chaetodipus penicillatus*
	Merriam's kangaroo rat	*Dipodomys merriami*
	Ord's kangaroo rat	*Dipodomys ordii*
	Banner-tailed kangaroo rat	*Dipodomys spectabilis*
Cricetidae	White-throated wood rat	*Neotoma albigula*
	Muskrat	*Ondatra zibethicus*
	Northern grasshopper mouse	*Onychomys leucogaster*
	Southern grasshopper mouse	*Onychomys torridus*
	Brush mouse	*Peromyscus boylii*
	Cactus mouse	*Peromyscus eremicus*
	White-footed mouse	*Peromyscus leucopus*

TABLE 6.1. *(continued)*

Family	Common name	Scientific name
Cricetidae *(cont.)*	Deer mouse	*Peromyscus maniculatus*
	Fulvous harvest mouse	*Reithrodontomys fulvescens*
	Western harvest mouse	*Reithrodontomys megalotis*
	Plains harvest mouse	*Reithrodontomys montanus*
	Arizona cotton rat	*Sigmodon arizonae*
	Fulvous cotton rat	*Sigmodon fulviventer*
	Yellow-nosed cotton rat	*Sigmodon ochrognathus*
Muridae	House mouse*	*Mus musculus**
Erithizontidae	Porcupine	*Erithizon dorsatum*
Castoridae	Beaver	*Castor canadensis*
Cervidae	Mule deer	*Odocoileus hemionus*
	White-tailed deer	*Odocoileus virginianus*
Tayassuidae	Collared peccary	*Pecari tajacu*
Procyonidae	Ringtail	*Bassariscus astutus*
	White-nosed coati	*Nasua narica*
	Northern raccoon	*Procyon lotor*
Mephitidae	Common hog-nosed skunk	*Conepatus mesoleucus*
	Hooded skunk	*Mephitis macroura*
	Striped skunk	*Mephitis mephitis*
	Western spotted skunk	*Spilogale gracilis*
Mustelidae	Northern river otter	*Lontra canadensis*
	Long-tailed weasel	*Mustela frenata*
	American badger	*Taxidea taxus*
Canidae	Domesticated dog*	*Canis familiaris**
	Coyote	*Canis latrans*
	Gray wolf	*Canis lupus*
	Gray fox	*Urocyon cinereoargenteus*
	Kit fox	*Vulpes macrotis*
Felidae	Domesticated cat*	*Felis silvestris**
	Bobcat	*Lynx rufus*
	Jaguar	*Panthera onca*
	Puma	*Puma concolor*
Ursidae	Black bear	*Ursus americanus*
	Grizzly bear	*Ursus arctos*

Note: Documented by captures, observations, and occurrence records (64 native species [6 extirpated] and 3 non-native species) based on Woolsey (1987), Duncan (1988), and Hass (2001).

TABLE 6.2. Mammal species with range maps that overlap the San Pedro riparian corridor.

Not documented by captures, observations, or occurrence records (20 species).

Family	Common name	Scientific name
Phyllostomatidae	Mexican long-tongued bat	*Choeronycteris mexicana*
	Southern long-nosed bat	*Leptonycteris curasoae*
Vespertilionidae	Spotted bat	*Euderma maculatum*
	Allen's big-eared bat	*Idionycteris phyllotis*
	Silver-haired bat	*Lasionycteris noctivagans*
	Hoary bat	*Lasiurus cinereus*
	Southern yellow bat	*Lasiurus ega*
	Southwestern myotis	*Myotis auriculus*
	California myotis	*Myotis californicus*
	Small-footed myotis	*Myotis ciliolabrum*
	Fringed myotis	*Myotis thysanodes*
	Long-legged myotis	*Myotis volans*
	Yuma myotis	*Myotis yumanensis*
	Western pipistrelle	*Pipistrellus hesperus*
Molossidae	Pocketed free-tailed bat	*Nyctinomops femorosaccus*
	Big free-tailed bat	*Nyctinomops macrotis*
Heteromyidae	Silky pocket mouse	*Perognathus flavus*
Muridae	Northern pigmy mouse	*Baiomys taylori*
Antilocapridae	Pronghorn	*Antilocapra americana*
Felidae	Ocelot	*Leopardus pardalis*

Source: Based on Duncan (1988).

local environmental conditions likely influence species richness. Historical factors (both natural and anthropogenic) have profound effects as well.

The tendency towards higher species richness at lower latitudes has received considerable attention but cannot be easily explained, in large part due to the host of factors (climate, geography, etc.) that vary in parallel with latitude (Schluter and Ricklefs 1993). The relationship between species richness and climate appears much clearer—conditions that favor biological production in terrestrial ecosystems (warm temperatures and abundant precipitation) often correlate with high species richness (Richerson and Lum 1980). High levels of productivity may not always maximize species richness, however; the relationship appears to be far more complex than a simple linear association. Other factors such as landscape position, seasonality, and species interactions affect both primary production and species richness (Whittaker 1975, Huston 1994, Brown 1995, Schade et al. 2003), leading to neutral or even inverse relationships between productivity and species richness, depending on the scale of observation. Thus, many studies have found that species richness increases with increasing production when species are sampled at large spatial scales. However, when sampled at smaller scales, species richness often peaks at intermediate levels of productivity.

Habitat heterogeneity refers to the number of different habitat types within a given area. Topographic and geomorphic variation combine with variation in temperature and water availability to provide diverse abiotic conditions, which, in turn, result in different habitat types. Because different species often favor particular habitats, a diversity of habitats results in high beta richness and has been cited as one of the reasons for the high levels of species richness in mountainous regions and riparian habitats (Simpson 1964, Sabo et al. 2005). Habitat complexity refers to vegetation structure within a given habitat, for example, the canopy, understory, and shrub- and ground-level vegetation in a forest make it a complex habitat. This complexity increases alpha richness, the number of species in a given habitat, by providing more niches for species to occupy (MacArthur and MacArthur 1961, Pianka 1967).

Disturbance and environmental stress (sensu Menge and Sutherland 1987) can increase both local and regional species richness (Connell 1978, Huston 1994). Increases in alpha richness at intermediate disturbance levels are thought to represent a trade-off between harsh abiotic conditions eliminating all or most species in a given area when disturbance is frequent (or stress levels are high) and competitive pressure from superior competitors crowding out other species in an area when disturbance is infrequent (or stress levels are low). Beta richness also increases as a result of disturbance, due to the creation of spatial heterogeneity across a landscape (Pollock et al. 1998) and through time. Disturbance resets the successional trajectory of a patch; spatially and temporally variable disturbance (as is characteristic of riparian habitats) results in patches of different seral stages, each containing a different suite of species. For example, riparian areas along the San Pedro River contain habitats ranging from mature cottonwood/willow forests to sparsely vegetated ground; different plant and animal species utilize the different seral stages, increasing the overall richness of the riparian zone.

As described by MacArthur and Wilson (1967), species richness on islands depends on a balance between colonization and extinction, both of which relate to island size and distance to the mainland. Small, isolated islands such as Hawaii are expected to have lower species richness, whereas large islands close to a continent, such as Cuba, are expected to have higher species richness. The theory of metapopulation dynamics extends this thinking to different habitat types in a terrestrial landscape (Hanski 1999). According to the theory, the size and location of habitat patches within a landscape influence the alpha, beta, and gamma richness of a region. Increasing patch size results in larger populations that are more resistant to extinction. Patch location relative to other patches influences dispersal and the probability of recolonization following extinction events. Together, these two geographic factors play an important role in determining species richness patterns for a given region (Brown 1971, Tonn and Magnuson 1982, Rosenzweig 1995).

Influences on Mammal Species Richness in the San Pedro Watershed

Several of the factors described above combine to produce the high mammal species richness of the San Pedro watershed. These factors include region size, biogeographic location, climate, water availability, primary productivity, habitat heterogeneity, disturbance, and edge habitat use. The factors can be summarized into three broad categories: geography, environmental gradients, and non-equilibrium processes. Any given factor may fall under more than one of these headings; we therefore use them more for heuristic than for classification purposes.

GEOGRAPHY

This category includes factors such as region size, biogeographic setting, and climate. The San Pedro watershed is large (ca. 12,000 km^2), well connected to other species-rich regions, and has a warm temperate climate (see chap. 11). The watershed lies at the intersection of four biogeographic zones (the Sonoran Desert to the west, the Chihuahuan Desert to the east, the Rocky Mountains to the north, and the Sierra Madres to the south) and includes species from each zone. Long summers and relatively mild winters allow mammals with tropical affinities, such as peccaries, coatis, opossums, and various bats, to extend their range northward into the San Pedro watershed (Hoffmeister 1986). Although precipitation is modest, rainfall peaks during late summer, maximizing biological production. Thus, regional conditions resemble those that correlate with high species richness. In sum, the San Pedro watershed is a large, well-connected region with a favorable climate.

ENVIRONMENTAL GRADIENTS

This category includes primary productivity and habitat heterogeneity, both of which are driven, in large part, by streamflow regime and gradients in water availability. In the San Pedro watershed, spatio-temporal variation in water availability is the principal factor constraining primary and secondary productivity (Huxman et al. 2004). Inflowing groundwater sustains permanent stream water and high productivity in some reaches, while in other reaches the stream flows only seasonally. In all reaches, high seasonal and annual variation in precipitation, flood magnitude, and aquifer recharge (typical of dry regions) creates sharp temporal fluctuations in water availability on the river floodplains and terraces.

Mammal species have evolved physiological and behavioral mechanisms for dealing with variation in resource availability. Heteromyids (e.g., kangaroo rats, pocket mice) have physiological adaptations that allow them to cope with little or no drinking water in their diets (Schmidt-Nielsen 1964, Tracy and Walsberg 2002); moreover, many desert mammals (e.g., ground squirrels, pocket gophers, kangaroo rats, pocket mice, woodrats, etc.) have evolved food-caching behaviors that allow them to endure periods of low resource levels (Vander Wall 1990). Some mammals, such as bats, ground squirrels, deer mice, kangaroo rats, and pocket mice, hibernate, aestivate, or undergo periods of inactivity, often associated with seasonal accumulation of fat (Hudson 1967). Finally, larger, more mobile mammals shift their movement and foraging patterns to areas of relatively higher resource abundance (Valenzuela and Ceballos 2000). For example, coatis in tropical dry forests shift their activity to arroyo forest during dry periods, where arthropods and water are more readily available, and may do the same along the San Pedro. Skunks, foxes, and other carnivores along the San Pedro also shift their diet and movement patterns, tracking the availability of seasonally or spatially abundant foods (Soykan and Sabo, forthcoming).

The vegetation types within the San Pedro watershed vary in plant species composition as well as in amount of cover in different height strata (see chap. 1). Vegetation influences mammal populations by providing food, affecting movement patterns, and providing protection from predators. Since each species has different environmental requirements, many mammals show affinities for particular vegetation types (tables 6.3, 6.4). Along the San Pedro River, five carnivores (bobcat, kit fox, puma, and two species of *Mephitis* skunks) used vegetation types in proportion to their availability, while six others (badger, coyote, gray fox, hog-nosed skunk, coati, and raccoon) showed some habitat affinity, with their sign being found more often than expected in certain habitats (Hass 2001). Some rodents also show habitat affinities. For example, cricetids (e.g., deer mice, wood rats) tend to utilize habitats with dense ground cover characteristic of sacaton grassland, saltcedar, or marsh habitats (Stamp and Ohmart 1979, Duncan 1988, Ellison and Van Riper 1998), while heteromyids (e.g., kangaroo rats, pocket mice) generally prefer more open vegetation cover characteristic of upland environments (Stamp and Ohmart 1979, Duncan 1988).

Water availability, in particular distance to groundwater, is a principal factor that determines vegetation types along the San Pedro River (see chap. 1). Mammal species vary in their habitat affinities; because of this, spatial variation in water table depth, and the resulting habitat heterogeneity, increases

TABLE 6.3. Common small mammal habitat associations along the San Pedro River.

Values represent the relative density of small mammals for each habitat type (number per 300 trap nights for rodents; number per 100 trap nights for shrews). Shaded cells indicate densities greater than expected based on a uniform distribution among habitats.

	Vegetation association							
	243.30	223.21	233.221	223.231	143.142	143.141	153.20	300.00
Common name	MSH	CW	TAM	MQ	SAC/MQ	SAC	DS	AG
Desert pocket mouse	0.0	0.3	7.0	2.8	1.0	0.0	0.9	0.0
Bailey's pocket mouse	0.0	0.0	0.0	2.0	0.0	0.0	0.1	0.0
Merriam's kangaroo rat	0.0	0.0	5.0	1.8	2.0	0.0	8.5	0.0
Plains harvest mouse	1.5	0.0	0.0	0.0	1.6	0.0	0.0	0.0
Western harvest mouse	3.5	0.7	1.0	1.6	4.8	8.5	0.3	0.0
Fulvous harvest mouse	0.5	0.0	1.0	0.0	0.0	0.0	0.1	1.2
Cactus mouse	1.5	2.2	6.0	4.0	0.0	0.0	2.4	0.0
Deer mouse	6.5	6.2	11.0	0.6	8.6	1.0	2.8	5.4
White-footed mouse	28.0	3.7	18.0	0.2	0.0	8.5	0.2	0.0
Northern grasshopper mouse	0.0	0.0	3.0	1.6	0.2	0.0	0.5	0.0
Southern grasshopper mouse	0.0	0.0	0.0	0.0	0.2	0.0	0.9	0.0
Arizona cotton rat	6.0	0.0	0.0	1.2	10.6	6.0	0.0	14.4
White-throated wood rat	1.0	1.8	4.0	3.2	1.2	0.0	3.1	0.0
Desert shrew	NA	24.6	11.8	8.0	11.4	5.9	1.0	NA
Number of Species Captured*	9	9	10	11	10	5	14	4
Number of Plots Sampled	2	6	1	5	5	2	12	5

Source: Based on Duncan 1988 and Duncan and Corman 1991.

Notes: Sampling did not incorporate detection probabilities and was not uniform among habitats—thus, these data must be interpreted with caution. MSH = Chihuahuan Interior Marshland Biome, CW = Cottonwood-Willow Series, TAM = Tamarix-Mixed Deciduous Association, MQ = Mesquite Association, SAC/MQ = Sacaton-Mesquite Association, SAC = Sacaton Association, DS = Chihuahuan Desertscrub Biome, AG = Agricultural Land.

*Total number of small mammal species for each habitat type, including several uncommon species not included in this table.

beta richness in the San Pedro watershed. The wettest habitats along the San Pedro River are marshlands adjacent to perennial water. Five small mammal species (deer mouse, western harvest mouse, plains harvest mouse, Arizona cotton rat, and white-footed mouse) had high abundance in San Pedro marsh habitat, with one (white-footed mouse) being substantially more abundant in marsh than in other habitats (Duncan 1988). Raccoon sign was detected more often than expected in cottonwood/willow forest (Hass 2001), a mesic, floodplain habitat, and desert shrews were two to three times more abundant in cottonwood/willow forests than in tamarisk (saltcedar), sacaton, or mesquite habitats (Duncan and Corman 1991). Several other small mammals (white-footed mouse, cactus mouse, deer mouse, white-throated wood rat) were abundant in cottonwood/willow habitats, but were not considered to be habitat specialists (Duncan 1988). Two small mammal species, desert pocket

TABLE 6.4. Mammalian carnivore habitat preferences along the San Pedro River

Values represent the proportion of total signs for that species encountered in each vegetation type. Shaded cells indicate values greater than expected based on availability.

	Riparian forest	Grassland	Desert scrub	Mountain woodland
Mephitis skunks	0.27	0.17	0.39	0.17
Hog-nosed skunk	0.12	0.19	0.47	0.23
Badger	0.07	0.40	0.40	0.13
Coati	0.11	0.00	0.11	0.78
Northern raccoon	0.47	0.20	0.29	0.04
Kit fox	0.21	0.21	0.50	0.07
Gray fox	0.14	0.15	0.39	0.32
Coyote	0.09	0.30	0.51	0.11
Bobcat	0.31	0.17	0.38	0.14
Puma	0.34	0.16	0.29	0.21

Source: Hass 2001.

Note: Vegetation type was classified according to the predominant vegetation within a 10 m circle surrounding the carnivore sign. Riparian forest included cottonwood/willow/ash gallery forest >5 m high. Mesquite bosques were categorized as desert scrub. Grassland refers to non-riparian grassland habitats found above the river terrace or in the foothills of the Huachuca Mountains.

mouse and southern grasshopper mouse, had substantially higher abundance along the San Pedro in patches of saltcedar/mixed deciduous trees than in other habitat types (Duncan 1988). Along the Rio Grande, both of these rodents have an affinity for saltcedar patches (Stamp and Ohmart 1979).

River terraces of the San Pedro are vegetated by sacaton grassland, mesquite woodland, mesquite savanna, or mixed herbaceous assemblages on abandoned farm fields and are commonly several meters above and 100 to 200 m from the low-flow channel. Several rodents (Bailey's pocket mouse, plains and western harvest mouse, and Arizona cotton rat) and carnivores (coyote, hog-nosed skunk, gray fox, coati, and badger) (fig. 6.1a) were detected more often on terraces than in floodplain habitats (Duncan 1988, Hass 2001). Within the terraces, Bailey's pocket mouse was most common in mesquite habitats (Duncan 1988), though this species was trapped in a variety of habitat types on the Verde River (Stamp and Ohmart 1979), suggesting that habitat affinities can be context dependent. The western harvest mouse was almost twice as abundant in sacaton grasslands as in any other habitat on the San Pedro, consistent with its affinity for dense grassland habitats along other rivers (Falck et al. 2003).

Upland habitats, characterized by Chihuahuan desert scrub (upper basin) or Sonoran desert scrub (lower basin), are located tens to hundreds of meters from the active channel. Due to their unique vegetation and climate,

Fig. 6.1. Two earth-dwelling mammals of the San Pedtro riparian corridor: American badger (top; 6.1a) and Botta's pocket gopher (bottom; 6.1b). Photo credits: Robert J. Luce.

they support a different suite of organisms. Two mammal species, Merriam's kangaroo rat and the northern grasshopper mouse, were most common in the Chihuahuan desert scrub habitats studied by Duncan (1988).

Mammal species of desert riparian corridors vary in their reliance on surface water. Wide-ranging carnivores such as puma and black bear probably depend on the river to rehydrate while moving between mountain areas (Hass 2001). Moreover, they utilize washes as movement corridors through inhospitable desert habitats. A number of bat species forage regularly over water (Hinman and Snow 2003). Peccary, raccoon, gray fox, bobcat, and puma and other large mammals breed along the river (Hoffmeister 1986, Duncan 1988, Hass 2001), but it is not known to what extent this is due to the surface water directly, the effects of water on riparian vegetation type and productivity, or both. Additionally, perennial water sources are important for a number of small mammals such as deer mice and desert shrews (Duncan 1988, Duncan and Corman 1991).

Due to its high habitat heterogeneity, the San Pedro watershed contains a great deal of edge habitat; that is, habitat boundaries separating two distinct vegetation types (e.g., the boundary between mesquite bosques and desert scrub). Ecological boundaries have long been known to affect the distribution and abundance of organisms, and different species can increase or decrease near edges (Ries et al. 2004).

Numerous studies have demonstrated that resource subsidies across ecological edges can affect population and community structure in both habitats (Rose and Polis 1998, Polis et al. 2004). Sabo and Power (2002) found that aquatic insects increased lizard abundance on riparian cobble bars. In turn, by increasing the abundance of predators, a resource subsidy may depress local prey populations that live in the subsidized habitat. For example, Murakami and Nakano (2002) found that aquatic insects increased near-river bird abundance; the birds also foraged on terrestrial insects, decreasing their abundance in near-river habitats.

Although few empirical examples exist, predator movement between habitats has the potential to reduce prey populations, perhaps even leading to prey exclusion from small habitat patches (Oksanen 1990, Cantrell et al. 2001). The abundance of carnivores along the San Pedro River makes it a likely location for this phenomenon, termed spillover predation. Species such as gray fox and striped skunk often live near the river but forage widely across the landscape, potentially reducing prey populations in desert scrub habitats located adjacent to the riparian corridor (Soykan and Sabo, forthcoming).

NON-EQUILIBRIUM PROCESSES

This category, which includes disturbance and subsequent increases in habitat heterogeneity, deals with spatio-temporal variation in the environment and its effects on species richness. Floods cause large changes in surface water depth and distribution and scour vegetation, seeds, and litter from the floodplain. In so doing, floods exert a strong influence on mammal community structure along the San Pedro and other free-flowing desert rivers (Andersen and Nelson 1999). Obligate riparian mammals such as beaver have behavioral and morphological adaptations to the flood regime (Andersen et al. 2000). Large or volant mammals generally have large home ranges and can escape flooded areas. In contrast, small mammal species that facultatively use riparian environments and that have small home ranges may be less tolerant of flooding and thus less able to exploit riparian environments (Andersen et al. 2000). In response to flood events, certain species, such as cactus and white-footed mice, utilized the canopy as a temporary refuge, though non-arboreal small mammals such as Ord's kangaroo rat experienced high mortality during flood events (Andersen and Nelson 1999, Andersen et al. 2000). Andersen and Nelson (1999) attributed the absence of certain species from cottonwood/willow forests on the lower Colorado and Bill Williams Rivers to an inability to move sufficiently far during periods of flooding.

On rivers such as the lower San Pedro, where flood regimes are characterized by high intensities and frequencies, open areas can cover large portions of the floodplain. This can decrease the available habitat for small mammal species that require dense vegetation cover as protection from predators (Andersen et al. 2000). However, this resetting of the successional process increases habitat heterogeneity in the riparian zone, creating a patchwork of seral stages (Pollock et al. 1998). If species are differentially adapted to exploit the various patches of riparian vegetation, then flooding can lead to high beta richness in the riparian zone. Additionally, woody debris resulting from flood events has been found to increase small mammal species richness, due to lower temperatures within the debris piles and/or protection from predators (Steel et al. 1999). Flood debris also provides materials needed for wood rat midden construction (Ellis et al. 1997, Ellison and Van Riper 1998). Thus, through its indirect effects on vegetation structure (patch dynamics, large woody debris accumulation), flooding profoundly influences mammal species richness and abundance along the San Pedro River.

Fire is another disturbance factor in riparian zones such as the San Pedro that develops seasonally dry understory vegetation. Some small mammals avoid direct heat by burrowing into the soil, while larger and more mobile animals flee the burning area (DeBano et al. 1998). Post-fire changes in vegetation structure influence mammal populations for many years. For example, fire in sacaton grasslands was followed by reduced numbers of green vegetation-feeding cotton rats (in response to foliage reduction) and increased numbers of seed-eating pocket mice and kangaroo rats (in response to abundant seed production by annual forbs; Bock and Bock 1978).

Livestock grazing is yet another form of disturbance, the cessation of which (on some reaches of the San Pedro) probably has increased the abundance of those small mammals that require dense vegetation. The substantial increase of plant cover at low height intervals that followed the removal of livestock on the upper San Pedro (Duncan 1988, Krueper et al. 2003) likely has substantially increased the abundance of small mammal species that prefer cover characteristic of grassland or riparian woodland habitats. The increase in herbaceous understory plants may also have increased the number of white-tailed deer (M. Fredlake, personal communication). In contrast, grazing cessation decreased the abundance of small mammal species that prefer more open vegetation structure, such as those characteristic of upland communities (Boeer and Schmidly 1977, Hayward et al. 1997, Andersen and Nelson 1999). Dense vegetation may also focus medium/large mammal movements onto well-worn paths (C. Hass, personal communication), potentially affecting foraging patterns and species interactions.

Habitat Modification by Mammals

The relationship between habitat and mammal species richness is often viewed as unidirectional, with habitat as the independent variable and species richness as the dependent variable. However, mammals can substantially influence their surroundings. This biotic feedback includes the dam-building activities of beaver, selective foraging by herbivores and granivores, soil disturbance due to burrowing, and seed dispersal.

DAM-BUILDING

Beaver are considered ecological engineers and keystone species in that they have disproportionate effects on ecosystems for their body size and abundance level (Paine 1966, Naiman et al. 1988). Fifteen beaver were released on the upper San Pedro River from 1999 to 2000, resulting in 6 to 12 colonies totaling 50 to 70 beaver, distributed among perennial and intermittent reaches (Snow and Olding 2003, M. Fredlake, personal communication). Preliminary findings on the San Pedro suggest that surface water is increasing in proximity to beaver dams, though there is no information yet on effects of beaver on stream bank vegetation or wildlife (Snow and Olding 2003). Creation of beaver dams in other riparian systems has been found to reduce the slope of the stream channel, reduce streamflow velocity, accelerate sediment deposition, and elevate water tables (Hair et al. 1978, Apple 1985). These changes favor wetland graminoids (grasses and grass-like plants), willows, and other plants that tolerate saturated soils. Beaver further modify vegetation structure by utilizing willow, cottonwood, and seepwillow for food and dam construction and by selectively foraging on aquatic plants such as cattails (Boeer and Schmidly 1977). In response to changes in hydrology and vegetation, species richness of both aquatic and terrestrial avian and mammalian species tends to increase after beaver reintroduction (Hair et al. 1978, Apple 1985). On the San Pedro, beaver activities may increase the abundance of mammals such as the fulvous cotton rat that depend on perennial water or wet meadows, or provide habitat for muskrats should they be reintroduced (Ceballos 1985, Duncan 1988, Ellis et al. 1997).

HERBIVORE AND GRANIVORE FORAGING

Selective foraging by herbivores and/or granivores impacts plant community structure and ecosystem processes (Brown and Heske 1990). For example, the exclusion of kangaroo rats from experimental plots in the Chihuahuan Desert resulted in a threefold increase in tall perennial and annual grasses

(Brown and Heske 1990). Black-tailed prairie dogs may act as a keystone species in semi-arid areas of the Southwest, suppressing competitively dominant woody vegetation such as mesquite (Weltzin et al. 1997). In so doing, the prairie dogs maintain certain areas in a grassland state. Early attempts to eradicate prairie dogs from grazing lands in the San Pedro watershed and other parts of the Southwest (see chap. 11) may have been one of the many factors that contributed to the increased abundance of mesquite and other woody plants in desert grassland habitats (Weltzin et al. 1997).

Herbivory by native mammals can be a significant determinant of tree seedling and sapling survivorship in floodplain environments (Andersen and Cooper 2000). Mammals that reduced tree survivorship included beaver, rabbits, and deer. The effects of mammal herbivory were contingent upon flow regime, with trees along regulated rivers declining more than those along unregulated rivers. As a driving mechanism, Andersen and Cooper (2000) suggest that river regulation allows small mammal populations, previously constrained by flooding, to expand in near-river habitats. Long-term monitoring along unregulated rivers such as the San Pedro would be useful to determine whether population size of some mammals increases to high levels during climatic periods characterized by reduced flood magnitude.

SOIL DISTURBANCE

The burrowing activities of small- and medium-sized mammals affect many ecosystem processes and properties. Prairie dogs, such as the black-tailed that once were abundant along the San Pedro, alter nutrient cycling and the rates and pathways of water flow through upland and riparian soils (Weltzin et al. 1997). Similarly, pocket gophers and their kin (still abundant along the San Pedro) fundamentally change physical and biotic processes, making them subterranean ecosystem engineers (Reichman and Seabloom 2002) (fig. 6.1b). By excavating and moving soil, pocket gophers change nitrification and mineralization rates, soil bulk density, and moisture regimes. These activities influence physical and biological processes such as the mixing of soil horizons, soil erosion, plant species richness, and biomass. The burrows and mounds of kangaroo rats have likewise been shown to impact biotic and abiotic habitat properties (Brown and Heske 1990).

SEED DISPERSAL

Mammals are important dispersal agents for numerous plant species and fungi through their consumption and subsequent egestion of seeds and spores and their inadvertent transport of seeds that attach to fur via barbs or hooks (Drezner et al. 2001; and see chap. 4). Rabbits disperse large quantities of seeds in grassland ecosystems with significant effects on the composition and structure of the soil seed bank (Malo and Suarez 1996) and may play a

Fig. 6.2. Gray fox (top; 6.2a) and collared peccary (bottom; 6.2b). Photo credits: Robert J. Luce.

similar role in the San Pedro watershed. Seed-caching rodents such as pocket mice, kangaroo rats, and deer mice also disperse seeds (Vander Wall 1997). Along the San Pedro River, mesquite pods and other large seeds, such as hackberry, are consumed or cached by numerous mammals including coyotes, foxes, skunks, peccaries, coati, black bear, and various rodents (Hoffmeister 1986, C. Hass, personal communication, C. Soykan, personal observation) (fig. 6.2). However, the impacts of mammalian seed dispersal for plant and fungal community structure have yet to be studied in this system.

Human Influences

Human activities influence the mammals of the San Pedro watershed both directly (e.g., hunting and road kill) and indirectly (e.g., alteration of hydrologic regime, land conversion, and other types of habitat modification). Early human impacts on mammals likely were restricted to hunting and habitat modification (agriculture, fire; Wilson 1995). With the arrival of European settlers came further changes in hunting regimes, hydrology, and habitat (Davis 1982, Glennon 2002). Increased hunting and trapping pressure eliminated or reduced the populations of numerous mammals including grizzly bears, gray wolves, and beaver. The loss of top carnivores can have cascading effects on lower trophic levels (Crooks and Soule 1999) but has not been investigated in the San Pedro watershed.

Prairie dogs also once thrived on both sides of the border, but by 1930 only a few remnant colonies could be found in Arizona, although some persist in Sonora. Many ranchers intensely disliked prairie dogs and insisted on their removal because their burrows caused problems for horses and they competed with cattle for grass. As discussed above, the loss of prairie dogs may have contributed to the spread of mesquite into grasslands, altering vegetation structure at the watershed level.

Along with these changes in biota, the San Pedro River went from a wide, slow-flowing series of ciénegas during the 1800s to a narrow, fast-flowing channel bordered by dense forests in the 1900s. Although the role of human actions in this historic river entrenchment is still debated (see chaps. 12 and 13), these changes likely had major impacts on the mammal fauna in the region, with some declining and others benefiting. In particular, mammals associated with marshes and riparian grasslands likely declined, as did those dependent upon upland grassland habitats. Conversely, arboreal species likely benefited from the expansion of the floodplain forests, while adaptable generalists such as deer mice may have benefited from human activities such as irrigated agriculture and ranching (Duncan 1988). Some mammals, such as the collared peccary and Virginia opossum, seem to have expanded their range and increased in abundance in Arizona, possibly as a result of human activities or in response to climate change and associated changes in vegetation (Hoffmeister 1986).

Animals introduced by humans to the San Pedro such as feral dogs, cats, and house mice may affect riparian mammals directly through predation (Hall et al. 2000) or indirectly via competition. Likewise, introduced plant species may affect them both directly, by changing resource availability, or indirectly, by changing vegetation structure (e.g., saltcedar trees grow in denser stands and produce finer litter than do cottonwood and willow trees, thereby influencing small mammal community composition; Ellis et al. 1997). Although few studies have examined the direct or indirect effects of non-native species on mammals along the San Pedro, their impact appears to be minor at this point in time (C. Soykan, personal observation).

More recently, major reaches of the river have been designated as preserves, and efforts have been undertaken to restore perennial riparian vegetation to abandoned fields, shifting the nature of human impacts. Road kill may have replaced hunting as the primary source of mortality for medium and large mammals along the river. Although no data exist for the San Pedro, a nearby study conducted in Saguaro National Park East suggests that thousands of mammals die annually due to road kill (Kline and Swann 2001).

Outside of protected areas, urban sprawl has altered habitat and increased human-wildlife conflicts (McKinney 2002). Urban and suburban sprawl threaten to intersect wildlife corridors that connect the river with protected lands in the surrounding mountains (Hass 2001). The river, its tributaries, and washes form important movement corridors for mammal species, providing cover, food, water, and easy passage through less-hospitable desert scrub habitats. By disconnecting the river from its tributaries or fragmenting reaches along the river, sprawl may reduce movement among suitable habitat patches, resulting in increased risk of local extinction due to reduced potential for recolonization. Furthermore, reduced movement opportunities would impede the natural recolonization of the watershed by extirpated species such as otter and jaguar that have source populations nearby.

Human population growth and associated water use has already altered stream hydrology (see chap. 15). If present trends continue, increasing portions of the river will cease to flow permanently, resulting in a loss of surface water—a vital resource for many of the mammal species that live in the region. Declining water tables would also alter riparian vegetation structure, with effects on mammal communities. At the same time, some portions of the river (including portions of the lower basin; see chap. 19) are expected to gain perennial streamflow, potentially offsetting the loss of water and riparian habitat elsewhere. However, the lower basin reaches are more arid than those of the upper basin, with less rain falling during the hot summer months. Longitudinal (upstream-downstream) studies of mammal abundance, as well as comparisons of mammal abundance between reaches with perennial vs. intermittent streamflow, would be valuable in helping us to understand how mammal populations may change over time along the San Pedro River in response to changes in stream hydrology.

Summary and Conclusion

The high mammal species richness of the San Pedro watershed can be attributed to a variety of factors including biogeographic position, climate, water availability, primary productivity, vegetation composition, structure, and heterogeneity, disturbance regimes, and species interactions. For simplicity, we have grouped these factors together into three categories—Geography, Environmental Gradients, and Non-equilibrium Processes—and discussed them using specific examples. Overall, existing data support our hypotheses, but much work remains to be done. Future efforts should focus on the specific mechanism(s) responsible for generating and/or maintaining species richness patterns. A fruitful approach may involve comparing species richness gradients from riparian to upland habitats, as suggested by Sabo et al. (2005). A hierarchical perspective of species richness may provide another useful method, bringing together mechanisms that operate at different spatio-temporal scales into a single predictive model (Sarr et al. 2005).

Just as habitat characteristics influence mammal communities, so too do mammals influence the habitats in which they live. Dam building by beavers and burrowing by rodents represent ecosystem engineering activities that significantly alter the physical and biological attributes of a habitat. Likewise, selective foraging by herbivores and granivores and seed dispersal can have important impacts on plant communities. Nevertheless, humans have had a far greater impact on the San Pedro watershed and its biota than any other mammal. Activities such as hunting, agriculture, and grazing were the first wave of human impacts, but these have been replaced by groundwater pumping and urbanization, which together represent the greatest, and most permanent, current threats to the mammals of the San Pedro. If left unchecked, groundwater pumping will lower water tables, leading to a loss of environmental gradients, while urbanization will isolate the watershed from surrounding regions. A reduction or loss of two out of the three factors hypothesized to account for the San Pedro's high mammal species richness would provide an interesting, albeit somber, test of our ideas. Ultimately, anthropogenic disturbance may prove to be the most important factor influencing mammal species richness along the San Pedro.

LAURA HANNON,
LESLIE RIES,
AND KATHY S. WILLIAMS

Terrestrial Arthropod Communities along the San Pedro

THREE CASE STUDIES

SEVEN

Introduction

In the arid landscapes of the southwestern United States, riparian areas support abundant and diverse arthropod faunas. As in other terrestrial ecosystems, arthropods play vital and varied ecological roles in riparian habitats. Insects, arachnids, and other arthropods function as predators, parasites, herbivores, pollinators, decomposers, and scavengers (Price 1984, Wilson 1987, K. Williams 1993) and thus influence plant productivity, facilitate nutrient cycling, and help regulate the population levels of other invertebrate groups (Price 1984, K. Williams 1997, Ellis et al. 2001). Riparian arthropod communities are also critical components of vertebrate food webs. Many passerine birds, for example, depend on arthropods in riparian corridors as prey during either their migration or breeding season (Rosenberg et al. 1991, Yard et al. 2004; and see chap. 8).

Because of their complex life histories and small body size, terrestrial arthropods may be especially sensitive to environmental conditions. Key environmental characteristics for arthropods include vegetation structure, plant community composition, habitat heterogeneity, and amounts and types of available floral resources (Murphy et al. 1983, Strong et al. 1984, Ehrhardt 1985, Kremen 1992, Holl 1995, 1996, Fleishman et al. 1999, Chovanec and

Waringer 2001). Many insects respond to vegetation structure because of its effect on light, temperature, and moisture microclimates (Willmer 1982, 1983, Warren 1985, Weiss and Murphy 1990, Kremen 1992, Holl 1996). Insects that feed on plant tissue and fluids frequently have specialized relationships with their host plants (Strong et al. 1984). Nectar-feeding insects vary widely in their degree of specialization on host plants. However, the larval stages of some of the major floral-visiting taxa (e.g., bees, butterflies, and moths) are often dependent on a limited number of host plant species (Ehrlich and Raven 1965, J. Scott 1986, Michener 2000). These herbivorous insect groups may, therefore, be expected to show strong responses to the composition of riparian plant communities (Nelson and Andersen 1994, Fleishman et al 1999). In addition, the mobility and short generation times of many insects mean that their distributions and density can quickly reflect changes in riparian habitat.

In preceding chapters, we have seen how plant communities in the San Pedro riparian corridor are shaped by hydrological and geomorphological conditions. Variation in surface water flow, groundwater level, flood regime, and stream morphology all influence the species composition and physical structure of riparian vegetation. A common theme of the research presented in this chapter is the response of arthropod communities to the variation in structure and composition of riparian vegetation found along the San Pedro River. We summarize data from studies of three arthropod groups in the San Pedro River corridor: adult butterflies (Lepidoptera), bees (Hymenoptera: Apoidea), and the assemblages of arthropods found on riparian vegetation. We conclude with an overview of relationships between these arthropod groups and major characteristics of riparian habitat and thoughts on directions for further research.

Butterflies

The San Pedro riparian corridor has an extremely rich butterfly fauna. Approximately 100 species have been recorded throughout the corridor (R. Bailowitz, personal communication). This number represents approximately 15 percent of the 750 species present in all of North America. Butterflies are quite popular with the general public, and many butterfly species are highly specialized in their habitat requirements, making them an excellent focus for monitoring of management actions (Ehrlich 2003). In this section, we show that butterfly populations are strongly associated with the presence of riparian habitats, as well as influenced by variation in the characteristics of those habitats.

During a three-year study period (1999–2001), more than 1,000 (10 × 10 m) survey plots were established in 16 separate study areas throughout the upper San Pedro within the boundaries of the San Pedro Riparian National

Fig. 7.1. Sleepy orange butterflies (Pieridae: *Eurema nicippe*) taking in water and nutrients from moist sand along the upper San Pedro River after a monsoon-season flood. Photo credit: Kerrith McKay.

Conservation Area (SPRNCA). One additional area was located in the upper end of the lower San Pedro. The full design is described in Ries (2003). Surveys were carried out in three lateral zones along the river corridor: floodplain, terrace, and adjacent desert scrub. Adult butterfly and flowering plant surveys were conducted three or four times each year in each plot during the monsoon and post-monsoon period (late July through early October), when adult butterflies are most abundant (Bailowitz and Brock 1991).

USE OF RIPARIAN HABITAT BY BUTTERFLIES

Butterflies have many resource requirements that influence their habitat choices (Dennis et al. 2003). Adult females must find suitable host plants on which to lay their eggs, and males must find females with which to mate. Both sexes of most species must regularly gain energy and nutrients, either through floral nectar (the most common source), rotting fruit, dung, or tree sap (J. Scott 1986). In addition, males of several butterfly species typically congregate at moist areas to gain nutrients (Dennis et al. 2003) (fig. 7.1).

Riparian areas are essential habitat for the majority of vertebrates in Arizona and New Mexico (Chaney et al. 1990, Krueper 1996b), and riparian habitats also appear to be critical for butterfly species in the lowlands of the arid Southwest. Along the upper San Pedro, butterfly communities are much richer in the riparian corridor (floodplain and terrace) than in the desert scrub (fig. 7.2a). Within the riparian corridor, butterflies are more abundant and species rich on the terrace than in the floodplain. Most butterfly species prefer open habitat; therefore, within the riparian zone, they tend to avoid areas with dense canopy or shrub cover. This is true in the floodplain (fig. 7.2b) as well as the terrace (fig. 7.2c). Although the desert scrub is open in structure, most butterfly species avoid this vegetation zone, and none seem to prefer it

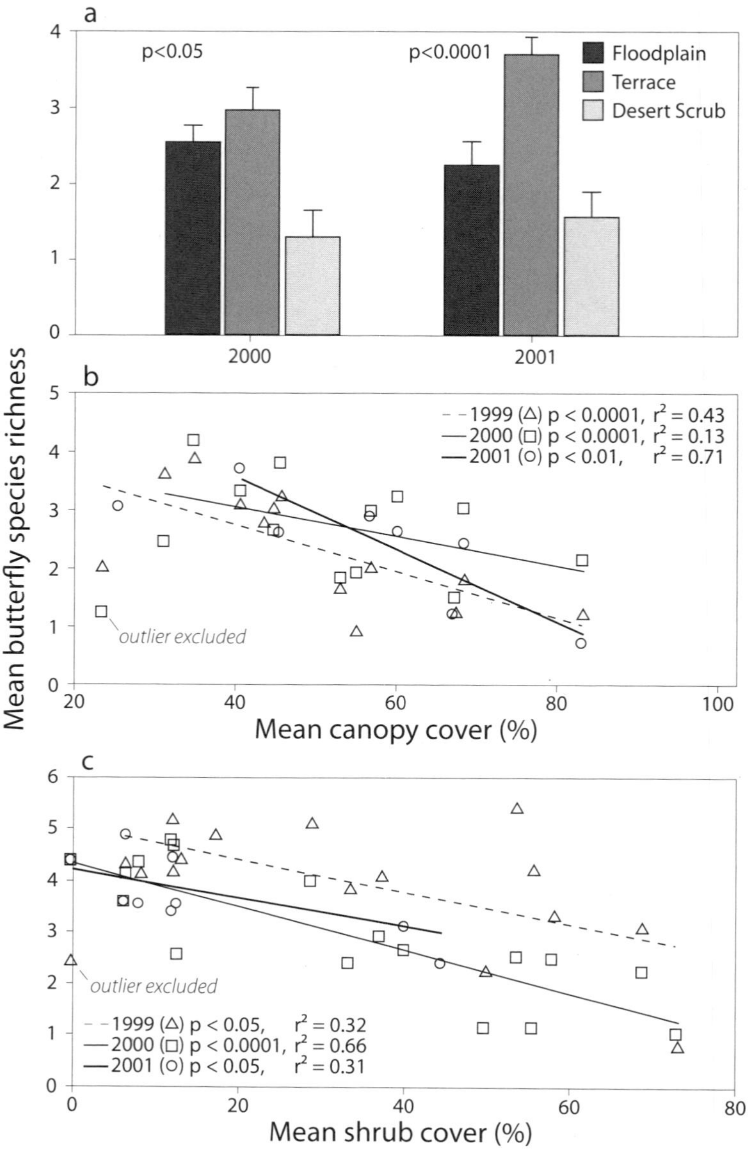

Fig. 7.2. Mean butterfly species richness per 100 m^2 plot: (a) in each zone of the riparian corridor, (b) in relation to canopy closure in the floodplain, and (c) in relation to shrub cover in the riparian terrace. Results are presented separately for each year of the study. Significance values are based on ANOVA (a) and linear regression (b, c).

(Ries 2003). It is important to note that adult butterflies are highly mobile. So, while we describe habitat preferences for individual butterfly species, no species was restricted to its preferred habitat, and most species were found in all habitat types at some density.

While the floodplain tends to have lower butterfly richness (fig. 7.2a) and abundance than the terrace, some butterfly species showed their highest abundance in the floodplain. In some cases, this pattern can be linked to the spatial distribution of host plants and other critical resources. For instance, the viceroy (*Limenitis archippus*) is rarely seen outside stands of its preferred host plant, Goodding's willow (*Salix gooddingii*). The viceroy also sometimes lays its eggs on cottonwood (*Populus* spp.), another phreatotypic tree limited to the floodplain zone. In addition to larval host plants, viceroys appear to require surface moisture to persist in riparian habitats. In observations along the lower Colorado River, Nelson (2003) found that the presence of larval host and nectar source plants was not sufficient to attract viceroys to restored areas of riparian vegetation. Viceroys occurred only at sites where the natural flow regime had been restored, thereby creating patches of wet soil where males could ingest water and minerals.

Other factors, such as microclimate preferences, may contribute to the association of other butterfly species with floodplain habitat. The pipevine swallowtail (*Battus philenor*) uses Watson's pipevine (*Aristolochia watsonii*) as its host plant on the San Pedro. While the pipevine plant is found throughout the riparian corridor, *B. philenor* reaches its highest abundance in the floodplain. The red satyr (*Megisto rubricata*) is often associated with woodlands (J. Scott 1986, Bailowitz and Brock 1991) and, along the San Pedro River, is most commonly seen in floodplain forest. As a final example, the question mark (*Polygonia interrogationnis*) is a woodland species usually found in mountainous areas (Bailowitz and Brock 1991). While this species has rarely been seen on the San Pedro, when found it usually occurs in heavily wooded areas along the river.

The terrace generally has the highest number of butterfly species (fig. 7.2a), and many species reach their highest abundances there (Ries 2003). However, it is difficult to classify butterfly species as to their association with the terrace per se, because of the great variability in vegetation structure found there. Terrace vegetation on the upper San Pedro ranges widely from grassland, to relatively open mesquite savanna, to closed-canopy mesquite forest or bosque. Since butterfly distributions are influenced by vegetation structure, butterfly surveys on the terrace were stratified into three habitat types: sacaton grasslands, mesquite savanna, and mesquite forest.

Some butterfly species showed a distinct preference for grasslands within the terrace zone (Ries 2003). The distribution of these species appeared to be relatively independent of, or at least not primarily driven by, the distribution of host plants. The white checkered-skipper (*Pyrgus albescens*), for example, was almost exclusively found in grasslands, despite the fact that its host plants, mallow species in the genera *Sphaeralcea* and *Sida*, were found in both mesquite savanna and desert scrub habitat. Another species clearly associated with grasslands was the southern dogface butterfly (*Colias cesonia*). This species' host plant, prairie clover (*Dalea* spp.), was not found along the river. Therefore, grasslands along the San Pedro could not have been breeding habitat for *C. cesonia* but may have been used as movement corridors. The bordered patch (*Chlosyne lacinia*) was more frequently found in mesquite savanna than mesquite forest habitat, despite the fact that its preferred larval host and nectar plant, *Viguiera dentata*, was common in both mesquite savanna and forest. Other butterflies were common in grasslands and mesquite savanna but avoided the mesquite forests. These species included the western pygmy-blue (*Brephidium exilis*), the orange sulphur (*Colias eurytheme*), and the cloudless sulphur (*Phoebis sennae*). Although many species avoided those areas of the riparian corridor with higher canopy or shrub density (figs. 7.2b, 7.2c), some species were abundant throughout the riparian corridor and adjacent desert scrub. These wide-ranging species included the queen (*Danaus gilippus*) and the sleepy orange (*Eurema nicippe*).

THE INFLUENCE OF PLANT DIVERSITY

Nectar resources. Butterfly species richness is often associated with plant diversity (Erhardt 1985, Simonson et al. 2001), because each butterfly species depends on particular types of host and nectar plants for reproduction and adult survivorship. (See Ries [2003] for a list of larval host and preferred nectar plants for San Pedro butterfly species.) Butterfly species richness was positively correlated with flowering plant richness (fig. 7.3a) in all three years of the study (although approaching significance only in 2001). However, this relationship was relatively weak, explaining only 12 to 17 percent of the variation, generally lower than that explained by canopy or shrub cover (figs. 7.2b, 7.2c). In one year, floral richness was highest in our desert scrub plots (Ries 2003), yet butterfly species richness was consistently lowest in the desert scrub plots.

The relationship between floral and butterfly species richness is weak, likely because plant species differ widely in their value to butterflies. Indeed, some flowering plants such as nightshades and ground cherries (Solanaceae) are not visited for nectar by butterflies (although some plants in this family may be visited for their pollen by certain bees). Other groups, including many members of the composite (Asteraceae) family, are visited by a wide variety of butterfly species (J. Scott 1986). Plant species that attracted the greatest number of butterfly species on the San Pedro were most likely to be encountered in the floodplain or terrace (fig. 7.3b). This distribution of attractive nectar resources is one possible explanation for the high butterfly richness in the riparian zone compared to the adjacent desert scrub (fig. 7.2). In fact, four of the five most attractive plant species were most common in the terrace, where butterfly species richness was highest.

Another factor that may have weakened the relationship between flower and butterfly richness is the variability in nectar volume among flowers. Within a given area, butterflies are often most abundant near available nectar, and the presence of nectar can be a more accurate predictor of butterfly abundance than the presence of flowers alone (C. Schultz and Dlugosch 1999). Nectar volume and quality can vary widely among different flower species, among individual flowers of the same species, and even within an individual flower throughout the day (Potts et al. 2003, Dafni 2005). Therefore, when we

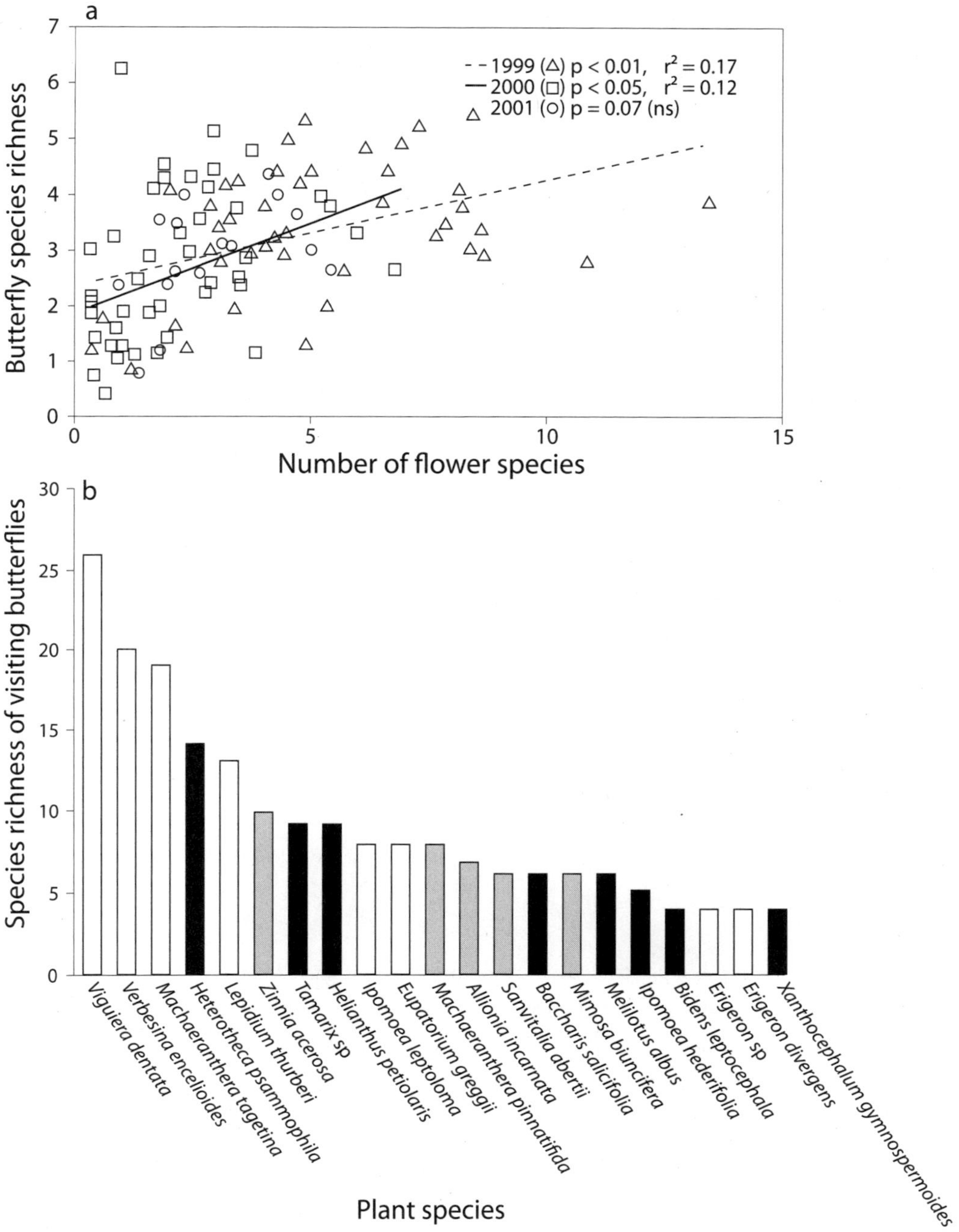

Fig. 7.3. The influence of floral diversity on butterfly species richness. (a) There was a significant relationship (significance values based on regression in SAS v. 8.1) between the number of different nectar source plants and butterfly species richness. Regressions performed separately for each year; sample sizes were 47 plots in 1999, and 46 plots in 2000. (b) The number of butterfly species recorded visiting each flower species varied. Flower species with three or fewer visiting butterfly species are not shown. The primary habitat association of each plant species is indicated by the color of the bars: black = floodplain, white = riparian terrace, gray = desert scrub upland.

surveyed floral resources by counting open flowers, we were not necessarily measuring the availability of nectar for foraging butterflies.

Some butterfly species, such as *C. cesonia*, *C. eurytheme*, and *Pontia protodice*, appeared to be nectar generalists, using a variety of nectar sources. In contrast, several butterfly species seemed to specialize on one or a few flower types. For instance, the large butterflies *B. philenor* and *P. sennae* preferentially visited flowers in the genus *Ipomoea* with long, tubular corollas, while common sootywing (*Pholisora catullus*) preferred small flowers that grow low to the ground. The most frequently visited flower species in our study area was the perennial toothleaf goldeneye (*V. dentata*), visited by 26 butterfly species during our surveys (fig. 7.3b). Goldeneye was also the preferred nectar plant for two common butterflies on the terrace, the bordered patch (*Chlosyne lacinia*) and the queen (*Danaus gilippus*).

Of course, these results reflect the timing of our surveys; our observations provide a snapshot of relationships between butterfly and flower species during the monsoon and post-monsoon periods. Plants that flower during other times of the year are undoubtedly also valuable nectar sources for butterflies. For instance, mesquite (*Prosopis velutina*) and saltcedar (*Tamarix* sp.) had already stopped flowering when our surveys began in the early monsoon season. Although butterflies are less abundant in the pre-monsoon period, these flowering tree species are likely important nectar sources during that time. Our surveys also missed another important floral species, rubber rabbitbrush (*Ericameria nauseosa*), which attracts many butterfly species throughout the riparian corridor along the San Pedro in the late fall (Bailowitz and Brock 1991).

Larval host plants. Plant species also vary in their use as larval host plants for butterflies. As with nectar source plants, some families, such as Solanaceae, are rarely, if ever, used as larval host plants by butterflies along the San Pedro. Other families, such as Fabaceae and Asteraceae, contain many larval host species (J. Scott 1986). Most butterfly species are much more specific in their larval host plant use than in their nectar plant requirements. For instance, the pipevine swallowtail is the only butterfly species whose larvae can eat and grow on Watson's pipevine, and this plant species is the only known host for pipevine swallowtails in the San Pedro River Valley. This exclusive relationship is due to the unique ablility of *B. philenor* caterpillars to sequester the poisons present in *Aristolochia* spp. These poisons make both pipevine caterpillars and adults unpalatable to predators (Fordyce 2000).

Other butterflies use a variety of larval host plants, usually within one or two families, but sometimes across many plant families (Ehrlich and Raven 1965). For instance, the dainty sulphur (*Nathalis iole*) is a host generalist on the family Asteraceae. Other butterflies that theoretically have broad host plant requirements actually use only a few host plant species in any one locality. The alfalfa sulphur (*C. eurytheme*) is a host generalist on the family Fabaceae. However, *C. eurytheme* likely uses cultivated alfalfa (*Medicago sativa*) as its main host plant along the San Pedro. *C. eurytheme* has been strongly associated with agricultural habitats in other western riparian areas along the lower Colorado River (Nelson and Andersen 1999) and in the Walker River basin (Fleishman et al. 1999).

Of the 14 common butterfly species for which likely host plants along the San Pedro could be identified, most potential host plant species were most common in either the riparian floodplain or terrace (Ries 2003). This pattern, along with the prevalence of preferred nectar resources along the river corridor and the access to wet soil, suggests that riparian habitat is critical to most species of the lowland butterfly community.

Bees

BEE SPECIES RICHNESS IN THE UPPER SAN PEDRO

All indications are that the upper San Pedro River corridor supports a large and diverse bee fauna. Until recently, the San Pedro River Valley has received relatively little attention from bee ecologists and taxonomists. As a result, the bee fauna of the area is still being discovered and described. The only exotic species detected so far is the European honeybee *Apis mellifera*; all other bee species are native to North America. To date, we have recorded 250 bee species in 47 genera and six families. This working species count, however, is certain to underestimate total bee species richness in the area because of spatial and temporal limitations in our collecting. Bee assemblages are known to be extremely variable spatially and temporally, showing high rates of species turnover interannually, seasonally, and among sites (Griswold et al. 1998, R. Minckley et al. 1999, N. Williams et al. 2001). In general, records of bee species richness are very sensitive to sampling intensity, especially the number of habitat types and seasons covered.

Results presented here are based on fieldwork along the upper San Pedro River between the town of St. David and the U.S.-Mexico border 62 km to the south (Hannon 2005). The majority of our sampling effort was directed at bee communities in agricultural and mesquite woodland habitats on the terrace, and most survey work occurred during the monsoon season (July to September). However, we also conducted less intensive sampling in other riparian and desert scrub habitats and collected bees from dominant flowering plants during the pre-monsoon early summer.

Because of its high diversity of insect-pollinated angiosperms (see chap. 5) and location at the junction of the Chihuahuan and Sonoran biomes, the upper San Pedro Valley would be expected to host a unique bee fauna. Southeastern Arizona is known to include both the northern and southernmost extents of the geographic ranges of other invertebrate species, e.g., butterflies (Bailowitz and Brock 1991). In identifying bee specimens from the SPRNCA, it was not uncommon to find major expansions of known ranges and new U.S. records. We also encountered previously undescribed species of some species-rich genera. One genus of pollen specialist bees provides a good example. The genus *Anthophorula* has its center of diversity and abundance in northern and central Mexico (Michener 2000), and the 12 *Anthophorula* species detected at our San Pedro study sites included 2 species previously known only from the Mexican states of Sonora and Chihuahua (Krombein et al. 1979). We also encountered 2 undescribed species: one new species of *Anthophorula* and another new species of *Brachynomada*, a genus of parasitic bees that lay their eggs in the nests of *Anthophorula* (T. L. Griswold, personal communication, Michener 2000).

USE OF RIPARIAN HABITAT BY BEES

Like butterflies, bees are most often associated with open and sunny habitats. Unlike most butterfly species, however, bees are abundant in arid desert scrub habitat, finding nectar and pollen source plants and suitable sites for ground nests in the dry, sandy soil (Michener 2000). Perhaps because of the affinity of bees for arid environments, the use of riparian habitats by native bee species is not usually emphasized. However, riparian zones are actually the primary habitat for certain species, and for many native bees, riparian habitat is one among several habitat types that provide critical resources. In our surveys of the bee fauna of the upper San Pedro riparian corridor, we found a diverse assemblage of bees using riparian habitats for foraging and/or nesting. Several species were closely tied to mesquite vegetation, which commonly dominates the terrace along the upper San Pedro.

Nesting sites. The large washes that drain into the main San Pedro River channel provide a high density of nest substrate for the large guild of bank-nesting native bees. Females build their nests horizontally into the cut banks of washes, making sturdy nest entrances and tunnels in the firm crust that forms when washes dry out after periodic flooding. Examples of bank-nesters include "digger bees" such as *Anthophora* sp. and *Melissodes* sp.; both genera consist of robust, furry bees with hind legs adapted for excavation (O'Toole and Raw 1991, Michener 2000). Bank-nesting taxa also include some smaller solitary and semi-social bees in the sweatbee family (Halictidae), some of which nest in large aggregations. Populations of these small Halictids are known to use and reuse the same bank nest site for several years (O'Toole and Raw 1991).

Another relatively large guild of native bees nest in old beetle holes in snags and dead tree branches. These species are part of the group called leafcutter bees; leafcutters are so named because of their distinctive behavior

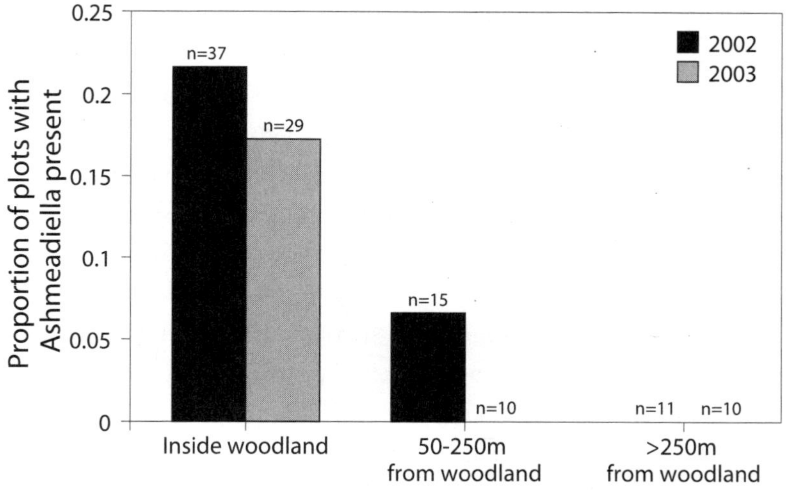

Fig. 7.4. Frequency of branch-nesting native bees in agricultural habitats on the riparian terrace, relative to distance from the nearest mesquite (*Prosopis*) woodland. In each year of the study, the proportion of plots where *Ashmeadiella* spp. were detected in pan trap samples declined with increasing distance from mesquite woodland habitat. Data are from July sampling periods (the season of peak activity for *Ashmeadiella*) in 2002 and 2003. The sample sizes indicate the number of plots surveyed in each distance class. (See Hannon 2005 for details of methods.)

of using small discs of leaf material to separate the larval cells in their nests (Michener 2000). In our study sites along the upper San Pedro, leafcutter bees frequently nested in dead branches of mature mesquite trees and standing mesquite snags. They may well also use dead and beetle-eaten parts of cottonwood and other floodplain tree species, such as Goodding's willow or elderberry (*Sambucus nigra* L. ssp. *cerulea*). Along the upper San Pedro, the distinct circular holes made by leafcutter bees are common in the leaves of cottonwood trees in the floodplain, as well as in the leaves of certain shrubs on the terrace.

Small species in the genus *Ashmeadiella* (Megachilidae: Osmiini) were the most common members of the leafcutter guild in our San Pedro surveys. Among the habitats we surveyed in the terrace—which included agricultural fields, hedgerows, and mesquite forest—*Ashmeadiella* density was highest in mesquite forest. In addition, there was an apparent effect of distance from mesquite vegetation in the distribution of *Ashmeadiella* on the terrace. In the summers of 2002 and 2003, we used pan traps to sample bees in riparian mesquite forest and adjacent agricultural habitats. (See Hannon [2005] for a detailed description of methods.) The frequency of these bees in agricultural habitats declined with increasing distance from mesquite forest (fig. 7.4). While there are multiple possible explanations for the spatial distribution of *Ashmeadiella* spp. on the terrace, it seems likely that these bees are concentrated in and near mesquite woodland because of proximity to nest sites.

Pollen and nectar resources. In the upper San Pedro riparian corridor, mesquite woodland provides rich foraging habitat for native bees. However, flower resource availability varies widely over the course of the growing season. There are two main peaks in available flower resources and abundance of foraging bees—one peak in the late spring to early summer, and one in the late summer to early fall (following the two major annual rainfall pulses). Mesquite trees and understory shrubs, such as graythorn (*Ziziphus obtusifolia*) and wolfberry (*Lycium andersonii*), produce most of their flowers during the late spring and early summer. Herbaceous composites, particularly the perennial composite goldeneye and the annual composite mesa tansyaster (*Machaeranthera tagetina*), dominate the late summer bloom in mesquite forest understory.

The abundant resource of mesquite blossoms in the early summer attracts a large assemblage of bee species, as well as other insect pollinators such as wasps, beeflies, beetles, and butterflies. Over the course of fieldwork along the upper San Pedro, we observed 30 different bee species visiting mesquite. This tally comes from relatively non-intensive sampling; more than 60 bee species were collected on mesquite in one season at a Sonoran Desert site (Simpson et al. 1977), and at least 52 species have been identified in one season of an ongoing study of mesquite visitors in Clark County, Nevada (Messinger, personal communication). In a broad survey of invertebrate assemblages associated with mesquite throughout the Southwest, Simpson et al. (1977) found that bees comprised the largest group among the invertebrates visiting mesquite flowers and that solitary bees were the most effective pollinators of mesquite.

Among the visitors to mesquite flowers on the upper San Pedro were several bee taxa that narrowly specialize on mesquite pollen. For example, we observed three species of *Perdita* (Andrenidae: Panurginae) exclusively visiting mesquite flowers. *Perdita* is a genus of small-bodied pollen specialist bees, which has its highest abundance and species diversity in the North American Southwest (Michener 2000). While each *Perdita* species narrowly specializes on only one or a few pollen hosts, as a group, bees in the genus *Perdita* collect pollen from a wide range of plant genera and families (Michener 2000). The mesquite bloom along the upper San Pedro also attracted numerous pollen specialists in two other bee families, Colletidae and Megachilidae.

Like the flowers of the mesquite overstory, the flowers of understory shrubs are also an attractive nectar and pollen resource for bees in the early summer. Bee visitors to the native shrub graythorn included one pollen specialist species in the genus *Perdita*, as well as numerous generalist species feeding on the readily available nectar in the flat, open flowers. Flowers of wolfberry bushes also attracted large assemblages of floral visitors. No specialists on wolfberry pollen were observed, but certain large solitary bees in the genera *Centris* and *Anthophora* were strongly associated with blooming wolfberry. Females of these genera collected both nectar and pollen from wolfberry flowers.

In the late summer, blooming goldeneye plants fill the entire herbaceous layer of some areas of riparian mesquite bosque along the upper section of the San Pedro River. In late-summer bee sampling in 2002 and 2003, we found that some bee species were closely tied to mesquite woodland habitats because of their specialization on goldeneye pollen. For example, the tiny bee *Dufourea boharti* (Halictidae: Rophitinae) specializes on the pollen of the Heliantheae subfamily of composites, which includes goldeneye. Over the course of our study, this species was never observed visiting flowers other than goldeneye and was only rarely detected in habitats other than dense mesquite woodland. Like the branch-nesting bees we monitored, *D. boharti* may not stray far from mesquite woodland. The frequency of *D. boharti* in trap samples dropped off at distances greater than 250 m from the nearest woodland, and they were not observed to forage on goldeneye flowers occurring outside of mesquite woodland habitat (Hannon 2005). These results suggest that *D. boharti* would be sensitive to patchiness or fragmentation of riparian mesquite forest. As with bee species generally, flight range size, nesting habitat specificity, pollen host plant specificity, and perhaps microclimate tolerance could all shape the spatial distribution of this small bee in the riparian corridor.

FLORAL RESOURCES AND THE STRUCTURE OF NATIVE BEE ASSEMBLAGES

In the summers of 2002 and 2003, we surveyed bees and flowering plant species in 24 (2 m x 40 m) plots in riparian mesquite woodland inside the SPRNCA preserve. Each plot was surveyed three times over the course of the summer—in July, August, and September. During each survey, in addition to capturing bees and recording the flower species each bee was visiting, we also recorded any insect-pollinated flower species inside the plot area (see Hannon 2005 for methods).

We found that greater flower species richness was associated with greater foraging bee species richness (fig. 7.5). This general pattern of a positive correlation between diversity of nectar and pollen sources and richness of pollinating insects is common to other plant-pollinator community studies (e.g., Banaszak 1996, Petanidou and Ellis 1996, Potts et al. 2003).

As might be expected, the relationship between flowers and bee assemblages was more complex than a broad correlation between flower and bee species richness. Different flowering plant species attract different assemblages of foraging bees, and individual flower species vary considerably in their ability to attract a diverse assemblage of visitors and to provide resources for pollen specialists. To look in more detail at the relationships between flowering plant and bee communities in riparian mesquite woodland (both

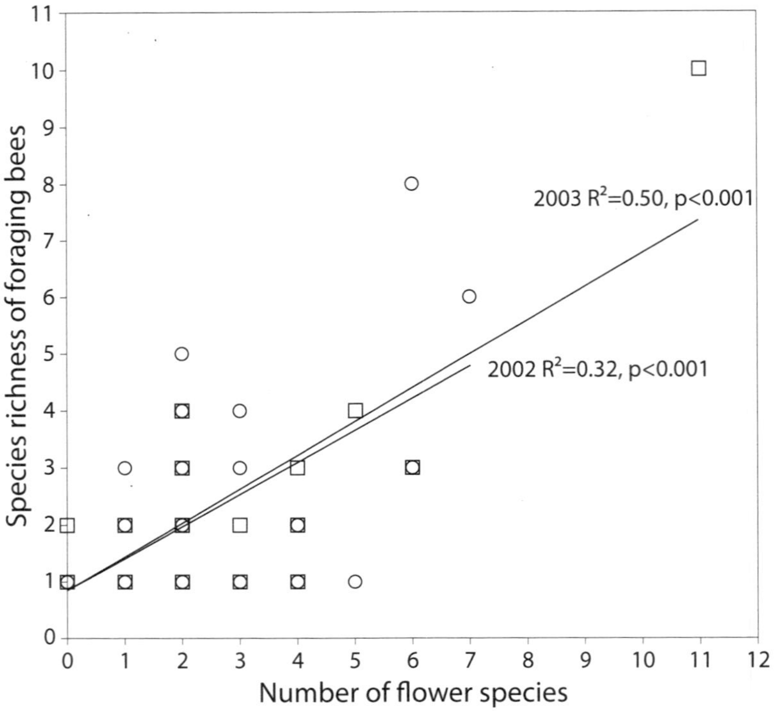

Fig. 7.5. Relationship between bee diversity and flower resource diversity in riparian mesquite (*Prosopis*) woodland. Surveys were conducted in July, August, and September 2002 (total survey n=38; circle symbols) and 2003 (n=52; square symbols). The species richness of foraging bees was strongly positively correlated with flower species richness in each survey year (2002 $R^2 = 0.32$, $p<0.001$; 2003 $R^2 = 0.50$, $p<0.001$).

bosque and savanna), we compared the number of bee species foraging on each flower species found in our plots (fig. 7.6). We also tallied the number of visiting bee species according to their degree of pollen specialization, characterizing them as either generalists or specialists based on information in the literature and from flower visitation data at the San Pedro study sites. (See Hannon [2005] for a full list of specialist/generalist designations and pollen host plants for bee species on the San Pedro.) For this analysis, we focused on the shrub and herbaceous plant species that have their peak blooms during the monsoon.

We found that flowering species in the mesquite understory showed a wide range of use by bees, with some species attracting large numbers of generalists and specialists, and others only attracting one or two species (fig. 7.6). Goldeneye attracted the greatest diversity of foraging bee species and supported a relatively large proportion of pollen specialists (specialists on sunflower Heliantheae, for example; see fig. 7.7). Two shrub species, graythorn and wolfberry, and one ruderal annual, common fleabane (*Erigeron divergens*), were also among the flower species most used by bees. As in other bee and flower community studies in the Southwest, globemallow (*Sphaeralcea* spp.)

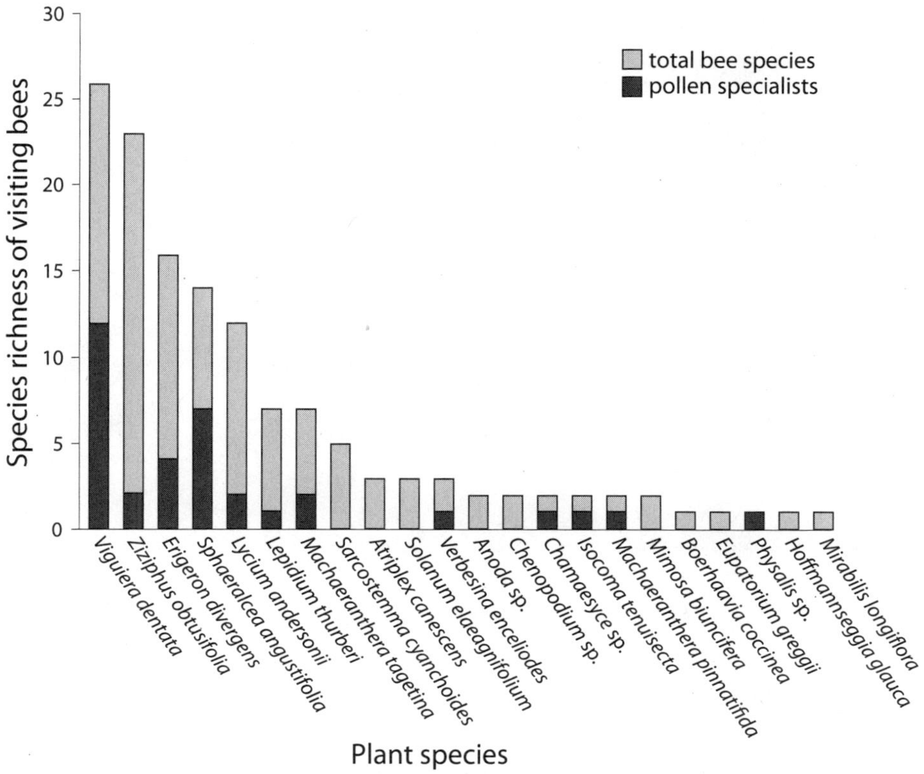

Fig. 7.6. Species richness of bees visiting various flower species in mesquite (*Prosopis*) woodland understory, and the proportion of pollen specialist bee species among flower visitors. Goldeneye (*Viguiera dentata*; Asteraceae) attracted the greatest number of foraging bee species and supported a relatively large number of pollen specialists.

attracted a high proportion of Malvaceae pollen specialists (Messinger and Griswold, unpublished data).

IMPORTANCE OF RIPARIAN HABITAT FOR LOCAL BEE POPULATIONS

In the San Pedro Valley, bee communities have not yet been systematically surveyed in different geomorphic zones. Therefore, the degree of overlap in bee community composition among riparian areas and adjacent uplands remains an open question. Another unknown is the degree to which local bee populations use resources in the riparian zone vs. the adjacent desert scrub. From analysis of the pollen loads on female bee specimens, we know that individuals of some medium-bodied bee species can gather resources in both riparian and desert scrub zones. In a 2002 pilot study of the foraging behavior of *Anthophora californica* and *Anthophora vallorum* at sites along the upper San Pedro and upper Santa Cruz Rivers, we found that some individual females had collected pollen both from plants in the desert scrub uplands (e.g., creosote bush, *Larrea tridentata*) and plants growing on the terrace (e.g., wolfberry) in a single foraging bout (Hannon, unpublished data). Theoretically, many other bee species could also access resources in both

Fig. 7.7. Female native bee (Apidae: Eucerini) collecting pollen from goldeneye (*Viguiera dentata*), a common component of the riparian mesquite (*Prosopis*) woodland flora in the upper San Pedro Valley. Photo credit: David Wood.

riparian and upland zones. While there is little specific information on home range size for most bees, the commonly hypothesized flight range of 1 km for solitary bee species (Cane 2001) would easily span from the floodplain to the desert scrub in most reaches of the upper San Pedro. Some known foraging distances for native bees (up to 6 km recorded for some bumblebees; Saville et al. 1997, Walther-Hellwig and Frankl 2000) far exceed the width of the entire riparian corridor along the San Pedro River.

Assemblages of Arthropods on Riparian Vegetation

Insects, spiders, and other members of the phylum Arthropoda living on riparian vegetation play major roles in local food webs, as consumers of plant tissue and as food for many other invertebrate and vertebrate animals. Despite their ecological importance, far less work has been done to describe the taxonomy and ecology of these arthropod assemblages than for the more conspicuous bee and butterfly faunas. Several characteristics of riparian habitat, including plant species composition, site hydrology, and seasonal patterns of precipitation, likely affect the abundance and composition of arthropod assemblages on vegetation. To measure the effects of each of these factors, arthropod assemblages were surveyed on a suite of plant species in San Pedro riparian forest and grassland sites with varying degrees of water permanence, both before and after monsoon rains.

Surveys were conducted in floodplain forests where streamflow was perennial (surface flow present every month; 3 sites in 1999 and 6 sites in 2000 and 2001) and where flow was intermittent (surface flow not present every month; 6 sites in 1999 and 10 sites in 2000 and 2001). Standardized sweep-sampling techniques were used to collect arthropods off four dominant floodplain tree and shrub species: willow (primarily *S. gooddingi*), cottonwood (*Populus fremontii*), seepwillow (*Baccharis salicifolia*), and saltcedar. All four species were sampled at each site, and sampling intensity was the same for each species. The trees and shrubs sampled were all growing in the floodplain, within 100 m of the river channel. Sweep-samples were also collected from big sacaton (*Sporobolus wrightii*) at 12 sites in 1999. Sacaton grassland sites were located on terraces and high floodplains, usually more than 200 m from the river channel. Of the 12 sacaton grassland sites, 6 had relatively shallow groundwater (June groundwater depths of less than 4 m) and 6 had groundwater at more moderate depths (4 or more m). Floodplain forest and grassland sites were sampled during the pre-monsoon (June–July) and monsoon/post-monsoon (August–October) periods (hereafter referred to as pre-monsoon and post-monsoon).

While it would be ideal to identify all collected individuals to species, the time required is often prohibitive when conducting broad surveys of multiple arthropod orders. In addition, many types of insects and spiders remain undescribed and unnamed. However, useful information can be gathered relatively rapidly without identifying individuals to the level of species (Oliver and Beattie 1997). Many insects can be assigned to an ecological or functional assemblage by identifying them to taxonomic order or family (Kremen 1992, Beattie and Oliver 1999, Goldstein 1999, Haddad et al. 2001).

In our studies along the San Pedro River, arthropods were sorted to the taxonomic level necessary to place them into functional groups (K. Williams 1993, 1997, 2000). All specimens were sorted to order, and to family or lower when possible. Detritivores were present only in very low numbers and were therefore not included in this analysis. Here, only the Herbivore and Predator assemblages will be discussed. The Herbivore assemblage included Lepidoptera (moths and butterflies), Homopteran suborders Sternorrhyncha and Auchenorrhyncha (planthoppers and aphids), Orthoptera (grasshoppers), herbivorous Coleoptera (beetles), and small Heteroptera ("true bugs"). The Predator assemblage included spiders (Araneida), predaceous Heteroptera, Neuroptera (lacewings and owlflies), Mantoidea (mantids), and predaceous beetles, as well as parasitic Hymenoptera (wasps) and Diptera (flies). The samples included taxa that were not necessarily specialists on the plant species from which they were collected. Most were probably generalists, predators or parasites of arthropods living on the sampled plants, or species simply moving through the habitat.

EFFECTS OF SEASON AND SITE HYDROLOGY

Season. The population dynamics of herbivorous and predaceous arthropods on riparian vegetation reflected monsoon rainfall patterns. Arthropod abundances consistently increased after the onset of the monsoon rains, in both

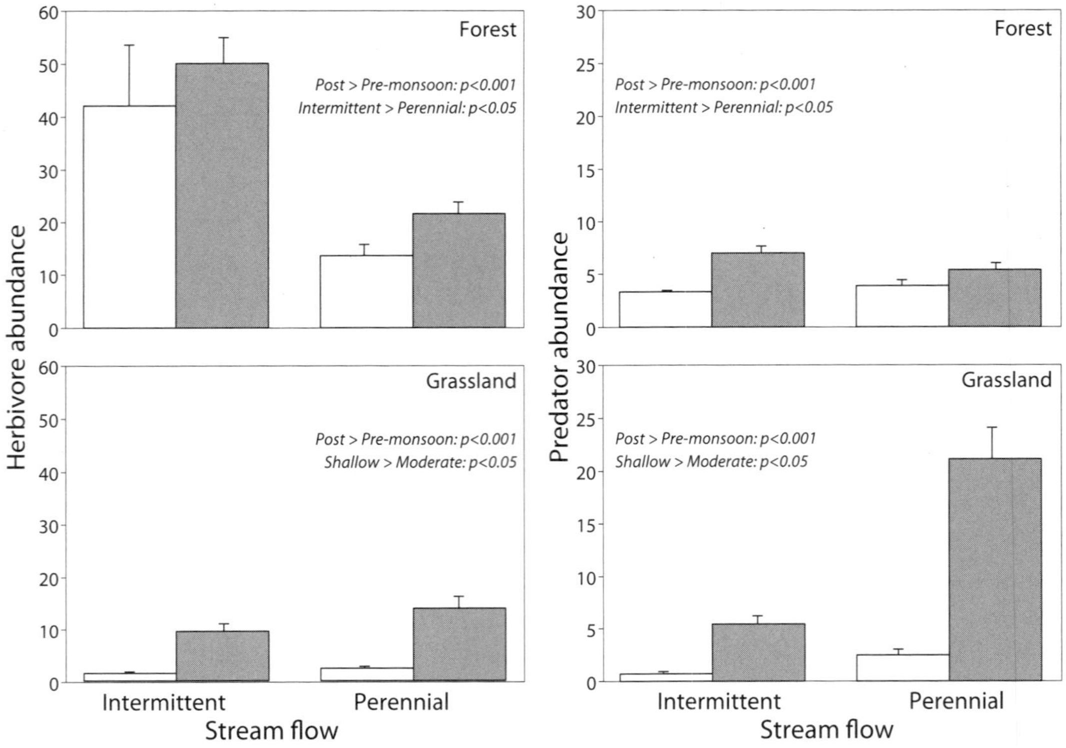

Fig. 7.8. Herbivore and Predator arthropod assemblages collected from dominant vegetation at riparian floodplain forest and grassland sites, before monsoon rains (open bars) and after monsoon rains (filled bars), and at sites with different hydrological conditions. Values represent mean number of individuals per sample across 3 years + 1 s.e. for forest sites, 1 year + 1 s.e. for grassland sites. Values are presented as arithmetic means, while statistics were performed on log-transformed data; significance values represents results of 2-way ANOVA (season x hydrology class). See text for methods and sample sizes.

floodplain forest and riparian grassland habitats (fig. 7.8). The seasonal pattern is similar to the responses of herbaceous riparian vegetation (see chap. 5) and mirrors increases in vegetation biomass, diversity, and nutritional quality associated with enhanced growing conditions during and after the monsoon rains and floods.

The seasonal increase in arthropod abundance was more pronounced in the grasslands than in the floodplain forests. This result likely reflects the differing fluvial zones of the forest and grassland sites, as well as differences in water sources and seasonal growth patterns between woody and herbaceous plants. While the floodplain trees are obligate phreatophytes with year-round access to shallow groundwater, sacaton is a facultative phreatophyte that increases in biomass, productivity, and nutritional quality as it shifts from deeper water sources (e.g., groundwater) to shallower water sources (e.g., precipitation) following the monsoons (see chap. 1). Arthropods may be responding to these seasonal changes.

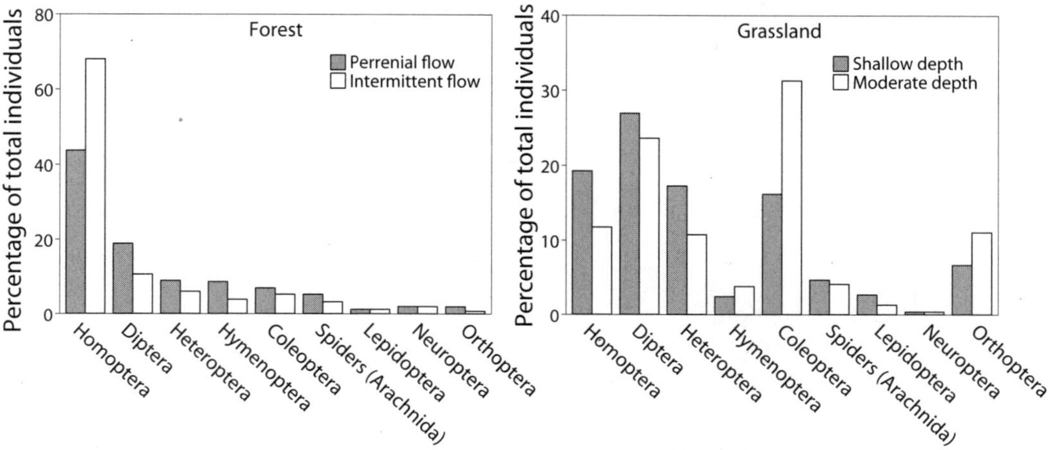

Fig. 7.9. Composition of arthropod communities on dominant vegetation at riparian floodplain forest and grassland sites characterized by different streamflow durations or groundwater depths during the post-monsoon season in 1999. Values represent the percentage of total arthropod collection within each habitat and hydrology class.

Site hydrology. In grassland habitat, Herbivore and Predator assemblages were more abundant at sites with shallower groundwater, both before and during the monsoon seasons (fig. 7.8). This was not surprising. A high degree of water stress at the drier grassland sites could have negative effects on both plant growth and food quality for herbivores. Likewise, plants at the wetter grassland sites would likely have experienced less water stress and could have supported denser herbivore populations and thus more predators.

In floodplain forest, in contrast, the herbivore assemblage was consistently more abundant at drier river sites than at sites with perennial surface flow (fig. 7.8). Homopteran and heteropteran fluid-feeders, and lepidopteran adults and caterpillars, were all more abundant at forest sites in the intermittent-flow reaches, with this pattern most pronounced in Homoptera (leafhoppers and aphids). These herbivorous insects also showed their greatest post-monsoon increases at intermittent flow sites. Homoptera represented the greatest proportion of the post-monsoon community in riparian forest collections (fig. 7.9), and the majority of these homopterans were leafhoppers collected from saltcedar. Thus, the responses of leafhopper populations on saltcedar contributed most to the pattern of high herbivore abundance in areas with intermittent compared to perennial flows during the post-monsoon period.

One possible explanation for the counterintuitive results of greater herbivore abundance at drier forest sites, and for proportionally greater post-monsoon increases in herbivore abundance at those sites, is increased food quality of plants with intermittent access to water. For a variety of woody and herbaceous plants, moderate levels of drought stress and recovery can result in mobilization of soluble nitrogen or reduction of plant defensive

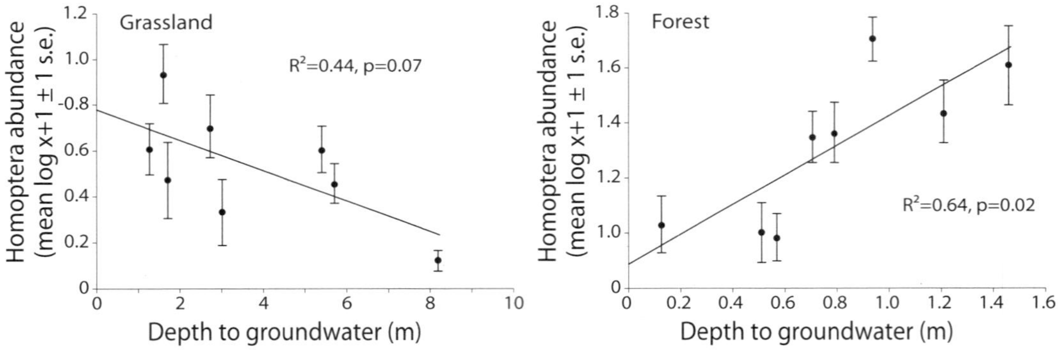

Fig. 7.10. Relationship between groundwater depth and Homoptera abundance on dominant vegetation at riparian grassland and floodplain forest sites. Data are from the post-monsoon season of 1999. Values represent mean Homoptera abundance at sites for which groundwater level measurements were available. Lines show results of linear regression analysis.

chemicals, improving food quality for phytophagous insects (White 1984, Mattson and Haack 1987, Louda 1988, A. Thomas and Hodkinson 1991, Louda and Collinge 1992, Shure et al. 1998).

Homopterans and groundwater depth. Because floodplain forest samples were dominated by Homoptera, and because homopterans appeared to a major factor in the forest herbivore response to hydrological conditions, we examined the effects of groundwater separately for this group. In both riparian grassland and floodplain forest, post-monsoonal Homoptera abundances were correlated with depth to groundwater. Populations of Homoptera showed different relationships to water availability in the two riparian zones. As would be expected if water availability improved food availability or quality for herbivores, homopteran abundance at grassland sites increased as depth to groundwater decreased (i.e., water availability increased; fig. 7.10). However, at floodplain forest sites, Homoptera showed the opposite pattern, with abundance increasing as depth to groundwater decreased (i.e., as sites became drier; fig. 7.10).

These opposing patterns suggest that different mechanisms are operating in forest and grassland sites, and without detailed studies of homopteran performance and plant quality in this system, we can only speculate as to the causes. At the more xeric end of the range of riparian corridor habitats (open sacaton grassland on the terrace), plants growing at drier sites could experience extreme water stress that lowers their food quality when groundwater is relatively deep. A positive relationship between homopteran abundance and the addition of water to host plants has been observed in various settings (Louda and Collinge 1992), including grasslands (Lavigne and Campion 2001). At the mesic end of the range of riparian habitats (gallery forest in the floodplain), the low groundwater levels could produce a mild degree of water stress on host plants that improves plant nutritional quality for phloem-feeding homopterans. Moderately or temporarily drought-stressed plants may have greater concentrations of soluble nitrogen (amino acids, amides) and therefore support better insect growth, survival, and reproduction (White 1984, Louda and Collinge 1992). The high abundance of Homoptera at drier floodplain forest sites also could be related to the characteristics of the plant community surrounding the sampled tree and shrubs. Drier river reaches have higher proportions of saltcedar trees in floodplain forests (see chap. 1), which could promote the establishment of denser populations of the leafhoppers and aphids that feed on saltcedar.

ARTHROPOD COMMUNITY DIFFERENCES AMONG DOMINANT PLANTS

There were broad differences in arthropod assemblages between sacaton grass and woody plant species. The arthropods collected from trees and shrubs in floodplain forests showed less evenness (in terms of diversity patterns) than those collected from sacaton grassland (fig. 7.9). Homoptera were by far the most abundant taxa at the floodplain forest sites, whereas individuals were distributed more equitably among taxa at sacaton grassland sites. Grasshoppers and beetles were collected more frequently from terrace sacaton grasses than from any of the woody plant types in floodplain forests.

Relative abundances of insect taxa and functional groups also differed among tree and shrub species within floodplain forest habitat (fig. 7.11). The strongest pattern was the high abundance of herbivores on saltcedar compared to the native riparian tree and shrub species. In both the pre-monsoon and post-monsoon seasons, saltcedar consistently supported the highest number of herbivore individuals among the four woody plant species sampled. This result was due to the large numbers of Homoptera (planthoppers), including the host-specific *Opsius stactogalus*, found on saltcedar. Overall, Fremont cottonwood had the lowest abundance of herbivores among the four plant species sampled. Seepwillow and Goodding's willow had similar herbivore densities; both species supported significantly fewer herbivores than saltcedar, but significantly more herbivores than cottonwood.

The finding of high herbivore abundance on saltcedar along the San Pedro is similar to the results of studies on many other southwestern rivers (see chap. 8). Often, saltcedar has been found to support higher insect abundance, but lower insect diversity, than native riparian tree and shrub species (L. Stevens 1976, 1985). Arthropod assemblages on saltcedar are commonly dominated by the specialist leafhopper *O. stactogalus* (Stevens 1985, Yard et al. 2004).

It was interesting to find that cottonwood had the lowest herbivore abundance of the four woody plant species sampled. In a study on the lower Colorado River, Fremont cottonwood supported lower arthropod abundances than other riparian trees including mesquite and Goodding's willow

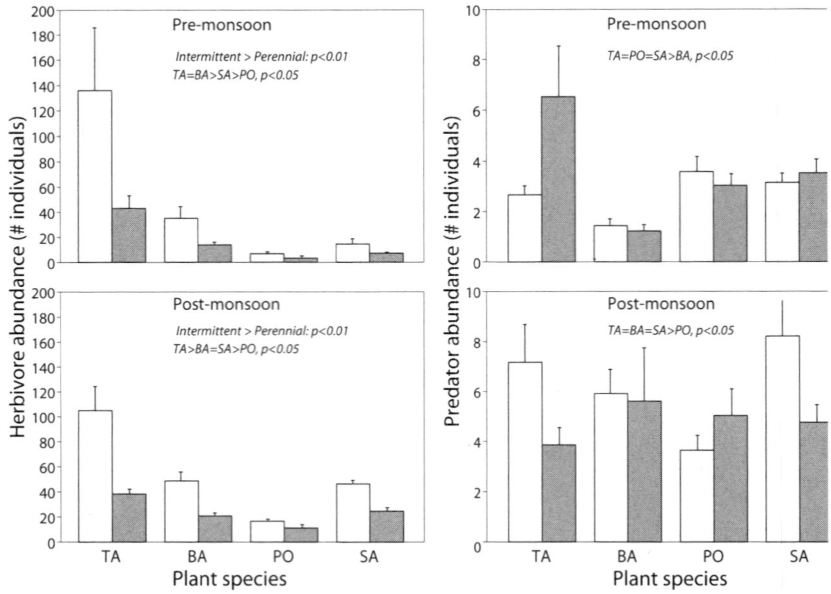

Fig. 7.11. Herbivore and Predator assemblages collected from four woody plant species at riparian floodplain forest sites, before and after monsoon rains, at sites with intermittent surface flows (open bars) and perennial flows (filled bars). TA = *Tamarix* sp., BA = *Baccharis salicifolia*, PO = *Populus fremontii*, and SA = *Salix gooddingii*.

(McGrath and van Riper 2005). However, studies in other southwestern river systems have found dense and species-rich insect communities on *Populus* foliage (Floate and Whitham 1995, Dickson and Whitham 1996, Wimp et al. 2005).

While mean herbivore abundance on cottonwood was significantly lower than on seepwillow or willow along the San Pedro, the magnitude of the difference was not great. The observed differences may be due to spatial and temporal limitations in the cottonwood sampling. Arthropod assemblages on low-canopy foliage accessible for sweep-sampling may not have been representative of those in the upper canopy where moisture and temperature conditions may differ. In addition, the richness of arthropod communities on *Populus* generally peak early in the season (G. Wimp, personal communication), prior to our collections. Those insects active during the initial leafing out of cottonwood trees along the San Pedro would have been missed in the summer sampling regime.

Synthesis: Relationships between Arthropod Communities and Riparian Habitat

EFFECTS OF SEASON

One common thread in our observations of terrestrial arthropod communities on the San Pedro was a strong seasonal pattern of increased abundance following the onset of the monsoon rains. This effect was quantified for arthropod assemblages living on riparian vegetation and observed, though not directly measured, for butterflies and bees. Butterflies seem to reach their highest numbers each year during the monsoonal late summer. Bees seem to show two major peaks in abundance annually, one in spring and one in late summer, but certainly the monsoon months represent a high point in bee abundance and activity.

In addition to effects of season, interannual variability in arthropod abundance and community composition can be quite significant. We have focused here on patterns consistent across years, but it is also important to note that the population levels of individual arthropod species can vary widely and idiosyncratically from year to year (see Denno and McClure 1983, Cappuccino and Price 1997).

EFFECTS OF VEGETATION STRUCTURE AND COMPOSITION

Each of the arthropod groups we studied responded to the structure of riparian vegetation. The abundance, richness, and taxonomic composition of these arthropod assemblages all differed to some degree between open and wooded habitat. Homoptera (leafhoppers) were present on both riparian grassland and forest vegetation but clearly dominated the assemblages living on tree and shrub species in floodplain forests. Wooded habitats in the riparian corridor (both cottonwood/willow and mesquite stands) supported lower numbers and fewer species of butterflies than open habitats. The composition of butterfly assemblages also differed somewhat between wooded and open areas, likely due to an interaction between microclimate tolerances and host and nectar plant distribution. Bee community composition also differed between open and wooded habitat within the terrace, apparently reflecting response of individual bee species to the distribution of available nest sites and flower resources.

Plant species composition was important in shaping the abundance and taxonomic composition of arthropod assemblages on riparian vegetation. Dominant plant species in the riparian floodplain showed significant variation in total arthropod abundance and in the relative proportions of insect orders in their arthropod assemblages. This variation in arthropod load could have implications for the value of riparian plant communities as foraging habitat for the many insectivorous birds that rely on riparian habitats along the San Pedro River for feeding and nesting (see chap. 8 and Morrison et al. 1990).

Fine-scale variation in the composition and diversity of flowering plants within riparian habitats also influenced the distribution of pollinating insects. We found a positive relationship between local flower diversity and bee and butterfly species richness. This relationship was stronger for bees than butterflies, possibly because bees show a greater degree of specialization in their use of flower resources. For bees, flowers represent both sources of energy for adults (nectar) and material for feeding offspring (pollen). While individual bee species do take nectar from some flower species more than others, obligate specialization occurs only in the collection of pollen. In general, butterflies would be expected to show less specialization than bees in their flower visitation since they collect only one type of floral resource: nectar for adult energy requirements.

The value of riparian habitat for foraging bees and butterflies no doubt depends both on the diversity of floral resources and the species composition of those resources. The variability among flower species in attractiveness for foraging bees and butterflies was clear when we charted the number of species visiting individual flowering plant species growing on the terrace. Certain flower species were simply more attractive than others, and in the case of native bees, some flower species uniquely attracted groups of pollen specialists.

There was wide overlap in the groups of flowering plant species most used by bee and butterfly communities on the San Pedro. On the terrace (where we have the most data to make comparisons), several composite flowers were highly attractive to both pollinator groups (figs. 7.3, 7.6). This observation has some implications for the habitat restoration efforts taking place in the abandoned agricultural fields on the San Pedro terraces. The data suggest that certain plant species, e.g., globemallow and yellow composites such as goldeneye and golden crownbeard (*Verbesina encelioides*), could be planted to attract diverse bee and butterfly communities to restored riparian habitats.

EFFECTS OF HYDROLOGY AND GEOMORPHOLOGY

Stream hydrology and geomorphology have both direct and indirect effects on terrestrial arthropod communities. In the research summarized here on bees, butterflies, and arthropod assemblages on riparian vegetation, we see strong evidence for the indirect effects of hydrology on arthropod communities. These indirect effects occur through the structuring of riparian plant communities, including the division between grassland and wooded habitat, and the species composition of dominant plants. We also see more direct effects of hydrology on arthropods in the responses of herbivorous insects on riparian vegetation (especially homopterans) to between-site differences in groundwater availability. These responses may occur through the effects of local hydrology and geomorphology on the productivity, nutrient, and/or water stress status of riparian plants.

Fine-scale variation in hydrology was not a focus of either the bee or butterfly research; therefore, we can only speculate on more direct influences of hydrology on these insects. However, known aspects of the biology of pollinators and flowering plants suggest some ways in which hydrology could influence the distribution and behavior of these insects. For example, local variation in groundwater depth and flood inundation may shape the amount and quality of floral rewards available to pollinators. It has been shown experimentally that the amount of water available to individual plants can shape floral characteristics to which pollinators respond, e.g., nectar volume and flower size (Carroll et al. 2001).

The carving of landforms and movement of nutrients by seasonal flooding are also important for arthropods living in the riparian corridor. For some native bees, dry washes and cut banks along the main river channel provide nesting sites, potentially making the riparian corridor more habitable for these bee taxa. In addition, bees, wasps, and butterflies all ingest water and salts from moist soil (Michener 2000, Dennis et al. 2003) and appear to be attracted to the nutrient-rich mud deposited in washes and along the main river channel by monsoon floods. Therefore, availability of surface moisture may limit the distribution of some terrestrial insects (e.g., Nelson 2003). Of course, insects with aquatic lifestages, such as dragonflies, mayflies, and caddisflies, are dependent on the presence of surface water.

Directions for Further Research

Basic gaps remain in our knowledge of major insect groups along the San Pedro River, including important pollinators such as bees, wasps, flies, beeflies, moths, and beetles. We lack even approximate species lists for most of these taxa. While we have accumulated a preliminary species list of native bees for the upper San Pedro, we still require a more complete fauna description. Moth diversity is potentially very high along the San Pedro, as more than 1,400 species of macro-lepidopteran moths are known to occur in southeastern Arizona (http://nitro.biosci.arizona.edu/zeeb/butterflies/mothlist.html). However, moth collecting in southeast Arizona has largely been concentrated in mountains and uplands, with relatively little work done in riparian areas (B. Walsh, personal communication).

In comparison to other insect groups in the San Pedro region, butterflies have been relatively well studied. The resident butterfly fauna has been described, and habitat affinities are known for many species. The same is true for dragonflies, as dragonfly distributions in the San Pedro and other riparian habitats of southeastern Arizona have received greater attention in recent years (R. Bailowitz, personal communication). There is still a need, however, for additional information on the resource requirements of individual butterfly and dragonfly species in order to better predict their responses to riparian habitat change. These autecological studies could be targeted at species that have the potential to be key pollinators or predators, or serve as indicators of riparian habitat quality.

Other important areas of investigation include links between riparian and upland arthropod faunas and the contribution of riparian areas to regional species richness. In their review of animal and plant diversity patterns in riparian areas and surrounding landscapes, Sabo et al. (2005) found that riparian habitats generally supported distinct assemblages of species but did not consistently support higher numbers of species than upland habitats. In contrast, along the upper San Pedro, butterfly diversity was significantly higher in the riparian corridor than in adjacent desert scrub uplands. It would be instructive to do this sort of analysis with other arthropod taxa. As agricultural land is a significant part of the San Pedro riparian landscape, agricultural habitats should also be included in any measurements of the area's arthropod diversity.

In light of predicted changes in the relative abundances of dry vs. wet reaches and saltcedar vs. cottonwood/willow forest patches along the San Pedro (see chaps. 3 and 19), responses of arthropods to water availability and dominant vegetation type deserve special attention. Lastly, it is critical to study the impact of arthropods on other trophic levels within riparian systems. For example, more data are needed on the effects of hydrological conditions on the arthropods that serve as prey for many songbirds.

In sum, these data would make it possible to model the responses of arthropods and their predators to future changes in riparian habitat along the San Pedro River and allow for better estimates of the importance of riparian areas to local and regional arthropod diversity. Ultimate goals would be to assess the impact of arthropod dynamics on riparian ecosystem function and elucidate connections between land use, hydrology, and the dynamics of riparian plant and arthropod communities.

L. ARRIANA BRAND,
DAVID J. CERASALE,
TERRELL D. RICH, AND
DAVID J. KRUEPER

Breeding and Migratory Birds

PATTERNS AND PROCESSES

EIGHT

Introduction

Desert riparian woodland is one of the rarest habitat types in North America (Knopf et al. 1988). While covering less than 1 percent of the land area in the southwestern United States, desert riparian forest supports some of the highest species richness and abundance totals of terrestrial vertebrates in North America (Szaro 1980, Rice et al. 1983, Krueper and Corman 1987, Knopf et al. 1988). Approximately 75 percent of the breeding bird species in the Southwest are classified as facultative riparian and over 50 percent as obligate riparian (Johnson et al. 1977). Avian use of riparian systems consistently far exceeds that of uplands in arid and semi-arid regions (Stevens et al. 1977, Wauer 1977, Johnson and Haight 1985, Szaro and Jakle 1985).

Avian diversity is high in desert riparian areas largely due to high vegetative diversity and complex habitat structure (Szaro 1980, M. Scott et al. 2003). On the San Pedro River, avian diversity is also high due to its geographic location at the interface between the Sonoran and Chihuahuan Deserts with biological influence from the Sierra Madre and Rocky Mountains (see chap. 6).

This chapter describes avian diversity on the San Pedro and discusses the processes that maintain diversity levels. First, we put the San Pedro riparian corridor into regional context for birds by comparing species richness and total densities with other southwestern rivers. Next, we discuss the ecological mechanisms, including anthropogenic influences, that affect avian diversity on the San Pedro. Birds vary in their requirements for water, food, shelter, nest substrate, and protection from predators (Anderson and Ohmart 1977, Szaro 1980); thus, we assess the roles of vegetation composition, structure, and arrangement; food availability; parasitism and predation; and various natural and anthropogenic disturbances on particular bird species. We treat breeding and migrating birds separately due to the large seasonal differences in mechanisms affecting each (Rotenberry et al. 1979, Szaro 1980; see table 8.2 for scientific names and classification of San Pedro bird species according to migratory status). Because of the extensive vegetation shifts that have occurred on many southwestern rivers, we emphasize bird response to differences between saltcedar and cottonwood/willow habitats.

Breeding Birds of the San Pedro River

BREEDING BIRD DIVERSITY LEVELS

Species richness estimates during the breeding season on the upper San Pedro are about 2 times higher than reported estimates from other southwestern rivers, within vegetation types (table 8.1). In total, the San Pedro hosts over 100 species that breed in the riparian corridor from May to August (Krueper 1999). On the San Pedro, as for the other rivers, bird density and richness are high in all wooded vegetation types, with values tending to be highest in the cottonwood/willow forests (table 8.1).

VEGETATION COMPOSITION AND STRUCTURE

Floodplain habitats. Vegetation structure, including the distribution of foliage at varying heights, is a key factor affecting avian community composition in desert riparian systems (MacArthur and MacArthur 1961, Anderson, Higgins et al. 1977, Szaro 1980, Mills et al. 1991, Rich 2002). The considerable difference in structure among San Pedro riparian vegetation types, including that between tall cottonwood/willow forests and shorter, shrubbier saltcedar, is an important mechanism producing differences in composition and diversity of breeding birds.

Some species build nests at heights greater than 4m above ground and thus require high canopy vegetation. Such species on the San Pedro include Gila Woodpecker, Cassin's Kingbird, Western Wood-Pewee, Gray Hawk, Vermilion Flycatcher, Great-horned Owl, Lesser Goldfinch, Yellow Warbler, and Summer Tanager (Ehrlich et al. 1988) (fig. 8.1). This canopy-nesting guild specializes in cottonwood/willow vegetation across southwestern rivers (Anderson and Ohmart 1977, Hunter, Ohmart et al. 1987).

Upper canopy also plays an important role for mid-story nesting spe-

TABLE 8.1. Estimates of avian species richness (number of species) and total bird density (number of birds per 40 ha) ± SE (standard error), where reported, for birds during the breeding season, spring migration, and fall migration, across three vegetation types in southwestern riparian systems.

River system	Cottonwood	Mesquite	Saltcedar	References
		Breeding season—Species richness		
Upper/mid San Pedro	60 ±1	61 ±1	52 ±2	Brand et al.
Middle Rio Grande	25	*	17	Ellis 1995
Lower Rio Grande	27	*	28	Engel-Wilson and Ohmart 1978
Lower Colorado	35	22	25	Anderson et al. 1977
Pecos	19	22	25	Hunter et al. 1988
		Breeding Season—Total bird density		
Upper/mid San Pedro	431 ±22	314 ±20	367 ±27	Brand et al. in review
Lower Rio Grande	708	*	486	Engel-Wilson and Ohmart 1978
Lower Colorado	336	323	216	Anderson et al. 1977
Pecos	445	75	298	Hunter et al. 1988
		Spring Migration—Species richness		
Upper San Pedro	58	*	46	D.C. Cerasale, unpubl. data
Middle Rio Grande	27	*	18	Ellis 1995
Lower Rio Grande	17	*	30	Engel-Wilson and Ohmart 1978
Lower Colorado	40	30	19	Anderson et al. 1977
		Spring Migration—Total bird density		
Upper San Pedro	1,468 ±92	*	766 ±56	D.C. Cerasale, unpubl. data
Lower Rio Grande	475	*	330	Engel-Wilson and Ohmart 1978
Lower Colorado	172	193	71	Anderson et al. 1977
		Fall Migration—Species richness		
Middle Rio Grande	31	*	27	Ellis 1995
Lower Rio Grande	26	*	25	Engel-Wilson and Ohmart 1978
Lower Colorado	41	28	27	Anderson et al. 1977
		Fall Migration—Total bird density		
Lower Rio Grande	749	*	178	Engel-Wilson and Ohmart 1978
Lower Colorado	262	195	177	Anderson et al. 1977

*No information available.

cies that build nests less than 4 m above ground. The upper canopy shades and buffers high summer temperatures while also increasing prey base (Anderson, Higgins et al. 1977, Anderson and Ohmart 1977, Hunter, Ohmart et al. 1987, Taylor 2003). Examples of birds that nest in the mid-story beneath cottonwood/willow canopy on the San Pedro include Ladder-backed Woodpecker, Black-chinned Hummingbird, Abert's Towhee, Yellow-breasted Chat,

TABLE 8.2. Scientific names of bird species named in text and/or tables.

Species name	Scientific name	Primary season
Abert's Towhee	*Pipilo aberti*	B, M
Ash-throated Flycatcher	*Myiarchus cinerascens*	B, M
Bank Swallow	*Riparia riparia*	M
Barn Swallow	*Hirundo rustica*	B, M
Bell's Vireo	*Vireo bellii*	B, M
Bewick's Wren	*Thryomanes bewickii*	B, M
Black Phoebe	*Sayornis nigricans*	B, M
Black-chinned Hummingbird	*Archilochus alexandri*	B, M
Black-headed Grosbeak	*Pheucticus melanocephalus*	B, M
Black-tailed Gnatcatcher	*Polioptila melanura*	B, M
Black-throated Gray Warbler	*Dendroica nigrescens*	M
Black-throated Sparrow	*Amphispiza bilineata*	B, M
Blue Grosbeak	*Passerina caerulea*	B, M
Blue-gray Gnatcatcher	*Polioptila caerulea*	M
Botteri's Sparrow	*Aimophila botterii*	B, M
Brewer's Sparrow	*Spizella breweri*	M
Brown-crested Flycatcher	*Myiarchus tyrannulus*	B, M
Brown-headed Cowbird	*Molothrus ater*	B, M
Bullock's Oriole	*Icterus bullockii*	B, M
Canyon Towhee	*Pipilo fuscus*	B, M
Cassin's Kingbird	*Tyrannus vociferans*	B, M
Cassin's Sparrow	*Aimophila cassinii*	B, M
Cassin's Vireo	*Vireo cassinii*	M
Cedar Waxwing	*Bombycilla cedrorum*	M
Chihuahuan Raven	*Corvus cryptoleucus*	B, M
Chipping Sparrow	*Spizella passerina*	M
Common Black-Hawk	*Buteogallus anthracinus*	M
Common Ground-Dove	*Columbina passerina*	B, M
Common Raven	*Corvus corax*	B, M
Common Yellowthroat	*Geothlypis trichas*	B, M
Dusky Flycatcher	*Empidonax oberholseri*	M
Eastern Meadowlark	*Sturnella magna*	B, M
Gambel's Quail	*Callipepla gambelii*	B, M
Gila Woodpecker	*Melanerpes uropygialis*	B, M
Gray Flycatcher	*Empidonax wrightii*	M
Gray Hawk	*Buteo nitida*	B, M
Great Horned Owl	*Bubo virginianus*	B, M
Great-blue Heron	*Ardea herodias*	B, M
Green-tailed Towhee	*Pipilo chlorurus*	M
Hammond's Flycatcher	*Empidonax hammondii*	M

TABLE 8.2. *(continued)*

Hermit Thrush	*Catharus guttatus*	M
Hooded Oriole	*Icterus cucullatus*	B, M
House Finch	*Carpodacus mexicanus*	B, M
House Sparrow	*Passer domesticus*	B, M
House Wren	*Troglodytes aedon*	M
Indigo Bunting	*Passerina cyanea*	B, M
Killdeer	*Charadrius vociferus*	B, M
Ladder-backed Woodpecker	*Picoides scalaris*	B, M
Lark Bunting	*Calamospiza melanocorys*	M
Lark Sparrow	*Chondestes grammacus*	M
Lazuli Bunting	*Passerina amoena*	B, M
Lesser Goldfinch	*Carduelis psaltria*	B, M
Lincoln's Sparrow	*Melospiza lincolnii*	M
Loggerhead Shrike	*Lanius ludovicianus*	B, M
Lucy's Warbler	*Vermivora luciae*	B, M
MacGillivray's Warbler	*Oporornis tolmiei*	M
Mallard	*Anas platyrhynchos*	B, M
Mourning Dove	*Zenaida macroura*	B, M
Nashville Warbler	*Vermivora ruficapilla*	M
Northern Beardless-Tyrannulet	*Camptostoma imberbe*	B, M
Northern Cardinal	*Cardinalis cardinalis*	B, M
Northern Flicker	*Colaptes auratus*	B, M
Northern Mockingbird	*Mimus polyglottos*	B, M
Northern Rough-winged Swallow	*Stelgidopteryx serripennis*	B, M
Orange-crowned Warbler	*Vermivora celata*	M
Pacific-slope Flycatcher	*Empidonax difficilis*	M
Phainopepla	*Phainopepla nitens*	B, M
Pine Siskin	*Carduelis pinus*	M
Plumbeous Vireo	*Vireo plumbeus*	M
Pyrrhuloxia	*Cardinalis sinuatus*	B, M
Red-naped Sapsucker	*Sphyrapicus nuchalis*	M
Red-winged Blackbird	*Agelaius phoeniceus*	B, M
Ruby-crowned Kinglet	*Regulus calendula*	M
Rufous Hummingbird	*Selasphorus rufus*	M
Savannah Sparrow	*Passerculus sandwichensis*	M
Scott's Oriole	*Icterus parisorum*	B, M
Sharp-shinned Hawk	*Accipiter striatus*	M
Song Sparrow	*Melospiza melodia*	B, M
Southwestern Willow Flycatcher	*Empidonax traillii extimus*	B, M
Spotted Towhee	*Pipilo maculatus*	M
Summer Tanager	*Piranga rubra*	B, M
Townsend's Warbler	*Dendroica townsendi*	M

(continued)

TABLE 8.2. *(continued)*

Turkey Vulture	*Cathartes aura*	B, M
Verdin	*Auriparus flaviceps*	B, M
Vermilion Flycatcher	*Pyrocephalus rubinus*	B, M
Vesper Sparrow	*Pooecetes gramineus*	M
Violet-green Swallow	*Tachycineta thalassina*	M
Virginia's Warbler	*Vermivora virginiae*	M
Warbling Vireo	*Vireo gilvus*	M
Western Kingbird	*Tyrannus verticalis*	B, M
Western Meadowlark	*Sturna neglecta*	M
Western Tanager	*Piranga ludoviciana*	M
Western Wood-Pewee	*Contopus sordidulus*	B, M
White-breasted Nuthatch	*Sitta carolinensis*	B, M
White-crowned Sparrow	*Zonotrichia leucophrys*	M
White-winged Dove	*Zenaida asiatica*	B, M
Willow Flycatcher	*Empidonax traillii*	M
Wilson's Warbler	*Wilsonia pusilla*	M
Yellow Warbler	*Dendroica petechia*	B, M
Yellow-billed Cuckoo	*Coccyzus americanus*	B, M
Yellow-breasted Chat	*Icteria virens*	B, M
Yellow-rumped Warbler	*Dendroica coronata*	M

Note: This list represents species that were relatively common during the breeding season and/or migratory periods modified from Krueper (1999). Species are classified by primary season of occurrence: the North American summer breeding period (B), during spring and/or fall migratory periods (M), or during both breeding and migratory periods (B, M). Please note that species classified here as occurring during breeding or migratory periods can also be present on an annual basis (i.e., resident) on the San Pedro. For a complete list of bird species and a more complete discussion of primary season of occurrence on the upper San Pedro, see Krueper (1999).

and Western Yellow-billed Cuckoo, the latter a candidate for listing under the Endangered Species Act (J. Hughes 1999, Halterman 2002) (fig. 8.1). The San Pedro likely sustains the largest remaining population of Western Yellow-billed Cuckoo (Halterman 2002).

Whereas over 15 species are more abundant in cottonwood compared with other San Pedro riparian vegetation types, 3 species (Yellow-breasted Chat, Abert's Towhee, and Brown-headed Cowbird) (fig. 8.1) are most abundant in monotypic saltcedar stands. Abert's Towhee and Yellow-breasted Chat are shrub-nesting species that are distributed widely across the wooded riparian corridor on the San Pedro, including cottonwood understory, but they have highest abundance in saltcedar thickets. Brown-crested Flycatcher and Black-chinned Hummingbird nest in cottonwood understory but also have relatively high abundance in saltcedar. Black-chinned Hummingbird uses a wide variety of woody species to nest, including saltcedar and broadleaf trees on the San Pedro (L. A. Brand, unpublished data). Brown-crested

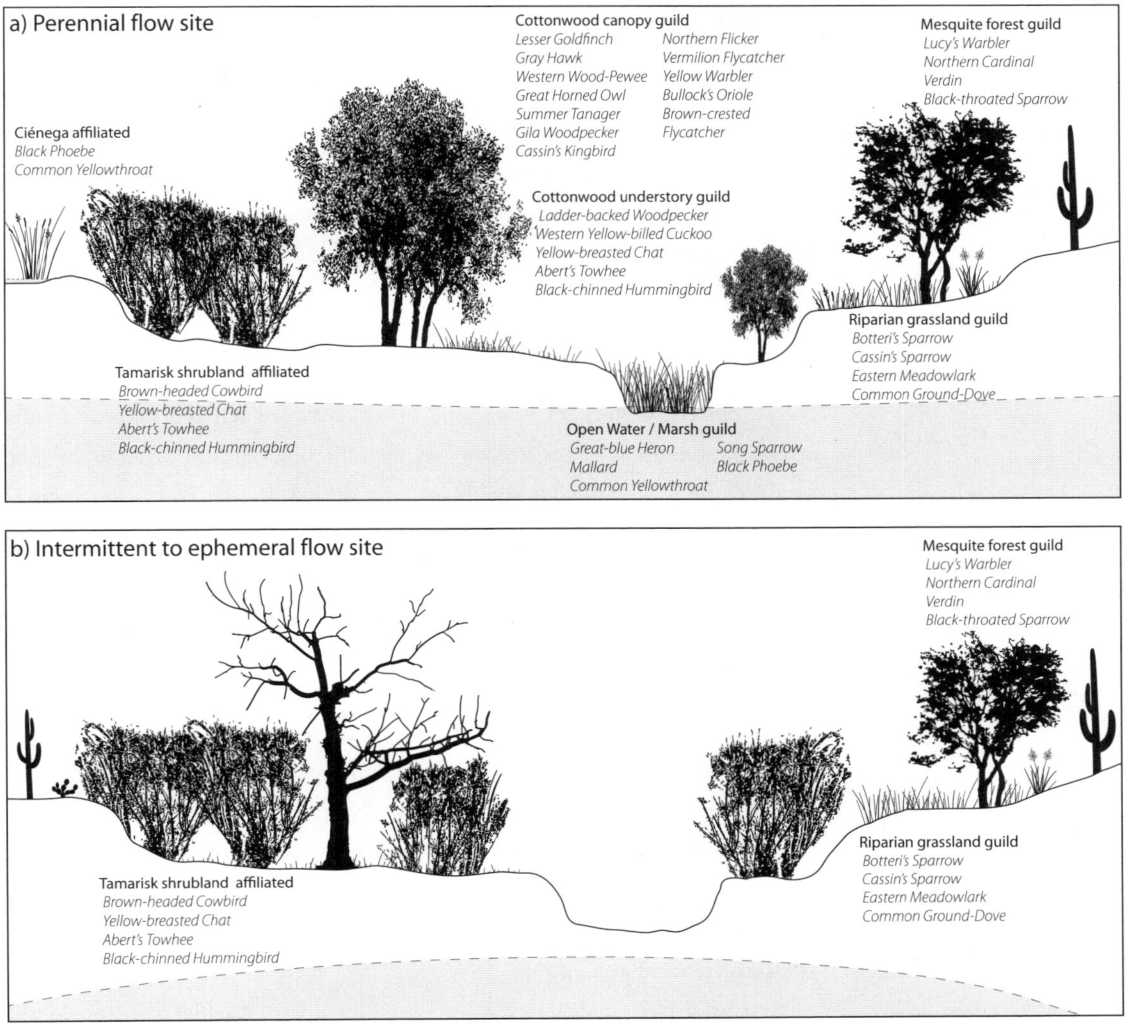

Fig. 8.1. Habitat affiliation of common breeding bird species of the San Pedro River in (a) perennial reaches and (b) intermittent-ephemeral reaches. The species listed are ones that had highest densities and/or highest spatial coverage in the indicated vegetation type, based on Brand et al. (in review), Brand et al. (2008), and Krueper and Corman (1987). Illustration credit: Mike Buffington.

Flycatchers nest in cavities; saltcedar does not provide them suitable nesting habitat, though they use cottonwood snags (standing dead trees) amongst saltcedar stands in the driest reaches (L.A. Brand, unpublished data). The Brown-headed Cowbird is an obligate brood parasite that lays its eggs in the nests of many species including Abert's Towhee, Yellow-breasted Chat, Yellow Warbler, and Bell's Vireo.

On some sections of the San Pedro River, saltcedar grows as an understory component in cottonwood forests. The addition of this vegetation layer may increase nest-site potential for certain species, including those adapted to nesting in young cohorts of broadleaf riparian plants. On the Rio Grande, cottonwood/willow canopy with saltcedar understory had more breeding birds than cottonwood with no understory (Leal et al. 1996), and on the lower Colorado River, the presence of scattered cottonwood trees amidst saltcedar stands increased the nest-site potential for a number of species (Anderson, Higgins et al. 1977). On the San Pedro, many understory nesting species were observed nesting in saltcedar shrubs below the cottonwood canopy (L. A. Brand, unpublished data). The endangered Southwestern Willow Flycatcher has not been documented to successfully breed on the upper San Pedro since 1977 (Krueper 1999), but the high density of small stems provided by saltcedar, combined with a dense surrounding forest matrix and close proximity to water, increased nest-site potential for this species on the lower San Pedro (Paradzick 2005; and see chap. 19).

Terrace habitats. Encroachment of woody vegetation into grassland habitats has been extensive in southwestern deserts (Turner et al. 2003) and riparian zones, but the effects on avian communities have garnered little scientific attention. On areas of the Pecos River that historically supported grassland, establishment of saltcedar shrublands increased density of species such as Yellow-breasted Chat and Spotted Towhee (Hunter et al. 1988, Livingston and Schemnitz 1996).

Species closely associated with sacaton grassland on the San Pedro include Botteri's and Cassin's Sparrows, Eastern Meadowlark, and Common Ground-Dove (fig. 8.1). Given their low structural diversity, grasslands generally support only a few, dominant bird species (most of which nest on the ground; Wiens 1973, Ehrlich et al. 1988). Sacaton grasslands adjacent to cottonwood/willow are also frequently used by canopy nesting birds such as Cassin's Kingbird and Vermilion Flycatcher to forage (Brand et al. 2008). Widely distributed species such as Blue Grosbeak, Western Kingbird, Mourning Dove, and Northern Mockingbird are also found in riparian grassland (Brand et al. 2008).

Species that specialize in mesquite include Lucy's Warbler, Verdin, Northern Cardinal, and Black-throated Sparrow (fig. 8.1), though the later species is most abundant in desert uplands (Brand et al. 2008). Lucy's Warbler, Verdin, and Northern Cardinal nest in the mid-story between 2 and 3.5 m. Mesquite also provides important foraging habitat for many species and increases the productivity of Gray Hawks nesting in adjacent cottonwood/willow (Bibles 1999).

Overall on the lower and upper San Pedro, species richness in mesquite is much higher than in sacaton grassland and is equal to that in cottonwood (Brand et al. 2008). These findings contrast with Krueper and Corman (1987), who found species richness to be lower in mesquite than cottonwood; they surveyed over an annual basis and only in the upper basin, where mesquite tends to be shorter and shrubbier.

In summary, the main vegetation types on the terrace, mesquite woodland, and sacaton grassland, are each important for maintaining certain species including those (e.g., Lucy's Warbler, Gray Hawk, Botteri's and Cassin's

Sparrows) that are Partners in Flight Watch List species due to restricted ranges and/or specialized habitat requirements (Rich et al. 2004). Overall, however, there are substantially fewer breeding bird species obligate to mesquite and grassland compared with cottonwood (fig. 8.1). Collectively, the heterogeneity provided by the various vegetation types in the different hydrogeomorphic zones of the San Pedro riparian corridor maintains a highly diverse bird community.

HYDROLOGIC REGIME

Some areas along the San Pedro have perennial surface water flow, while others are substantially drier, with intermittent to ephemeral flow that occurs primarily following rains (see chap. 15). There is a strong relationship between hydrologic regime (depth to groundwater, groundwater fluctuation, and surface water permanence) and floodplain vegetation composition and structure (see chap. 1). For example, cottonwood and willow trees need shallower depth to groundwater compared with saltcedar and mesquite and are most abundant in wetter river stretches. As such, hydrologic regime is an important driver of avian community composition, via its impact on vegetation (fig. 8.1).

In addition to requiring a certain vegetation structure, many bird species require surface water. A number of the species that are cottonwood-affiliated on the San Pedro, such as Song Sparrow and Common Yellowthroat, are associated with mesic habitat throughout their range. These species are likely responding to the available surface water in the perennial to intermittent flow sites that also sustain cottonwood/willow forests. Great Blue Herons, an aquatic species that forms rookeries in cottonwood forests, require available surface water (Wauer 1977, Krueper and Corman 1987). Several other aquatic birds (e.g., Mallard, Killdeer, and other shorebirds and wading birds) breed on the San Pedro and depend on its surface water. Wastewater ponds or other created ponds also provide habitat for some of these species (Ohmart et al. 1985, Krueper 1999).

Ciénega habitat is important for aquatic-associated species including Black Phoebe and Common Yellowthroat (fig. 8.1). The latter species was strongly associated with ciénega habitat, where it constituted over 84 percent of the total breeding bird density (Krueper and Corman 1987). However, ciénegas have low overall breeding bird species richness compared with other vegetation types on the San Pedro (Krueper and Corman 1987).

PREDATION AND PARASITISM

Nest predation and parasitism can substantially affect population dynamics of breeding birds and were significant sources of nest mortality for bird species breeding on the San Pedro (Brand et al. in review). Common nest predators likely include raccoon (*Procyon lotor*), cricetid rodents (e.g., *Neotoma* spp.), snakes (e.g., *Thamnophis* spp.), owls (e.g., Great-horned Owl), and Chihuahuan Raven. The main nest parasite is the Brown-headed Cowbird. In some riparian systems, declines or local extirpation of desert riparian bird species such as Bell's Vireo and Yellow Warbler have been attributed to Brown-headed Cowbird parasitism (Phillips 1968, Anderson, Higgins et al. 1977, Wauer 1977). On the San Pedro, Brown-headed Cowbirds are more abundant in saltcedar than in cottonwood/willow vegetation, which may result in increased nest parasitism for certain host species (Brand et al. in review). Livestock grazing can favor saltcedar (see chap. 1), and linkages between land use, saltcedar, and cowbirds warrant further investigation. Abundance of cowbirds on the lower Colorado River increased in association with river channelization and conversion of adjacent land to agriculture (Anderson, Higgins et al. 1977).

When studying breeding birds, it is important to assess whether abundance is a good measure of habitat quality (Van Horne 1983, Bock and Jones 2004). Evolutionarily novel habitats, such as those created by introduced plant species, can show disconnects between abundance and productivity (Battin 2004, Bock and Jones 2004). An ecological trap occurs when a low-quality habitat (e.g., with low recruitment or survival rate) is preferred over a high-quality habitat (i.e., with high recruitment or survival rate), which can lead to population extinction, though empirical examples of ecological traps are relatively rare (Battin 2004, Robertson and Hutto 2006). Because of these concerns, an assessment of nest success (an important component of recruitment rates) was done for Abert's Towhee, Yellow-breasted Chat, and Arizona Bell's Vireo, all of which had relatively high abundance in saltcedar (Brand et al. 2008). While nest success of Abert's Towhee and Yellow-breasted Chat was similar across vegetation types, Arizona Bell's Vireo had substantially higher parasitism rates and lower nest success in saltcedar and mesquite compared with cottonwood/willow. Further assessment of recruitment, survivorship, parasitism rates, and population status is particularly important for the Arizona Bell's Vireo across its range due to low observed nest success and similarity with the endangered Least Bell's Vireo (*Vireo bellii pusillus*) of southern California that is highly vulnerable to both parasitism and habitat loss (Brown 1993, Robinson et al. 1995, Averill-Murray et al. 1999).

VEGETATION ARRANGEMENT AND HETEROGENEITY

The spatial distribution of vegetation patches across the landscape plays an important role in habitat selection for many desert riparian breeding birds (Anderson, Higgins et al. 1977, Szaro 1980). The presence of a mix of plant communities on the landscape allows birds to utilize resources from multiple habitat types. For example, species nesting in cottonwood, such as Gray Hawks, also use adjacent areas, such as grassland, mesquite, and agriculture, to forage (Carothers et al. 1974, Bibles 1999), and species that nest in saltcedar may benefit from supplemental foraging in cottonwood or mesquite (Cohan et al. 1978).

Effects of patch size on avian community structure have not been studied on the San Pedro. However, studies of distance to edge show substantially more species on the San Pedro to have neutral edge response (versus a positive or negative response) compared with species occupying forested areas across the United States (Brand 2004). The high percentage of neutral edge responses on the San Pedro may reflect adaptation of desert riparian birds to the typical narrowness of southwestern riparian corridors and embedded vegetation patches, along with the ability to utilize nesting and/or food resources from multiple habitat types.

FOOD AVAILABILITY

Food availability is a key factor structuring avian communities. Most species breeding along the San Pedro are insectivores, and we focus on food resources for this guild. Several studies in desert riparian systems have compared arthropod abundances between saltcedar and other vegetation types. Some empirical studies (Cohan et al. 1978) and some reviews (Di Tomaso 1998, DeLoach et al. 2000) conclude that saltcedar habitats have less arthropod abundance and diversity than cottonwood/willow or mesquite vegetation, and this may explain why bird species that were abundant in saltcedar on the lower Colorado were generally granivores or omnivores such as Northern Mockingbird, House Sparrow, White-winged Dove, and Common Ground-Dove. Additionally, sticky exudate from saltcedar leaves may discourage insectivorous birds from foraging in saltcedar (Cohan et al. 1978). Exudate levels may vary among habitats with different salinity levels, and whether exudate levels are lower in non-saline habitats such as the San Pedro River (see chap. 14) remains to be investigated.

In contrast, other studies find arthropod densities to be equal or higher in saltcedar than in other woody riparian habitats. No difference in arthropod density occurred between cottonwood and saltcedar vegetation along the Rio Grande (Mund-Meyerson 1999), and some arthropod orders had higher frequencies in saltcedar than mesquite along the Colorado River (Yard et al. 2004). On the San Pedro, small arthropods were more abundant on saltcedar than on Fremont cottonwood or Goodding's willow, and large arthropods were equally abundant among all three tree species; these patterns were consistent across seasons, but there were interactions with site streamflow regime (see chap. 7).

Thus, there is considerable variability in insect abundance across riparian habitats. Comparisons of arthropod abundance among habitat types and drainages should be viewed with caution, however, because factors such as sampling method, season, streamflow regime, time since flood disturbance, livestock grazing, and adjacent vegetation may significantly affect estimates. In addition, avian diets and nutritional requirements change seasonally. Observed abundances of proposed prey items do not always represent available food, and more research is needed to compare food availability between vegetation types from the perspective of an individual bird (Hutto 1990).

Availability of prey for insectivorous birds also may be influenced by diversity and composition of plants in the herbaceous ground layer and by availability of surface water. Along the San Pedro River, species richness of butterflies, an important prey for birds, increased both as floral diversity increased and as forest canopy declined (see chap. 7). Availability of prey in riparian areas may also be influenced by the presence or lack of standing water and moist soils, since some life cycle stages of many riparian insects depend on water. Thus, studies of food availability for insectivorous birds should consider the collective roles of the available surface water or moist soils, understory herbaceous layer, mid-story shrub layers, and upper canopy tree layers.

FLOOD AND FIRE DISTURBANCE

Flood and fire disturbance have both short- and long-term effects for breeding birds. A large flood event on the Bill Williams River that reduced canopy foliage caused significant declines in species such as Yellow-billed Cuckoo, Gila Woodpecker, and Summer Tanager that depend on cottonwood/willow to nest (Hunter, Anderson et al. 1987). Significant flood events on the Bill Williams, lower Colorado, and San Pedro Rivers have also had detrimental short-term impacts on species vulnerable to nest inundation (species that nest near to the ground or close to the river channel) such as Abert's Towhee, Yellow-breasted Chat, and Song Sparrow (Brown and Johnson 1985, Hunter, Anderson et al. 1987). Likewise, fire can destroy nests and nesting substrate and thereby impact bird populations. Over the long term, however, periodic floods and fires allow for the establishment of young, dense patches of riparian forest and for development of a diverse mix of woody and herbaceous plants; these factors, in turn, maintain nest substrate and food base for many riparian bird species.

LIVESTOCK GRAZING

Livestock grazing generally has less effect on bird communities than does the total loss of riparian vegetation due to agricultural land conversion (Cohan et al. 1978, Szaro 1980, Leal et al. 1996). However, intensive graz-

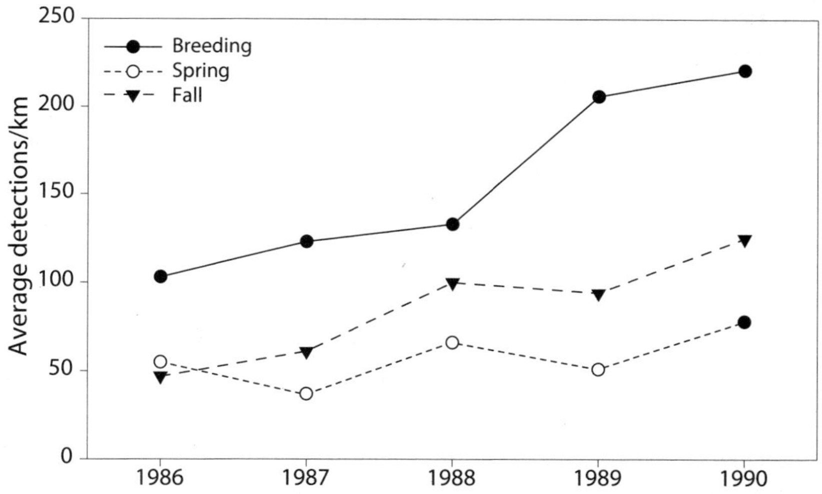

Fig. 8.2. Average bird detections per km of transect, for all species with adequate sample size, by year and season following the removal of cattle from the upper San Pedro River. All three trends are significant over the five-year period (breeding $P<0.001$, fall migration $P=0.005$, spring migration $P=0.037$) based on a one-sample t-test of the slope of the regression equation. Full description of methodology is in Krueper et al. (2003).

ing of desert riparian areas has had major impacts on desert riparian birds, with the effects largely mediated by the changes in cover and composition of the herbaceous and woody vegetation (Ames 1977, Szaro 1980). In late 1987, cattle were removed from a portion of the upper San Pedro River. The subsequent vegetation recovery and response of breeding bird populations were pronounced (fig. 8.2). Herbaceous vegetation increased substantially in cottonwood and mesquite communities after cessation of livestock grazing, and many bird populations increased in tandem (Krueper et al. 2003). Of 61 bird species with sufficient data, 87 percent increased or stayed the same, with only 13 percent decreasing. The average number of individuals detected on surveys increased 114 percent over the five-year period (103 to 221 mean detections/km from 1986 to 1990; fig. 8.2). Regional and continental population trends for species suggest that population responses on the San Pedro were not simply part of coincidental temporal changes in other environmental factors (Krueper et al. 2003).

Food availability is an important mechanism influencing bird response to livestock grazing. A particular vegetation type may benefit a particular bird species depending on its diet, foraging substrate, and foraging mode (e.g., whether it forages in the air column or on the ground). While all trophic guilds (insectivores, omnivores, granivores, nectarivores, carnivores, and scavengers) increased after cattle removal, insectivores showed the strongest increases (Krueper et al. 2003). This response probably was due to increased food resources associated, in part, with increased herbaceous vegetation.

Migratory Birds of the San Pedro River

MIGRATORY BIRD DIVERSITY LEVELS

The importance of southwestern riparian systems to migrating birds is well documented (Johnson et al. 1977, Stevens et al. 1977, Hehnke and Stone 1979, Emmerich and Vohs 1982, Skagen et al. 2005). Important migration corridors in the Southwest include the Rio Grande and the Colorado River, as well as smaller, isolated riparian oases (Skagen et al. 1998, Skagen et al. 2005). Over 200 species use the San Pedro riparian corridor to move between their breeding and wintering grounds, constituting one of the most important corridors for migrating birds in North America (Krueper 1999). Both species richness and density during spring migration were substantially higher on the San Pedro compared with other riparian systems, with density of over 35 birds per ha in cottonwood/willow habitats (table 8.1). Across southwestern drainages, species richness and total bird density tends to be highest in cottonwood/willow compared with other wooded vegetation types (table 8.1).

Birds are under selection to limit migration time and arrive on the breeding or wintering grounds in good condition (Alerstam and Lindström 1990, Alerstam and Hedenström 1998, Weber et al. 1998). While successful migration may be influenced by migrants' ability to avoid predation and minimize the impacts of competition, the ability to acquire sufficient quality and quantity of food to meet energetic needs may be the most important constraint (Moore et al. 1995). Thus, stopover habitat that meets the needs of migrating birds is important (Moore 2000).

VEGETATION COMPOSITION AND FOOD AVAILABILITY

Food availability may be the most important resource for birds during migration (F. Moore 2000). Limited numbers of frugivores migrate through the San Pedro (Skagen 1995), and little is known about their use of different vegetation types. Mistletoes occur commonly in cottonwood and mesquite (see chap. 1) and occasionally in saltcedar (Haig 1996) and may be important for frugivores such as Cedar Waxwing. Migrant nectarivores, including Black-chinned and Rufous Hummingbirds, have been detected in a variety of habitats including cottonwood/willow and saltcedar along the middle Rio Grande (Ellis 1995) and upper San Pedro (Cerasale 2004; table 8.3), but little is known about variability in their food resources. Some migrating omnivores, such as Chipping Sparrow and Brown-headed Cowbird, are present in large numbers in saltcedar, while others, such as Song Sparrow, are more abundant in cottonwood/willow vegetation (Yong and Finch 2002). Habitats adjacent to the riparian zone, such as agricultural fields with high cover of fecund herbaceous plants, may influence habitat selection by granivores and omnivores (Hehnke and Stone 1979). Relatively small numbers of carnivores, such as Cooper's Hawk, migrate along the San Pedro (Skagen 1995, Cerasale 2004), and little information regarding habitat associations of these raptors is available.

Most of the migrants along the San Pedro are insectivorous. Within this group, many warblers, such as Yellow Warbler and Wilson's Warbler, are most common in cottonwood/willow vegetation on the San Pedro (Cerasale 2004) and the Rio Grande (Kelly et al. 2000, Yong and Finch 2002). Several species, including Ash-throated Flycatcher, Bell's Vireo, Common Yellowthroat, and Yellow-breasted Chat, are abundant in saltcedar during migration, with some of these detected most frequently in this habitat type on the San Pedro (table 8.3) and Rio Grande (Kelly et al. 2000, Yong and Finch 2002). The Southwestern Willow Flycatcher uses saltcedar and cottonwood/willow during migration (Yong and Finch 2002). Habitat use during migratory periods by other endangered species, such as the Western Yellow-billed Cuckoo, remains unknown.

During spring migration (peaking in April–May), two studies reported that arthropod abundance (DeLay et al. 1999; Rio Grande) or arthropod biomass (Cerasale 2004; San Pedro) was higher in cottonwood/willow than in saltcedar vegetation, perhaps reflecting the earlier leaf flush of the former. Stopover habitat use by insectivorous migrants was more evenly distributed among riparian vegetation types during fall than spring migration along the middle Rio Grande (Ellis 1995, Yong et al. 1998). This may be due to less variation in food resources among vegetation types during the highly productive post-monsoon period (Delay et al. 1999) or to the large influx of immature birds forcing migrants to use more of the available habitat types (Yong et al. 1998).

REFUELLING RATES

Refuelling rates can affect habitat selection and migration success and thus affect fitness attributes of migrants (Alerstam and Lindström 1990, Weber et al. 1998). Migrants that refuel quickly can arrive at breeding or wintering grounds earlier and obtain higher-quality territories or higher status in social hierarchies (Møller 1994, Cristol 1995, Lozano et al. 1996). Measurement of refuelling performance can provide valuable insight into avian stopover ecology and aid in conservation and management efforts to protect stopover habitat. Yet, most research has ignored avian performance in southwestern riparian zones during stopover (but see Yong et al. 1998, Yong and Finch 2002, Cerasale 2004).

Despite having higher abundance in cottonwood/willow compared with saltcedar (table 8.3), Wilson's Warblers along the upper San Pedro had higher refuelling rates in saltcedar than in cottonwood as measured by blood compounds (Cerasale 2004). This effect occurred despite findings of lower

TABLE 8.3. Number of detections (mean ± SE) per census point in saltcedar and cottonwood/willow for species detected during spring migration 2003 along the upper San Pedro River.

Species	Cottonwood	SE	Saltcedar	SE	P[1]
Carnivores					
Common Black-Hawk	0.00	0.00	0.07	0.07	0.326
Common Raven	0.07	0.07	0.07	0.07	1.000
Gray Hawk	0.00	0.00	0.20	0.20	0.072
Turkey Vulture	0.07	0.07	0.00	0.00	0.326
Frugivores					
Cedar Waxwing	0.07	0.07	0.00	0.00	0.326
Phainopepla	0.00	0.00	0.07	0.07	0.326
Granivores					
Abert's Towhee	3.20	0.46	0.67	0.21	<0.001
Common Ground-Dove	0.07	0.07	0.07	0.07	1.000
Gambel's Quail	0.00	0.00	0.07	0.07	0.326
House Finch	1.73	0.43	3.00	0.56	0.061
Lesser Goldfinch	1.00	0.34	0.53	0.19	0.240
Mallard	0.53	0.26	0.13	0.13	0.176
Mourning Dove	1.00	0.40	0.33	0.23	0.162
Northern Cardinal	0.13	0.09	0.20	0.14	0.699
Pyrrhuloxia	0.07	0.07	0.00	0.00	0.326
White-winged Dove	0.60	0.24	1.20	0.30	0.481
Insectivores					
Ash-throated Flycatcher	0.13	0.09	0.40	0.19	0.216
Brown-crested Flycatcher	0.13	0.09	0.00	0.00	0.153
Bell's Vireo	0.87	0.29	1.60	0.50	0.212
Bewick's Wren	4.20	0.53	1.73	0.46	0.002
Black-tailed Gnatcatcher	0.07	0.07	0.00	0.00	0.326
Black Phoebe	0.20	0.11	0.07	0.07	0.299
Black-throated Gray Warbler	0.07	0.07	0.00	0.00	0.326
Bullock's Oriole	0.93	0.30	0.40	0.13	0.115
Cassin's Kingbird	1.13	0.27	0.60	0.24	0.151
Cassin's Vireo	0.07	0.07	0.00	0.00	0.326
Common Yellowthroat	3.60	0.71	1.93	0.64	0.093
Gray Flycatcher	0.07	0.07	0.00	0.00	0.326
Hammond's Flycatcher	0.13	0.09	0.07	0.07	0.559
House Wren	0.20	0.14	0.00	0.00	0.178
Ladder-backed Woodpecker	0.40	0.16	0.20	0.14	0.367

TABLE 8.3. *(continued)*

Species	Cottonwood	SE	Saltcedar	SE	P[1]
Insectivores *(cont.)*					
Lucy's Warbler	2.53	0.51	2.53	0.73	1.000
Northern Beardless-Tyrannulet	0.33	0.13	0.13	0.09	0.493
Northern Rough-winged Swallow	0.80	0.34	0.67	0.21	0.742
Orange-crowned Warbler	0.13	0.13	0.00	0.00	0.326
Pacific-slope Flycatcher	0.07	0.07	0.00	0.00	0.326
Ruby-crowned Kinglet	0.73	0.25	0.07	0.07	0.011
Summer Tanager	2.80	0.39	0.27	0.12	<0.001
Vermilion Flycatcher	3.33	0.80	0.33	0.33	<0.001
Verdin	0.27	0.15	0.07	0.07	0.242
Violet-green Swallow	0.00	0.00	0.07	0.07	0.326
Virginia's Warbler	0.07	0.07	0.00	0.00	0.326
Warbling Vireo	0.40	0.24	0.00	0.00	0.100
Western Kingbird	0.07	0.07	0.00	0.00	0.326
Western Tanager	0.00	0.00	0.07	0.07	0.326
Western Wood-Pewee	0.27	0.15	0.00	0.00	0.032
White-breasted Nuthatch	0.33	0.13	0.00	0.00	0.013
Wilson's Warbler	3.28	0.59	1.67	0.36	0.025
Yellow-breasted Chat	1.93	0.30	2.67	0.58	0.273
Yellow Warbler	14.67	1.97	0.20	0.11	<0.001
Nectarivores					
Black-chinned Hummingbird	1.40	0.35	1.13	0.24	0.532
Omnivores					
Black-headed Grosbeak	0.07	0.07	0.00	0.00	0.326
Black-throated Sparrow	0.20	0.11	0.00	0.00	0.072
Brewer's Sparrow	0.07	0.07	0.00	0.00	0.326
Brown-headed Cowbird	1.20	0.26	2.67	0.68	0.068
Chipping Sparrow	1.13	0.36	1.00	0.49	0.828
Gila Woodpecker	0.93	0.33	0.07	0.07	0.016
Green-tailed Towhee	0.07	0.07	0.13	0.09	0.559
Lark Sparrow	0.00	0.00	0.13	0.13	0.326
Lincoln's Sparrow	0.33	0.13	0.00	0.00	0.013
Song Sparrow	2.93	0.53	0.27	0.15	<0.001
White-crowned Sparrow	0.07	0.07	0.07	0.07	1.000

Sources: Modified from Cerasale 2004. Avian diets during migration based on Skagen (1995).

Note: Significance values from 2-sample independent t-tests.

[1]Detection probabilities were not measured and significance values do not account for multiple comparisons.

arthropod abundance in saltcedar than cottonwood/willow vegetation during spring migration (Cerasale 2004). These results lead to questions of what ecological mechanisms (i.e., food availability, predation, competition) might be controlling habitat selection by Wilson's Warblers and other migrants along the San Pedro.

PREDATION

Predation constitutes a significant mortality risk during migration. Predation risk must be balanced against energy requirements and affects refueling performance and habitat selection (F. Moore and Aborn 2000). Information about predation on birds migrating through southwestern riparian systems, including differences in predation rates among riparian vegetation types, is limited.

COMPETITION

The concentration of migrants with similar food preferences and high energy demands during stopover may heighten competition for food (F. Moore and Yong 1991). Population density-dependent mass gain in warblers suggests that exploitation competition may be an important mechanism affecting habitat selection along the Rio Grande (Kelly et al. 2002). Exploitation competition occurs when food is limited and each migrant is consuming as much food as possible, suppressing food availability and causing fattening rates to decrease. Interference competition also may be an important mechanism controlling fattening rates and habitat selection during migration. Interference occurs when the focus of competition shifts from the primary resource (i.e., food) to access to that resource. It is unknown if arthropod prey are a limiting resource along the San Pedro, but aggression among insectivorous species has been observed (Skagen 1995, Cerasale 2004). An abundance of migrants may increase these interactions, forcing some individuals to move from preferred habitats into adjacent habitats to refuel.

VEGETATION ARRANGEMENT

Habitat selection and performance of migrants among different-sized patches of riparian habitat have received little attention. However, Skagen et al. (1998) found that relative abundances of most spring migrants did not differ among small oases compared with intermediate and large continuous riparian patches. Overall, effects of riparian patch size and arrangement on the stopover ecology of migrants in the Southwest requires further scrutiny.

LIVESTOCK GRAZING

Compared with breeding birds, less information is available on migrating bird response to livestock grazing. Overall, the effects of livestock removal

TABLE 8.4. Five-year population trends (1986 to 1990) following cessation of livestock grazing and their significance for bird species with adequate sample sizes during spring and fall migration on the upper San Pedro River.

Common name	Fall trend	P	Spring trend	P
Scott's Oriole	NA	NA	0.34	0.00
Red-winged Blackbird	NA	NA	11.41	0.09
Blue-gray Gnatcatcher	NA	NA	0.39	0.15
Hooded Oriole	NA	NA	2.85	0.22
Pine Siskin	NA	NA	1.03	0.97
Indigo Bunting	31.10	0.02	NA	NA
Lazuli Bunting	25.91	0.01	3.00	0.04
Common Ground-Dove	20.50	0.00	4.43	0.15
Northern Beardless-Tyrannulet	20.46	0.01	1.96	0.50
Common Yellowthroat	16.50	0.00	2.75	0.18
Yellow-billed Cuckoo	16.50	0.02	1.60	0.18
Dusky Flycatcher	12.08	0.11	1.26	0.39
Yellow-breasted Chat	10.74	0.00	3.05	0.02
Western Tanager	9.89	0.03	2.82	0.30
Black-chinned Hummingbird	8.77	0.03	1.61	0.08
Brewer's Sparrow	8.36	0.00	1.09	0.00
Rufous Hummingbird	5.20	0.00	NA	NA
Cassin's Sparrow	4.81	0.16	1.68	0.31
Hermit Thrush	4.71	0.21	6.58	0.20
Lincoln's Sparrow	4.14	0.01	1.53	0.00
Gray Flycatcher	4.14	0.00	3.49	0.09
Yellow Warbler	4.03	0.00	2.28	0.00
Black-throated Gray Warbler	4.03	0.00	1.36	0.66
Townsend's Warbler	3.88	0.00	1.92	0.17
Western Wood-Pewee	3.78	0.00	NA	NA
Blue Grosbeak	3.61	0.00	0.81	0.56
Black-headed Grosbeak	3.47	0.01	2.27	0.34
MacGillivray's Warbler	3.20	0.00	15.89	0.08
Bullock's Oriole	3.17	0.00	1.19	0.03
Vermilion Flycatcher	2.88	0.00	2.41	0.01
Warbling Vireo	2.78	0.05	11.17	0.03
Red-naped Sapsucker	2.74	0.06	NA	NA
Brown-headed Cowbird	2.62	0.01	0.57	0.26

(continued)

TABLE 8.4. *(continued)*

Common name	Fall trend	P	Spring trend	P
Ash-throated Flycatcher	2.57	0.01	1.49	0.31
Green-tailed Towhee	2.48	0.02	1.49	0.31
Plumbeous Vireo	2.34	0.02	1.51	0.39
Pacific-slope Flycatcher	2.30	0.00	48.82	0.04
Gray Hawk	2.29	0.03	1.01	0.98
Summer Tanager	2.24	0.01	0.07	0.01
House Wren	2.16	0.00	2.65	0.22
Ruby-crowned Kinglet	2.14	0.00	1.72	0.09
Botteri's Sparrow	1.74	0.37	1.37	0.67
Bank Swallow	1.61	0.46	NA	NA
Wilson's Warbler	1.60	0.22	3.92	0.00
White-crowned Sparrow	1.60	0.00	0.45	0.05
Hammond's Flycatcher	1.54	0.00	1.89	0.10
Spotted Towhee	1.51	0.80	NA	NA
Brown-crested Flycatcher	1.39	0.00	NA	NA
Northern Mockingbird	1.34	0.81	0.06	0.00
Orange-crowned Warbler	1.33	0.57	0.73	0.54
Killdeer	1.26	0.00	0.43	0.05
Vesper Sparrow	1.17	0.87	0.54	0.31
Sharp-shinned Hawk	1.16	0.88	NA	NA
Yellow-rumped Warbler	1.13	0.00	1.11	0.63
Nashville Warbler	0.96	0.96	NA	NA
Northern Rough-winged Swallow	0.95	0.96	0.72	0.75
Chipping Sparrow	0.92	0.61	2.42	0.22
Willow Flycatcher	0.82	0.86	NA	NA
Lark Sparrow	0.77	0.48	0.02	0.01
Turkey Vulture	0.58	0.00	0.38	0.14
Loggerhead Shrike	0.40	0.05	NA	NA
Lark Bunting	0.07	0.04	0.26	0.01
Violet-green Swallow	0.06	0.00	NA	NA
Barn Swallow	0.05	0.01	NA	NA
Savannah Sparrow	0.03	0.00	NA	NA

Source: Krueper and Rich, unpublished data.

Notes: Spring and fall trend greater than 1.00 indicates increasing trends, and less than 1.00 indicates decreasing trends. Full description of methodology used to calculate and test trends in Krueper et al. (2003). Species are sorted by decreasing strength of the fall trend and then by the significance (P) of that trend. NA = species did not occur in that season.

on spring and fall migrants along the upper San Pedro appear to be similar to those during the breeding season (Krueper and Rich, unpublished data), with significant increases in bird density (fig. 8.2). Over a five-year period, the average number of individuals detected increased 42 percent in spring (55 to 78 mean detections/km) and 166 percent in fall (47 to 125 mean detections/km; fig. 8.2). Of 51 species in spring, 86 percent increased or stayed the same while 14 percent decreased; of 60 species in fall, 90 percent increased or stayed the same, and 10 percent decreased (table 8.4). As a group, insectivorous birds showed the strongest increases in abundance (table 8.4). Among the species that increased were Yellow Warbler, Wilson's Warbler, Lincoln's Sparrow, Brewer's Sparrow, Vermilion Flycatcher, Yellow-breasted Chat, Warbling Vireo, Bullock's Oriole, Pacific-slope Flycatcher, and Lazuli Bunting (table 8.4). Several other species, including Turkey Vulture, Loggerhead Shrike, Violet-green Swallow, Barn Swallow, Savannah Sparrow, and Lazuli Bunting, decreased in at least one season (table 8.4).

The difference in seasonal trends, with greater relative increase in fall on the upper San Pedro River following livestock removal, may be chiefly a result of available food resources. Summer monsoon rains and floods drive a pulse of late summer and fall insect production (see chap. 7), and food availability differences between grazed and ungrazed areas may be most pronounced during this season. The seasonal differences also may reflect food-resource differences between young riparian forest vegetation, which are more fully leafed out during the fall migration (Farley et al. 1994). Near-channel woody vegetation at the time of the San Pedro study was one to four years old (Krueper and Rich, unpublished data) and thus likely had high stem density and leaf cover that provided abundant substrates for insect production (Hutto 1989).

Summary and Conclusion

The San Pedro riparian corridor supports a diverse and abundant community of breeding and migrating birds. Each of the main vegetation types present, including cottonwood/willow forests, saltcedar shrublands, mesquite forests and shrublands, sacaton grasslands, and ciénega marshlands, has a different vegetation structure and resource base and thus sustains a different, but overlapping, assemblage of bird species.

Along the San Pedro River over the past century, there has been a shift from riparian grassland to mesquite bosque, although both types have been extensively cleared for agriculture. While substantially fewer breeding birds are obligate to mesquite and grassland compared with cottonwood/willow, mesquite and grassland both play important roles for certain species. Mesquite and grassland also provide cross-boundary subsidies for a number of riparian birds that utilize floodplain habitats.

Saltcedar became abundant in dry and heavily grazed reaches over the past half-century, whereas cottonwood/willow forests developed in the wetter reaches, with differing effects on bird communities. Stands of saltcedar show lower species richness than cottonwood/willow for breeding and migratory birds. While saltcedar maintains high densities of some shrub-nesting birds, cottonwood/willow stands are needed to maintain high densities of canopy-nesting birds, including Gray Hawk, and understory-nesting birds that require upper canopy, such as Yellow-billed Cuckoo. Cottonwood/willow forests are also associated with increased surface water permanence that supports a variety of wading, swimming, and marsh-associated birds.

The San Pedro is an important corridor for migratory birds. Cottonwood/willow provides higher-quality habitat for some migrating species, but for other species, saltcedar provides adequate stopover habitat, particularly during spring migration. Avian response to a mixture of cottonwood/willow or mesquite and saltcedar vegetation has yet to be thoroughly investigated.

Livestock have been removed from many reaches of the San Pedro River. Cessation of livestock grazing has led to increased abundance for breeding and migratory birds. The increases are likely due, in part, to increased insect availability following the regrowth of vegetation.

The San Pedro River sustains several rare and threatened bird species including the Western Yellow-billed Cuckoo and Southwestern Willow Flycatcher. These birds have declined regionally due to loss of suitable riparian habitats in the desert Southwest. The San Pedro also hosts many species that currently are common but have declined on other rivers and are vulnerable to further habitat loss. For example, relatively common bird species on the San Pedro, such as Yellow Warbler and Great Blue Heron, have been extirpated from the lower Rio Grande due to loss of cottonwood/willow habitat. Habitat conditions that maintain these and other bird species on the San Pedro include perennial streamflows to sustain mesic forest types such as cottonwood/willow, as well as periodic floods to sustain tree recruitment and stimulate growth of the herbaceous plants that provide forage for insectivores and other bird guilds.

PHILIP C. ROSEN

Reptiles and Amphibians

NINE

Introduction

The amphibians and reptiles (herpetofauna) of the San Pedro River present us with an opportunity to observe a high-diversity assemblage of species under relatively undisturbed conditions. Yet, the river is not without major human-caused changes affecting the herpetofauna. This chapter outlines the species diversity and habitat relationships of the "herps" and explores reasons for natural patterns in the assemblage and for changes in the herpetofauna that may have occurred or are ongoing.

Species Diversity and Riparian Affinities

The San Pedro River riparian corridor supports the characteristic southeastern Arizona lowland riparian reptiles and amphibians (table 9.1; see Rosen et al. 2005) along with many facultative riparian species with upland (desert grassland and desert scrub) affinities. It is a species-rich assemblage. Along the roughly 130 mi reach of the San Pedro River in the United States plus the portion in Mexico, there are 55 known species of herpetofauna including 4 non-native species, with a probable total of 63 species including 5 non-natives (table 9.2; and see Rosen 2005). This chapter focuses on the U.S. portion of the river, since little organized sampling has been done in the Sonoran part.

TABLE 9.1. Riparian-obligate (R-O) and preferential-riparian (PR) species of herpetofauna of the San Pedro River and its riparian environs in the United States.

Common name (Scientific name)	Relationship to riverine system	Riparianness	Landscape occurrence pattern
Amphibians[1]			
Sonoran Desert Toad (*Ollotis alvaria*)	Breeds on floodplain	OR-O	Valley-Canyon
Great Plains Toad (*Anaxyrus cognatus*)	Breeds on floodplain	OR-O	Valley
Green Toad (*Anaxyrus debilis*)	Breeds on floodplain	OR-O	Valley
Red-spotted Toad (*Anaxyrus punctatus*)	Breeds on floodplain	OR-O	Canyon-rocky areas
Woodhouse's Toad (*Anaxyrus woodhousii*)	Stream breeder	R-O	Valley
Canyon Treefrog (*Hyla arenicolor*)	Stream breeder	OR-O	Canyon-rocky areas
American Bullfrog (NN) (*Lithobates catesbeianus*)	Aquatic obligate	R-O	Valley
Chiricahua Leopard Frog (*Lithobates chiricahuensis*)	Aquatic obligate	R-O	Valley-Canyon
Lowland Leopard Frog (*Lithobates yavapaiensis*)	Aquatic obligate	R-O	Valley-Canyon
Couch's Spadefoot (*Scaphiopus couchii*)	Breeds on floodplain	PR	Valley
Mexican Spadefoot (*Spea multiplicata*)	Breeds on floodplain	OR-O	Valley
Turtles[2]			
Spiny Softshell (NN) (*Apalone spinifera*)	Aquatic obligate	R-O	Valley
Sonoran Mud Turtle (*Kinosternon sonoriense*)	Aquatic obligate	R-O	Valley-Canyon
Ornate Box Turtle (*Terrapene ornata*)	Abundant on floodplain	PR	Valley
Pond Slider (NN) (*Trachemys scripta*)	Aquatic obligate	R-O	Valley

TABLE 9.1. *(continued)*

Common name (Scientific name)	Relationship to riverine system	Riparianness	Landscape occurrence pattern
Lizards[3]			
Giant Spotted Whiptail * (*Aspidoscelis burti stictogramma*)	Requires woodland and thornscrub	R-O	Canyon-Valley
Sonoran Spotted Whiptail (*Aspidoscelis sonorae*)	River-associated at lower elevations	PR	Canyon-Valley
Zebra-tailed Lizard (*Callisaurus draconoides*)	River-associated at higher elevations	PR	Valley
Greater Earless Lizard (*Cophosaurus texanus*)	Locally riparian-obligate	PR	Canyon-Valley
Madrean Alligator Lizard (*Elgaria kingii*)	Riparian when non-montane	R-O	Canyon-Valley
Clark's Spiny Lizard (*Sceloporus clarkii*)	Riparian at lower elevations	R-O	Valley-Canyon
Southwestern Fence Lizard (*Sceloporus cowlesi*)	Riparian at lower elevations	R-O	Valley
Northern Tree Lizard (*Urosaurus ornatus*)	High riparian abundance	PR	Ubiquitous
Snakes[4]			
Banded Sand Snake (*Chilomeniscus cinctus*)	Xeroriparian-associated	PR	Valley-Canyon
Ring-necked Snake (*Diadophis punctatus*)	Riparian at lower elevations	R-O	Valley-Canyon
Sonoran Whipsnake (*Masticophis bilineatus*)	Riparian-associated at lower elevations	PR	Canyon-rocky areas
Smith's Black-headed Snake (*Tantilla hobartsmithi*)	Riparian at lower elevations	R-O	Valley-Canyon
Black-necked Gartersnake (*Thamnophis cyrtopsis*)	Aquatic obligate	R-O	Canyon-rocky areas
Mexican Gartersnake (*Thamnophis eques*)	Aquatic obligate	R-O	Valley
Checkered Gartersnake (*Thamnophis marcianus*)	Aquatic/ephemeral water obligate	R-O	Valley

Notes: Several ephemeral water-breeding toads were likely largely restricted to the floodplain prior to human habitat modification (see text) and are indicated as originally riparian-obligate (OR-O) here. All species listed are confirmed to occur along the river. Non-native species are indicated by NN.

*Not verified by voucher specimen or photograph.

[1] 3 aquatic, 2 stream breeding, 5 originally floodplain dependant

[2] 3 aquatic, 1 preferential

[3] 4 riparian, 4 preferential

[4] 3 aquatic, 2 riparian, 2 preferential

TABLE 9.2. Herpetofauna of the San Pedro River and its riparian environs in the United States.

English name	Scientific name	River reach				
		Lower	Benson	Upper	**Entire U.S. Reach**	Upper Valley
Amphibians						
Tiger Salamander (NN)	*Ambystoma tigrinum*		E		E	E
Sonoran Desert Toad	*Ollotis alvaria*	4	1	TEC	5	3
Great Plains Toad	*Anaxyrus cognatus*	E	1	1	2	2
Green Toad	*Anaxyrus debilis*		14	?	14	1
Red-spotted Toad	*Anaxyrus punctatus*	TEC	?	?	TEC	3
Woodhouse's Toad	*Anaxyrus woodhousii*	1	2	67	70	7
Canyon Treefrog	*Hyla arenicolor*	2(r)	?(r)	?(r)	2(r)	
American Bullfrog (NN)	*Lithobates catesbeianus*	6	PCR	1	7	E
Chiricahua Leopard Frog[1]	*Lithobates chiricahuensis*			1(ex)	1(ex)	E
Lowland Leopard Frog	*Lithobates yavapaiensis*	7	1	13	21	E
Couch's Spadefoot	*Scaphiopus couchii*	TEC	E	10	10	8
Mexican Spadefoot	*Spea multiplicata*	?	E	PCR	PCR	32
Turtles						
Spiny Softshell (NN)	*Apalone spinifera*	1	?	TEC	1	
Desert Tortoise	*Gopherus agassizii*	?(r)			?(r)	
Sonoran Mud Turtle	*Kinosternon sonoriense*	1	2	14	17	
Ornate Box Turtle	*Terrapene ornata*	1	PCR	15	16	PCR
Pond Slider (NN)	*Trachemys scripta*	1	E	TEC	1	
Lizards						
Giant Spotted Whiptail	*Aspidoscelis burti stictogramma*	?	?	TEC	TEC	
Little Striped Whiptail[2]	*Aspidoscelis inornata*			1(?)		
Sonoran Spotted Whiptail	*Aspidoscelis sonorae*	4	E	TEC	4	E
Tiger Whiptail	*Aspidoscelis tigris*	8	3	6	17	2
Desert Grassland Whiptail	*Aspidoscelis uniparens*	PCR	2	32	34	17
Zebra-tailed Lizard	*Callisaurus draconoides*	3	4	2	9	1
Western Banded Gecko	*Coleonyx variegatus*	1	E	1	2	E
Greater Earless Lizard	*Cophosaurus texanus*	7	E	2	9	E
Eastern Collared Lizard	*Crotaphytus collaris*	1(r)	?(r)	1(r)	2(r)	E
Madrean Alligator Lizard	*Elgaria kingii*	1	E	2	3	1
Long-nosed Leopard Lizard	*Gambelia wislizenii*	1	E	1	2	E
Mediterranean Gecko (NN)	*Hemidactylus turcicus*	E	E		E	E
Lesser Earless Lizard	*Holbrookia maculata*	1	E	5	6	49
Texas Horned Lizard	*Phrynosoma cornutum*		2	4	6	9

TABLE 9.2. *(continued)*

English name	Scientific name	River reach: Lower	River reach: Benson	River reach: Upper	River reach: Entire U.S. Reach	Upper Valley
Lizards *(cont.)*						
Greater Short-horned Lizard	*Phrynosoma hernandesi*			1	**1**	1
Regal Horned Lizard	*Phrynosoma solare*	PCR	4	3	**7**	PCR
Clark's Spiny Lizard	*Sceloporus clarkii*	5	4	10	**19**	4
Desert Spiny Lizard	*Sceloporus magister*	8	1	3	**12**	2
Southwestern Fence Lizard	*Sceloporus cowlesi*	21	2	8	**31**	4
Northern Tree Lizard	*Urosaurus ornatus*	29	1	21	**51**	12
Common Side-blotched Lizard	*Uta stansburiana*	12	?	1(?)	**12**	
Gila Monster*	*Heloderma suspectum*	1	1	1	**3**	5
Snakes						
Glossy Snake	*Arizona elegans*	3	1	1	**5**	E
Banded Sand Snake	*Chilomeniscus cinctus*	2			**2**	
Ring-necked Snake	*Diadophis punctatus*	E	1	1	**2**	E
Chihuahuan Hook-nosed Snake	*Gyalopion canum*	(1)	1	2	**3**	E
Mexican Hog-nosed Snake	*Heterodon kennerlyi*		?	1	**1**	4
Desert Night Snake	*Hypsiglena chlorophaea*	1	2	1	**4**	3
Common Kingsnake	*Lampropeltis getula*	1	2	3	**6**	8
Plains Threadsnake	*Leptotyphlops dulcis*		2	TEC	**2**	1
Western Threadsnake	*Leptotyphlops humilis*	E	1		**1**	?
Sonoran Whipsnake	*Masticophis bilineatus*	?	?	1	**1**	2
Coachwhip	*Masticophis flagellum*	5	4	1	**10**	10
Saddled Leaf-nosed Snake	*Phyllorhynchus browni*	3			**3**	
Gopher Snake	*Pituophis catenifer*	3	2	4	**9**	11
Long-nosed Snake	*Rhinocheilus lecontei*	3	7	2	**12**	13
Mountain Patch-nosed Snake[3]	*Salvadora grahamiae*			?	**?**	E
Western Patch-nosed Snake	*Salvadora hexalepis*	2	3	4	**9**	5
Western Groundsnake[3]	*Sonora semiannulata*			?	**?**	2
Smith's Black-headed Snake	*Tantilla hobartsmithi*	8	4	1	**13**	E
Plains Black-headed Snake	*Tantilla nigriceps*	?	2	1	**3**	E
Yaqui Black-headed Snake[3]	*Tantilla yaquia*			?	**?**	?
Black-necked Gartersnake	*Thamnophis cyrtopsis*	?(r)	?(r)	?(r)	**E(r)**	1
Mexican Gartersnake	*Thamnophis eques*	?	(ex?)	6	**6**	E
Checkered Gartersnake	*Thamnophis marcianus*	6	5	22	**33**	5

(continued)

TABLE 9.2. *(continued)*

		River reach				
English name	Scientific name	Lower	Benson	Upper	Entire U.S. Reach	Upper Valley
Snakes *(cont.)*						
Lyre Snake	*Trimorphodon biscutatus*	1	1	2	4	E
Sonoran Coralsnake*	*Micruroides euryxanthus*	5	2	2	9	2
Massasauga*[1]	*Sistrurus catenatus*			(ex)	(ex)	1
Western Diamondback*	*Crotalus atrox*	2	1	9	12	6
Mohave Rattlesnake*	*Crotalus scutulatus*	2	3	1	6	14
Tiger Rattlesnake*	*Crotalus tigris*	TEC			TEC	
Total museum records found		174	89	292	555	251
Vouchered total		40	34	44	55	35
Expected species total		50	48	53	63	57
Non-native species		4	4	3	5	3

Source: Reprinted from Rosen (2005).

Notes: Numbers in the table are museum voucher specimens found in a nationwide search. Records interpreted as within ≤2 miles of the riparian bottomland are included. The Upper reach extends from the international border to Curtiss Flats, the Benson reach is from Curtiss Flats to Pomerene, and the Lower reach is from Pomerene to the Gila River. TEC and PCR refer to additional records by Troy Corman (1988 and personal communication) and P. C. Rosen (unpublished notes); E indicates a species that is expected to occur, with (r) indicating rock-dwelling species that have been or are likely to be found very locally along the river where appropriate habitat abuts the riparian zone. Species less likely, but possible, are indicated with a question mark (?). Non-native species are denoted by (NN), and dangerously venomous ones by (*). The specimen of *Gyalopion canum* from the lower basin was lost (Mayne 1985 and personal communication). The table also provides a listing for the upper San Pedro Valley outside the river environs.

[1]Apparently extirpated in the U.S. reach but might still occur in Mexico.

[2]See text. The local subspecies of this form is the Arizona Whiptail (*A. i. arizonae*), sometimes regarded as a full species.

[3]Not confirmed in the river herpetofauna but may occur in upper basin in Mexico, which has not been extensively surveyed.

The only riparian ecosystem in Arizona with similar herpetological species richness is the Santa Cruz River (Santa Cruz and Pima counties), with about 68 documented species. Ciénega Creek, a smaller tributary to the Santa Cruz, which has fewer total species, also has fewer introduced species and retains the closest approximation of the original lowland ecological situation (Rosen et al. 2005). By way of contrast, the Colorado River below Lake Mead supported about 46 species of herpetofauna (Vitt and Ohmart 1978, Stebbins 2003), and the lower Gila River below Phoenix supported about 45 (Rosen et al. 2005).

Thirty-five percent of the San Pedro herpetofauna are riparian-obligate species, i.e., species whose regional occurrence depends upon the riparian corridor. This value is similar to estimates for the entire North American desert herpetofauna (37 percent; Lowe 1989). With an additional nine or so species that are preferential-riparian (i.e., more abundant in the riparian zone than the surrounding uplands), around 49 percent of the herpetofauna of the San Pedro River Valley is riparian associated. This value is lower than the 70 percent figure for riparian biota cited by Naiman et al. (1993) in a global review, and much lower than the riparian-obligate proportion for birds in southwestern lowland systems (Johnson et al. 1977).

Of the riparian-obligate species along the San Pedro, nine, or 14 percent, of the herpetofauna are aquatic species that depend on perennial water. This aquatic group includes three turtles, two snakes, and three amphibians (table 9.1). Additionally, seven amphibians at least historically depended on the stream and bottomland for breeding, and at least six "squamate" reptiles (lizards plus snakes) are largely riparian-restricted in the San Pedro River Valley. The amphibians of this region are generally riparian-obligate species; yet some of them, especially summer-breeding toads, have become widespread across surrounding upland as cattle ponds have proliferated and are now less dependant upon the river floodplain than they were originally.

Most reptile species in arid regions are more closely affiliated with terrestrial rather than aquatic ecosystems. The San Pedro reptile assemblage includes high- to mid-elevation upland species that follow their more mesic habitats downgradient along the riparian corridor into landscape environs that are otherwise too hot and dry for them. Many of the locally riparian-obligate herpetofauna that also occur in mountain canyons or other highland environments do not require perennial water but rather require the dense vegetation and high humidity and soil moisture that, at low elevations, are found only along the river.

Rare and Common Species

Many species in the San Pedro herpetofauna have declined in abundance. These include the native ranid frogs (Chiricahua and Lowland Leopard Frogs), the stream-breeding Woodhouse's Toad, Desert Grassland Box Turtle, and the Sonoran Mud Turtle (Corman 1988). Aquatic snakes, including the Mexican Gartersnake, were once abundant but now also have become rare (Rosen and Schwalbe, unpublished reports).

Conspicuous, active, desert generalist snakes with wide geographic distributions, such as the Coachwhip, Western Diamondback Rattlesnake, and Gopher Snake, tend to dominate the observed snake fauna of the San Pedro and other lowland riparian areas (Corman 1988, Rosen 2005, Rosen et al. 2005). Small, secretive snakes, such as the arthropod-eating Smith's Black-headed Snake and snake-eating Sonoran Coral Snake and Ring-necked Snake, are diverse (totaling about 10 species along the San Pedro) and doubtless are more abundant than sightings indicate (see Rosen 2005). Abundant lizards include the Desert Grassland Whiptail and the arboreal Clark's Spiny, Southwestern Fence, and Tree Lizards. The Lesser Earless Lizard is locally and regionally abundant in grassy uplands adjacent to the river (Corman 1988, Rosen 2005).

Ecological Mechanisms

Too little is specifically known about the ecology of most of the herpetofauna to fully explain the mechanisms producing observed patterns of abundance, distribution, and temporal change. Detailed field studies, quantitative monitoring records, and associated experimental observations are sorely needed. Nevertheless, some general ecological explanations can be offered here. As is the case for many animal taxa (see chaps. 6 and 8), the diversity of environments and vegetation types along the river corridor (see chap. 1) contributes to the high species diversity of the herpetofauna.

Diversity and composition are also enriched by the additions of species found by going longitudinally downstream in elevation. Both upland and riparian conditions change from the uppermost basin in the United States to the lower reaches. Associated with these changes and elevation, there are species transitions (table 9.2; usually gradual replacements of one species by a related and ecologically similar one) in horned lizards, whiptails, and likely other taxa as the plant communities become increasingly open and arid and the character of the stream itself becomes more desert-like and prone to flood scour. Although these examples involve upland, rather than aquatic or riparian species, there was originally a similar transition from the now-extirpated Chiricahua Leopard Frog (in the upper portion of the upper basin near Hereford) to the Lowland Leopard Frog (which was at Lewis Springs but now persists mainly, and perhaps exclusively, downstream of Benson).

MACROHABITAT OVERVIEW

No species of the southwestern herpetofauna depends primarily on cottonwood/willow forests. The high vertical structure provided by these trees is more accessible to mobile species like birds that may compete with, and prey on, reptiles and amphibians. Riparian gallery forests do, however, support substantial diversity and abundance of trunk- and ground-dwelling lizards and other herpetofauna that thrive in the highly productive and relatively moist environment. Corman (1988), in a study along the San Pedro, reported 19 herpetofaunal species as included in, but not restricted to, cottonwood/willow riparian environments.

Notably, however, it is the less-mesic environments, such as mesquite bosque, mesquite scrub, semi-desert grassland, and desert scrub, that tend to have the highest herpetofaunal species diversity (Jones 1981a, Corman 1988). For the San Pedro River, habitats with mesquite had the highest herpetofaunal species richness (27 species), closely followed by Chihuahuan desert scrub (26 species; Corman 1988); highest relative abundance (i.e., pitfall trap success) was found in "mixed grass-mesquite" and "sacaton-mesquite." Only 11 species were characterized as aquatic. These findings, along with Corman's species list (43 species; see Hanson 2001, pp. 180–183), agree well with broader habitat survey results at Ciénega Creek (Rosen et al. 2005, Rosen and D. J. Caldwell, unpublished data) and San Pedro River (Rosen 2005).

Ciénegas generally support the strongest populations of the arid Southwest's most aquatic herpetofauna. Exceptions to this include two primarily stream-breeding toads, the Woodhouse's Toad and the Arizona Toad (*Anaxyrus* [formerly *Bufo*] *microscaphus*), and the Narrow-headed Gartersnake (*Thamnophis rufipunctatus*), only one of which—the Woodhouse's Toad—occurs in the San Pedro.

MACROHABITAT PATTERNS BY SUBGROUP

Macrohabitat patterns vary among subgroups of the herpetofauna. Reptiles often reach maximal diversity in rich desert scrub environments (Jones 1988a); lizards are especially successful in arid environments that consist of stands of small trees with an abundance of low to medium shrubs and dense herbaceous and vine-thicket cover (personal observations). Amphibians, in contrast, achieve highest diversity in the cottonwood/willow association (K. Jones 1981a). Prior to the widespread construction of cattle ponds and road embankments, most (probably all; see Lowe 1989) amphibians in the San Pedro River Valley depended primarily on the river and its floodplain. A group of species, including Couch's and Mexican Spadefoot Toads and two or more species of bufonid "true toads" (family Bufonidae), breed on the river floodplain. In the Southwest, as elsewhere in the North American Desert (Lowe 1989), these ephemeral-water-breeding anurans join permanent-water aquatic species to produce unique regional aquatic herpetofaunas. These endemics were mostly dependent on lowland river corridors, level floodplains of un-incised larger washes, and playas (shallow, seasonal lakes on very flat, broad valley floors). Along with the Mexican Gartersnake and Lowland Leopard Frog, which are characteristic of lowland streams and ciénegas, the summer rainpool-breeding frogs and toads represent the most endemic and riparian corridor–dependent group of herps in southwestern lowland environments like the San Pedro.

MICROHABITAT

Herpetofaunal diversity, composition, and abundance are influenced by microhabitat features such as soil, rock, and debris substrata; understory and overstory vegetation density and type; and the presence of ephemeral or perennial water. Key substrata for riparian reptiles in the Southwest include brush piles, flood debris, leaf litter, open sand, and rocks (Jones and Glinski 1985). Small rocks and shrubs are important as cover for the Side-blotched Lizard (Warren and Schwalbe 1985), as are cobble bars for body-warming of Western Fence Lizard (*Sceloporus occidentalis*) in a riparian ecosystem in California (Sabo 2003).

The spadefoots and toads may depend on the river for its moist and loosely consolidated soils (which facilitate burrowing) and on floodplain scour pools (for breeding habitat). Key microhabitat features for them are temporary summer puddles and pools and short-lived streamflow segments created by seasonal rain runoff. These waters are characterized by the near or complete absence of fish and low-to-moderate abundance of predatory insects. Adaptations permitting exploitation of these "short hydroperiod" environments are explosive breeding immediately following major rainfall events, high foraging activity and hence growth rates of tadpoles, and short time from egg to metamorphosed toadlet. Along a hydroperiod gradient from ephemeral to perennial, species replacements occur, with tadpole traits matching the habitat. No single species can master all the different hydroperiod-determined environments, and no one environment could sustain all the species.

Many microhabitat features have not been adequately investigated and may be underappreciated. For example, near-ground vegetation structure and ground surface irregularities probably have importance, as do standing and fallen dead wood, treeholes, cracked trunks and hollow boughs, and specific-sized rock crevices. These structures provide the substrata that allow individuals to avoid overheating and to escape predation, thereby permitting safe, and thus efficient, foraging. Overall environmental productivity, and especially that close to the ground, creates abundant food resources for herpetofauna, and this creates the opportunity to obtain sufficient energy and nutrients in a variety of different ways. This is at the root of local species diversity.

Recent Changes in Southwestern Riparian Ecosystems

CHANGES IN WATER AVAILABILITY

Our knowledge of life stage-specific and seasonal refuge habitats for southwestern amphibians and reptiles is limited. Even basic facts such as the duration of aquatic larval stages remain poorly known for most species, and the effects of habitat quality, predation, competition, over-wintering, and flood- or drought-refuges all remain essentially blank spots in our knowledge base for most of the herpetofauna.

Although hydroperiod is a major force that structures assemblages of amphibians (Collins and Wilbur 1979, Sexton and Phillips 1986, Werner and McPeek 1994, Relyea and Werner 2000) and defines which species flourish in a region, this issue has received little research or management attention in the Southwest. Mean and minimum larval periods are only sketchily known, even for many North American anurans (= frogs and toads, collectively). On the San Pedro, there are species like Couch's Spadefoot that can go through the larval stage from egg to toadlet as quickly as 7.5–8 days (Newman 1989), allowing them to breed in puddles as well as in pools on the floodplain. Other species take longer. For example, most bufonid toads are thought to take about 20–40 or more days (see Lannoo 2005). Native leopard frogs live primarily in perennial water and require at least two to three months to reach the froglet stage. Non-native bullfrogs need three and one-half months to two years to attain this same stage and thus are highly dependant upon permanent water and limited flood scour for successful establishment (Sartorius and Rosen 2000).

Humans affect southwestern wetland hydroperiods for summer-breeding toads in two critical ways. First, some of our actions have reduced water availability and shortened hydroperiods. Many streams have become incised, and this has decreased channel-floodplain connectivity, reducing the availability and hydroperiod of periodically inundated off-channel breeding pools. Additionally, most desert lowland rivers have been desiccated to greater or lesser extents, eliminating perennial water species and influencing the abundance of others. Sabo and Power (2002) demonstrated that aquatic insect emergence from a river can be critical for the feeding and energetics of terrestrial lizards near the stream edge, at least in areas with relatively sparse vegetation. In drier reaches of the San Pedro, streamside vegetation is less abundant (see chap. 1), and this hydrologic condition may result in fewer insects being available to sustain lizards.

Second, human actions have lengthened the hydroperiod in some areas. We have created a myriad of stock ponds and other water-retaining features across environments where such medium- to long-hydroperiod water sources were historically scarce. In many instances, the hydroperiod is so lengthened that predation and competition increase and can eliminate species adapted to the less-intensive species interactions usually associated with short hydroperiods. By providing refuges from drought and scour, humans have helped bullfrogs flourish and thus have greatly exacerbated their impacts on native herpetofauna.

CHANGES IN RIPARIAN VEGETATION: DECLINES IN MARSHLANDS

Over the long term, the San Pedro River channel has incised, and the riparian vegetation has shifted from primarily herbaceous species to primarily woody vegetation (see chaps. 1 and 12). The great reduction of ciénegas, with their pools, moist ground, and myriad of microhabitat features, has likely had major impacts on many species. Both species of leopard frogs, the Sonoran Mud Turtle, Slevin's Bunchgrass Lizard, Mexican Gartersnake, Desert Grassland Box Turtle, Madrean Alligator Lizard, Giant Spotted Whiptail, and Ring-necked Snake all may have been more common along the San Pedro River prior to the historic period of river incision, when ciénega-like conditions and sacaton grasslands predominated.

Box turtles thrive in environments where they may seek out cool sites in hot weather and where insects, earthworms, and other foods are abundant on or near the ground surface. Reduction of the ciénegas, combined with heavy grazing of adjacent grasslands, probably contributed to markedly reduced population sizes of the Desert Grassland Box Turtle (*Terrapene ornata luteola*; a Southwestern endemic) in parts of southeastern Arizona.

The Mexican Gartersnake is a lowland reptile primarily dependent on ciénegas in Arizona (Rosen and Schwalbe 1988). It prefers dense cover (Rosen 1991), characteristic of ciénegass, and when water conditions change and introduced species invade, it is likely replaced by the less aquatic, but otherwise similar, Checkered Garter Snake (Rosen and Schwalbe 1988, 1998, 2001). The Mexican Gartersnake exemplifies species that are "suffering from a thousand cuts" rather than primarily a single cause of decline: these causes include habitat degradation (including desiccation of stream reaches); habitat change (including conversion of flowing waters to ponds and deep pools where exotics thrive); loss of its main prey (native leopard frogs, which are eliminated by bullfrogs and introduced fish) and reduction of other important prey such as medium-sized native fishes (which occurs when predatory non-native fishes invade); and direct predation by bullfrogs. The current existence of the Mexican Gartersnake on the San Pedro is tenuous.

CHANGES IN RIPARIAN VEGETATION: INCREASES IN WOODLANDS

As suggested above, most species of herpetofauna in Arizona are not dependant on cottonwood/willow forests but do extensively use resources provided by mesquite woodlands. Arboreal species, especially spiny lizards and the Tree Lizard, have likely benefited from the increase in woody riparian vegetation along the San Pedro during the past century, but most other species have probably been less positively affected.

The Giant Spotted Whiptail provides an exceptional and complicated example. This large whiptail lizard's size (up to 17 inches total length) gives it "thermal inertia," which allows it to forage deep into the relatively cooler shade of densely vegetated environments after first warming itself in the sunnier spots (Asplund 1974). It is associated with Sonoran Desert region thornscrub (semi-arid, dense, subtropical brushland), but early habitat degradation has largely restricted it to canyon habitats in Arizona (Rosen et al. 2002). This species should benefit from the development of dense scrublands and gallery forests where grasslands once stood, as in the San Pedro. Yet the rapid pace and frequency of ecological changes associated with human land use act against it for a remarkable reason: it is not a "weedy species," as some other local whiptails have been called (Wright and Lowe 1968). Several of these occur in the San Pedro region, and two are all-female (parthenogenetic) species, which produce only female eggs (see Wright 1993). These all-female species colonize rapidly (only one is needed, rather than a male and female), and their populations can expand much more rapidly than other lizards, because every offspring is a female that can produce eggs and offspring of its own. The two parthenogenetic species—the Desert Grassland Whiptail and Sonoran Spotted Whiptail—numerically dominate along most streams in semi-arid southern Arizona today, including the San Pedro. When environments fluctuate greatly (as has occurred along the San Pedro in response to factors including historically devastating livestock overgrazing, agricultural disturbance, and the rapid shift from grassy to forested riparian environments), successful colonizers with high potential for population growth (i.e., "weeds"), like the all-female whiptail species, are at an advantage (Wright and Lowe 1968). This may explain why the Giant Spotted Whiptail occurs in only a few lowland riparian areas and suggests that the intense ecological interactions (such as competition) it faces from the parthenogenetic species will slow its spread into newly improved habitat (Rosen et al. 2002). On the San Pedro it has been reported only once (Corman 1988).

CHANGES IN RIPARIAN VEGETATION: SHIFTS IN TREE COMPOSITION

Over recent decades, non-native saltcedar (a small riparian tree) has become established on parts of the San Pedro River (see chap. 5), with little-studied impacts on the herpetofauna. Although some authors have speculated that saltcedar degrades habitat for amphibians and reptiles, particularly specialized species (DeLoach et al. 2000), there is a dearth of published information on the subject (Lovich and DeGouvenain 1998). Jakle and Gatz (1985) found low abundance and diversity in closed thickets of saltcedar along the flow-regulated Gila River compared to Sonoran desert scrub, and Griffen et al. (1989) reported low abundance and diversity in Athel tamarisk stands in Australia. In contrast, P. Warren and Schwalbe (1985) reported high diversity in open stands of desert woodland, including those vegetated by saltcedar. Jones (1988b) and Jones et al. (1985) found higher herpetofaunal diversity at the free-flowing Hassayampa River in central Arizona, which supports mixtures of cottonwood/willow, mesquite, saltcedar, and other vegetation types, compared to more anthropogenically altered sites, with dense stands of saltcedar, on the flow-regulated lower Salt and lower Gila Rivers. These results suggest that dense saltcedar monocultures provide poor habitat for amphibians and reptiles. On the San Pedro, saltcedar is usually in open stands or is a component of a floristically diverse riparian corridor, and in these circumstances, its significance is unknown. It might be having negative, positive, or little detectable effect on habitat quality for the herpetofauna.

CHANGES IN VEGETATION: UPLAND AND RIPARIAN GRASSLAND CONVERSION

Degradation of grasslands, and their conversion to woody scrub, has been widespread in the San Pedro River Valley (Turner et al. 2003) and may have eliminated some reptile species. The San Pedro River Valley experienced a major and relatively sudden transition from arid grassland to scrubland in response to livestock grazing, drought, and fire reduction during the late 1800s (Turner et al. 2003). Grassland species of long tenure, such as the Little Striped Whiptail, may have declined and disappeared rapidly when the landscapes became denuded (Wright and Lowe 1965). However, Rosen et al. (1998) argued that original locality data for this species may have been in error, and the issue of its natural distribution and subsequent regional decline remains to be convincingly resolved. It still persists, although it is declining, in parts of New Mexico that underwent similar grazing impacts (C. W. Painter, personal communication, 2002).

The Massasauga Rattlesnake is a Plains grassland species that was first reported in Arizona from the Fort Huachuca parade grounds. Its mapped occurrence on the upper San Pedro (Lowe et al. 1986), though based on unvouchered observations, suggests that it contracted toward the river bottomlands, and probably it has subsequently been extirpated from the valley. This small rattlesnake has become progressively less widespread in southeastern Arizona based on its recorded occurrences about 100, 50, and 10 years ago (A. T. Holycross, unpublished data). The likelihood is high that suitable habitat existed near the river in the upper San Pedro, continuing into the mid- to late twentieth century near the U.S.-Mexico border in the vicinity of Palominas and Hereford. Perhaps the species could yet be relocated, or re-established, during the present century.

Another species that may have disappeared when the grassland was razed is Slevin's Bunchgrass Lizard (*Sceloporus slevini*). H. Smith et al. (1998) demonstrated that this often-montane species is sensitive to drought and grazing at lower elevations. Its occurrence in the Sonoita Grasslands (in Pima and Santa Cruz counties) suggests that suitable habitat existed in grasslands near the San Pedro River. At Empire Ciénega (Pima County, to a minimum elevation of 1613 m), populations have continued to thrive in moderately grazed sacaton, but the population is centered on ciénega-stream margins in close proximity to sacaton-mesquite savannah bottomlands. The near-total loss of ciénega, along with overgrazing and replacement of sacaton grasslands by agricultural fields, may have eliminated this specialized lizard from the San Pedro, though it might do well if introduced there now.

CESSATION OF LIVESTOCK GRAZING

Cattle grazing has demonstrable impacts on herpetofaunal abundance (Jones 1981b, Szaro et al. 1985, Bock et al. 1990, Fleischner 1994), and parts of the San Pedro have seen heavy grazing, followed by recovery after its cessation (Krueper et al. 2003). Grazing can reduce forage for insects, creating resource scarcity for lizards and some small snakes. It may also reduce forage for small mammals, thus decreasing the food supply for many other snakes that feed chiefly on lizards and rodents. Such reductions in energy and nutrient supply would reduce growth and reproductive output, assuming these animals are resource limited rather than primarily cover- and predator-limited.

Grazing can decrease or eliminate key substrata, such as low ground cover of grass, forbs, small shrubs, and leaf litter, that facilitate predator escape and permits safe, efficient foraging (Szaro et al. 1985). It is likely that for most species both forage reduction and cover loss are important impacts that interact via behavioral mechanisms: foraging activity is likely to decline when risk is higher and rewards are lower. On the San Pedro, the species most likely to be affected by overgrazing are those that depend on marshland vegetation or dense stands of mesic riparian grassland such as sacaton-dominated grasslands. Examples include Slevin's Bunchgrass Lizard, Mexican Gartersnake, Desert Grassland Box Turtle, and Madrean Alligator Lizard. Species that have high energy demands and require dense cover such as Giant Spotted Whiptail, Sonoran Whipsnake, and Coachwhip also may be grazing sensitive. Cessation of grazing also, of course, can have indirect effects by causing population changes in a species' predators and competitors. However, there are no ongoing studies monitoring and experimentally exploring how such changes affect the herpetofauna.

Although many arid-land lizard species decline in abundance under heavy grazing (Jones 1981b), cessation of grazing in the San Pedro Riparian National Conservation Area, with its resultant proliferation of riparian vegetation, has actually reduced the open habitat utilized by other species. Some, such as the Zebra-tailed Lizard and Greater Earless Lizard, are predicted to decline in the upper basin in response to vegetation re-growth. Others, such as the horned lizards, might decline close to the river but persist in open vegetation elsewhere in the valley landscape. Some of the species that are negatively affected by cessation of grazing in the upper basin will remain abundant in the lower basin, with its more open, desert environment.

INTRODUCED SPECIES EFFECTS

For aquatic herps, the introduction of non-native species has had marked impacts via mechanisms of competition, predation, and disease transmission. In addition to eating and competing with native frogs and garter snakes, the introduced bullfrog may act as a disease vector and reservoir (Mazzoni et al. 2003) for emerging diseases that affect the native leopard frogs (Bradley et al. 2002). However, at low and moderate elevations like those on the San Pedro, observations by many herpetologists in Arizona suggest that disease impacts on frog populations may be limited in severity (Hale et al. 2005, Rosen, unpublished data, and personal communications from D. Caldwell, D. Hall, T. Jones, V. Miera, M. Schlaepfer, M. Sredl, and D. Swann).

Predation and competition by bullfrogs, introduced predatory fishes, and crayfish may be major driving forces in declines of most native ranid frogs and garter snakes in Arizona, partly via interactions with habitat modification by humans (Rosen and Schwalbe 2002). The recent range-wide declines and local extirpations of several species, including the Mexican Gartersnake, Lowland Leopard Frog, and Chiricahua Leopard Frog, which have declined or gone extinct in the upper basin of the San Pedro (Corman 1988, Clarkson and Rorabaugh 1989, Rosen and Schwalbe 2001), are correlated with increases in non-native bullfrogs, centrarchid fishes (basses and sunfishes), catfishes, and the northern crayfish (*Orconectes virilis*). These non-native animals eat adults, juveniles, and larvae of aquatic amphibians and reptiles. Some, especially crayfish, exert competitive effects through their broad impacts on aquatic plants, insects, water quality, and substratum (Fernandez and Rosen 1996). Impoundments—either in-stream or near-stream ponds—often support exotic species, most of which are better adapted to such modified habitat than they are to the rigorous drying and flooding regimes that characterize southwestern streams (Meffe 1984, Minckley and Meffe 1987, Sartorius and Rosen 2000).

Summary

The San Pedro River supports a species-rich herpetofauna that is characteristic of several lowland riparian corridors in southeastern Arizona. There are records for 55 species and an expected total of 63 species, including 5 non-natives. This diversity reflects changes in macrohabitats along the 130 mi length of stream and changes that occur laterally from the river to the upland as well as marked local habitat diversity within the river and adjoining bottomland. Similar to some other taxonomic groups (including plants), aquatic and riparian obligates comprise 35 percent of the total species number. Herpetofaunal species along the river and in riparian habitats range from aquatic specialists like ranid frogs to widespread North American desert generalists like the Western Diamondback Rattlesnake. Many of the regional endemic species that occur on the San Pedro are aquatic, including various kinds of toads that breed in temporary summer waters on the floodplain. The biotic communities with the most species and highest abundances for most classes are not the wettest ones, but rather mesquite- and sacaton-dominated and desert scrub habitats. Also important to herpetofaunal abundance are oft-overlooked microhabitat features, such as cracks and crevices, brush piles, dense low vegetation and thickets, puddles, and ephemeral streamflows. Species interactions such as competition and predation emerge as complex interactions with each other and with microhabitat features, but with a few exceptions remain little understood for the herpetofauna.

Human activities have produced some marked changes in the river's herpetofauna, particularly affecting those species that utilize marshes, mesic meadow environments, and grasslands. Humans have altered natural hydroperiods, and these have had known and inferred impacts on the herpetofauna. Incision of the river channel is thought to have reduced the hydroperiod of temporary pools used by several toad species for breeding. In contrast, construction of stock ponds has yielded longer hydroperiods on surrounding upland pastures, benefiting some of these same species, but it has also, in the case of permanent ponds, enhanced populations of introduced species that negatively impact native fishes, amphibians, and reptiles.

A few species of grassland-associated reptiles have probably been extirpated from the San Pedro, including the Massasauga Rattlesnake and possibly the Bunchgrass Lizard and Little Striped Whiptail. A higher proportion of aquatic species has been impacted, and the Chiricahua and Lowland Leopard Frogs and Mexican Gartersnake, which were formerly prominent species, have disappeared from all or large parts of the river. These species seem most strongly affected by competition and predation by the introduced American Bullfrog and by introduced crayfish and large predatory fishes. Habitat features or changes that favor larger areas of deep, still water, and that ameliorate the effects of flood scour, are expected to exacerbate the impacts of these non-native species.

Natural laboratories like the San Pedro River with relatively intact ecological functions can provide an essential observational basis against which we may design, contrast, and evaluate experimental research systems, human-created landscape systems, and intensively managed habitat we might use as recovery systems for some imperiled species.

JEROME A. STEFFERUD,
PAUL C. MARSH,
SALLY E. STEFFERUD,
AND ROBERT W.
CLARKSON

Fishes

TEN

HISTORICAL CHANGES AND AN IMPERILED NATIVE FAUNA

Introduction

The history of fishes in the San Pedro River basin is an exemplar of the discovery and subsequent extirpation of native fishes throughout southwestern North America (Miller et al. 2005, Minckley and Marsh forthcoming). The San Pedro River and tributaries were historically occupied by a fish fauna comprised of seven minnows (family Cyprinidae), four suckers (Catostomidae), a pupfish (Cyprinodontidae), and a livebearer (Poeciliidae), several of which were endemic to the Gila River basin (table 10.1). This fauna ranged in size from the diminutive desert pupfish and Gila topminnow to the huge Colorado pikeminnow.

The San Pedro River Valley was one of the earliest routes of exploration into Arizona, and the river figured prominently in early ichthyological investigations in the desert Southwest. Six species—spikedace, speckled dace, loach minnow, Sonora sucker, flannelmouth sucker, and desert pupfish—were discovered and originally described based on specimens obtained from the San Pedro and Babocomari Rivers (Minckley 1985). Additionally, razorback sucker, roundtail chub, Gila chub, longfin dace, and desert sucker were present.

TABLE 10.1. Records verified by specimens (X) and probable occurrences due to existence of later records (O) of native and non-native fishes in San Pedro River main stem from before 1840 through 2005.

	Years of occurrence									
Species	pre-1840	1841–1860	1861–1880	1881–1900	1901–1920	1921–1940	1941–1960	1961–1980	1981–2000	2001–2005
NATIVE TAXA										
Colorado pikeminnow (*Ptychocheilus lucius*)	X									
Razorback sucker (*Xyrauchen texanus*)	X	X								
Flannelmouth sucker (*Catostomus latipinnis*)	O	X								
Gila chub (*Gila intermedia*)	O	O	O	O	X					
Desert pupfish (*Cyprinodon macularius*)	O	X	O	O	O	O	X			
Speckled dace (*Rhinichthys osculus*)	O	O	O	O	X	O	O	X		
Roundtail chub (*Gila robusta*)	O	X	O	O	O	O	O	X		
Loach minnow (*Tiaroga cobitis*)	O	X	O	O	O	O	O	X		
Spikedace (*Meda fulgida*)	O	X	O	O	O	X	X	X		
Gila topminnow (*Poeciliopsis occidentalis*)	O	O	O	O	O	O	X	O		
Sonora sucker (*Catostomas insignis*)	O	X	O	O	X	X	X	X	X	X
Longfin dace (*Agosia chrysogaster*)	O	X	O	O	X	X	X	X	X	X
Desert sucker (*Pantosteus clarki*)	O	X	O	O	X	X	X	X	X	X
NON-NATIVE TAXA										
Common carp (*Cyprinus carpio*)				X	O	O	O	X	X	X
Black bullhead (*Ameiurus melas*)						X	O	O	X	X
Green sunfish (*Lepomis cyanellus*)						X	O	X	X	X
Western mosquitofish (*Gambusia affinis*)							X	X	X	X
Rainbow trout (*Oncorhynchus mykiss*)								X		
Goldfish (*Carassius auratus*)								X	O	X
Fathead minnow (*Pimephales promelas*)								X	X	X
Yellow bullhead (*Amerius natalis*)								X	X	X
Channel catfish (*Ictalurus punctatus*)								X	X	X
Bluegill (*Lepomis macrochirus*)								X	X	
Largemouth bass (*Micropterus salmoides*)								X	X	X
Brook trout (*Salvelinus fontinalis*)								X		
Threadfin shad (*Dorosoma petenense*)								X		
Red shiner (*Cyprinella lutrensis*)									X	X
Redear sunfish (*Lepomis microlophus*)									X	

Information from Arizona State University (2002), W. Jackson et al. (1987), J. Stefferud and Stefferud (2003), and R. W. Clarkson (unpublished data).

TABLE 10.2. Post-1990 distribution of native and non-native fishes in tributaries to San Pedro River.

Fishes/Water	Aravaipa Creek[1]	Bingham Ciéega	Redfield Canyon	Buehman Canyon	Hot Springs Canyon[2]	Paige Canyon	Babocomari River[3]	Lewis Spring	El Sauz	Rservoir near lg. Zaragoza	Cave Canyon[4]	Los Alisos	Los Fresnos	San Rafael	Las Nutrias	La Cieneguita
NATIVE TAXA																
Longfin dace	X		X	X	X		X		X		X	X	X			
Gila chub			X		X		X						X?			X?
Sonora sucker	X		X		X		X									
Speckled dace	X		X		X											
Desert sucker	X				X		X									
Loach minnow	X															
Roundtail chub	X															
Spikedace	X															
Gila topminnow							X									
NON-NATIVE TAXA																
Western mosquitofish	X	X					X	X	X			X		X	X	X
Quitobaquito pupfish							X[5]									
Green sunfish	X		X			X	X	X	X		X					X
Largemouth bass							X		X	X						
Fathead minnow	X						X			X						
Black bullhead						X	X									
Bluegill							X									
Channel catfish	X															
Red shiner	X															
Yellow bullhead	X															

Sources: Information from Arizona State University (2002), Varela-Romero (1992), K. Young and Lopez (1995), Gori (1995), Campoy-Favela (1996), Girmendonk and Young (1997), Voeltz and Bettaso (2003), J. Rorabaugh, D. Duncan, USFWS, personal communication, and unpublished data of authors.

Note: Tributaries are arranged in order of entry into the San Pedro River, from downstream to upstream.

[1]Includes tributaries Turkey Creek, Deer (Hell Hole) Creek, and Oak Spring Canyon.

[2]Includes tributaries Hookers Hot Springs, and Bass, Double R, and Wildcat Canyons.

[3]Includes tributaries Post and O'Donnell Canyons, Turkey Creek, and Finley Tank.

[4]Includes tributaries Lone Mountain and Sycamore Canyons, and Joaquin and Bear Creeks.

[5]In Finley Tank on National Audubon Society's Appleton-Whittell Research Ranch.

During the past 150 years, native fishes have declined and disappeared from the San Pedro River basin and other drainages in the Southwest. Of the original 13 species in the San Pedro River, only 2 consistently remain in the main stem, and an additional 7 persist in tributaries (table 10.2). Losses are a result of geomorphological alteration due to natural and human-induced factors and introduction and spread of non-native fishes (R. R. Miller 1961, W. Minckley 1987). Natural (e.g., earthquakes, precipitation patterns, climate change) and human-caused factors (e.g., mining, grazing, construction of roads, trapping of beaver (*Castor canadensis*), surface water and groundwater use, urbanization) have caused the channel to incise, surface flows to diminish, and aquatic habitats to shrink and become less diverse. Non-native species introduced for food or sport, forage for sport species, public health, or by accident or illegally are established and have largely replaced the native species in many streams. Since the late nineteenth century, at least 15 non-native fishes have been either deliberately stocked into San Pedro River or invaded from downstream waters. Eleven of those now have viable populations in the San Pedro River and its tributaries.

In this chapter we review the ways in which natural and human-induced factors, compounded by additions of non-native fishes, have decimated the native fish fauna of the San Pedro River basin. We emphasize that whereas habitat improvement is necessary, control of non-native fishes is the paramount action needed to initiate restoration of the native ichthyofauna.

Native Fishes of the San Pedro River Basin

The basin's ichthyofauna is well documented through collections and museum records (Minckley and Marsh forthcoming). U.S. and Mexican boundary surveys during 1849–1855 collected fishes in the San Pedro and Babocomari Rivers that were new to science (Baird and Girard 1853, 1854). Documentation continued in 1904 by F. W. Chamberlain (Minckley 1999b) and into the early to mid-twentieth century during the Hubbs-Miller expeditions (Miller et al. 1991). W. L. Minckley's arrival in Arizona in 1963 initiated investigations of ecology and biology of ichthyofauna of Arizona in general (Minckley 1973), and Aravaipa Creek in particular (Stefferud and Reinthal 2005). More recently, monitoring provided information on distribution and status of the fishes in the main stem and tributaries (Gori 1993, 1997, J. Stefferud and Stefferud 2003, Marsh and Kesner 2004). Investigations in the Mexican portion of the drainage have added to this knowledge base (Varela-Romero 1992, Varela-Romero et al. 1992).

All members of the native assemblage were riverine oriented, but some favored quiet habitats of pools, side channels, ciénegas, and backwaters. W. L. Minckley (1987) divided native fishes into four broad categories based on habitat requirements and associations:

I. Tending to live in large, eroding rivers and associated floodplain habitats;
II. Tending to inhabit perennial, moderate- or small-sized streams of variable erosiveness;
III. Occupying spring-fed or river-associated, aggrading habitats such as backwaters, cutoff pools, or stream margins; and
IV. Ubiquitous and/or variable in habitat use, including occurrences in spatially intermittent systems.

CATEGORY I FISH

"Big river" species (Colorado pikeminnow, flannelmouth sucker, and razorback sucker) were most likely confined to the main stem San Pedro River. Except during spawning forays, their habitat needs and body size probably restricted them from occupying tributaries, although young individuals may have opportunistically exploited smaller habitats. All are long lived and can achieve large to very large size, and all disappeared from the San Pedro River early in recorded history.

Colorado pikeminnow. Colorado pikeminnow is endemic to the Colorado River basin and historically ranged freely among larger rivers from Wyoming to the delta in Mexico. Colorado pikeminnow persists as wild populations only in the upper Colorado River basin above Glen Canyon Dam (Osmundson and Burnham 1998, Mueller and Marsh 2002). High dams block their spawning migrations, and populations of predacious non-native fish have hindered recruitment (Minckley 1991). In the San Pedro River, Colorado pikeminnow occurred upstream to about Fairbank, but were extirpated by the late nineteenth century (Miller 1961), and from the entire Gila River basin by the 1960s (Minckley and Marsh forthcoming). Colorado pikeminnow was an important food source in the region both historically and prehistorically, and Anglo explorers reported using "salmon trout" from the San Pedro River in the mid-1800s for food (Minckley 1965, Minckley and Alger 1968, Davis 1982). Lower San Pedro River incised in the late nineteenth century (see chap. 12), and the disappearance of pikeminnow may have been in response to subsequent habitat changes or loss of strong currents associated with geomorphologic changes caused by incision (Miller 1961, Minckley 1987).

Colorado pikeminnow is North America's largest minnow and the largest native fish in Arizona, attaining lengths approaching 2 m and weights of 45 kg (Osmundson et al. 1997). Individuals exceeding 1 m and 7 kg now are rare, likely a result of blockage by dams of migration routes into warmer, lower basin waters (Mueller and Marsh 2002). Longevity can exceed 40 years (Scoppettone 1988, Osmundson et al. 1997). It is the top predator among native fish in the Colorado River basin and after achieving juvenile size feeds almost exclusively on other fishes. The species can make migrations of several hundred kilometers to spawn in high-gradient canyon-bound habitats (Tyus and Karp 1990, Harvey and Mussetter 1993). Spawning areas are a complex of deep pools, eddies, and rapid-velocity water over cobble substrates, and spawning typically occurs near the end of spring runoff (Propst 1999). Post-hatch larvae drift downstream up to hundreds of kilometers before settling in backwaters to grow (Tyus 1990, Tyus and Karp 1990).

Flannelmouth sucker. Flannelmouth sucker, another Colorado River basin endemic, historically dispersed unimpeded among larger rivers of the Colorado River system. The species remains in the Colorado River basin above Lake Powell and in the Grand Canyon but was extirpated from Colorado River below Lake Mead and the Gila River basin before 1970 (Bezzerides and Bestgen 2002, Mueller and Marsh 2002). A repatriated population has persisted in the Colorado River below Davis Dam, Arizona-Nevada, since 1976 (Mueller and Wydoski 2004). Flannelmouth sucker is documented from the San Pedro River by only one specimen, collected in 1846 between Benson and Fairbank; its abundance at that time is not known but was likely low (Minckley 1987). The species probably suffered the same fate as Colorado pikeminnow and razorback sucker in the San Pedro River, as strong currents and deep pools diminished due to geomorphological and hydrological changes (Miller 1961).

Flannelmouth suckers reach 50 cm or more and live 15 to 30 years (Scoppettone 1988). Adults often are found in tributary mouths of larger rivers, which they ascend in springtime to reproduce (Minckley et al. 1991, Douglas and Marsh 1998, Weiss et al. 1998). Spawning is in early spring with multiple males attending a single female, and eggs are deposited over gravel (Weiss et al. 1998). As with most catostomids, larvae drift downstream and congregate in shallow slack-water areas (Childs et al. 1998). Adult flannelmouth suckers can be found in shallow to deep habitats but prefer deep pools with hard substrates (Bezzerides and Bestgen 2002). Flannelmouth suckers are omnivorous, consuming chironomids, copepods, phytoplankton, and organic detritus (Childs et al. 1998).

Razorback sucker. Razorback sucker is another endemic that formerly moved freely among larger streams of the Colorado River system (Mueller and Marsh 2002). It was used as food by aboriginal peoples and was common in the San Pedro River in the 1800s, where it was marketed for food as "buffalo fish" in nearby Tombstone (Minckley and Alger 1968, Davis 1982). It was extirpated from the San Pedro River by the 1860s, either due to mine spillage (Minckley 1999a) or loss of a sizeable, permanent river with strong currents (Miller 1961).

Razorback sucker can grow to 75 cm and exceed 50 years of age (Minckley 1983, McCarthy and Minckley 1987, Marsh et al. 2003). It is mostly a slack-water specialist, where it filter-feeds on small invertebrates and detritus from the water column (Marsh 1987). Spawning is in swift water over sand and gravel substrate. No nest is built and no parental care is given (Minckley 1983, Tyus and Karp 1990).

Recovery efforts. Colorado pikeminnow was listed as endangered in 1967 (USFWS 1991a) and razorback sucker as endangered in 1991 (USFWS 1998). Critical habitat, designated in 1994 for both, does not include areas within the San Pedro River basin. Recovery plans are available for Colorado pikeminnow and razorback sucker (USFWS 2002a, 2002b). Flannelmouth sucker was a candidate for listing under the Endangered Species Act of 1973 and is included in a multi-agency, range-wide conservation agreement (USFWS 1994a, Utah Dept. of Natural Resources 2004).

Recovery efforts in Arizona for Colorado pikeminnow have included the stocking of hundreds of thousands of mostly larvae and fingerling-size individuals into the Salt and Verde Rivers beginning in the 1980s, and for the razorback sucker, reintroductions into the main stream and larger tributaries of the Gila River both above and below the San Pedro River. This effort has achieved little success (Hyatt 2004), primarily due to predation on stocked individuals by non-native fishes (Marsh and Brooks 1989). An introduced population of razorback sucker was present in an artificial off-channel pond near Dudleyville (Desert Fishes Team 2003). In Arizona, recovery efforts for flannelmouth sucker have been limited to the 1976 successful repatriation of a population in the Colorado River below Davis Dam (Mueller and Wydoski 2004). There have been no recovery efforts for any of these three species in the San Pedro River basin (Desert Fishes Team 2006).

CATEGORY II FISH

Roundtail chub, loach minnow, and spikedace occupy rivers and moderate- to small-sized, hard-bottomed tributaries. In the San Pedro River basin, the three species are now only in Aravaipa Creek.

Roundtail chub. Roundtail chub historically was widespread in the Colorado River drainage (Miller et al. 2005, Minckley and Marsh forthcoming). In the Gila River basin, it now occupies only about 20 percent of its historical range (Voeltz 2002). Roundtail chub was last collected from the San Pedro River in 1931 (Arizona State University 2002) but persists in Aravaipa Creek (unpublished data of authors). Its disappearance from the San Pedro River was likely due to loss of pools (Miller 1961, Minckley 1987).

Roundtail chub is a medium-sized cyprinid that can grow to 40 cm or more (Bestgen 1985) and live 20 years or more (Scoppettone 1988). It inhabits pools, eddies, and swift waters below rapids and riffles. They are opportunistic and show no apparent preferences in foods, consuming a variety of insects, algae, gastropods, crayfish, fish, and even small lizards (Vanicek and Kramer 1969, Neve 1976, Schreiber and Minckley 1981, Bestgen 1985, Rinne 1992). In the absence of Colorado pikeminnow, roundtail chub when large occupies the "top carnivore" position (Minckley 1973, Schreiber and Minckley 1981). Spawning occurs in May when chubs deposit eggs over clean gravel at the head of a riffle. Reproductive success is positively influenced by large flood events, which rejuvenate spawning and nursery areas by churning substrates and removing fine sediments. Abundance of age-1 roundtail chubs is typically higher the year after large winter-spring events than in years following periods with small or no floods (Bestgen and Propst 1989, Brouder 2001).

Loach minnow. Loach minnow is endemic to the Gila River basin, including the upper San Pedro River in Mexico (Miller et al. 2005, Minckley and Marsh forthcoming). The species is extirpated from greater than 75 percent of its historical range and in the San Pedro River basin remains only in Aravaipa Creek (USFWS 1985a). Loach minnow persisted in the upper San Pedro River until the 1950s (Miller and Winn 1951) and until 1961 near Charleston (Arizona State University 2002). Sedimentation of riffle habitats may have contributed to its loss (Minckley 1987), as perhaps did elevated water temperatures (Widmer et al. 2006).

Loach minnow is a small (<60 mm), short-lived (2 years or less), current-loving species, primarily inhabiting interstices of gravel and rubble in shallow, well-defined stream riffles (Rinne 1989, 1992, Propst and Bestgen 1991). Foods are predominantly ephemeropteran nymphs and simuliid larvae (Schreiber and Minckley 1981). Spawning occurs in riffles, where eggs are emitted by the female and then fertilized, retrieved, and affixed in clumps to the underside of rocks by the male (Johnston 1999, Childs 2004). Under experimental conditions, loach minnow exhibited thermal stress when water temperatures rose above 30°C (Widmer et al. 2006).

Spikedace. Spikedace is endemic to the Gila River basin and occurred in the San Pedro River basin into Mexico (Miller et al. 2005, Minckley and Marsh forthcoming) but is extirpated from over 90 percent of its historical range (USFWS 1985b). It was common in the San Pedro River until the 1950s and was last documented near Charleston in 1965 (Minckley 1999a), its disappearance likely due to drying of significant portions of the river (Minckley 1987). In the basin, it persists only in Aravaipa Creek. This species is arguably the most endangered fish in the Gila River basin due to its specialized habitat preferences, need for relatively high baseflows, and vulnerability to competitive and/or predatory non-native fishes (USFWS 1994b). Elevated stream temperatures also may be exacting a toll (Carveth et al. 2006).

Spikedace is small (<70 mm), short-lived (ca. 2 years), and occupies flowing pools generally less than 1 m deep over sand and gravel bottoms below riffles or in eddies. Spawning occurs over sand-gravel substrates with no parental care given (Barber et al. 1970, Propst et al. 1986, Rinne 1991, 1992). Foods are primarily ephemeropteran nymphs and dipteran larvae, but winged adults of these groups and trichopterans also are taken (Schreiber and Minckley 1981, Barber and Minckley 1983). In experimental conditions, water temperatures above 28°C caused thermal stress in spikedace (Carveth et al. 2006).

Recovery efforts. Loach minnow and spikedace were listed as threatened in 1986, and uplisting to endangered is warranted (USFWS 1994b). Critical habitats for both species have been designated and withdrawn twice (1994 and 2000) and were again designated in 2007 (USFWS 2007). Aravaipa Creek was included in all designations for both species, and the 2000 designations included long reaches in both the upper and lower San Pedro River and Redfield, Bass, and Hot Springs Canyons. The current designations include Aravaipa Creek and the San Pedro River between Aravaipa Creek and Gila River. Spikedace and loach minnow have recovery plans, which have been inadequately implemented (USFWS 1991b, 1991c, Desert Fishes Team 2006).

Few on-the-ground activities have occurred for conservation of roundtail chub (Desert Fishes Team 2003). It is included in a multi-agency, range-wide conservation agreement that sets out plans for actions to benefit the species (Utah Dept. of Natural Resources 2004). In Arizona, roundtail chub is a sport species with special angling regulations. Concerns about its decline in the Gila River basin resulted in a petition requesting federal listing, which was denied due to non-biological reasons (USFWS 2006).

All three species have benefited from construction of barriers on Aravaipa Creek to prevent upstream invasion by non-native fishes (Desert Fishes Team 2003). No other recovery actions specifically for any of these species have occurred in the San Pedro River basin.

CATEGORY III FISH

Species that typically occupied relatively stable habitats, such as deep pools in streams, ciénegas, springs, cutoff channels or spatially unstable habitats such as stream margins, floodplain marshes, and backwaters in the San Pedro River basin, included Gila chub, desert pupfish, and Gila topminnow. Gila chub is still present in tributaries of the San Pedro River. Desert pupfish and Gila topminnow were extirpated by the mid- to late twentieth century but have been stocked into a few isolated waters in the basin.

Desert pupfish. Desert pupfish was widespread among main river backwaters, springs, ciénegas, and slow-flowing streams in lower elevations of the Gila River basin, including the Colorado River delta and its overflow sinks, and Salton Sink, California (Miller et al. 2005, Minckley and Marsh forthcoming). It was extirpated from the Gila River drainage but has been repatriated to a few isolated waters therein (Voeltz and Bettaso 2003). Desert pupfish was first collected in the San Pedro River in a marshy canyon near Benson in the mid-1800s. It was not found again until 1950, when it was detected in Mexican headwaters of the San Pedro River above a dam, where it was soon extirpated (Miller 1961, Minckley 1987). Reasons for its extirpation are probably tied to the draining of marshy areas and ciénegass that accompanied arroyo cutting (Miller 1961, Hendrickson and Minckley 1984).

Desert pupfish rarely exceed 30 mm and probably do not live longer than 2 years. Reproductive males are brightly colored blue, black, and yellow-orange and are highly territorial. Dominant males gather on a patch of silt-free bottom and try to lure females to spawn. Although diminutive in size, pupfish are highly pugnacious and aggressively defend oviposition sites from both smaller males and other species (Loiselle 1994). They forage primarily on small invertebrates and algae picked off the substrate and occasionally off their own eggs and young (Schoenherr 1988). They can resist almost any natural environmental extreme known in aquatic systems of the Sonoran Desert, tolerating water temperatures between 7° and 45°C, salinities from fresh water to twice sea water (68 ppt), and oxygen from saturation down to 0.1–0.4 mg/liter (Lowe et al. 1967).

Gila topminnow. The range of Gila topminnow extends south from the Gila River basin to the Río Mayo in Sonora. Historically it was one of the most common species below 1500 m elevation in the Gila River drainage, where it inhabited springs, streams, ciénegass, and margins of main rivers (Miller et al. 2005, Minckley and Marsh forthcoming). Loss of range began early in the twentieth century due to lowering water tables and arroyo cutting (Hendrickson and Minckley 1984). Introduction and spread of non-native fishes, particularly western mosquitofish in the 1930s–1940s, accelerated the decline and is the primary reason for its endangerment (Meffe et al. 1983, Meffe 1985, Minckley 1999b). Currently, fewer than a dozen natural populations remain in the United States, with all but one confined to the Santa Cruz River basin (Voeltz and Bettaso 2003).

In the San Pedro River basin, there were two known natural occurrences of Gila topminnow: a collection from San Pedro River in 1943 about 3 km downstream from Aravaipa Creek (Barber and Minckley 1966) and a population found in 1978 in the outflow of an artesian well 13 km southeast of Mammoth that was extirpated when the pipe was capped the following year

(McNatt 1979). Absence of Gila topminnow from collections from the San Pedro River remains an enigma, as the entire river is well within the elevational range of the species, and suitable habitats would have been present, notwithstanding incision and sedimentation (Minckley 1987).

Gila topminnow is a small (males <35 mm, females <50 mm), live-bearing fish, and longevity is less than 1 year (Schoenherr 1974). It feeds on a variety of small plants and macroinvertebrates. Reproduction may occur any time water temperatures are suitable but typically is during spring through summer. Brood size varies with female size, nature of habitat, and season. Females can store spermatozoa for several months and are capable of superfetation, where two or more groups of embryos develop simultaneously at different developmental stages, with births occurring at about 21-day intervals. Broods consist of 14–49 embryos (Schoenherr 1977). Sexual maturity is as early as two months, and up to 10 broods per year can be produced under laboratory conditions (Schultz 1961).

Gila chub. Gila chub was widespread and locally common in suitable habitat throughout the Gila River basin, where it is endemic (Minckley and Marsh forthcoming). It occurred in the upper San Pedro River until the early 1900s (Minckley 1987), but much of its habitat was lost and only remnant populations in tributaries persist today (DeMarais 1986, Weedman et al. 1996). Reasons for its loss in the San Pedro River are tied to loss of deep pools in ciénega habitats (Minckley 1987). Gila chub were relatively common in the San Pedro River main stem for nearly 50 years after roundtail chub essentially disappeared, which may indicate that habitat changes in the late 1800s resulted in replacement of large and deep habitats favored by roundtail chub with smaller, slower habitats favored by Gila chub (Minckley, deceased, personal communication). Recent surveys in Mexico have failed to find Gila chub in locations where it occurred in the late 1990s (Rorabaugh, Duncan, USFWS, personal communication).

Gila chub is smaller and thicker-bodied than the closely related roundtail chub and is characteristic of deeper pools in small streams, ciénegas, and springs. Females grow to 25 cm, whereas males seldom exceed 15 cm. Life span is four years or more. The species is omnivorous with a significant amount of insects consumed. Reproduction takes place throughout much of the year except the coldest months, and young are found from early spring through autumn. Gila chub is reclusive, hiding in deep water among roots and other cover (Griffith and Tiersch 1989, Minckley and Rinne 1991).

Recovery efforts. Recovery plans are available for desert pupfish and Gila topminnow (USFWS 1984, 1993). Desert pupfish was listed as endangered in 1986 with critical habitat, which did not include waters within the San Pedro

River basin. In the San Pedro River basin, pupfish were stocked into artificial ponds near Dudleyville in the early 2000s and continue to thrive there, but other reintroductions were unsuccessful (Voeltz and Bettaso 2003).

Gila topminnow was classified as endangered in 1967 without critical habitat. During the 1970s–1980s, hundreds of natural and artificial habitats were stocked with this species in a recovery attempt, but most sites failed (Voeltz and Bettaso 2003). Other than an introduced population in O'Donnell Canyon (stocked in 1974 and not found again until 2003 [J. Stefferud and Stefferud 2004]), sites stocked in the San Pedro River basin were unsuccessful (Voeltz and Bettaso 2003). In 2005, desert pupfish and Gila topminnow were stocked into three small springs in Aravaipa Canyon (D. Duncan, USFWS, personal communication), where they currently persist.

Gila chub is listed as endangered due to habitat loss and establishment of non-native fishes throughout much of its range. Critical habitat in the San Pedro River basin includes portions of O'Donnell, Bass, Hot Springs, and Redfield Canyons, and Turkey Creek (USFWS 2005a). It was unsuccessfully stocked into Gardner Canyon in 1988 (Weedman et al. 1996). O'Donnell Creek was treated with a piscicide in 2002 to remove green sunfish, and salvaged Gila chub repatriated (Blasius 2002), where they persist in low numbers. In 2005, Gila chub were reintroduced to Turkey Creek (D. Foster, AGFD, personal communication).

CATEGORY IV FISH

These are primarily generalist species that occupy a variety of habitats in perennial or intermittent streams. Speckled dace, longfin dace, Sonora sucker, and desert sucker are widely distributed in the Gila River basin. Longfin dace and desert sucker persist in the San Pedro River main stem and tributaries, but Sonora sucker and speckled dace are restricted to tributaries. All are experiencing range-wide declines in distribution, and all at one time were on the Endangered Species Act candidate list (USFWS 1994a).

Speckled dace. Speckled dace is a bottom-dwelling minnow typical of swiftly flowing riffles and runs in tiny headwater creeks to large rivers. It ranges widely across most of the western United States and is represented in all major drainages of Arizona (Minckley et al. 1986, Minckley and Marsh forthcoming). However, like the rest of Arizona's native fishes, speckled dace has suffered local declines in distribution and abundance (Fagan et al. 2005). Speckled dace in the San Pedro River were considered rare by 1904. A combination of channel entrenchment, loss of riparian vegetation, rising water temperatures, and declining streamflows likely caused its disappearance from the river soon thereafter (R. R. Miller 1961). The species persists in the San Pedro River basin only in Aravaipa Creek and Bass, Hot Springs, and Redfield Canyons.

Speckled dace is small (<60 mm), with a life span of 2 to 3 years. Whereas closely related loach minnow inhabits interstices of rubble bottoms, speckled dace typically occupies the water column immediately above those substrates (Rinne 1992). Speckled dace spawns in response to summer rains and other substrate-disturbing events, and spawning occurs in gravel riffles where females deposit eggs into nests excavated by the male (John 1963, Mueller 1984). Foods are predominantly ephemeropteran nymphs and dipteran larvae (Schreiber and Minckley 1981).

Longfin dace. Longfin dace is among the most common native fishes in the San Pedro River basin (J. Stefferud and Stefferud 2003, Marsh and Kesner 2004) and other lower-elevation (<2000 m) streams of the Gila River basin (Miller et al. 2005, Minckley and Marsh forthcoming). Its love for shallow, sandy-bottomed streams allows it to persist in areas where most other species (native or non-native) do not (Miller 1961). Despite this adaptation, longfin dace has disappeared from many stream segments in the Gila River basin, especially large rivers (Fagan et al. 2005), and its range in the San Pedro River basin has declined (Stefferud and Stefferud 2003). Longfin dace has the habit of moving both up- and downstream into formerly dry stream reaches during flood events. Although mortality typically results, this behavior can occasionally establish new populations. In addition to the main stem river, longfin dace persists in Babocomari River; Aravaipa Creek; Buehman, Hot Springs, Redfield, and Cave Canyons in Arizona; and El Sauz, Los Alisos, and Los Fresnos in Sonora. Longfin dace was extant in O'Donnell Creek as late as 1997, but none were found during renovation efforts in 2002. However, during post-treatment surveys in 2004, a single longfin dace was found (D. Foster, AGFD, personal communication); its current status is uncertain. Green sunfish have invaded its range in Mexico (J. Rorabaugh, D. Duncan, USFWS, personal communication), and its future there is problematic.

Longfin dace is small (<75 mm) and short-lived (2 to 3 years). It typically occupies moderately shallow waters with moderate velocities over sand-gravel substrates (Rinne 1992). It is omnivorous, consuming both algae and aquatic invertebrates according to availability (Fisher et al. 1981, Schreiber and Minckley 1981). Reproduction can occur throughout the year (Minckley and Barber 1971, Kepner 1982) but typically is in spring to late summer in sandy-bottomed, quiet-water areas along stream margins, where actions by spawning individuals excavate saucer-shaped depressions into which eggs are deposited.

Sonora sucker. Sonora sucker is endemic to the Gila and Bill Williams Rivers drainages of Arizona, New Mexico, and Sonora (Miller et al. 2005, Minckley and Marsh forthcoming). It disappeared from the San Pedro River above

Benson by 1950, likely a result of the filling of deep pools with sand (Miller 1961), and is now functionally extirpated from the lower river also (Marsh and Kesner 2004), its presence supported only by dispersal from Aravaipa Creek. The species occurs in Aravaipa and O'Donnell Creeks, and Hot Springs and Redfield Canyons.

Sonora sucker is large and robust (up to 80 cm, 2 kg) and frequents deep pools in medium-sized streams and rivers between 300 and 2000 m elevation, where it primarily consumes a variety of benthic invertebrates (Schreiber and Minckley 1981, Clarkson and Minckley 1988). Spawning occurs in gravel riffles in late winter or early spring, similar to desert sucker with which it occasionally hybridizes (Clarkson and Minckley 1988). Spawning is typical of cyprinoids, with two or more males and a larger female swimming in a tight circle until all individuals pause and emit gametes. Release of gametes is usually accompanied by agitation of substrates by spawners' fins, which may clean the gravel and bury eggs within the substrate (Reighard 1920, W. Minckley 1981). Larvae of Sonora sucker comprise a major component of stream drift in Gila River basin waters (Bestgen 1985, Remington 2002). Adults were used as food by aboriginal inhabitants (W. Minckley and Alger 1968).

Desert sucker. Desert sucker tends to occupy smaller, higher-elevation streams compared with Sonora sucker, but the two species are broadly sympatric in the Gila and Bill Williams Rivers drainages (Minckley and Marsh forthcoming). Desert sucker remains common in these basins but is locally extirpated from some river and tributary reaches (Fagan et al. 2005). The species persists in the San Pedro River and tributaries but is declining in range and abundance in the main river (J. Stefferud and Stefferud 2003, R. W. Clarkson, unpublished data).

Desert sucker attains >40 cm in length. Habitat is riffles with laminar flow over gravel substrates (Rinne 1992). It is largely herbivorous, scraping algae and detritus off rock surfaces in riffles and runs with specialized cartilaginous sheaths on its upper and lower jaws (Schreiber and Minckley 1981, Clarkson and Minckley 1988). Spawning of desert sucker is similar to that described for Sonora sucker (J. A. Stefferud, unpublished data). As with Sonora sucker, a significant life history feature of desert sucker is its proclivity to enter stream drift as larvae (Bestgen 1985, Remington 2002).

Recovery efforts. None of the four species is federally or state-listed; however, all were recommended for listing due to their declining range-wide status and persistence of threats (USFWS 1994a). Conservation activities in the San Pedro River basin have included repatriation of Sonora sucker into O'Donnell Creek after it was renovated (Blasius 2002) and repatriation of longfin dace into Turkey Creek in 2005, from where it had disappeared after 1993 (D. Foster, AGFD, personal communication). As with many of the other native fishes, these species have benefited from barrier construction in Aravaipa Creek (Desert Fishes Team 2004).

Arrival and Spread of Non-native Fishes

The introduction and spread of non-native fishes is of local, regional, and national significance in natural resource management and conservation biology (USFWS 1980, Fuller et al. 1999, W. Minckley and Marsh forthcoming). Non-native fishes can alter populations and assemblages of native species, as well as ecosystem functioning—a capacity that is well demonstrated in the San Pedro River basin.

At least 15 non-native fishes have been introduced into the San Pedro River basin during the past 100+ years (table 10.1). There now are established populations of nearly a dozen non-native fish species in the main river and tributaries. These species were brought to Arizona mostly from other parts of the United States and a few from Europe (Minckley and Meffe 1987), and their intended destination likely was not the San Pedro River, nor was conservation of native fishes a consideration when they were brought to the region.

Earliest transplants of non-native fishes were to establish reliable food sources for humans; thus, common carp was brought to the Colorado River drainage in the late 1800s and later stocked into ponds at St. David (Minckley 1987). Over the years, other non-native fishes were widely distributed by state or federal resource management agencies, or private parties, to develop or improve recreational angling by adding new sport species or by providing forage for sport fish. Other species were introduced accidentally in shipments of desired species (Fuller et al. 1999), and a few were likely brought in surreptitiously. The presence of western mosquitofish is a result of public health officials assuming native fishes were ineffective at controlling mosquitoes, even though Gila topminnow was an efficient insectivore (Childs 2006).

Among other non-native animal taxa that were brought in or invaded the San Pedro River, several likely affect native fish stocks, including northern crayfish (*Orconectes virilis*), American bullfrog (*Rana catesbeiana*), and spiny softshell turtle (*Trionyx spiniferus*). These species adversely affect fishes through direct predation, habitat alteration, inhibition of spawning, or competition for food and other resources (Boyd 1975, Fernandez and Rosen 1996, Robinson et al. 1998, Bryan et al. 2002, Carpenter 2005).

Mpoame and Rinne (1983) detected no non-native parasites in fishes from Aravaipa Creek. Since then, great Asian tapeworm (*Bothriocephalus opsarichthy-*

dis) has spread to fish in all river basins of Arizona, including the San Pedro (Arizona Game and Fish Department, unpublished data), and is a chronic threat to native fish health (Clarkson et al. 1997). Another non-native, the external parasitic anchor worm (*Lernaea cyprinacea*), is established in the San Pedro River and can adversely affect native fishes in heavy infestations (A. Robinson et al. 1998).

Decline and Loss of Native Fish

The large-scale change in biodiversity of aquatic fauna of the San Pedro River basin in the recent past is a complex, cumulative, and synergistic interaction of human-induced and naturally occurring stochastic factors. Streams are a product of their watersheds, and almost any change within the watershed can affect the stream and its inhabitants. Time frames are highly variable, ranging from immediate (e.g., water diversion) to decades or centuries (e.g., vegetation removal on ridges of upper watersheds). Effects of all change are modified, magnified, diminished, accelerated, or decelerated by interactive factors in complex patterns (see chap. 13). But viewed simply, decline of native fishes in the San Pedro River can be attributed to (1) alteration and destruction of aquatic habitats, and (2) introduction and spread of non-native fishes and other aquatic species (Miller 1961, Minckley and Deacon 1968, Minckley 1987, Deacon and Minckley 1991, Minckley and Rinne 1991, Marsh and Pacey 2005).

Southwestern fishes persisted for millennia in the severe climate and highly fluctuating river systems of southwestern North America. Over that period, geologic and climatic changes likely caused major local and regional alterations in aquatic habitats (W. Minckley 1983, W. Minckley et al. 1986), but the fauna survived into the historic period. However, natural and human-induced events that have occurred since the arrival of European man have significantly affected the ichthyofauna in the San Pedro River basin.

CHANGES IN AQUATIC HABITATS

Prior to the early nineteenth century, most river systems in the Southwest were perennially or seasonally connected by surface water, and fish could freely move between the various watercourses (Rinne 1993). Historical accounts of the San Pedro River suggest that prior to the 1850s the upper river was not incised but meandered through marshy areas and pond complexes constructed by beavers, and the banks of the lower river were plentifully timbered with cottonwood and willow (see chaps. 12 and 19). Aquatic habitats in the San Pedro River and tributaries would, depending on gradient, likely have been a series of large deep pools separated by reaches of shallow, rapidly flowing runs and riffles, interspersed with a diverse mix of shallow to deep, slow- to fast-moving aquatic habitats, including marshes, ciénegas, backwaters, cutoff channels, stream margins, eddies, springs, canyons, and other forms of riverine habitats.

A decrease in the diversity of aquatic habitats probably began in the late 1600s when domestic livestock arrived in the San Pedro River Valley (see chap. 11). The presence of surface water and lush vegetation would have encouraged cattle to spend an inordinate amount of time in the riparian area, where they would have trampled the streambanks, consumed the vegetation, and altered aquatic habitats, particularly those associated with stream margins and in shallow reaches (see Belsky et al. 1999). Trapping-out of beaver colonies beginning in the early nineteenth century and into the twentieth century would have led to the loss of their flow- and channel-stabilizing dams, particularly degrading pool habitats. Channel incision in the late 1800s following an earthquake and several very large floods exacerbated by land uses would have caused massive changes in aquatic habitats, essentially changing the river from one with a diverse mix of aquatic habitats to one of wide and shallow homogenous habitats (Miller 1961, Minckley 1987; and see chap. 12). During this period, Colorado pikeminnow, razorback sucker, and flannelmouth sucker, very large fish that needed deep water and connected habitats, disappeared from the San Pedro River.

During and following the period of river incision, human activities along the river and on the uplands continued to contribute to alteration of aquatic habitats. Mine spills made the water toxic (Eberhardt 1981), grazing continued to destabilize streambanks and prevent reestablishment of vegetation, and timbering, roads, and agricultural development exacerbated sediment movement off the uplands (see chap. 12). Excessive erosion can overwhelm a river's capacity to process sediment, which results in the depth of pools being reduced, coarser substrates being covered and filled with fine sediments, and lateral channel erosion being increased (D. Wood et al. 1990), causing a reduction in abundance, biomass, and biodiversity of native fish assemblages (Shields et al. 1994). Loss of native species in the San Pedro River continued in the first half of the twentieth century, as speckled dace, roundtail chub, Gila chub, and desert pupfish disappeared, likely a result of the homogenization of diverse habitats caused by degradation of the stream channel (Miller 1961, Hendrickson and Minckley 1984, Minckley 1987) and perhaps elevated stream temperatures (Widmer et al. 2006, Carveth et al. 2006).

The mid- to late 1900s saw the advent of deep wells punched into the aquifer of the San Pedro River and extensive urbanization of the uplands (see chaps. 11 and 15). Pumping from deep wells decreased the extent and amount of surface flow in the river and many of its tributaries and restricted the amount of habitat available for fish. In addition to depletion of groundwater, urbanization affected the San Pedro River by changing runoff patterns, increasing erosion and sedimentation, and increasing pollutants (see chaps. 11 and 17). The increase in human population brought with it demands for "improved" recreational angling opportunities in the San Pedro River, which were answered with introduction of a suite of non-native sport fishes. During this period, loach minnow, spikedace, and Gila topminnow disappeared, and the presence of Sonora sucker was severely reduced in the main river (Minckley 1987).

Today, populations of native fish in the Southwest are isolated from each other by stream channels that are fragmented by dams and diversions, and long reaches sucked dry by groundwater pumping (Rinne 1993). The San Pedro River remains undammed, but diversions and groundwater pumping have depleted its surface flow (see chaps. 12 and 19). Aquatic habitat in the San Pedro River is mostly shallow runs over sand with infrequent large pools, and surface water is intermittent and interrupted. Areas of double or braided channel are common. The substrate is deeply embedded with sand and fine gravel, and many deeper water habitats are filled with sand (Velasco 1993, J. Stefferud and Stefferud 2003). Similar conditions exist in the tributaries where human-caused impacts have occurred.

INTRODUCTION OF NON-NATIVE FISHES

Native fishes were profoundly affected by additions of non-native fish to the San Pedro River basin. Non-native species can alter population, community, and ecosystem structure and function through direct predation, habitat alteration, competition for food and other resources, changes in behavior of native fishes, and other effects (Taylor et al. 1984, Moyle and Light 1996, Fuller et al. 1999, Marchetti and Moyle 2001, Marchetti et al. 2004). Precisely how they cause decline and replacement of native fauna is a complex mix of individual, cumulative, and synergistic actions, but the result is unequivocally a loss of native species.

Native fishes evolved in a relatively simple community that included only one obligate piscivore and were influenced by, and adapted to, extreme conditions of periodic droughts and floods typical of the region. They occupied a free-flowing river system characterized by heterogeneous habitats that ranged from mainstream runs and riffles to sluggish pools and ciénegas. Use of available habitats was relatively well partitioned among species (Minckley 1981, Rinne 1992, Childs et al. 1998).

The native fauna persisted for millennia in part because local populations lost to stochastic events were either repopulated by fish from unaffected areas within the basins of the interconnected San Pedro, Gila, and Colorado Rivers or by individuals that survived during periods of extreme conditions by moving temporarily into other habitats (Rinne 1993). Conditions in the

San Pedro River today make natural recolonization of local population losses unlikely (Minckley 1999b, Marsh et al. 2000, Clarkson et al. 2005).

Differences between native and non-native fishes in behavioral patterns, competitive ability, and habitat use help explain the changes that have occurred within the fish community. Native species have behavioral adaptations that allow them to withstand extreme flows, both droughts and floods (Meffe and Minckley 1987, Minckley and Meffe 1987). Few native fishes are particularly cover-oriented, preferring instead to move into the water column away from perceived dangers or disturbances. All have relatively simple life histories. Most provide no or only minimal preparation or parental protection of spawning sites or care of young (Kodric-Brown 1977, Schoenherr 1977, Constantz 1980, Kepner 1982, Minckley and Rinne 1991, Rinne 1992, Weiss et al. 1998, Johnston 1999, Johnson 2002, Childs 2004, Marsh and Pacey 2005). Differences in piscivory and predator avoidance are important factors in native–non-native fish relationships. Most native fishes are predator-naive and co-existed with only Colorado pikeminnow and occasionally roundtail chub as predators. Many show no fear or flight reaction in the presence of predators (Johnson and Hines 1999, Johnson 2002).

In contrast, most non-native fishes are from the species-rich environment of the Mississippi River system where intense competition and habitat-use constraints kept populations in dynamic balance such that co-evolved species were able to co-exist (Pflieger 1975, Trautman 1981, Ross 2001). These species evolved in predominately lentic habitats with deep, slow-moving pools, and where flooding increased volume and discharge but resulted in little increase in velocity or turbulence (Minckley and Meffe 1987). Most species prefer deep pools with abundant in-channel or bank-associated cover and have complex life histories (Unmack and Fagan 2004, Marsh and Pacey 2005). Parental care is given in the forms of nest building and aggressive protection of young. When these species were imported to Arizona's rivers, they were released from population-limiting factors such as interspecific competition and highly specific resource partitioning, which allowed them to flourish (Minckley and Douglas 1991, Douglas et al. 1994).

Most non-native fishes in the San Pedro River basin are piscivorous and capable of feeding on all life stages of the natives. Agonistic behaviors of non-native predators induce energetically costly responses by native species that can result in decreased growth, fecundity, swimming ability, and alteration of habitat use. These extra energy outlays may indirectly contribute to higher levels of mortality (Tollrian and Harvell 1999), especially to young-of-year during winter (Thompson et al. 1991).

The large predators (largemouth bass, green sunfish, yellow and black bullheads) are probably the most destructive to native fishes (Marsh et al.

1989). Abundance and distribution of non-native species vary substantially throughout the basin, although the trend in both parameters is upward. Most non-natives are relatively long-lived (5 to 10+ years), and their adult populations do not fluctuate greatly. But the impact of non-native species on natives is highly disproportionate to their relative abundance in the assemblage (Tyus and Saunders 2000), and they may in fact preferentially consume native catostomids (Schwemm and Unmack 2001). Although native fish may be numerically predominant, each predator will consume many individual prey and can have a significant impact on total abundance of prey species and perhaps even survival of entire taxa in the system (Leslie et al. 2005, Velez et al. 2005).

Even the non-predatory non-native species can impact native populations. They are highly competitive for space and foods and will displace native species into less-desirable habitats (Taylor et al. 1984, Brown and Moyle 1991, Douglas et al. 1994). Of the smaller omnivores, red shiner has been implicated as a factor in decline of spikedace (Abarca-Gonzales 1989, Douglas et al. 1994) and can impact recruitment of native fish (Gido et al. 1999). Fathead minnow and red shiner prey on larvae of native fishes (Ruppert et al. 1993, Brandenburg and Gido 1999). Common carp increase water turbidity by dislodging plants and rooting around in the substrate, and they prey on eggs of other fish species (Taylor et al. 1984).

Flood-dominated hydrographs of erosive tributary streams may contribute to the security of native assemblages there and help keep non-native species from establishing. Streams that are isolated from the main river and have hydrographs and habitats closer to historical conditions are more likely to resist invasion by non-native species (Marchetti and Moyle 2001). However, although characteristic floods still occur and are known to differentially displace many non-native species, they usually do not eliminate non-native populations (Meffe and Minckley 1987, Minckley and Meffe 1987). For example, Aravaipa Creek, the largest tributary, retains seven native fish species and has only limited populations of six non-natives. However, despite large floods, non-natives are increasing in distribution and abundance and may be causing gradual loss of natives (Eby et al. 2003). History demonstrates that over time, we can expect all tributaries unprotected by natural or constructed barriers to upstream fish movement to become dominated by non-native fishes, with few if any native fish remaining (Clarkson et al. 2005).

The ubiquitous, persistent, and increasing presence of non-natives in the San Pedro River basin is a significant impediment to long-term survival of the native fishes. Where non-native fish become established, natives invariably wane or disappear, as has occurred in the San Pedro River and its tributaries (Unmack and Fagan 2004, Marsh and Pacey 2005).

Conservation Actions

Recovery plans for the listed fishes outline a suite of conservation actions that could be applied in the San Pedro River (Desert Fishes Team 2006) and which would benefit the entire assemblage of native species, listed or not. Additionally, numerous restorations of native fishes into both the main river and tributaries have been recommended in state and federal management plans, but the few that have been implemented were limited to tributaries. Attempts to control non-natives similarly have been limited, and restricted to tributaries.

Land acquisition and conservation in the San Pedro River basin by governmental agencies, The Nature Conservancy (TNC), or other public and private entities have been the most significant activities from a general conservation perspective. Significant reaches of stream in the San Pedro River basin are under conservation-oriented management, most notably on the lower and upper main river, Aravaipa, O'Donnell, and Turkey Creeks, Redfield and Hot Springs Canyons, Bingham Ciénega, and Los Fresnos. Passive benefits have included riparian restoration and amelioration of detrimental human activities such as irrigated agriculture and domestic livestock grazing (Krueper et al. 2003; and see chap. 19). Other actions within the basin have secondarily benefited native fishes. For example, mine tailings at Cananea were reconfigured to avoid pollution of the San Pedro River, roads and bridges at a number of sites were reconstructed to minimize or avoid channel impacts, and erosion control in ephemeral tributaries may reduce sediment inputs.

However, habitat protection has yet to demonstrate benefits to native fishes in the San Pedro River. There has been remarkably little in-stream habitat recovery despite striking improvements in riparian condition. Streamside vegetation has rebounded from earlier decimation, but aquatic habitats have not. Ciénegas and beaver ponds once characteristic of the river are rare, substrates are dominated by sand and fine gravel instead of a diversity of materials, riffles are scarce, long sandy runs are common, and infrequent deep pools that are not filled with sand are usually turbid. Other parameters of aquatic habitat types (e.g., depth, length, width, cover) have not substantially changed since the 1988 establishment of the San Pedro Riparian National Conservation Area on the upper river (Velasco 1993, J. Stefferud and Stefferud 2003).

Efforts to protect instream flow have occurred in tributaries, such as Aravaipa Creek, through legal recognition of instream flow rights (USBLM 1998a). Land acquisition has retired irrigation water rights and returned those flows to the river. Ways to prevent desiccation of upper portions of the river from groundwater pumping are being discussed (Hess 2005; and see chap. 21).

Construction of barriers on Aravaipa Creek is inarguably the single most important action to directly benefit native fish that has been taken in the watershed. Barriers do, of course, also stop upstream migration of native fish, so their benefits versus detriments need to be carefully considered in an ecosystem context. However, they are the only way available at present to protect existing populations of native fish, and they are widely used for this purpose, primarily for cold-water species but increasingly for warm-water assemblages also (Carpenter 2005). In Aravaipa Creek specifically, the effects of restricting any potential upstream migration of natives is far outweighed by the benefits of protecting the existing assemblage above the barrier.

Beaver were repatriated into the upper portion of the river in 1999 (USBLM 1998b), an action which will likely prove detrimental to native fishes, primarily due to the presence of non-native fish. The two native fish species remaining in the San Pedro River are primarily riffle-dwellers, a habitat type locally decreased by activities of beaver. The mechanism is straightforward: beavers construct dams, dams impound water to form pools or ponds that back up behind the dam, and flowing stream habitat that was there previously is inundated (Collen and Gibson 2001). Compounding this loss of habitat, the non-native predatory species present do well in pools, where they prey on native fish seeking refuge during drought or low-flow periods (Stefferud and Stefferud 2003).

There are few options for native fish management in the San Pedro River basin. Many of the fishes are extirpated locally and declining range-wide; the entire fauna is biologically imperiled. Most species are federally listed, but efforts to manage or recover these species are rare. The single most important conservation action to benefit native fishes—control or elimination of non-natives—has rarely been implemented, and other actions, while benefiting other resources, do little to improve conditions for native fishes. In fact, in the face of the continuing presence of detrimental non-native species, habitat improvements are likely to contribute to further losses of natives by enhancing conditions for non-natives as well.

Summary

The San Pedro River has experienced a staggering loss of its once-rich native fish fauna. Originally an assemblage of 13 native fish species, all but 2 have been replaced by a suite of non-native fishes in the main river. Perturbations to the watershed and channel and declining streamflows have contributed to the loss of native fishes, but it is the persistent, ubiquitous, and expanding presence of non-native fishes that precludes hope for recovery of the native fauna. Tributary streams retain 9 native species, but not all in any one stream. Even these streams are vulnerable to invasion by non-native fishes and are experiencing declines in number of native species present.

Conservation actions directed toward habitat protection or improvement by themselves are insufficient to recover native fishes. Even if ongoing and future water depletion and habitat alteration issues can be resolved, chronic presence and unceasing invasion of an abundance of non-native fishes in the main river, tributaries, ponds, and stock tanks may prevent even persistence of the remaining native fish fauna and rules out the likelihood of re-establishment of extirpated native species. The most urgent conservation need of native San Pedro River basin fishes is control or elimination of non-native fishes (Marsh and Pacey 2005, Clarkson et al. 2005).

The San Pedro River has been described as one of America's "most endangered rivers" and "last great places" (respectively, American Rivers 1999 and The Nature Conservancy 1991). Its true potential as a "last great place" can never be realized until its endangered native fish fauna is restored.

SECTION 3

The Geomorphic Template

BARBARA TELLMAN
AND GARY HUCKLEBERRY

The Land and the People

ELEVEN

The San Pedro basin has a long and rich human history. The goal of this chapter is to provide a general overview of human activities in the San Pedro River and its watershed. This history provides important context for understanding the changes, described in other chapters, that have occurred in the river's hydrology, geomorphology, and biotic communities.

Chronology of Settlement

EARLY HISTORY OF THE SAN PEDRO RIVER VALLEY

The first evidence of human occupation of the area comes from remains of hunting groups that lived in temporary camps starting about 12,000 YBP (years before present). Archaeological evidence points to human hunting of mammoths that lived there, along with other megafauna such as the saber-toothed tiger and giant ground sloth (Wright and Martin 1967, Haynes and Huckell 2007). After a gap in the archaeological record, evidence of human settlement in the area again appears, showing that people made a gradual transition from a hunting and gathering lifestyle to an agricultural one augmented by hunting and gathering at least 3,000 years ago. The northern parts of the river were influenced by Hopis and Zunis starting about 1100 AD, a period of drought throughout the Southwest when population shifts occurred

(Anyon 1999, Ferguson et al. 2004). These people lived alongside the Hohokam and Tohono O'Odham tribes already in the region. The Sobaipuri were well established by the time the first Spaniards arrived, using irrigated agriculture along the San Pedro River and tributaries such as the Babocomari.

SPANIARDS AND APACHES

The sixteenth and seventeenth centuries were times of major change. Spaniards arrived from the south about the same time Apaches arrived from the north at a time of upheaval in the region. The indigenous civilization throughout the area had declined by the mid-fifteenth century for reasons still debated, including drought, salination of irrigation systems, and overuse of resources. Spaniards introduced cattle, horses, goats, sheep, donkeys, wheeled vehicles, iron tools, and new crops. They constructed a major wagon road, the Camino Real, which traversed the upper basin going from the Rio Grande to the Colorado River and California. Spanish military encampments and missions were short-lived, except in the Mexican portion of the watershed (Officer 1987).

Apaches primarily occupied upland areas, with periodic travels into and through the lowlands mainly of the upper basin. Most of the Sobaipuri Indians (a Piman tribe), unable to compete with the Apaches, moved to the Santa Cruz basin in the 1760s at Spanish instigation. Apaches challenged Spanish and Mexican attempts to farm and ranch the region and did limited farming in addition to hunting and gathering native plants. At the time of American settlement, Apaches strongly resisted incursions into what they considered their territory. After the American Civil War when the American army mobilized throughout the American West, it took less than five years to end Western Apache domination of the region and to send the survivors to reservations in Arizona and elsewhere (Seymour 2001).

THE AMERICAN PERIOD

The first Americans in the area were beaver trappers, followed by "'49ers" moving though the area on their way to the California gold fields. War between the United States and Mexico ended in 1847 when the San Pedro was still part of Mexico. Just six years later, the United States purchased land from Mexico in the Gadsden Purchase, moving the border south to its present location. Significant products of the subsequent border surveys included biological inventories of the areas crossed (e.g., Emory 1858–1859). Following the American Civil War in the 1860s, the U.S. Army sent hundreds of troops to defeat the Apaches and make American mining and settlement possible. By the 1870s, prospectors had established claims in many parts of the San Pedro region. Arrival of the railroad in the mid-1870s increased the pace of development. In the United States, the Homestead Act (1862) and the Desert

Land Act (1877) opened lands for ranching and farming. As the land was surveyed, it became available for purchase at very little cost (Sheridan 1995).

By 1912, when Arizona became a state, gold and silver mining had declined, but copper mining was on the rise. Ranching, copper mining, and farming were the predominant land uses in both the Sonoran and Arizona parts of the valley. After World War II, urbanization began to predominate in the Fort Huachuca area but had little influence in the Sonoran and lower basin parts of the region.

History of Land Uses

BEAVER TRAPPING

Famed trapper James Ohio Pattie passed through the San Pedro area in the 1820s and named this watercourse "Beaver River," reportedly procuring 200 pelts near the Gila River confluence. In the late 1800s, European travelers, prior to floodplain entrenchment, commented on numerous beaver dams and associated ponds (Hastings 1959, Rodgers 1965:16, Bahre 1991). Beavers were nearly extirpated along the San Pedro River by 1900 as they were killed in an effort to drain local wetlands caused or enhanced by their dams, or because they interfered with irrigation systems. They persist today near the Gila River confluence and in the San Pedro Riparian National Conservation Area (SPRNCA), where they were introduced in the 1990s (see chap. 6).

RANCHING

Livestock grazing has a long history in the San Pedro River Valley starting in the 1600s (Dobyns 1981, Bahre 1991). In the early 1800s, the Mexican government established land grants for ranching in the upper basin. These ranches were eventually abandoned due to Apache deprivations, but large feral herds remained behind. In fact, the only violent encounter of the Mormon Battalion during their 1848 sojourn through Arizona was with a band of rampaging feral bulls in the upper San Pedro River Valley!

European ranchers brought in large herds of cattle to the upper and lower basins in the late 1800s (Hastings 1959, Hastings and Turner 1965, Bahre 1991) (fig. 11.1). During decades of heavy rainfall, thousands of cattle were brought to the area, and numbers peaked at different times during the last quarter of the nineteenth century and early twentieth century. When drought years occurred, and forage became sparse, most of the cattle died from starvation, after having eaten everything edible. This in turn led to a search in the early 1900s and again in the 1930s for forage grasses to revegetate the area, with emphasis on a few species imported from Africa (Tellman 2002).

Until recently, most river and tributary banks were grazed and also used for large cattle drives (Krueper et al. 2003). The overgrazing of the past has been significantly reduced, especially since passage of the Taylor Grazing

Fig 11.1. (top; 11.1a): Cattle grazing in the San Pedro on the Boquillas Ranch. No date, twentieth century. Courtesy of Bisbee Mining and Historical Museum Accession 80.141.34. (bottom; 11.1b): Cattle drive north from Mexico in the vicinity of Hereford. No date, 1900–1910. Courtesy of the Arizona Historical Society, Tucson, Buehman Collection 200360.

Act (1934) and development of permit systems for federal and state lands (Rowley 1985). Almost all ranching today in the area is conducted on ranches made up of an assemblage of public land leases and private land. Along the river itself, grazing is excluded from the SPRNCA, The Nature Conservancy (TNC) preserves, and other preserves. Within the watershed, grazing is excluded from Fort Huachuca and urban lands.

AGRICULTURE

Agriculture was another significant source of change in the San Pedro River basin. Irrigated agriculture began in the area more than 2,000 years ago. These early farmers tapped surface water during the summer rainy period, using brush diversion dams that were frequently washed out during floods, but easily replaced, and also utilized terracing to capture surface flow for crops. Spaniards did little farming in the area, except near presidios, especially along the Rio Nutrias, a major San Pedro headwaters stream. Apaches had small farms, harvesting summer runoff.

Virtually all the good locations for farming—broad fertile floodplain with good access to water—were utilized at some time. Irrigation has been essential throughout the region, and has long been the primary water use in the Southwest. When major floods widened and deepened the river channel, it became more difficult to construct dams and diversion ditches through which the water could flow by gravity to the fields.

In contrast to much of the West, modern-day farming along the San Pedro has been conducted by small family farmers, rather than large corporations. The most prominent crops have been forage crops and pasturage. In the early days, Fort Huachuca's many horses required large amounts of hay, so the army contracted with people who gathered wild hay or grew hay for market (Fort Huachuca 1976). Fruits and vegetables were important crops in the nineteenth century, when mining was at its height and importation difficult. St. David, in the upper basin, has been the center of a family farming area since the 1870s. Orchard crops, a wide variety of vegetables, and forage have been, and are, grown. The residents of St. David, and later Pomerene to the north, built diversion dams to capture San Pedro water for agricultural use. Farther downstream, from Redington to the Gila confluence, small family farms similarly have prevailed along the river, again mostly growing forage crops and pasture (Muffey 1938, Larson 1999).

Many of these agricultural fields have been abandoned and lie fallow or have been sold for other uses. In the early twenty-first century, TNC began to acquire farms along the river in order to retire water rights and is making efforts to establish perennial native grasslands on the fields (see chap. 19). Within and adjoining the SPRNCA, more than 2,000 acres (810 ha) of agricultural land has been retired (Fredlake et al. 1993).

MINING

Gold, silver, and other ores. Prospectors found silver ore in the Tombstone area in the late 1870s, and soon hundreds of prospectors flocked to the area, and Tombstone quickly became the largest town in Arizona. Since little water was available at Tombstone, ore processing occurred along the San Pedro River in towns such as Millville and Charleston. As mining proceeded deeper and deeper, however, water was encountered and soon became a serious problem. Huge pumps operated 24 hours a day, moving water into Walnut Gulch, a San Pedro tributary. The ore-processing facilities were moved away from the river partly to utilize this excess water. In 1910, however, the pumps failed and most mining ceased. This deep water source is under consideration today as a way to augment water supplies for the San Pedro (Cook 1987, Shillingberg 1999).

The first gold mining claim in the lower basin was at Tiger in 1881. During World War I, molybdenum and vanadium were extracted. The mine reopened in 1936 to produce lead and zinc, and during World War II it produced vanadium. Tungsten mines and processing facilities operated in the Huachucas and Whetstones in the first half of the twentieth century (Dale and Stewart 1960).

Abandoned mines are a poorly documented source of water pollution in the area (see chap. 17). Use of toxic material such as cyanide and arsenic in processing is well documented, but the long-term impacts on water quality are less well known. Some abandoned mines have become important roosting and nest places for bats (Snow and Castner 1996).

Copper mines. The major copper mining areas were at Cananea, Bisbee, San Manuel, and Winkelman, part of a large copper belt that stretches from Sonora to central Arizona. Large-scale copper mining in Cananea, Sonora, began in the 1880s, and by 1900, the mines and processing facilities, smelter, and the town itself were in full operation. Water came from nearby springs and both surface water and groundwater in both the San Pedro and Rio Sonora watersheds. By 1980 the Cananea District had the largest copper pits in North America. Rather than spend money to renovate an old polluting smelter, the owner, Grupo Mexico, closed it in 1999. By 2000 most mining had ceased. Most of the water supply and water impacts from mining occurred on the Rio Sonora side of the watershed, although some were in the San Pedro watershed.

Prospectors first staked claims in the Bisbee area in 1877, and open-pit mining began in 1917. Entire hills were removed to extract ore, extensively changing the landscape (Newkirk 1966, Epler and Dillard 1981). Mining ceased when Phelps Dodge closed the mine in 1974, and the pit and underground mine are now tourist attractions. Because water was scarce, the old smelter was replaced by a large smelter in nearby Douglas in the Río Yaqui watershed, where water was somewhat more plentiful. Phelps Dodge closed the smelter in 1987 because of the cost of meeting new air quality regulations.

Farther downstream, Magma Copper Company in 1950 purchased lands in the Tiger/Mammoth/San Manuel area and tapped into artesian wells near the San Pedro River from which they piped water 8 km to the town and underground mine. Since the ore was low grade, the resulting tailings ponds grew to more than 11.5 km^2 (Knoerr 1956, Canty 1991). The smelter emitted about 75,000 tons of sulfur annually until replaced in the 1980s. BMP, the most recent owner, closed the mine in 1999, except for the in situ leaching section. The deep mine shafts were left to flood.

Open-pit copper mines and processing facilities are at Hayden and Winkelman near the Gila River and San Pedro confluence. In 1883 the Ray Copper Company installed its first mine and smelter in the lower basin. Over the years this mine continued under different owners and was a major copper producer and employer for nearly a century. The current owner, Grupo Mexico, has land and water rights in the San Pedro and Gila Rivers that affect TNC's San Pedro Preserve (see chap. 23).

FUELWOOD CUTTING

Associated with mining was tree harvest in the watershed and riparian zones to supply mines with fuelwood and timber and the populace with firewood (Dobyns 1981). Wood roads were abundant and fanned out from Tombstone "like the veins of a leaf" (Bahre and Hutchinson 1985). Direct disturbance of the watershed through acquisition and transport of wood to fuel the mines increased erosion and sediment movement into the stream channels (Bahre and Hutchinson 1985). The demand for wood was so great that entire hillsides were denuded of all useable timber, although many trees including oak and mesquite resprouted after harvest. In the Cananea and Bisbee areas, deforestation led to especially severe erosion and flooding problems. Hundreds of check dams were built in the 1920s and 1930s in the hills surrounding Bisbee to help control flooding. Fort Huachuca, too, had an active check dam program (Newkirk 1966, Merrill 1981, Johnson 1988).

Hundreds of people were employed in small-scale charcoal-making operations during the last quarter of the nineteenth century. Large smelting operations contracted with entrepreneurs to obtain much of their supply. Since it took about one cord of wood to produce 35–40 bushels of charcoal, areas around the burning operations were quickly depleted. The charcoal burner would then move elsewhere or bring wood from farther and farther away (Ramenofsky 1984).

From 1860 to 1900, commercial lumbering occurred in the Huachucas, where sawmills processed thousands of trees (mostly Ponderosa pine) annually. The arrival of the railroad led to importation of lumber from elsewhere. Small-scale operations continued, but the great period of lumbering was over by 1900, and the forests have re-grown.

FIRE MANAGEMENT

Fires in the San Pedro River watershed occur in response to lightning strikes, typically in association with early monsoonal thunderstorms. Tree-ring scar studies in five mountain ranges in the San Pedro watershed show that before 1900 small fires occurred at least once per decade (Swetnam and Betancourt 1990, Wohl and Pearthree 1991, Danzer et al. 1996). These small fires tended to reduce fuel load so that large fires were unlikely. Following active fire suppression on the U.S. side of the border, large Arizona wildfires became infrequent but more intense when they did occur (Swetnam et al. 2001). The U.S. Forest Service installed its first fire lookout towers in the Huachucas before 1920 and adopted policies of suppressing fires as quickly as possible. New firefighting techniques and equipment made firefighting relatively efficient. Without frequent fires, fuel loads built up, culminating in massive crown fires in the Huachucas and the Catalinas at the turn of the twenty-first century (Biggs et al. 1999, Swetnam 1999, Swetnam et al. 2001, Allen 2002, Parker 2002).

Historic grassland fires are more difficult to document than forest fires, but they were frequent in early historic times. Again, fire suppression became the dominant policy, especially where the fires threatened human structures. But many ranchers have long advocated natural and controlled burns. As ranches have been divided into smaller and smaller units, allowing fires to burn is more difficult because there are more human structures to protect (Biggs et al. 1999).

WILDLIFE MANAGEMENT AND CONTROL

Increased population, ranching, and farming led to demands for control of wildlife considered detrimental. By 1900 Cochise County was offering bounties for wildlife such as mountain lions, coyotes, and bears as well as "nuisance animals" such as badgers. Fifteen years later, the United States government had developed a program for control of certain predators, rodents, and other unwanted wildlife, including species such as skunks that may spread rabies. The program succeeded in extirpating wolves, grizzly bears, jaguars, and prairie dogs and reducing populations of some felines, but coyote and rodent numbers continually recovered in spite of repeated efforts. Personal hunting was common, especially for deer, bear, and mountain lions. Establishment of a state game and fish agency and hunting regulations starting in the 1920s has limited hunting of species such as bear, turkey, and deer to what is considered a sustainable level (Tellman 2006).

ROADS AND RAILROADS

The U.S. government sponsored road surveys throughout the West to facilitate communication and travel, including the Army of the West, led by Col. Stephen Watts Kearny, that traversed the San Pedro River Valley in 1846 (Jackson 1952). The arrival of animal-drawn wagons with steel-rimmed wheels in 1846 played a role in watershed erosion by cutting soil surface ruts, which further incised during surface water runoffs (Dobyns 1981). Roads were built in, along, and across channels that soon became rutted, capturing and concentrating runoff. In addition to heavily traveled roads that paralleled the San Pedro River, a network of roads that connected mines and mills, isolated communities, ranches, and watering points was constructed. Railroads were established during the last quarter of the nineteenth century and facilitated transport to and from the mines (Myrick 1975). During major floods, road and railway bridges often failed and were replaced by much-wider ones. An unintended consequence of bridges has been adoption of the expansion cracks by bats.

URBANIZATION

The population of the Cochise County portion of the upper San Pedro basin has changed considerably in the past 150 years. The area experienced two major periods of rapid urbanization, one during the nineteenth-century peak of gold and silver mining (fig. 11.2) and another during the twentieth century as subdivision development in the lower basin increased with Fort Huachuca expansion and the area became a desirable retirement and commercial region (fig. 11.3a). Population has steadily increased in the Sierra Vista region since 1955 in spite of mine closings and a decline in Fort Huachuca population since the heyday of World War II. Population of the lower basin has consistently been much less than the upper basin for the last 150 years, and its population today is growing at a much slower rate (fig. 11.3b).

The towns. In the late nineteenth and early twentieth century, towns such as Fairbank and Tiger thrived, but they later became ghost towns as mining declined. Cananea, Bisbee, Tombstone, San Manuel, Hayden, and Winkelman all began as mining towns, usually designed as company towns. Cananea, with a population of about 15,000, was the largest town in the watershed for more than 50 years until the growth of Sierra Vista. Bisbee, San Manuel, and Tombstone succeeded in making the transition from mining towns to towns based on tourism, retirement, or commuting. Of the twin border towns of Naco, Arizona, and Naco, Sonora, the latter grew larger than the former with the growth of maquiladora (twin industrial plant) construction starting in the 1980s.

For more than 50 years, small settlements were located near the entrance to Fort Huachuca. During the Fort population peak of World War II, a small community of mobile homes, bars, and whorehouses grew up there. During the Korean War, the army again expanded and Sierra Vista was incorporated. This area is experiencing the highest rate of population growth in the watershed. Benson began as a Southern Pacific Railroad station in the 1880s and was an important transportation hub that for a while served three railroads. Population rates remained steady until establishment of nearby Kartchner

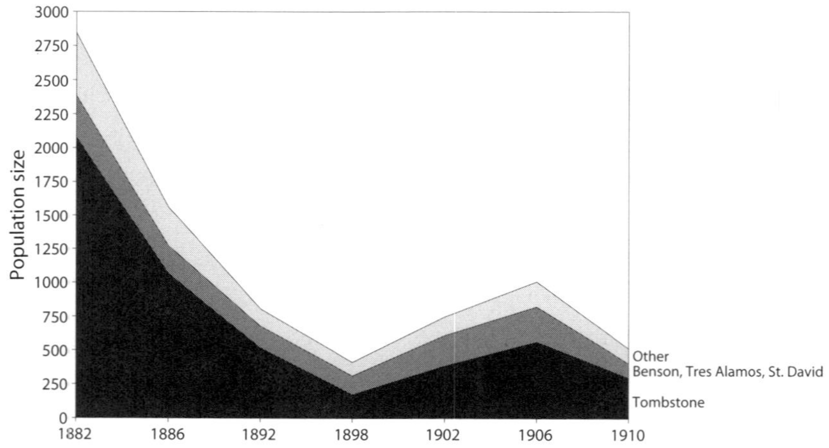

Fig. 11.2. Estimated adult male voting population of the U.S. portion of the San Pedro Valley in the late nineteenth and early twentieth centuries. Estimates derived from Great Registers of Voters. "Other" category includes Fort Huachuca, Huachuca Mountains, and miscellaneous mining towns and other towns.

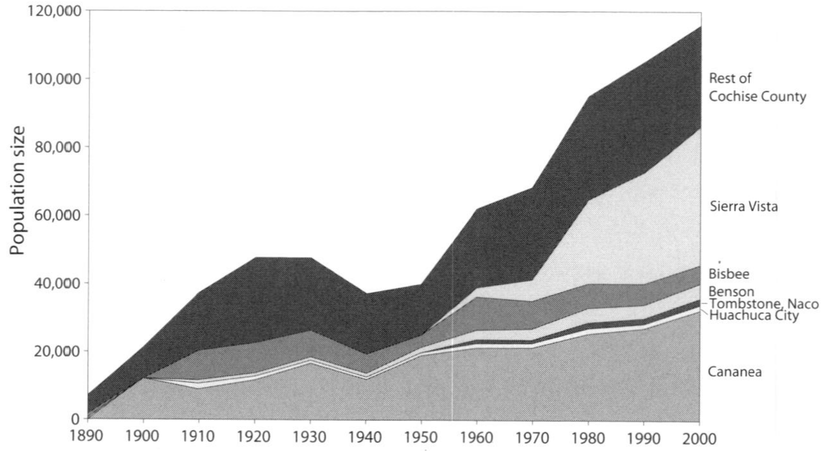

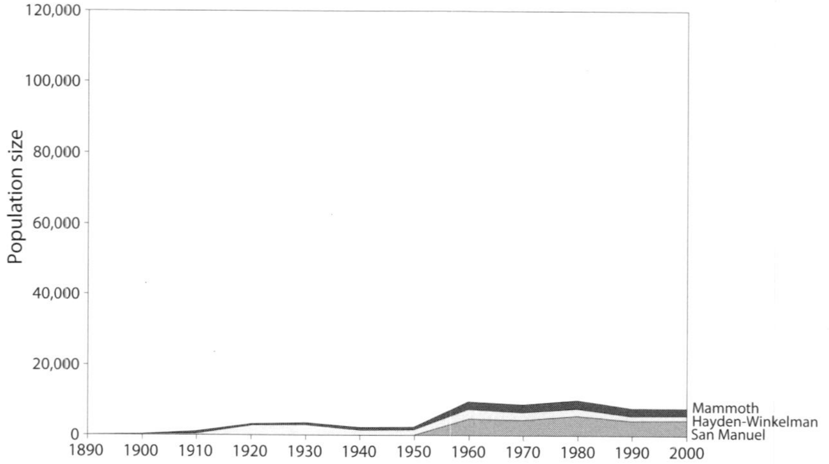

Fig. 11.3. Population of towns in the upper (top; 11.3a) and lower (bottom; 11.3b) San Pedro watershed in the twentieth century. Figures derived from U.S. Census.

Caverns State Park when the area was promoted for tourism and development. The nearby small town of Pomerene is an unincorporated farming community established in the early 1900s (Larson 1999); the Benson-Pomerene area currently is undergoing a population increase.

The region has several unincorporated communities, including Hereford and St. David, a family-oriented farming town founded in the 1880s as a Mormon settlement. Cascabel, the only community between Benson-Pomerene and San Manuel, is on an unpaved road that keeps this ranching area relatively isolated (Muffey 1938). Plans to build a railroad along the river in the 1870s were abandoned, as were subsequent plans for paved roads. In 2006, planners once again proposed a paved road from Benson to San Manuel, leading to concerns over the impacts the road and accompanying population growth would have on the river.

WATER USE AND MANAGEMENT

Conflicts over water rights in the San Pedro basin are documented as early as the 1880s when the first lawsuits were brought to court. The first big dam on the San Pedro was proposed in the 1880s, and a short-lived dam was actually built for mine processing. Demands for a dam increased in the 1920s and 1930s to increase water for agriculture, but the costs were too high (Griggs Engineering 1927, G. Smith 1942). In the 1950s, the U.S. Bureau of Reclamation investigated the feasibility of a dam at Charleston as part of the Central Arizona Project (CAP), a major water project that transports Colorado River water to central and southern Arizona (Johnson 1977) (fig. 11.4). The CAP was authorized in 1968, but the dam was never built.

Another plan for increasing water supply for human use involved phreatophyte removal. From the 1950s to the 1970s, saltcedar as well as native phreatophytes were cleared from several Arizona rivers, with the expectation of increased flow to the rivers and thus downstream for human use. The San Pedro River was studied, but large-scale projects were not implemented (Manes 1968).

Water supplies for towns. All of the mining towns have been dependent on groundwater both for mining and for domestic use. Cananea was unusual among all these towns in not having a central water supply system for all residents. Bisbee did not have adequate potable water supplies, so it developed a central water system dependent on wells in the Naco area. Tombstone's water comes from wells and a pipeline built in the 1880s that brings water from springs in the Huachuca Mountains 42 km across the San Pedro River Valley (Cook 1987).

In the greater Sierra Vista area, private water companies provide water from wells, and many residents have their own wells. The multiplicity of

Fig. 11.4. 1960s artist's rendering of what Charleston Dam and the reservoir would look like if constructed. The "photo" looks south into much of what is now SPRNCA. Photo source: U.S. Bureau of Reclamation, available at University of Arizona Special Collections Library (AZ15).

water providers offers challenges to developing water conservation programs. Benson has had a municipal water company since the 1920s, also using groundwater. Private wells are the norm along the rest of the river, except for San Manuel, which has a municipal water utility.

WASTEWATER TREATMENT AND REUSE

All of the towns in the watershed, with the exception of Cananea, have a community wastewater treatment system. In 1980 Tombstone became the most recent Cochise County town to build a modern wastewater treatment plant. Many residents in sparsely populated areas have their own septic tanks. Inadequate sewage treatment has created water quality problems in Cananea, Naco, and Benson.

Fort Huachuca began to use wastewater on its golf course during World War II and has continued to promote reuse and recharge. Sierra Vista operates a regional wastewater treatment plant, with the water discharged into the regional aquifer. This concentration of wastewater offers potential for coordinated reuse and recharge projects that are underway with a goal of prolonging the supply to protect the river while allowing for population growth (ASL 1995). The Naco wastewater plant has a reuse component, but in the rest of the area, all treated wastewater is discharged to washes or evaporation ponds.

Present Land Ownership and Management

The current border between the United States and Mexico was established by the Gadsden Purchase in 1853. For the first 100 years, the border was relatively open, with people and their cattle, as well as wildlife moving relatively freely between the two nations. Over the past 50 years, the United States has attempted to secure the border and has increasingly attempted to restrict entry, with varied success.

SONORA

Land ownership in the Sonoran part of the watershed is mostly divided between ejidos (communally owned ranches) and private lands. Most of this area formerly was part of a land grant that was subsequently bought by American mining entrepreneur and rancher, Col. William Greene. The ejidos developed from the Greene lands (Longwell 1974).

The Sierra de los Ajos is a largely public preserve in Sonora, Mexico, a portion of which is in the San Pedro watershed. Non-profit groups own land for preservation purposes along Rio Nutrias (see chap. 23).

UNITED STATES

Federal land. Land ownership and management in the San Pedro basin is a complex mixture of federal, state, and private land. The national forests (U.S. Department of Agriculture) operate under a multiple-use mandate to protect the watershed and forest resources and provide for appropriate human uses, including grazing, timber cutting, mining, and recreation. The Coronado National Forest is spread over 12 mountain ranges in southeastern Arizona, including portions of the San Pedro watershed in the Catalina, Huachuca, Rincon, Muleshoe, Galiuro, Winchester, and Dragoon Mountains. Most Bureau of Land Management (BLM—U.S. Department of Interior) property is mandated as multiple-use land. Grazing is the main land use of the BLM land in the watershed (J. Williams et al. 2003)

There are many riparian preserves along the San Pedro and its tributaries, ranging from public to private. Congress designated a portion of land along the river as the San Pedro Riparian Conservation Area (SPRNCA) in 1988, giving it protected status (see chap. 20). The area had been a Spanish land grant and subsequently an American ranch and was about to be rezoned for a large planned development. BLM also manages a riparian preserve in Aravaipa Canyon. The U.S. Bureau of Reclamation oversees Cook's Lake and is a partner in preserving the lower San Pedro River flows (see chap. 19).

Fort Huachuca (U.S. Department of Defense) occupies 30,000 ha in the Huachuca Mountains and on the uplands to the east. This land is used for military purposes and is not open for grazing, but some areas are available to the public for hiking and bird-watching. Portions of the area are managed to protect threatened and endangered species and their habitat.

The Coronado National Memorial (U.S. National Park Service) occupies 1,922 ha of grassland and oak woodland on the south end of the Hua-

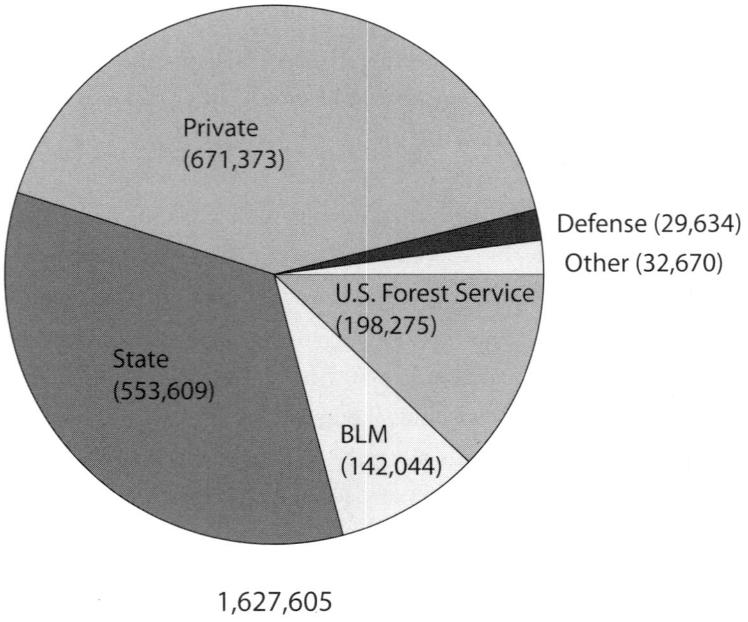

Fig. 11.5. Land ownership in Cochise County (in hectares) in 2000.

chucas. It is managed primarily for its historic value but also for habitat conservation.

State of Arizona. Approximately one-third of the land in Cochise County is held in trust by the Arizona State Land Department, which also holds lands downstream (fig. 11.5). Because state trust lands are constitutionally mandated to be managed for the economic benefit of public schools and other purposes, they may be leased or sold under certain conditions. Most leasing is for livestock grazing, although leases are available for such purposes as sand and gravel operations and mining. Nearly all state trust land is by law subject to sale to the highest bidder (Varady 1991). Over the past decade voters have rejected initiatives to allow for conservation uses. State property also includes Kartchner Caverns State Park, a large, wet cave in the Whetstone Mountains about 11 km south of Benson that hosts a breeding colony of some 1,000 cave myotis bats (*Myotis velifer*).

Natural Resource Conservation Districts (NRCDs) are legal state entities within the Arizona State Land Department whose mission is to "provide for the restoration and conservation of lands and soil resources of the state, the preservation of water rights and the control and prevention of soil erosion, and thereby to conserve natural resources, conserve wildlife, protect the tax base, protect public lands and protect and restore this state's rivers and streams and associated riparian habitats." NRCDs also receive assistance from the U.S. Natural Resources Conservation Service. The San Pedro River Valley includes parts of four NRCDs.

Private land. While the mountainous areas of the watershed are mostly under federal jurisdiction, nearly half of the land at the lower elevations is privately owned. Permissible private land uses are under the jurisdiction of the county boards of supervisors and, within cities, the appropriate city council, operating under state enabling legislation. A countywide comprehensive plan outlines permissible land uses, but rezonings and plan amendments are common in the county and the cities (Cochise County 2002). The lower basin with very little population pressure has less detailed planning efforts but operates under similar laws.

TNC and its partners have been pioneers in preserving land and water, starting with the Ramsey Canyon Preserve in 1974 and continuing to this day along the river and its tributaries from Sonora to the Gila River confluence (see chap. 19). Although the river flows through only a small remote part of Pima County, the county bought Bingham Ciénega along the lower San Pedro, which is managed by TNC.

Summary

Boom and bust cycles were common in the American West, and the San Pedro region was no exception. The silver mining boom of the late nineteenth century that made Tombstone the largest town in Arizona Territory lasted only about twenty years. Cutting of lumber and firewood left many hillsides bare. Grazing cycles prevailed with thousands of cattle thriving on the range, followed by droughts when thousands perished, leading to loss of ground cover, erosion, and introduction of non-native grasses for forage, but since the 1930s cattle numbers have remained relatively stable. The copper boom lasted nearly a century, but copper mining has declined in most of the region, although most mining towns survived supported by new economies. The most recent boom, rapid population growth in the upper basin, has led to decline of the aquifer and water supply for the river.

Some of the impacts on the river were short-lived, while others continue to this day and will continue long into the future. Mining pollution had major impacts for short periods, but the long-term impacts are mostly minimal. Hillsides have largely recovered from deforestation and overgrazing. Channel change from several sources has been long-lasting. The long-term impacts of the twentieth century population boom and large-scale groundwater pumping are unknown.

Attempts to conserve and restore the area began with National Forest formation in the early twentieth century. In the 1980s SPRNCA's creation occurred along with non-profit organization land acquisition and preservation efforts from the Sonoran region to the Gila confluence. Recovery of the river as a result of these efforts along with land use change has been dramatic in parts of the lower and upper basins.

RICHARD HEREFORD AND
JULIO L. BETANCOURT

Historic Geomorphology of the San Pedro River

TWELVE

ARCHIVAL AND PHYSICAL EVIDENCE

Introduction

The need to explain and manage arroyos, or water-carved gullies, in the western United States has been a dominant theme in American geomorphology since the turn of the twentieth century. To date, no single explanation satisfies widespread and almost synchronous arroyo formation around the turn of the century. Is this dramatic episode of erosion unique, or has it repeated itself both in kind and in magnitude during past millennia? Surprisingly, attempts to explain arroyos far outnumber efforts to characterize their initiation and subsequent history.

The San Pedro River is cited often in reference to historic arroyos (Bryan 1925, Antevs 1955, Hastings 1959, Hastings and Turner 1965, Martin 1963, Melton 1965, Rodgers 1965, Cooke and Reeves 1976, Dobyns 1981, Hendrickson and Minckley 1984), but neither the archival nor physical evidence has received more than cursory attention. Unlike the heavily urbanized floodplains along the Santa Cruz River at Tucson, floodplain surfaces and cutbank stratigraphy remain relatively unspoiled along the San Pedro River, particularly in its upper reaches.

When arroyos expanded into the upper San Pedro, they exposed the remains of mammoth in association with Clovis, notably at the Naco, Lehner, and Murray Springs sites. Investigation of these sites has led to an unusually complete record of late Quaternary alluvial history (Haynes 1968, 1987) that contrasts with our haphazard understanding of the more recent floodplain history. We correct for this oversight by evaluating both archival and physical evidence for floodplain evolution before and after historic arroyo cutting on the San Pedro.

In this study, we used archival evidence from the lower and upper basins, but field mapping was limited to the upper San Pedro. The primary objectives of the archival research were to describe general floodplain conditions before arroyo cutting and to establish timelines for major floods and cutting episodes. The physical evidence was marshaled to determine rates and causes of channel widening once the arroyo developed, as a prerequisite for understanding how alluvial channels might progress towards equilibrium after entrenchment (Hereford 1993).

Archival Evidence

Historical studies of environmental change must depend on documentary sources of variable quality. Standard observations made at regular time intervals, such as those obtained at a stream gage or weather station, usually are unavailable for the periods of interest; for example, neither weather nor discharge measurements exist for the San Pedro River during the critical period of arroyo initiation. In the Southwest, the field notes of the cadastral surveys made by the General Land Office consistently record the width of stream channels but mention channel depth only sporadically (both before and after arroyo initiation occurred). Were there significant differences between cross sections where channel depths are mentioned and where they are omitted, or is any reference to depth purely whimsical? Can we infer unincised floodplains where the surveyor failed to mention depth, as Bryan (1928a) did on New Mexico's Rio Puerco? In 1873, Theodore White, one of the first land surveyors in southern Arizona, surveyed the San Pedro River Valley from St. David to just below the Narrows. From the journals of itinerants, we know that the river was entrenched at St. David, Tres Alamos, and below the Narrows (fig. 12.1), with perpendicular banks 3 to 6 m deep as early as the 1850s. White failed to record any channel depths at these same localities (Cooke and Reeves 1976). Was White making a distinction between terraces formed during an earlier erosional episode and active channel depths, a distinction that escaped the itinerants?

Historical sources, such as newspapers, provide descriptions of extreme and rare episodes, most importantly floods. These accounts serve the environmental historian well, because degradation of alluvial stream channels occurs catastrophically during extreme flows. The erosional work done by floods often is described in great detail, as was the case with headcut migration in the Santa Cruz River Valley at Tucson in summer 1890 (Hastings

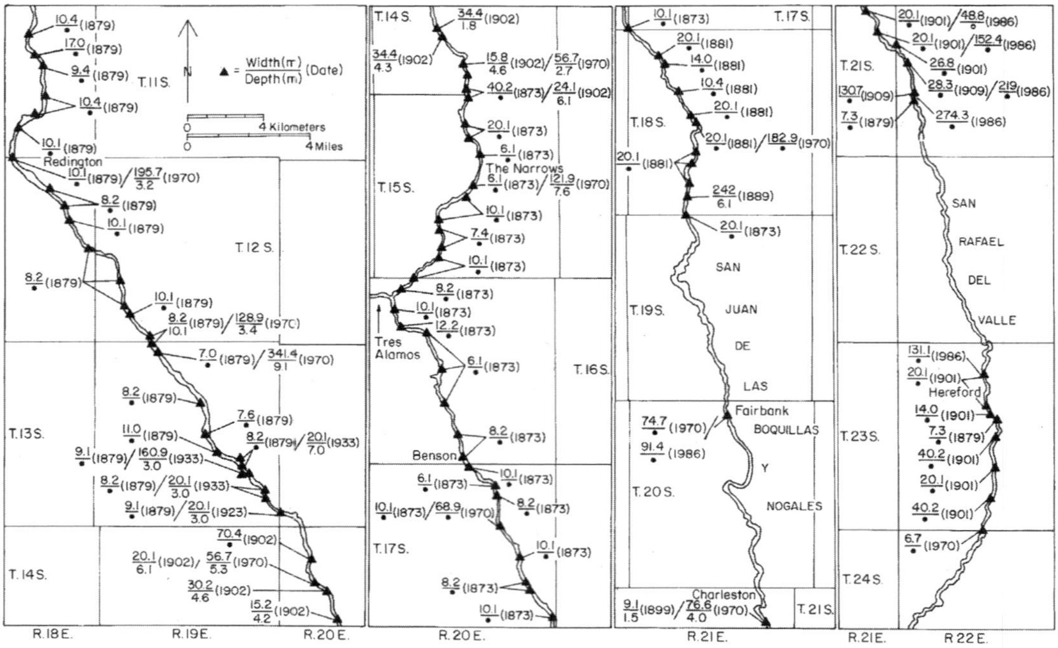

Fig. 12.1. Maps showing width and depth observations from cadastral surveys 1873–1933, amended from Cooke and Reeves (1976).

1959, Betancourt 1990). The degree of detail given in the accounts correlates well with distance to large settlements, and coverage of flood damage is patchy, giving the false impression that some reaches were more afflicted than others.

For the San Pedro River, we relied on a variety of primary and secondary sources. The earliest relevant observations are those related to administration of the Presidio of Santa Cruz de Terrenate, established in 1742 in the headwaters of the San Pedro and moved to Quiburi near Fairbank (fig. 12.2) in 1772 (Kessell 1966). Many of the documents pertaining to this presidio are contained in the Archivo General de Indias in Seville, Spain (Beers 1979). The next period for which documentation exists involves the early years (1820s–1830s) following Mexican Independence, when four land grants—the San Ignacio del Babocomari, the San Rafael del Valle, the San Juan de las Boquillas y Nogales, and the San Pedro (fig. 12.2)—were sought, surveyed, and approved (Mattison 1946). A fifth grant was ceded at Tres Alamos in 1852. For the post–Civil War period, we relied mainly on local newspaper accounts.

PRE-ENTRENCHMENT CONDITIONS

The inability to accurately portray pre-entrenchment conditions has plagued historic arroyo studies. The written record just prior to arroyo formation is patchy and incomplete. In the case of the San Pedro, gaps in written observations can be bridged by stratigraphic records from critical localities.

These records can help resolve two questions about pre-entrenchment

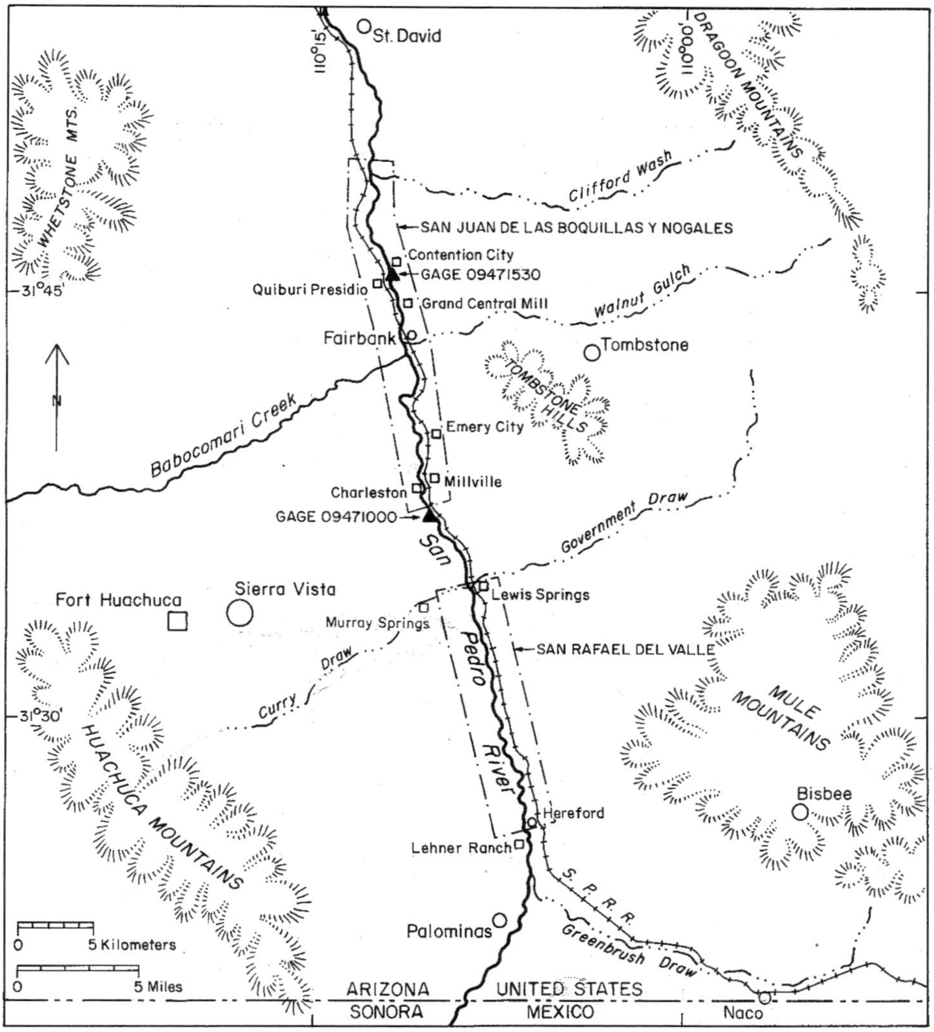

Fig. 12.2. Map of upper San Pedro River Valley showing locations of current and historical features. Illustration credit: Chuck Sternberg.

conditions: (1) which reaches of this interrupted stream had perennial surface flow and which did not, and (2) does evidence exist for unincised floodplains and contemporaneous discontinuous arroyos? In the case of discontinuous arroyos, the evidence could be ambiguous because the observer usually was unaccustomed to making subtle distinctions between inset and superimposed stratigraphic relations between alluvial deposits. Such a distinction is critical to geomorphic interpretation. Figure 12.3 illustrates the difference between inset and superimposed relations. These relations result from two or more cut-and-fill cycles in which the younger longitudinal gradient is the steepest. A superimposed relation is typical of the area upstream of Lewis Springs; an inset relation occurs locally from Charleston downstream to Fairbank (see fig. 12.2 for locations). A steep terrace rise near the river,

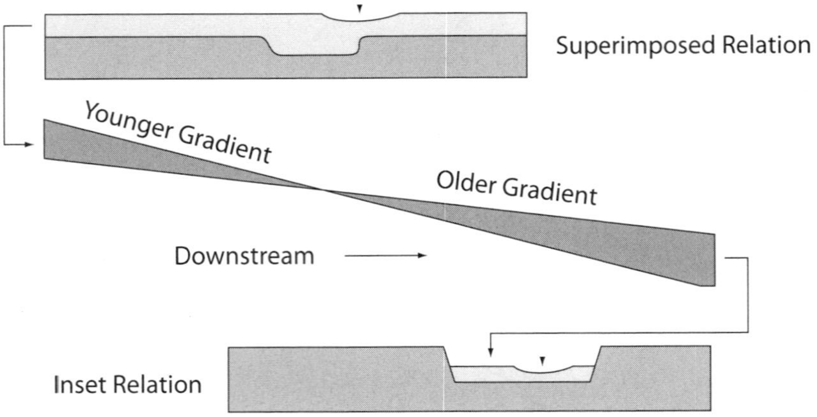

Fig. 12.3. Superimposed and inset stratigraphic relations and their geomorphic expression.

which perhaps had to be "cut down" to help wagons cross the river, could represent an earlier entrenchment, unrelated to current base level, and with an inset stratigraphic relation.

1600s and 1700s. The accounts of Kino, Manje, and Bernal (Karns 1954) in the 1690s and those of Velarde in 1716 contain few relevant observations about channel or flow conditions in the valley, other than the fact that irrigation was practiced on swampy land near Quiburi (near Fairbank).

1820s and 1830s. Attempts to settle the San Pedro increased in the 1820s after Mexican Independence, when several settlers filed for land grants in the bottomlands. This happened at a time when beaver ponds dotted the lower reaches of the San Pedro River Valley (Pattie 1905). The State of the West law, adopted by Mexico in 1825, limited grants to ranchers to 4 square leagues or *sitios* (ca. 2784 ha). The price of a sitio with running water was $60; the price for dry rangeland was $10. In 1832, Don Ygnacio Elias y Gonzales was issued title for eight sitios, six with water, along Babocomari Creek, where he ran about 40,000 head of cattle and a large herd of horses and mules (Christiansen 1983). He and Juan Nepomucino Felix were also granted four sitios with water along the San Pedro (the San Juan de las Boquillas y Nogales Land Grant) in 1833. The Boquillas Land Grant was a narrow strip of land on both sides of the river, from Charleston to just south of Fairbank. In 1833, Rafael Elias Gonzales was granted four sitios, again with water, along the San Pedro. This was the San Rafael del Valle Land Grant with its southern boundary between Hereford and the Lehner Ranch and extending north to Lewis Springs (fig. 12.2). Gonzales also received title to the San Pedro Land Grant, another four sitios along the San Pedro straddling the international boundary. A selling price of $60 for each sitio along the Babocomari and San Pedro Rivers suggests that relevant reaches of these streams were perennial in the 1830s. The San Pedro remains perennial today from Hereford to Fairbank, while the Babocomari contains two perennial reaches, one near the Brophy Ranch headquarters and another just downstream of the grant's eastern boundary (D. Brown et al. 1981).

1840s to 1860s. Accounts during this period generally indicate marshy and commonly treeless conditions throughout the upper San Pedro, with intermittent flow below Tres Alamos and the Narrows and discontinuous arroyos below the Narrows, at Tres Alamos, and near St. David (Hastings 1959, Hastings and Turner 1965, Dobyns 1981, Hendrickson and Minckley 1984). In 1849, Eccleston (1950) noted that below the Narrows, the river "is lined with a poor growth of swamp willow and other brush, so that it cannot be seen until you come within a few feet of it; then the bank is perpendicular." Five years later, in the same reach, Parke (1857:24–26) noted that: "The valley bottom is generally smooth and open, with the streambed curving through it, sometimes a few inches, and at others as much as fifteen feet below the surface of the meadow. At Tres Alamos, the stream is about fifteen inches deep and twelve feet wide, and flows with a rapid current over a light sandy bed, about fifteen feet below its banks, which are nearly vertical. The water here is turbid, and not a stick of timber is seen to mark the meanderings of its bed. In the gorge below (the Narrows) and in some of the meadows, the stream approaches more nearly the surface, and often spreads itself on a wide area, producing a dense growth of cottonwood, willows and underbrush, which forced us to ascend and cross the outjutting terraces. The flow of water, however, is not continuous." Hutton (1859) gave a similar description for the reach just below Tres Alamos in 1857: "The San Pedro has a width of about twelve feet and a depth of twelve inches, flowing between clay banks, ten or twelve feet deep, but below it widens out, and from beaver dams and other obstructions overflows a large extent of bottomland, forming marshes, densely timbered with cottonwood and ash." Another apparent arroyo just below St. David was described by Bartlett (1854) and Graham (1852) of the International Boundary Commission.

LATE NINETEENTH-CENTURY FLOODS AND ARROYO CUTTING

1870s to 1890s. Settlements along the San Pedro were first established in the 1870s, with the arrival of Mormons at St. David and the discovery of silver near Tombstone (Fulton 1966). In 1884, the anthropologist Adolph Bandelier visited ruins along the San Pedro and described the arroyos near Tres Alamos and St. David: "[At Tres Alamos] the river, now rendered muddy by the washings of the mines worked on its upper course near Contention and Charleston, runs in a cut which is from eight to twelve feet deep . . . [at St. David] the river runs in a cut with abrupt sides. This cut is 10 to 15 feet deep, and about 25 feet wide" (Bandelier 1892:475–478). This also agrees with McClintock's (1921) account that the first Mormon settlers encountered an entrenched channel of the San

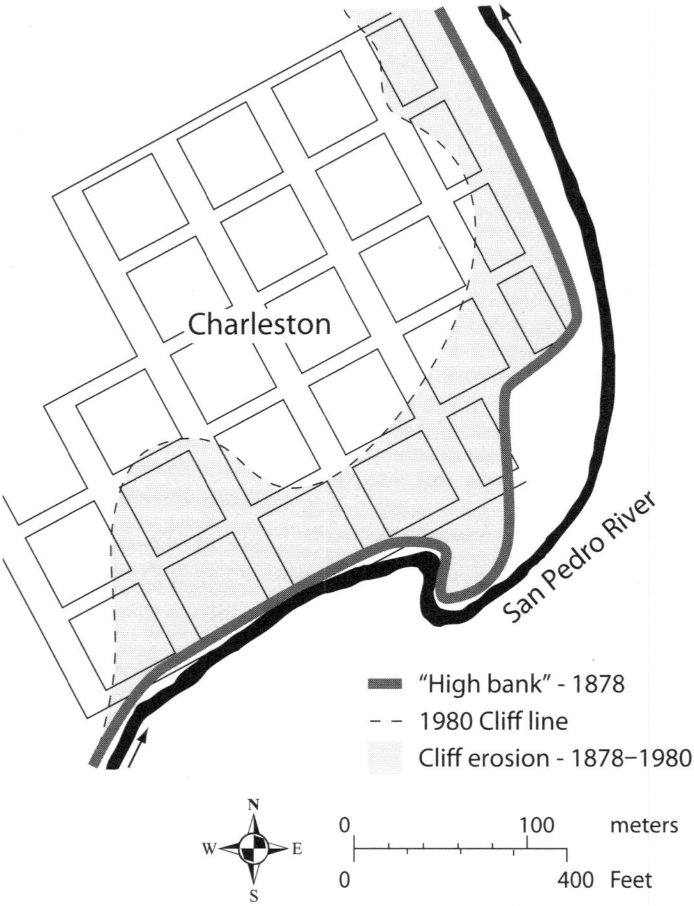

Fig. 12.4. Plat of Charleston in 1879 by A. J. Mitchell. Adapted from K. Tiller (1982).

Pedro below St. David in 1877. Hastings and Turner (1965) suggest that extensive mesquite thickets existed where the floodplain was entrenched. Mesquite also dominated in the lower reaches where the flow was intermittent.

In 1879, the town of Charleston (a planned community) and a mill site were founded on opposite sides of the San Pedro, with the intent of using the river's permanent flow for processing ore from the newly created Tombstone Mining District (fig. 12.4). Early photographs of Charleston (fig. 12.5) again beg the question about discriminating between superimposed and inset relations for steep "banks" bordering the San Pedro. Figure 12.5 is an upstream view of Charleston from Millville showing the position of the inner channel between two older terraces. The date of the channel cutting that produced these erosional terraces remains uncertain. However, major flooding at any time before establishment of Charleston could have formed these terraces and caused valley widening.

The first mention of active arroyo cutting is from the reminiscences of Mary Wood, published in the *Tombstone Epitaph* in 1929. Wood recalled that a flood in August 1881 destroyed the small dam near Millville, and the banks of the river were widened and deepened. The years 1881, 1882, and 1883 had

Fig. 12.5. (top; 12.5a): Photograph of Charleston taken by Carleton Watkins in 1883 looking southwest toward Huachuca Mountains. The San Pedro River runs from left to right through well-defined channel. (bottom; 12.5b): Same view in 1960. Photo credit: Rod Hastings, U.S. Geological Survey Desert Laboratory Photo Archive.

unusually wet summers, the only three consecutive years to produce more than 20 cm of rainfall in June–August at Tucson (Betancourt 1990). The wet summer of 1881 produced enough runoff to cause overflow of the active channel and erosion of the terrace on which Charleston was built. The 1883 photograph of Charleston indeed shows some evidence of recent undercutting of the west terrace (fig. 12.5).

Bryan (1925:342), in an often-cited statement, maintained that "the trench on San Pedro River was cut progressively headward between 1883, when the arroyo formed at the mouth of the river, and 1892, when the headwater fall cut through the boundaries of the Boquillas grant 200 km upstream." He cited a report then in preparation (Bryan et al. 1934), but this manuscript contains no further reference that would warrant an 1883 date for initiation of a headcut at the mouth, or that arroyo development progressed from mouth into the upper San Pedro in less than a decade (Hastings 1959, Rodgers 1965, Cooke and Reeves 1976). Bryan apparently thought that headcut migration was enhanced by increased longitudinal surface slope in each successive subbasin, contrary to normal steepening upstream exhibited by most streams. Bedrock outcrops at Charleston and the Narrows produce independent base levels in each subbasin. Over the short term, these outcrops should have restricted propagation of headcuts or coalescence of discontinuous arroyos from one subbasin to another.

Large floods also occurred on the San Pedro River in the summers of 1886 and 1887 and the summer and fall of 1890. The newspapers reported overbank flooding in 1886 and 1887, but no mention was made of channel erosion. Hastings (1959), however, cites testimony in a court case that the bed of the river near Tres Alamos was lowered 4 m between 1885 and 1889.

1890s to 1900s. Though in other reaches arroyos might have developed in the 1880s, 1890 does appear to mark the beginning of extensive degradation in the lower San Pedro. In August and September 1890, floods on the San Pedro and Santa Cruz Rivers received unusual attention in southern Arizona newspapers. On the San Pedro, most of the bridges were swept downstream. At Dudleyville, the San Pedro "caved within 5 meters of Cook's place," indicating extensive channel widening. On October 2, the *Arizona Daily Star* described deepening of the channel near Mammoth by 9 m.

In the winter of 1891, flooding again affected the San Pedro River Valley, eroding valuable land in some areas and silting other land downstream. Damaging floods occurred again in August 1893, 1894, and in July, August, and September 1896. Above-normal summer rains preconditioned the watershed to excessive runoff during a generalized storm in the fall of 1896. This storm produced the third greatest September–October rainfall at Tucson. In October 1896, streams originating in the Whetstones flooded the Benson area; near the mouths of these streams, channels were deepened by as much as 9 m. This storm persisted for two weeks and caused significant damage to settlements and farms along the San Pedro.

It was probably during the 1896 flood that a channel almost 244 m wide and 6 m deep developed at the northern end of the Boquillas Land Grant, as recorded in an 1899 survey (fig. 12.1). A survey in 1873 recorded a width of no more than a chain (ca. 20 m) in the same area (Cooke and Reeves 1976).

Yet, a channel only 9 m wide and 1.5 m deep defined the river's course at the southern end of the grant near Charleston in 1899. We speculate that in 1899 there was an active headcut somewhere in the 25 km reach between the northern and southern boundaries of the Boquillas Land Grant. In 1909, J. B. Wright recorded a channel width of 130 m at Lewis Springs (fig. 12.1), which may suggest that the headcut progressed to the northern end of the San Rafael del Valle Land Grant between 1899 and 1909, possibly during the floods in the winter of 1904–1905.

1910s and 1920s. According to several accounts, neither the main stem nor tributaries became entrenched upstream of Charleston until the 1910s. Ranchers at Hereford told Haury et al. (1959) that between 1910 and 1914 the river channel was narrow and only 0.5–1.0 m deep. The channel from Fairbank to Hereford, a reach of more than 32 km, was probably entrenched in less than 18 years. Haynes (1987) states that Curry Draw on the Murray Ranch became entrenched along the ruts of a wagon road in 1916.

Channel widening and further degradation in the San Pedro River Valley occurred in September 1926, when floods produced peak discharges of 2780 m^3s^{-1} at Charleston. This is three times greater than the next highest peak in the 69-year gaged record at Charleston from 1916 to 1987. The 1926 flood is one of the better-documented floods in the early twentieth century. Numerous occurrences of channel erosion at bridges were reported from the international boundary to the Gila. The river overflowed its 6-m-deep channel at Benson. At St. David, the channel, which was 18 m wide in 1918 and 46 m wide in 1922, widened to 107 m. The second largest gaged flow occurred in August 1940, but by then the channel could accommodate larger flows.

Over the length of the river, the areas of extensive channel widening are near Redington and Benson, where arroyos cut to the greatest depths in the floods of 1890–1926. Degradation and channel widening persist in these areas. Elsewhere, tributaries have been aggrading in recent decades, as have reaches of the main stem, particularly below Mammoth (near the confluence with the Gila) and above Benson.

Factors Contributing to Entrenchment

HUMAN SETTLEMENT

Overgrazing, trampling of springs and marshes by cattle, eradication of beavers, draining of marshes through ditch diversions, and fuel harvesting in the 1870s and 1880s may have preconditioned the watershed to arroyo cutting in the 1890s. Bahre and Hutchinson (1985) estimate that about 80,000 cords of fuelwood, including mesquite from the floodplains and oaks and junipers from the uplands, were consumed in the Tombstone Mining District between

1879 and 1886. By 1890, upland and floodplain vegetation had been seriously reduced by grazing and fuelcutting. A number of new ditches, which concentrated drainage and used ciénegas as their source, had been dug in the valley (Bryan et al. 1934, Rodgers 1965). Railroad construction involved lengthy embankments along the San Pedro, which may have impeded sediment contributions from the adjacent *bajadas*. More significantly, flow was constricted at bridges (Cooke and Reeves 1976, Dobyns 1981).

EARTHQUAKE

Another factor that may have preconditioned the valley to widespread arroyo cutting was the 1887 earthquake. On May 3, 1887, an earthquake rocked southern Arizona, northern Sonora, and northwestern Chihuahua (DuBois and Smith 1980). Hydrologic effects were noted within a 160-km radius of Bavispe (Dubois and Smith 1980). The upper San Pedro River Valley was within the fissured zone, with several reports of liquefaction from Charleston to Tres Alamos. A fissure 32 km long was reported along the San Pedro River north of Benson and issued a considerable stream of water. Some springs went dry, others doubled in flow, and there was a rise of 1 m in the flow depth of the San Pedro, this during the driest month of the year. The earthquake leveled Charleston, while at St. David, it alerted settlers to the presence of artesian water (Fulton 1966, Fulton and Bahre 1967, Tiller 1982).

According to Tevis (1954), a similar earthquake affected the San Pedro River Valley in 1800–1810. References to unincised floodplains in the 1850s suggest that if this earlier earthquake occurred, it had no large-scale effects on subsequent channel histories in the San Pedro River Valley. However, it cannot yet be discounted that geohydrological phenomena associated with the 1887 earthquake set the stage for arroyo initiation. The earthquake conceivably could explain the remarkable synchroneity of arroyo cutting throughout southern Arizona and northern Sonora. One might expect channel adjustment to a 32-km fissure in the floodplain or to the changed configuration of groundwater surfaces. The immediate withdrawal from artesian aquifers probably produced changes in head that might have accelerated rates of compaction by reducing buoyant forces. The same effect, perhaps not as catastrophic, can stem from pressure losses in artesian aquifers during extremely dry periods. Regardless, investigation of the possible links between the 1887 earthquake and subsequent channel trenching is long overdue. A first step would be to examine evidence for fissures in the 1937 aerial photos of the San Pedro River Valley, provided that arroyo cutting did not eliminate such evidence.

WATER TABLE FLUCTUATION

There is fair agreement that a period of major channel cutting in southern Arizona took place in the middle Holocene (Haynes 1968, Waters 1985,

Haynes 1987), but great discordance over the number and timing of cut and fill cycles during the late Holocene (Waters 1985). Waters (1985) argues that in the late Holocene, streams responded to geomorphic controls irrespective of regional climates. Haynes (1987) maintains that drought-induced fluctuations in regional water tables determined late Holocene cutting and filling. He also suggests that the historic arroyo is just the modern expression of frequent cutting and filling in the late Holocene, which would have happened eventually without human impact. Few would argue that simultaneous cutting and filling on ephemeral streams could lead to ambiguity in the alluvial record. However, there is also danger in assuming that zones of aggradation and degradation migrate systematically to flush sediment from the system, completely out of step with climatic trends (Patton and Schumm 1981). It would be equally difficult to discount the role of a falling water table in promoting arroyo cutting or that of a rising one in enhancing aggradation.

Floodplain Evolution after Arroyo Cutting in the Upper San Pedro: Physical Evidence

The channel and floodplain of the San Pedro River, through time, were mapped from Hereford to the northern boundary of the Boquillas Land Grant, an area that encompasses much of the San Pedro Riparian National Conservation Area (SPRNCA). The age of the various channel and floodplain deposits was estimated from analysis of aerial photography taken at five different times and scales (Soil Conservation Service, April 1937, 1:30,000; USGS, January 1955, 1:20,000; U.S. Air Force, October 1970, 1:55,000; Soil Conservation Service, October 1978, 1:25,000; U.S. Bureau of Land Management, September 11, 1986, 1:6,600). Ages were assigned by the first appearance of a particular deposit in the photographs. The area of the entrenched channel (here defined as the area between the walls of the post-entrenchment terraces) was mapped on sequential, stereoscopic small-scale aerial photography to evaluate rates of channel widening. The channel walls are readily identifiable in stereoscopic aerial photographs because the walls form a nearly vertical, continuous feature that separates two broad surfaces of different elevation.

PRE-ENTRENCHMENT ALLUVIUM

Late Holocene (4,000 YBP to present) alluvium, inset against the St. David Formation, can be divided into pre-entrenchment alluvium, which forms a terrace that occupies most of the inner valley, and post-entrenchment alluvium, which represents the active floodplain of the San Pedro River (fig. 12.6). Near Hereford, the pre-entrenchment alluvium forms a two-stepped terrace separated by 0.5 to 1.0 m of relief. Lenses of dark, carbonaceous sediments, or ciénega deposits, mark the former heights of the water table in the pre-entrenchment alluvium. The pre-entrenchment alluvium correlates with the

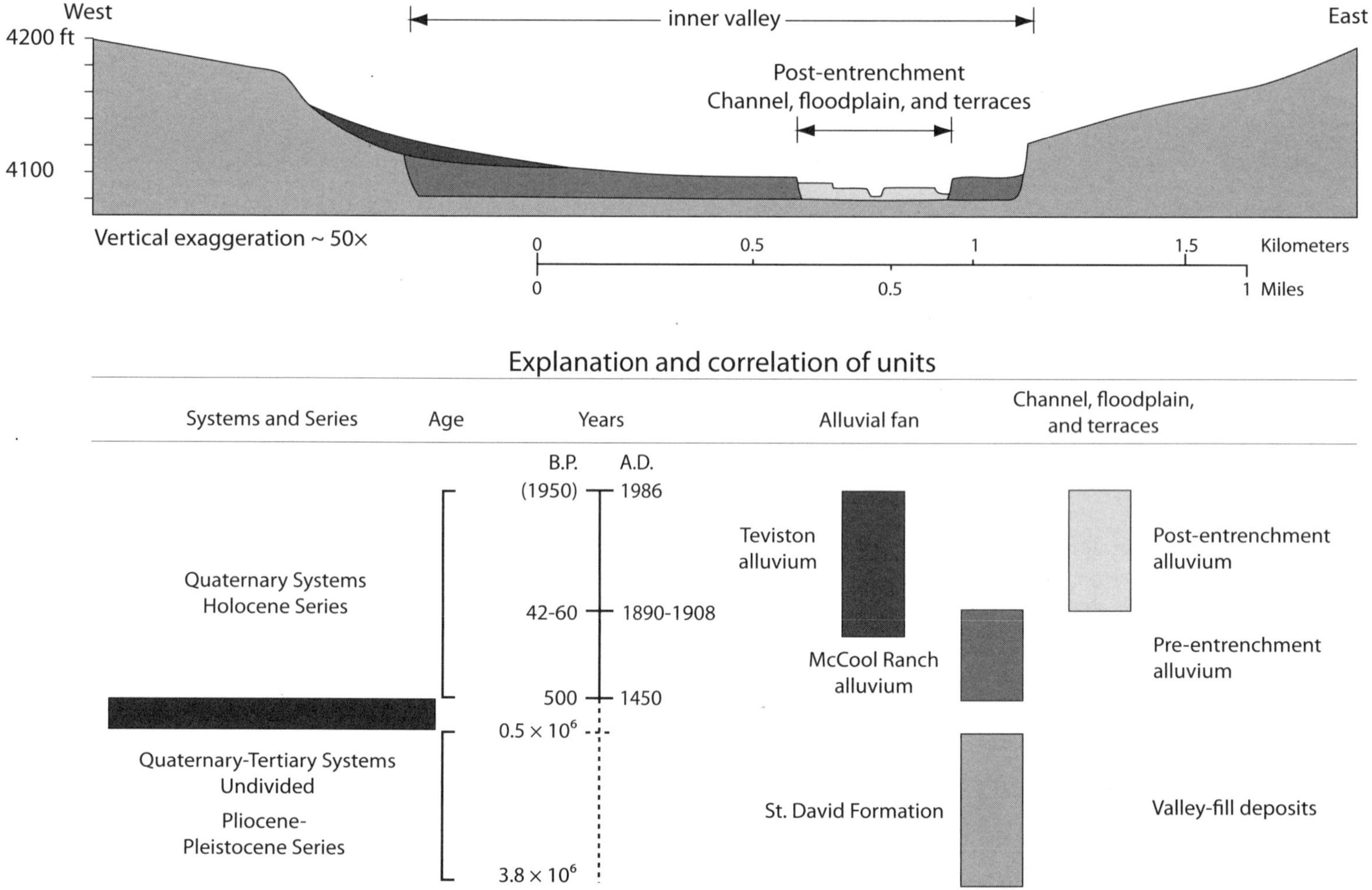

Fig. 12.6. Geologic cross section showing correlation of surficial deposits and geomorphology of the inner valley of the San Pedro River in the vicinity of Lewis Springs. The geologic relations, geomorphology, and deposits are typical of the upper San Pedro.

"Escapule Ranch formation" of Haynes (1987), which can be traced from Curry Draw into the inner valley near Lewis Springs. Near Hereford a sinuous abandoned channel (1 m deep and 10–20 m wide) on the lower terrace was probably the active channel of the San Pedro River before arroyo cutting.

POST-ENTRENCHMENT ALLUVIUM

From youngest to oldest, the post-entrenchment alluvium consists of the active channel, floodplain, and terrace of the San Pedro River, although alluvial fans have formed contemporaneously. The active channel is inset from 1 to 10 m below the pre-entrenchment terrace. Entrenchment in the SPNRCA is greatest below Lewis Springs where it ranges from 5 to 10 m deep. Upstream of Lewis Springs, the river is entrenched only 1 to 5 m below the pre-entrenchment terrace.

Deposition of the alluvial fans and sheetwash deposits began slightly before entrenchment of the San Pedro River. The deposits are cut by the entrenched channel of the San Pedro River, suggesting that deposition began before channel entrenchment. At Walnut Gulch, historic artifacts dating from the turn of the century occur at the basal contact, and artifacts are present locally within the Teviston alluvium. Deposits of similar age are also present in Curry Draw (Haynes 1987). In short, the Teviston alluvium and its correlatives in the inner valley resulted from tributary stream entrenchment and increased hillslope erosion that began before the entrenchment of the main channel.

RATE OF CHANNEL ENLARGEMENT

The spatial distribution of the post-entrenchment alluvium (fig. 12.7) indicates clearly that the area of the channel and floodplain have enlarged since initial entrenchment around the turn of the century. In an alluvial system with a strong component of lateral accretion such as the San Pedro River, progressively younger floodplains form as the channel migrates (see chap. 13). Channel migration simultaneously erodes the pre-entrenchment alluvium, while providing space for subsequent floodplain deposition. Two important questions emerge regarding this process: what is the rate of widening of the high-flow channel, and is the process complete? The process of channel widening is poorly understood. Thus, it is not known whether the widening process is self-limiting or controlled by external factors such as climate or land use.

Figure 12.8 illustrates expansion of the channel from pre-entrenchment to 1986 in a 2-km reach of the river beginning 3.2 km downstream of the Hereford Bridge. Channel area increased rapidly from entrenchment to 1955, but the rate of enlargement slackened since then as shown in figure 12.9, which illustrates the cumulative area of the entrenched channel as a function

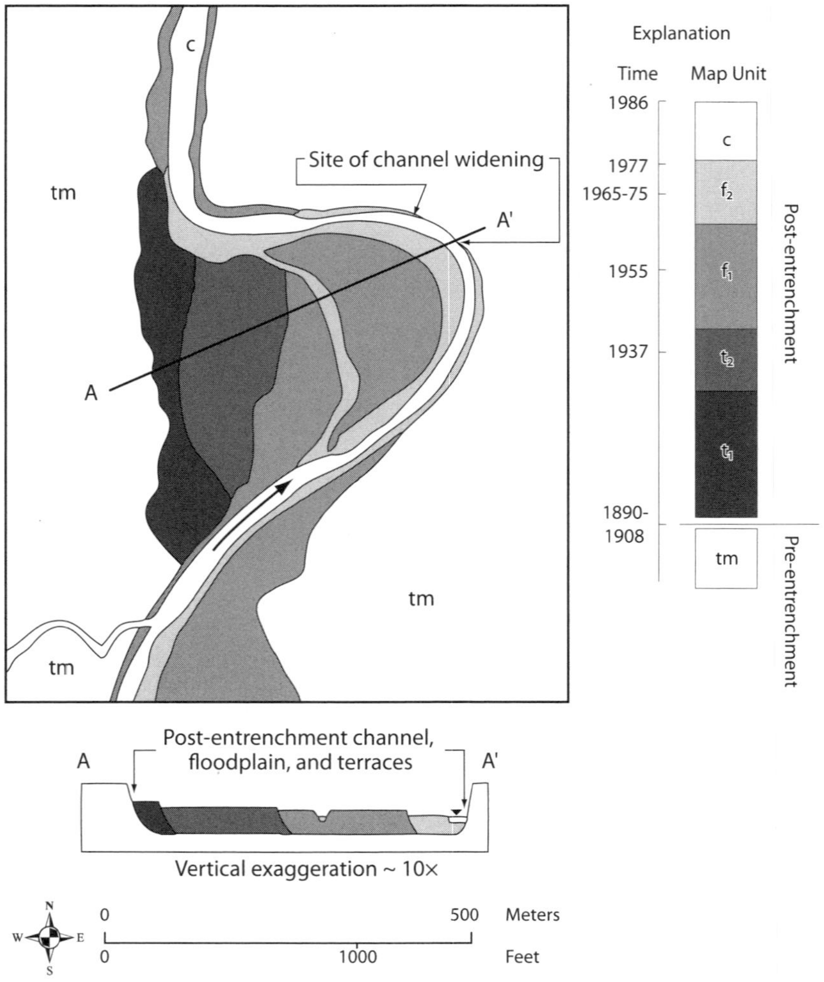

Fig. 12.7. Geologic map and cross-section of the post-entrenchment alluvium exposed on a point bar north of Hereford.

of time. The increase of channel area is approximately an exponential function of time and follows a "rate law," which describes the time-dependent adjustment of many disturbed physical systems (Graf 1988a). Considering the entire area and assuming that entrenchment occurred by 1900, the estimated rate of enlargement from 1900 to 1955 was 0.109 km^2 yr^{-1}, and from 1956 to 1986 the rate was only 0.024 km^2 yr^{-1}. Thus, the rate of channel enlargement has declined in recent years. This probably signifies stabilization of the channel and the end of significant widening.

FLOODS AND CHANNEL WIDENING

The morphology of the channel is controlled largely by the frequency of channel-forming floods (the control variable). The annual flood series at Charleston (see chap. 16) shows a clear pattern of relatively frequent large floods (defined as events in the upper quartile of all flows) during the first part of the twentieth century. Seventeen floods equal to or greater than the

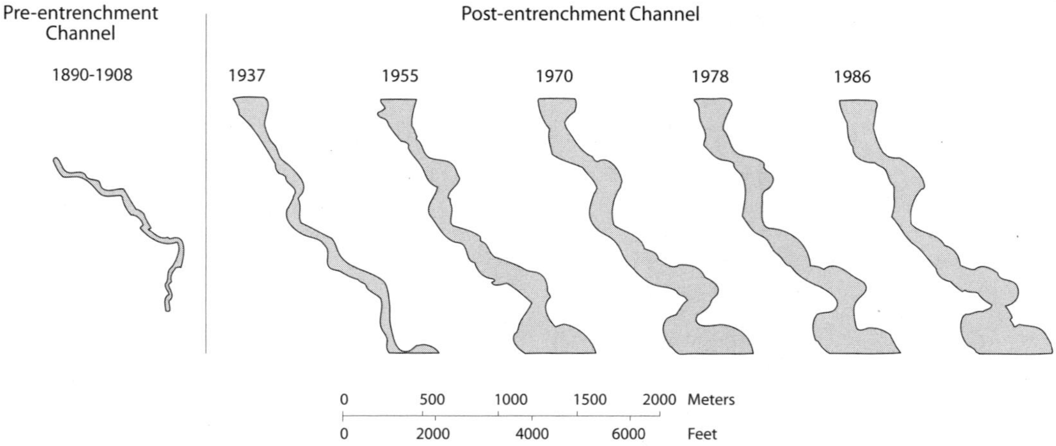

Fig. 12.8. Maps showing the pre-entrenchment channel and expansion of the post-entrenchment channel as compiled from sequential aerial photography since 1937, and from cadastral survey notes and plats at the turn of the century.

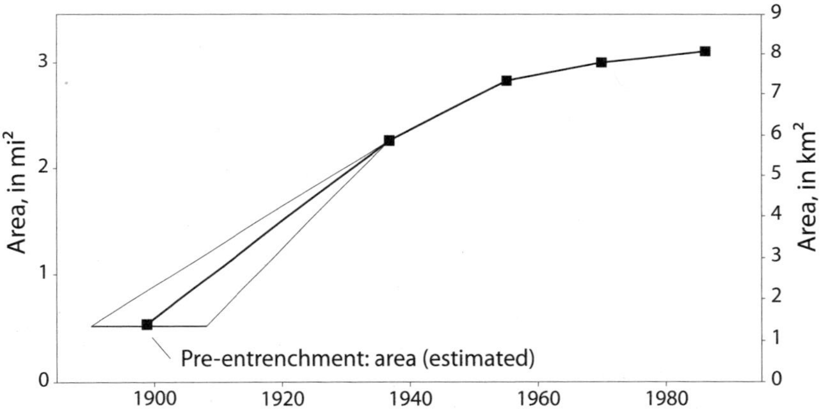

Fig. 12.9. Time series showing cumulative area of the entrenched channel. Channel expansion slowed appreciably by at least 1955.

75th percentile occurred between 1916 and 1955, an average rate of about one such flood every 2.4 years. This period includes the flood of record in 1926. Most of these large floods occur during the summer runoff season.

In contrast, only four floods larger than the upper quartile occurred from 1956 to 1987, an average rate of one such flood about every eight years. Only one of these four floods, the flood of October 9, 1977, was comparable in size to the largest floods of the earlier period. Reduced frequency of large floods on the San Pedro after 1955 runs counter to trends noted in adjacent watersheds, notably the Santa Cruz River (Webb and Betancourt 1992). Less-frequent large floods after 1955 probably stem in part from increased channel storage due to greater channel areas and sinuosities, which seem to have stabilized during the last five decades, as well as perhaps to increased revegetation of the watershed.

CHANNEL WIDENING AND EQUILIBRIUM

Widening of the San Pedro River channel could not continue indefinitely. Once the channel cross section is capable of transporting the water and sediment load of the post-entrenchment discharge regimen, it should stabilize and cease to widen. The negligible rate of channel enlargement since about 1955 indicates that the widening process has ended or slowed greatly (Hereford 1993). In terms of geomorphic equilibrium, the river system has adjusted to the entrenchment disturbance and has probably attained a new equilibrium with a quasi-stable channel configuration.

This transition from pre- to post-entrenchment equilibrium is analyzed diagrammetrically in figure 12.10. The effect of an increase in flooding is to increase the channel area after a reaction or lag time. Thus, the pre-entrenchment equilibrium was disturbed by a change of flood frequency probably beginning in the early 1880s, when destructive floods were first described in the upper San Pedro River Valley. An additional disturbance with unknown effect was the 1887 earthquake. The reaction time to these disturbances began about 1880 and lasted until entrenchment began between 1890 and 1908. The period of disequilibrium and rapid increase of channel area is the relaxation time, or the time it takes to attain a new quasi-stable equilibrium. The relaxation time was about 55 years, assuming that entrenchment began by 1900 and that the channel was essentially stabilized by 1955.

The relaxation time for channel stabilization was probably controlled by factors influencing the frequency of channel-forming floods. This variable is affected by feedback mechanisms, climate, and land use. The feedback is between vegetation and the expanding channel. As the channel expands, more room is provided for riparian vegetation, which has the effect of reducing peak-flood discharge (Burkham 1972). In addition, larger channel area increases transmission losses, compounding the influence of vegetation. This feedback process shortens the time to stabilization, because vegetation increases boundary shear stress, eventually minimizing further bank erosion. Climate directly controls flood frequency through rainfall variations and indirectly controls flood frequency through its effect on vegetation both within and out of the channel.

Changes in grazing practices and development of tributary water-retention structures probably shortened the time required for channel stabilization. Generally, these changes served to reduce runoff and peak flows. The number of cattle grazing in the upper basin decreased since entrenchment from a historic high of 36,000 cattle in 1890 to 7,500 by 1964, well within grazing capacity (Wagoner 1962, Rodgers 1965). In addition, numerous small water-

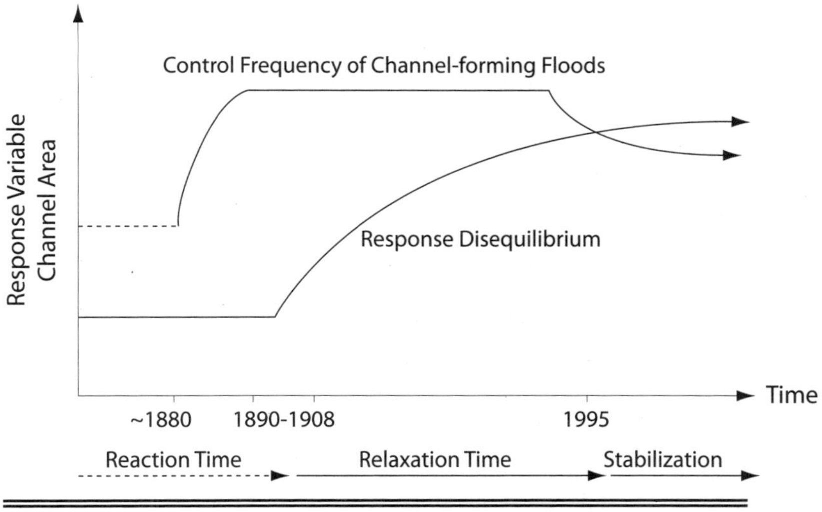

Fig. 12.10. Conceptualization of channel equilibrium in terms of control and response variables, based on Graf (1988a) and Knighton (1984).

retention structures have been built in small tributaries of the river. Although their overall effect is unknown, they were designed to reduce runoff.

Summary

The historical record suggests that in the mid-nineteenth century the San Pedro was a continuously perennial stream from its source near Cananea to just beyond the Narrows. Flow was interrupted (spatially intermittent) in the lower reaches, with the dry discontinuities outdistancing limited surface flow from groundwater outcroppings. Apparent discontinuous arroyos up to 6 m deep at St. David, Tres Alamos, and below the Narrows transitioned a short distance downstream into ciénegas dammed by beaver. Mesquite thickets occupied dry and incised reaches, while mostly treeless conditions characterized the unincised, marshy floodplains particularly in the upper basin. Treeless conditions could imply permanently saturated soils, where reducing conditions would limit tree growth and favor graminoids (Hendrickson and Minckley 1984).

The exact timing of arroyo initiation is still uncertain. Bryan's (1925) statement that arroyos started at the mouth in 1883 and progressed headward 200 km to the Boquillas Land Grant by 1892 cannot be substantiated. Though in other reaches arroyos might have developed in the 1880s, 1890 does appear to mark the beginning of extensive degradation in the lower San Pedro. Extensive erosion in the upper San Pedro apparently did not occur until the early 1900s. Newspaper accounts, survey records, and other written records describe extensive channel erosion in association with the series of large floods that occurred near the turn of the nineteenth century.

Today on the San Pedro River, post-entrenchment alluvium deposits occupy the lowest topographic level of the inner valley, which is 1 to 10 m below the pre-entrenchment terrace. A widespread, locally dense riparian forest has developed simultaneously with deposition of the post-entrenchment alluvium. The nature of post-entrenchment deposits imply an entrenched, meandering, low-sinuosity alluvial system. The post-entrenchment alluvial deposits are successively younger across the floodplain surface, indicating that the channel has widened since initial entrenchment. Channel area of the upper San Pedro increased rapidly from initial entrenchment until at least 1955; since 1955, channel area has increased only slightly. Peak-flood discharge of the San Pedro River declined substantially after 1955. Our conclusions are that the channel in the upper basin is largely stabilized and that equilibrium or near-equilibrium conditions exist.

GARY HUCKLEBERRY,
SHARON J. LITE,
GABRIELLE KATZ,
AND PHILLIP PEARTHREE

Fluvial Geomorphology

THIRTEEN

Introduction

Grounded in the disciplines of geography and geology, geomorphology seeks to understand the active processes and history responsible for the configuration of planetary surfaces. Fluvial geomorphology is a subdiscipline that focuses on flowing water as a surficial process. Geomorphologists pay considerable attention to flowing water because in most places of the world, and particularly in deserts, it is the primary agency by which landscapes are shaped. Whereas many physical and chemical processes act to break down rock, for most of earth's surface it is water that removes the weathered material and transports it away.

Given that most geomorphic changes are accomplished by flowing water, some of the most dynamic components of the landscape are streams, rivers, and floodplains. Fluvial systems gather water from a broad area and concentrate it along streams and rivers, producing extensive erosion and sediment transport along channels and adjacent floodplains. Floodplains are a very small part of the terrestrial landscape but are where much of the geomorphic work is accomplished. Stream networks are a primary means by which landscapes are sculpted, and they are constantly adjusting to changes in the environment to transport geological materials effectively.

Climate change and human changes to the watershed have affected the geomorphology of the San Pedro River in the past and continue to do so today. Dramatic climate changes may occur at timescales of 10^3–10^4 years in association with glacial-interglacial oscillations, whereas smaller changes in temperature and moisture occur at timescales of 10^1–10^2 years due to solar variation, volcanic eruptions, and ocean-atmosphere dynamics such as El Niño–Southern Oscillation and Pacific Decadal Oscillation. As long as there is climatic variability, there is an inherent tendency for rivers to downcut, backfill, and shift their channels; human disturbances may augment or dampen these climate-driven transformations. Regardless of cause, fluvial changes represent recurrent disturbances that affect the structure of riparian communities (Malanson 1993, Hughes 1997). Because geomorphic processes influence biotic components in floodplain ecosystems, effective management of riparian resources requires a consideration of geomorphic history and process.

This chapter places the San Pedro River in a geomorphic context with the aim of providing readers with a sense of the dynamic and nonlinear aspects of hydrologic processes and their impact on stream geomorphology and riparian resources. The emphasis is on both natural and human-caused processes. Following an overview of fluvial processes, we describe physical characteristics of the San Pedro River and its floodplain and discuss mechanisms by which river channels change in time and space, and the ecological implications. This is followed by a review of agents of change affecting San Pedro River geomorphology.

Fluvial Fundamentals

Despite an abundance of quantitative studies in fluvial geomorphology, a universal equation (or set of equations) defining the full mechanics of streamflow and sediment transport has yet to be defined, and may never be defined, given the complexity. Nonetheless, basic processes involved in streamflow, sediment transport, and channel formation are relatively well understood (Leliavsky 1955, Leopold et al. 1964, Morisawa 1968, Schumm 1977, Knighton 1998). One conceptual approach to understanding rivers and sediment transport is to view them as open systems whereby a balance is approached between the sediment to be transported and the water available to accomplish the task. Sediment is provided to the system by physical and chemical breakdown of rock that is then available for transport by gravity or some fluid agent (wind or water) from hillslopes to stream channels. Once in motion, sediment continues to move until gravitational forces exceed the tractive forces acting on its motion. If more grains settle than are translocated, then a sedimentary deposit is created. These sedimentary deposits accumulate through time and underlie floodplain surfaces.

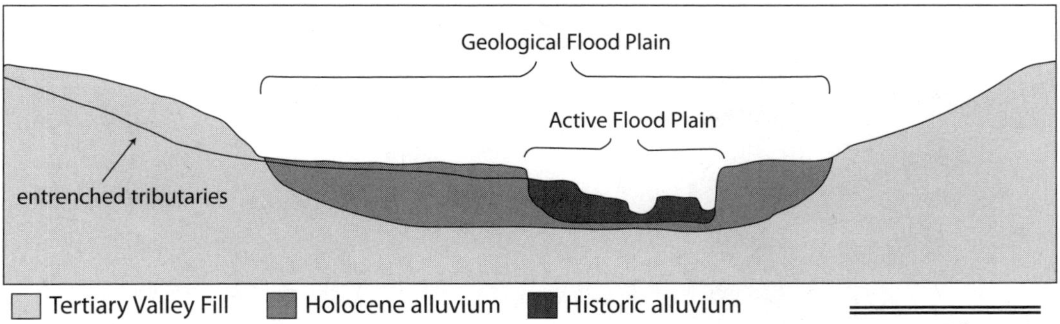

Fig. 13.1. Schematic cross section of San Pedro River floodplain and the post-entrenchment channel.

FLOODPLAINS AND TERRACES

Definitions. The concept of a floodplain is intuitively obvious: an area that periodically is inundated with floodwater. However, geomorphologists, hydrologists, and others often disagree with the boundaries of such areas (Graf 1988b) and the precise definition of "periodically." Many consider the floodplain to be surfaces adjacent to the active channel that are inundated when streamflow exceeds bank-full conditions. Others include surfaces adjacent to the channel composed of alluvium deposited during the current interglacial period, i.e., the last 11,000 years.

Surfaces that are currently too high above the modern channel to experience flooding given current channel conditions are not part of the floodplain but instead are stream terraces (Wolman and Leopold 1957). If the floodplain is deeply entrenched, such as much of the San Pedro River, the largest flows may be contained entirely within the inset channel (or series of channels). Given a definition of floodplain as geomorphic surfaces prone to frequent inundation, the San Pedro floodplain is limited to the entrenched zone (figs. 13.1, 13.2).

Prior to historic arroyo cutting (see chap. 12), the San Pedro River contained a relatively broad floodplain. Following arroyo cutting, surfaces that were once part of the floodplain became terraces. We include the pre-entrenchment floodplain surface with the entrenched zone to represent the geological floodplain and refer to the entrenched zone that contains the modern channel and adjacent surfaces prone to frequent inundation as the active floodplain.

Floodplain construction. Sediment is constantly added and removed from floodplains by moving water. Periods of net sediment storage are associated with floodplain construction. Given the complexity of streamflow and sediment transport, deposition is spatially and temporally variable, and there are many ways a floodplain can be constructed; these can be grouped into the two general processes of lateral and vertical accretion (Wolman and Leopold 1957, Morisawa 1968, Brakenridge 1988). Lateral accretion is characterized by the horizontal shifting of the main channel such that the outer edge of a bend migrates into

Fig. 13.2. Aerial photograph of San Pedro River floodplain at its junction with Aravaipa Creek. Highlighted features include 1. San Pedro River geological floodplain; 2. active (post-entrenchment) channel; 3. secondary flood channel; 4. Aravaipa Creek; 5. Putnam Wash; 6. mining on hillslopes; and 7. agriculture. Photo source: Arizona Regional Image Archive, University of Arizona.

existing alluvium while deposition occurs on the inside of the bend, often forming a gravel lag or point bar. Vertical accretion occurs as sediment-laden streamflow exceeds channel capacity and spreads onto adjacent floodplains.

In humid environments, channel flow is exceeded (resulting in overbank flow) on average every 1 to 2 years (Leopold and Wolman 1957). In arid environments, overbank flooding occurs more irregularly (see chap. 16) but nonetheless produces a vertical sequence of alluvial deposits, usually composed of fine sand, silt, and clay (see chap. 14). Through time, the height of the overbank deposits increases, resulting in a progressive increase in the size of discharge required to inundate adjacent surfaces. Eventually, the channel swings laterally and cuts into the stack of vertical accretion deposits, replacing them with channel sands and gravels. Alternatively, a large, erosive flood will strip the entire depositional sequence and leave a layer of gravels in its place (Nanson 1986). Both processes result in a mosaic of different-aged and -textured floodplain deposits that influence the age, structure, and composition of riparian plant communities (see chap. 1).

CHANNELS

Channel geometry is controlled by multiple factors including valley slope, discharge, sediment load, and floodplain vegetation. These factors influence how rivers erode and deposit sediment and thus influence the shape, size, and overall geometry of a river channel and its floodplain. Viewed from above, rivers generally fall into one of three plan-view shape categories: straight, meandering, and braided (Leopold and Wolman 1957). Straight channels are rare and generally occur where confined by bedrock or human structures.

Meandering channels. Meandering channels are frequently associated with low-gradient valleys, fine-grained alluvial sediment, and heavily vegetated banks (Schumm 1977). During relatively low to moderate flows, sediment accumulates on the inside of a meander, forming point bars, while erosion occurs on the outside of the meander; this causes the meander to shift horizontally across the floodplain transverse to valley slope, reducing the overall gradient of the channel. Meander loops also travel downvalley as the zone of maximum erosion is slightly offset downstream from the apex of the meander (Wolman and Leopold 1957). Meander cutoffs, semicircular abandoned segments of a stream channel, develop as meander loops become too large and unstable. If the meander cutoff retains water, it creates an oxbow lake.

Braided rivers. Braided rivers are characterized by multiple shallow, interconnecting channels passing among shallow sand and gravel bars. During high streamflow, bars are inundated and channels shift laterally. During periods of extended low flow, the sand and gravel bars can become covered with pioneer vegetation and eventually form islands within the braided floodplain. The channels are constantly filling with sediment forming sand/gravel bars that act to divert flow laterally creating new channels, driving rapid change in channel position and the number of channels (or anabranches).

Braided rivers are common where valley gradients are relatively steep and there is an ample supply of coarse sediment for transport. An abundance of sediment is yielded from floodplains that have a paucity of bank-stabilizing vegetation, and from arid or devegetated watersheds, perhaps through fire, deforestation, overgrazing, or mining. A highly variable discharge also appears important, given that braided rivers are most commonly found in dry environments where changes in streamflow can be rapid and dramatic (Graf 1988a).

Compound channels. Arid-region stream and river channels may have a combination of straight, meandering, and braided forms. Indeed, a common feature of floodplains in the desert Southwest is the compound channel (Graf 1988a), whereby a low-flow meandering channel is nested within a larger braided channel. Such forms are associated with fluvial systems that have frequent low flow contained within a single, meandering channel but also commonly experience larger floods whereby excess discharge is channeled through adjacent secondary and tertiary channels. As in braided channel forms, a highly variable discharge regime plays a critical role in creating and maintaining compound channels, but so does a smaller, semi-regular discharge.

CHANNEL PATTERNS IN TIME AND SPACE

Streamflow variance and channel change. Rivers may be uniformly meandering or braided, they may contain reaches that alternate between meanders and braids, or they may contain elements of both along a single reach. A river segment may also change its size and shape through time, particularly in response to large floods. Rivers in arid environments are particularly susceptible to radical changes in width and depth (Knighton and Nanson 1997), given their high variance in seasonal discharge (Graf 1988a; and see chap. 16). During large, high-energy floods, channels may downcut, widen, and shift laterally, often creating a channel much larger than what is needed to convey most flows. Changes in the form and course of the stream, such as in a meander cutoff, are referred to as avulsions.

Floods, vegetation, and channel change. The degree to which floodplains are altered through erosion depends on the magnitude and duration of a flood as well as the resistance of bank sediments to surface flow (Wolman and Miller 1960, Wolman and Gerson 1978, Costa and O'Connor 1995). If the interval between large floods is fairly long, the stream channel returns to a form more suited for smaller discharges, aided by vegetation that colonizes floodplains and sand/gravel bars (J. Friedman and Lee 2002). Along perennial stream reaches, the flood-widened channels soon re-narrow in association with vegetation establishment (J. Friedman et al. 1996a, 1996b). The recovery time in dryland streams is slower than for humid streams (Wolman and Gerson 1978).

Where the vegetation colonizes within a recently disturbed floodplain (e.g., within or outside the channel) can influence how the channel geometry will recover. Dense bank vegetation can result in narrower channels, while the presence of abundant channel-bed vegetation may result in channel widening (Huang and Nanson 1997). Regardless of the specific channel response, highly irregular streamflow and slow recovery times often preclude the formation of any equilibrium channel in dryland fluvial systems (Stevens et al. 1975, Graf 1988a).

Sediment yield and channel change. Vegetation type and precipitation pattern within the watershed both influence sediment yield. Arid-region catchments have low biomass and plant cover. Plant cover reduces runoff and erosion by minimizing raindrop splash energy on bare soil and slowing sheetwash so that more water infiltrates into the subsurface, while plant roots hold sediment in place. Shifts in precipitation and in plant cover both can have a significant effect on sediment yield in desert and grassland biomes (Langbein and Schumm 1958, Bull 1991) and can lead to significant changes in channel geometry.

San Pedro River Geomorphology

ENVIRONMENTAL SETTING

Geologic context. Most of the mountains in the San Pedro basin contain volcanic and metamorphic rock created during the Mid-Tertiary Orogeny 30–20 million years ago, whereas the mountains were formed by extensional forces during the Basin and Range Physiographic Province disturbance 15 to 8 million years ago (Shafiqullah et al. 1980, Menges and Pearthree 1989). Originally, the valleys were internally drained, as evidenced by alluvial fan deposits grading into extensive playa and lake deposits exposed along the valley margin, but eventually the basins filled and the San Pedro and Gila Rivers became integrated. Once this occurred, alluvial fans became the dominant depositional landform extending from mountains to floodplain. However, with drainage integration came episodes of valley entrenchment whereby the San Pedro River and its tributaries incised into basin fill, cutting a series of erosional fan terraces and pediments into margins of the valley, each veneered with stream gravels (Tuan 1962, Haynes 1987, Cochran and Richardson 2003). These alluvial fan and pediment surfaces extending upslope from the San Pedro River are collectively referred to as the valley piedmont.

Since the height of the last glaciation approximately 20,000 years ago, the river has cut a trench into the Tertiary basin fill along the axis of the valley (fig. 13.1). The trench is filled with post-glacial alluvium that underlies the geological floodplain of the San Pedro River. This includes deposits introduced by piedmont streams, commonly forming tributary alluvial fans at the margin of the geologic floodplain where rapid sedimentation occurs. The geologic floodplain and tributary fans have been periodically entrenched by the San Pedro River, most recently during the mid- to late 1800s and early 1900s, forming an inset, entrenched zone, i.e., the active floodplain (see chap. 12). In places, historic alluvium is beginning to fill the entrenched zone (Hereford 1993).

Upper and lower basins. The piedmont in the upper basin of the San Pedro is broad and only moderately dissected by tributary streams and slopes gently above the axial valley trench. In the lower basin, the axial trench increases in slope and is bounded by steeper and more dissected piedmont surfaces (Heindl 1952a, 1952b, Tuan 1962). This results in a higher drainage density (length of stream channel per unit land area) in the watershed.

The higher drainage density and slope of the bounding piedmont affects baseflow and the flood regime of the lower San Pedro River. Basins with high drainage densities display high drainage network efficiency and more rapid hydrograph response (Patton 1988). Runoff is more quickly conveyed through the network of tributary channels to the main trunk stream, increasing peak flows and also reducing infiltration on the piedmont, an important source of baseflow for the river. Increasing urbanization in the upper basin is likely to mimic the effect of increased drainage density through replacement of vegetated surfaces with impervious ones (Kepner et al. 2004).

Floodplain width. Overall, the width of the San Pedro River active floodplain increases downstream, except where locally constrained by geologic features. This downstream increase in width reflects the effects of large floods on bank cutting. The lower basin with its larger catchment area and greater drainage density has greater maximum flood size potential. Moreover, hydraulic measurements indicate that the middle and lower reaches of the San Pedro River channel have higher values of stream power per unit length of channel compared to the upper basin, thus increasing the river's ability to strip vegetation and erode sediment (Lite 2003). Whereas the width of the active floodplain is greater in the lower basin, the size of the main flow channel decreases downstream. Unlike the active floodplain, the main flow channel is formed by more seasonal runoff and baseflow, and regular streamflow decreases downstream due to infiltration.

Aquifer conditions. With the exception of a few short segments where bedrock emerges at the surface, the San Pedro River flows over Holocene alluvium inset within the axial trench. Both basin fill and Holocene alluvium serve as aquifers for the valley, with the Holocene alluvium very permeable but prone to seasonal fluctuations of the water table. In most places, the San Pedro River is influent, i.e., surface flow infiltrates down to the water table below the channel, resulting in reduced downstream discharge. However, there are perennial reaches where local water tables intersect the surface, usually near bedrock outcrops. The longest perennial reach is approximately 30 km long between Hereford and Fairbank in the upper basin (D. Brown et al. 1981; and see Introduction).

HISTORIC CHANGES

The San Pedro River and its tributaries have done their part in shaping the landscape in southeastern Arizona and northern Sonora: they have cut canyons, filled basins, and created terraced landforms for at least the last five million years (K. Bryan 1926, Tuan 1962, Lindsay et al. 1990). During this time, the region overall has been tectonically stable, and changes in deposition and erosion by the river and its tributaries are attributable mainly to climate change. Numerous glacial-interglacial cycles during the Quaternary altered the hydrology of the river and repeatedly reshuffled plant and animal communities in the valley and adjacent mountains. The river responded to these long-term changes in climate, and consequent effects on vegetation, fire frequency, and flood regime, by cycles of downcutting and backfilling, resulting in a dissected landscape in the basin and terraced landforms along the river.

In addition to long-term changes, there have been short-term stream dynamics occurring at timescales of decades to centuries. For example, incision of the San Pedro River channel in the mid-1800s and early 1900s radically altered floodplain geometry and vegetation (see chap. 12). These changes may be linked to smaller variations in climate—such as El Niño frequency, strength of the southwestern monsoon, and drought—that control flood regime, as well as human activities within the hydrologic basin.

The probability that incision along the San Pedro was caused by a combination of natural climate variability and human alterations of the landscape creates a problem for resource managers. Given the San Pedro River's history of change due to climatic variability, how does one separate natural from anthropogenic changes? Arroyos and gullies that formed throughout the San Pedro River Valley by the late 1800s were attributed by many to overgrazing; yet the San Pedro River and many other streams in southeastern Arizona had downcut and backfilled at least five times prior (Waters and Haynes 2001). Rivers like the San Pedro are complex, open systems that adjust channel size, shape, and configuration in response to changes in runoff and sediment yield from drainage basins. Such changes can have multiple causes, and it may not be possible to determine to what degree river metamorphosis is human induced.

Channel geometry. Dramatic changes in channel geometry during the last 150 years are well documented for the San Pedro and other rivers in the American Southwest (Burkham 1972, Cooke and Reeves 1976, Graf 1983, Hereford 1984, Huckleberry 1994a, Hooke 1996; and see chap. 12). These include

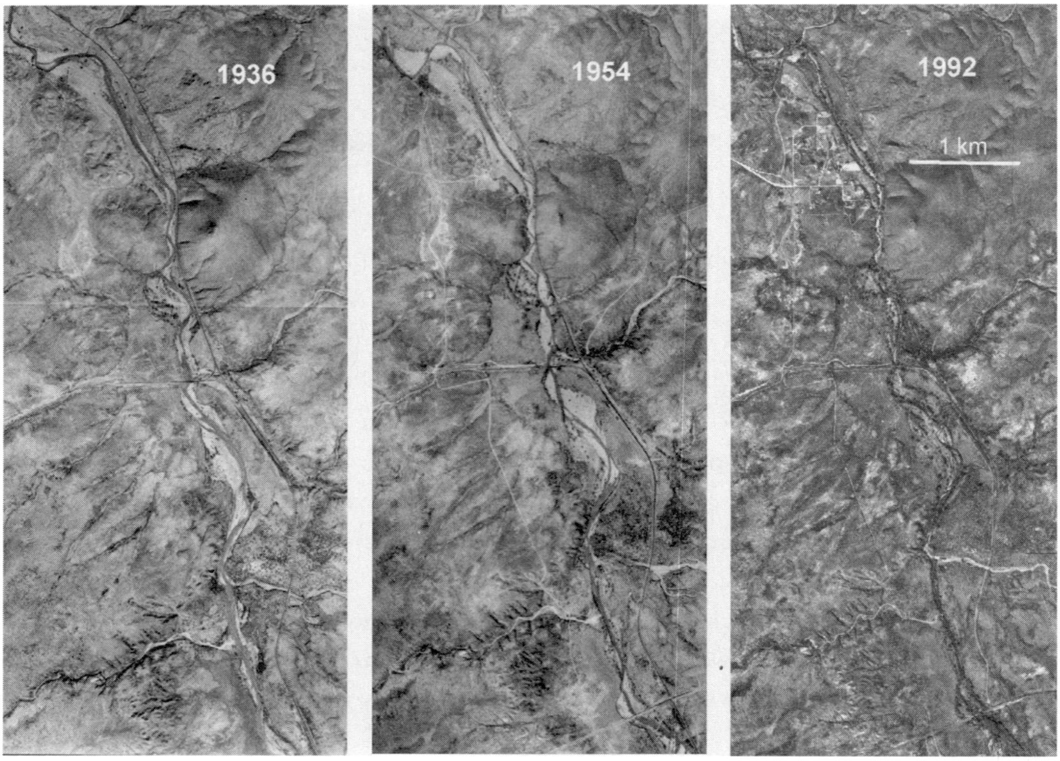

Fig. 13.3. Repeat aerial photography (1936, 1954, 1992) for the upper San Pedro River at Lewis Springs. Extensive areas of the active floodplain that were unvegetated in 1936 and 1954 were subsequently occupied by large stands of woody vegetation. Photos courtesy of Robert Webb.

floodplain transformations caused by vertical downcutting of the channel (i.e., arroyo formation) and fluctuations between meandering and braided patterns. These transformations affected riparian vegetation through erosion and consequent changes to water tables.

Prior to the mid-1800s, the San Pedro River flowed over an unincised surface and had a larger area prone to flooding. It contained a single meandering channel, and marshes were common (Hastings 1959, Hereford 1993, Huckleberry 1994b, Wood 1997; and see chap. 12). Following historic floodplain entrenchment, water tables lowered and marsh habitat declined (Hendrickson and Minckley 1984), but the subsequent channel widening also created the wide, braided channel conditions that facilitated the establishment of riparian forests of cottonwood (*Populus fremontii*) and willow (*Salix gooddingii*) (figs. 13.3, 13.4).

The incised channel of the San Pedro River has widened since initial arroyo development, especially during the largest recorded flood, that of September 1926. The result is an inset floodplain consisting of a post-entrenchment channel with numerous secondary flood channels and low gravel surfaces that in places is several hundred meters wide (fig. 13.3). The low-flow channel

a)

b)

Fig. 13.4. Repeat photography of San Pedro River near the U.S.-Mexico border at Palominas, looking downstream. (top; 13.4a): Photo taken 4/17/1930 by W. E. Dickinson (#1456, courtesy of U.S. Geological Survey Arizona District) reveals relatively unvegetated channel with exposed sand and gravel bars. (bottom; 13.4b): Photo taken 11/28/2000 by Dominic Oldershaw (survey stake 1954, courtesy of the U.S. Geological Survey Desert Laboratory Repeat Photography Collection) reveals a narrower channel with increased riparian vegetation.

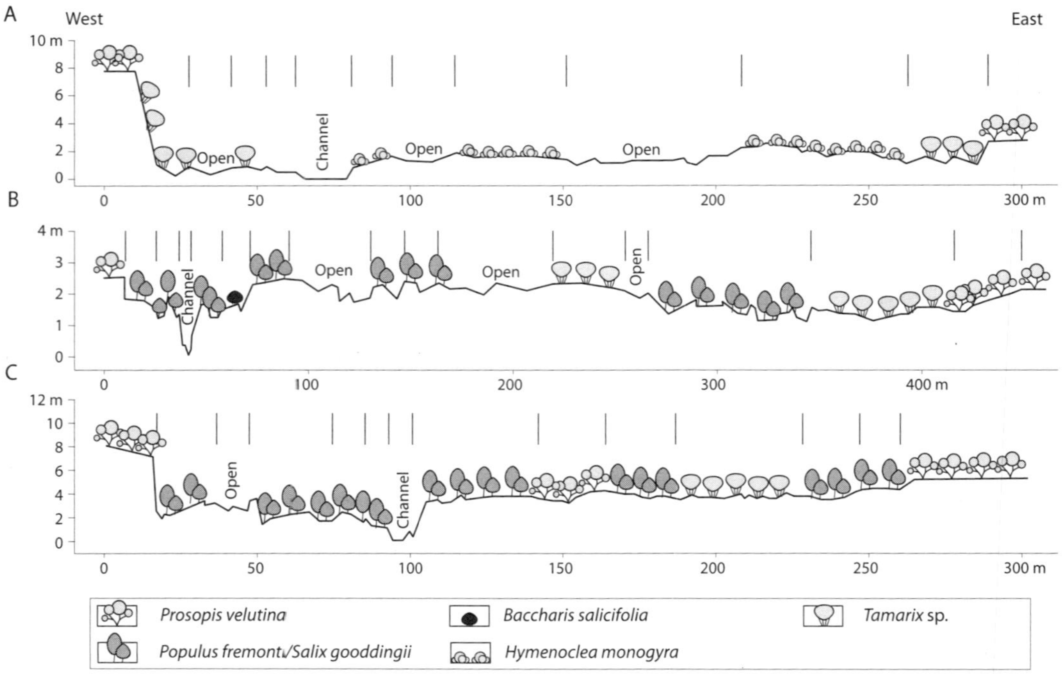

Fig. 13.5. Cross sections of the San Pedro River active floodplain with vegetation patch types (adapted from Lite 2003) at a reach in lower basin with highly intermittent flow (A), reach in lower basin with moderately intermittent flow (B), and reach in upper basin with perennial flow (C). The short vertical lines represent divisions between vegetation types and vegetation age classes. Illustration credit: Mike Buffington.

meanders but also has several braided sections and, in most reaches, forms a compound plan-view form. The low-flow channel has changed position due to meander migration and cutoffs (Wood 1997), as well as braided channel shifts. The result is an active floodplain with surfaces of different ages including gravel bars, swales (elliptical depressions formed by scour of sediment), and active and abandoned channels that contain different-textured alluvial soils and support a patchwork of riparian plant communities (fig. 13.5).

Floodplain development. As pioneer seedlings become established after floods on new bare sand and gravel surfaces, the new plant growth increases the hydraulic roughness of the floodplain and decreases flow velocities during floods. During the first half of the twentieth century, the channel was wide and braided and sparsely vegetated (Hereford 1993) (figs. 13.4, 13.5). Today, the active floodplain contains dense vegetation that acts as an impediment to the swift movement of large floods (Phillips et al. 1998).

By increasing hydraulic roughness and reducing peak discharge, dense floodplain vegetation facilitates the deposition of sediment, causing the floodplain to aggrade. Hence, there exists an alluvial cycle on the San Pedro River marked by channel downcutting followed by channel widening through lateral channel swings and bank collapse. As the channel widens, new alluvial surfaces are created for vegetation growth that help to trap sediment and facilitate the backfilling of the incised channel (Meyer 1989). This process can be accelerated if hillslopes are disturbed (e.g., see open-pit mining in fig. 13.2). As riparian vegetation colonizes and covers much of the active floodplain, the San Pedro River may eventually backfill much like it has several times during the Holocene. To what degree human activities will modify this cycle remains to be seen.

Agents of Change

Rivers are naturally prone to change. For the San Pedro River, we highlight four primary agents of geomorphic change, variously influenced by anthropogenic actions and often interrelated, that affect the form and behavior of the channel: changes in dominant streamflow, floods, fire, and animal populations (including cattle and beavers).

CHANGES IN STREAMFLOW

Today, portions of the San Pedro River Valley are experiencing unprecedented demands on water, while other reaches are being rewatered for conservation purposes. Changes in streamflow rates can have geomorphic consequences, but the degree to which a river channel responds may not be commensurate with the magnitude of the disturbance. For example, streams may be close to a threshold of change in channel geometry and will respond dramatically to small disturbances, while streams farther from such thresholds can withstand large disturbances and retain a stable configuration. Determining the proximity of a fluvial system to an internal threshold is difficult, and this precludes easy predictions of channel behavior for rivers like the San Pedro.

In general, stream banks are less stable, and streams have greater width/depth ratios, in areas where streamflow is ephemeral or intermittent (vs. perennial) and where the channel is composed of more easily erodable materials (Gordon et al. 1992). The reduction in cover of densely rooted wetland plants that occurs as streams become increasingly intermittent (see chap. 1) further contributes to bank instability. The reduction in bank vegetation by drought ultimately can lead to changes in channel pattern, such as a shift from a meandering to a braided form (Millar 2000).

Changes in streamflow also influence stream geomorphology by driving shifts in floodplain vegetation. As rivers such as the San Pedro become drier, cottonwood/willow forests shift to saltcedar shrublands and then to sparsely vegetated xeroriparian shrublands (see chap. 1); the associated changes in stem density and stem size lead to geomorphic changes. Stands with high density of large-stemmed shrubs such as saltcedar may result in increased hydraulic roughness and greater dissipation of energy and sedimentation in comparison to the more widely spaced, but larger, cottonwoods and willow stems (Hadley 1961, Graf 1980, Hupp 1992, Phillips et al. 1998). High density of woody stems can cause geomorphic adjustments including bank stabilization, channel downcutting, decreased channel migration, and channel narrowing (Graf 1978, Graf 1982, Birkeland 1996, Allred and Schmidt 1999, Tickner et al. 2001, Cooper et al. 2003). The resulting restricted and stabilized channel may not be able to adjust to changes in discharge or convey large floods, and flows that would previously have been contained within the channel will inundate portions of the floodplain (Blackburn et al. 1982). Greater study is needed to assess the full geomorphic effects of saltcedar presence on actively scoured rivers such as the San Pedro, and to determine if flood energy dissipation processes vary substantially along hydrologic gradients.

FLOOD REGIME CHANGES

Flood magnitude and frequency may increase in the future in response to intensification of the hydrological cycle associated with global warming (see chap. 3) and perhaps also to urbanization of the watershed. The size of a flood is important with respect to the amount of erosion that occurs: larger floods have greater stream power, and most sediment movement in dryland fluvial systems is accomplished by large floods (Baker 1977). Equally important to size is duration of flooding. Floods of medium to long duration with large peak stream power per unit area are considered to be most effective in altering floodplain geometry (Costa and O'Connor 1995) and can result in dramatic erosion and channel widening (Huckleberry 1994a).

In addition to direct effects, floods indirectly produce geomorphic changes by altering the density, age, and species composition of riparian vegetation (Auble and Scott 1998, Rood et al. 1998; and see chap. 1). Over time, as the pioneer trees grow and as new plant species colonize their understory, geomorphically significant characteristics such as stem density, stem size, rooting depth, and vegetation height change in tandem.

FIRE REGIME CHANGES

Wildfires influence fluvial geomorphic processes by altering vegetation, soil, and rock characteristics. When upland hillslopes burn, this can increase the amount of sediment supplied to river channels (Benda et al. 2003). Post-fire sediment is delivered from tributaries to higher-order channels in debris flows or in sediment-laden streamflows following storms (Wohl and Pearthree 1991). This punctuated sediment delivery can lead to channel aggradation, altered substrate particle size, and alluvial fan growth at tributary junctions and can create heterogeneous habitat and landforms in river and floodplain ecosystems (Miller et al. 2003). Fire-induced sediment pulses can also alter channel gradient. Because arid-region rivers such as the San Pedro transport most of their sediment load during infrequent large flows, sediment from tributary streams may remain in place for years, forming alluvial fans within the axial floodplain and altering slope-energy relationships within the channel until discharge is competent to remove it.

Fire in the riparian zone, in the short term, is similar to floods in destroying vegetation and thus decreasing flow resistance, bank strength, and nucleation of bar sedimentation and in increasing the supply of woody debris for overbank deposits (Hicken 1984, Huang and Nanson 1997). Reduced flow resistance in the floodplain could lead to higher flow velocities and to enhanced sediment transport and channel erosion. The longer-term effects of floodplain fires will depend upon the fire tolerance of riparian plants and their rate and style of post-fire revegetation (see chap. 1). The net effect may be reduced flow resistance in parts of the floodplain with high tree mortality and little resprouting, but heightened flow resistance and sedimentation in areas with dense stands of resprouted trees.

ANIMAL ABUNDANCE

Livestock. Livestock grazing has a long history in the San Pedro River Valley, stemming back to the 1600s (see chap. 11). Livestock grazing can affect floodplain geomorphology by inducing changes in riparian vegetation abundance and composition, but cause-and-effect relationships between grazing and geomorphic processes are complex. Dense herbaceous cover increases hydraulic roughness, slows water velocity, and facilitates sedimentation, and its removal through grazing can increase the erosion potential of floodplain surfaces. However, selective browsing by cattle can drive shifts from pioneer trees to pioneer shrubs (see chap. 1), potentially resulting in channel narrowing.
Grazing on surrounding hillslopes influences geomorphic processes by compacting soils and modifying runoff. Reduced infiltration leads to increased surface runoff; cattle trails then foster rill and gully formation by concentrating this runoff (Cooke and Reeves 1976). Erosion on hillslopes translates to increased sediment loads in streams and washes, leading to sedimentation which can favor disturbance-dependent vegetation. Other possible geomorphic effects include changes to stream width and depth; bank angle, height, and stability; and pool depth (Myers and Swanson 1995, Sidle and Sharma 1996, Belsky et al. 1999).

Beaver. North American beaver (*Castor canadensis*) are semiaquatic rodents that construct lodges and dams out of logs, branches, mud, and stone (see chap. 6). One primary geomorphic effect of beaver is enhancement of sedimentation, which follows from the reduction of water velocity due to dam construction (Butler 1995). As beaver ponds fill with silt, and water tables are raised, the areas become wet meadows (Ives 1942). During large floods, beaver dams are damaged or destroyed, but they do reduce conveyance and attenuate flood waves, thus reducing the erosive potential of large floods. In addition to dams, beavers construct canals and other diversions to flow either to assist the transport of logs or to maintain pond depth (Butler 1995). These features can divert water within the floodplain and ostensibly play a role in channel shifts and meander abandonment.

Some argue that the removal of beaver in the late 1800s played an important role in the entrenchment of floodplains throughout the Gila River watershed (Dobyns 1981, McNamee 1994). However, the effectiveness of beaver dams to slow the largest floods that have tremendous erosive power is questionable. Unfortunately, quantitative data demonstrating the hydraulic effects of beaver dams along arid streams are lacking, and thus it is difficult to determine the geomorphic response to beaver removal. Today, beavers are present along portions of the upper and lower San Pedro River, providing opportunities to measure the geomorphic significance of this species.

Conclusions

Rivers like the San Pedro are dynamic components of the landscape that constantly adjust channel size, shape, and gradient in response to changes in runoff and sediment. Vegetation plays an important role in this process by influencing patterns of runoff, surface stability, and sediment movement both within and outside of floodplains. As channels make adjustments, floodplains experience erosion and deposition: the former process destroying previously established vegetation and the latter process commonly creating new planting surfaces. During plant succession, differences in plant type, age, and structure alter the hydraulic conditions of the floodplain, thus affecting channel behavior. It is a two-way relationship between plants and rivers, and the dynamic between the two can be quite complex and difficult to predict.

One certainty, however, is change. Stratigraphic evidence indicates that the San Pedro River has altered its channel geometry several times over the last several thousand years. Over the last 150 years, the river channel has downcut, shifted laterally, changed between braided and meandering forms, and deposited sediment in the floodplain. Arroyo cutting removed numerous marshes within the floodplain but eventually created fresh disturbed surfaces for establishment of the large willow and cottonwood galleries that are the hallmark of the San Pedro and provide key habitat for migratory birds and other species.

Because fluvial systems are naturally prone to change due to climate variability and intrinsic geomorphic processes, it is difficult to quantify the degree to which humans have caused past and present transformations of the San Pedro River. Many of the geomorphic changes experienced by the San Pedro River during the last 150 years are undoubtedly linked in part to water depletion, overgrazing, deforestation, and introduction of plant species. Human actions continue to impact fluvial geomorphic processes in the river today through improved grazing management but also increased urbanization in the uplands; intentional burning and increased accidental fire ignitions; introduction of beaver and exclusion of livestock from some river reaches; and, depending on reach, reduced or increased groundwater pumping. More research is needed to monitor current geomorphic processes in rivers like the San Pedro to provide greater insight into current conditions and help forecast future changes in floodplain behavior under a variety of climate, land use, and population growth scenarios.

DOUGLAS M. GREEN,
JULIET C. STROMBERG,
AND RONALD L. TILLER

Riparian Soils

FOURTEEN

Introduction

Soils are the unconsolidated mineral and organic material on the immediate surface of the earth that serve as natural media for the growth of land plants (Gardiner and Miller 2004). Soils also can be regarded as complex biogeochemical systems that receive the bulk of photosynthetically fixed carbon via litter fall or root turnover (Coleman and Crossley 1996, Bardgett et al. 2001, Wardel 2002). Riparian soils, due to their position within the landscape, are the recipients of sediments and other materials from the watershed and are thus important regulators and transformers of energy and materials between terrestrial and aquatic ecosystems (Lowrance et al. 1984, Naiman and Decamps 1997, Hill and Cardaci 2004). Because of flood-induced deposition and erosion, the riparian soilscape is more temporally dynamic and spatially heterogeneous than that of adjoining uplands.

In this chapter we focus on soils of the San Pedro, drawing on local efforts and studies conducted on other desert rivers. We briefly describe soil taxonomy and review key factors in soil development. We then describe the soil characteristics of four zones within the riparian corridor—channel bars, backwater wetlands, floodplains, and terraces (old, no-longer-flooded surfaces).

Soil Formation and Taxonomy

Soils are a product of five factors: parent material, climate, relief (topography), organisms, and time (Jenny 1941, Hendricks 1985, Buol et al. 2003). These factors drive processes such as sedimentation, erosion, leaching, eluviation, illuviation, calcification, salination, precipitation and dissolution of minerals, oxidation and reduction, and physical weathering. Over time, these processes produce soils with a distinctive sequence of horizons and chemical and physical properties (or, in the case of many floodplain soils, the lack of a distinctive soil profile).

Riparian soils associated with the San Pedro River are primarily classified in the Entisol or Aridisol soil orders (Buol et al. 2003, USDA, NRCS 2003a). Entisols are young soils that have little or no evidence of development of pedogenic horizons (Soil Survey Staff 1999) and are common in floodplains throughout the desert Southwest (Muldavin et al. 2000). Lack of pedogenic horizons can be due to insufficient time for soil-forming processes to operate, low water-holding capacity of gravels or sands, erosion rate that equals or exceeds depositional processes, low fertility, or anaerobic environments that restrict biogeochemical processes; the first three are paramount on the San Pedro.

Aridisols are soils that occur in cool and warm deserts and show development of pedogenic horizons. Most terrace soils in the San Pedro River Valley are Aridisols, in the Calcid suborder (soils with accumulation of calcium carbonate in the profile) or Gypsid suborder (accumulation of gypsum, i.e., calcium sulfate). Other soil orders present in riparian areas of the San Pedro or nearby rivers include Inceptisols (soils of terraces lacking an illuvial horizon enriched with clay, iron, or organic matter), Vertisols (clayey soils in seasonally flooded wetlands such as playa lake beds), Mollisols (dark organic-rich soils of some riparian grasslands), and Histosols (in some graminoid-dominated wetlands)(Muldavin et al. 2000).

STREAM GEOMORPHOLOGY, WATERSHED POSITION, AND PARENT MATERIAL

The position of a stream in its watershed influences riparian soils in part by controlling the type and diameter of parent materials (Malanson 1993). Soils of valley systems such as the San Pedro are more diverse than those in headwater canyons because the riparian corridors have complex internal geomorphology and receive sediment from a variety of sources (L. Lewis et al. 2003). Alluvial sediments re-worked and deposited by the stream are the dominant parent material on valley floodplains. These sediment deposits are typically well sorted, with the long axis of the lenses roughly parallel to streamflow. Alluvial sediments also arrive from adjacent uplands. These typically overlie river-deposited sediments at the margins of the floodplain or on stream terraces. Underlying these stream-deposited and upland-deposited sediments are clays, silts, sands, and cobbles that fill the valley. These valley-fill sediments can be several hundred meters thick, having been laid down during previous geologic eras; they are important groundwater aquifers and are hydraulically connected with the river (see chap. 15).

FLOODS

San Pedro River floods transport and deposit considerable amounts of sediment. When the stream exceeds bank-full capacity and flows onto its floodplain, the hydrologic roughness of the floodplain causes stream velocities to slow and entrained sediments to settle as a layer on the surface (fig. 14.1) (see chap. 13). The sediment vertically accretes at an average rate of 1 to 3 cm per year (Clark 1987, Stromberg 1998b), although this rate is highly variable. Floods of long duration can erode the terrace walls and fill channels with deep layers of sediment (Minckley and Clark 1984, Stromberg et al. 1997). Following an extreme flood in the San Francisco River (upper Gila basin), for example, well-developed soils up to 2 m deep were eroded in places and replaced with riverwash (deep stratified sand, gravel, cobbles, and stones; Brock 1985). In contrast, small, monsoon floods typically deposit thin layers of silts and clays. Soil profiles characterized by vertical accretion have (1) distinct horizon boundaries with often sharply contrasting textures indicative of differing unit stream powers associated with different flooding events, (2) organic matter that decreases irregularly with depth, and (3) buried horizons.

Lateral accretion typically occurs on surfaces close to the stream channel in areas of high unit stream power, for example, the point bar of a stream meander. At these sites, soils opposite the point bar on the cut bank are undercut and collapse into the stream. The resulting newly derived parent material is transported and deposited on downstream point bars, often in a turbulent environment. Soil profiles formed from laterally accreted parent materials have (1) thick horizons containing coarse gravels, (2) organic matter content that decreases regularly or is homogenous with depth, and (3) absence of buried horizons.

Woody debris deposited by floods likely influences soil formation on floodplains of the San Pedro as it does on rivers elsewhere (Montgomery and Abbe 2006). Woody debris acts as a site of deposition for fine soil particles and may itself act as a source of nutrients and organic matter to the soil (see chap. 18). However, the extent to which woody debris influences soil formation in southwestern riparian systems is largely unknown.

TOPOGRAPHY AND HYDROLOGY

Desert riparian soils can be quite dry. On high floodplains and terraces, where groundwater is deep and inundation infrequent or absent, soil types develop that show little evidence of aquic (hydric) or oxyaquic conditions

Fig. 14.1. Horizontal layers characteristic of vertically deposited sediments. Note the contrasting layers denoting various flooding events. Verde River near Horseshoe Reservoir. Photo credit: Doug Green.

within the first meter or so (Muldavin et al. 2000). On lower floodplains, aquic or oxyaquic soil conditions can be present at soil depths of 50 to 150 cm or so, depending on surface elevation and depth to groundwater.

When soils are flooded or ponded long enough during the growing season to develop anaerobic conditions, hydric (aquic) soils develop (USDA, NRCS 2003b). Such soils can develop in groundwater-fed wetlands, along the edges of perennial streams, and in depressions in the floodplain such as channel cutoffs (Dick-Peddie et al. 1987). Gley, mottles, and blackened organic matter are redoximorphic features that are indicators of hydric soils. Gleying, the result of the biochemical reduction of iron, occurs in soil horizons with permanent saturation and anoxia and gives the profile a gray or bluish gray color. Mottles (often a bright rusty orange/red in a matrix of gray soil material) often occur in upper soil layers where saturated conditions alternate with dry periods, and indicate seasonally fluctuating water levels.

Oxyaquic soils develop where conditions are saturated but not reduced. Most southern Arizona groundwater is oxic (Winograd and Robertson 1982, Robertson 1991), and oxygen flow through the coarse-textured soils that prevail on floodplains of the San Pedro River generally is sufficient to prevent redoximorphic features from forming. Within this aerobic matrix, however, localized anoxic microsites can develop in buried pockets of organic matter in channel bar and floodplain soils (Holmes et al. 1996).

CLIMATE

Soils of many rivers of the desert Southwest have a thermic temperature regime, with mean annual temperature between 15° to 22°C at a depth of

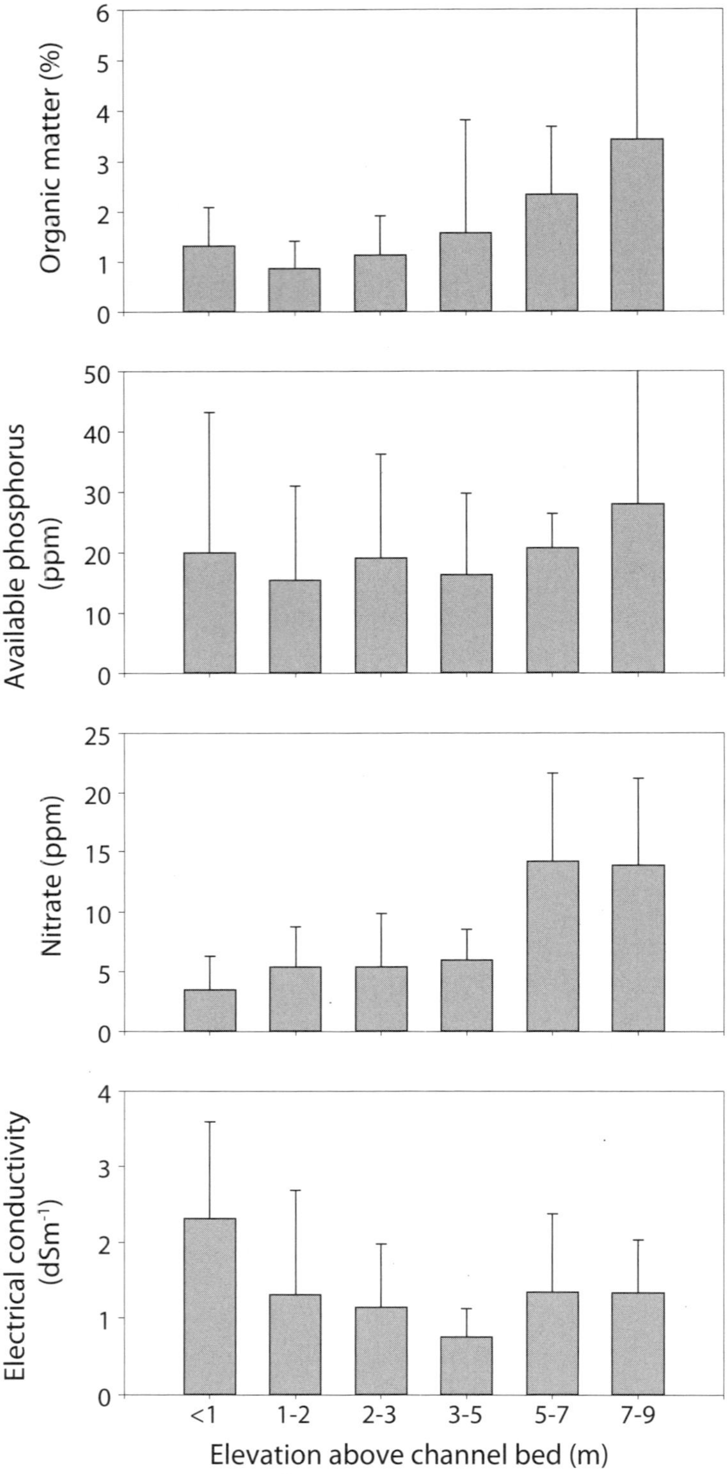

Fig. 14.2. Chemical and physical traits of riparian zone surface soils (top 15 cm) along a gradient of elevation above the channel bed. Values are means (plus one standard deviation) of soils collected from sites in the San Pedro Riparian National Conservation Area (unpublished data of authors).

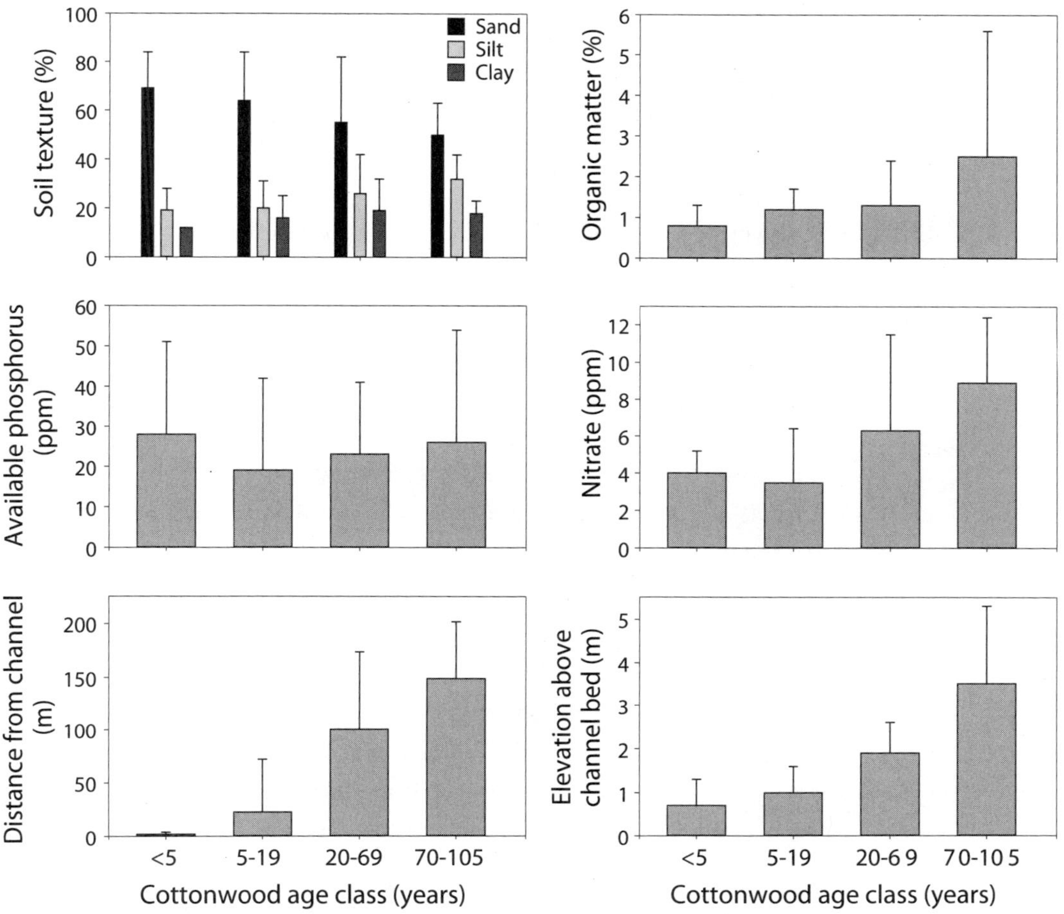

Fig. 14.3. Mean values of soil physical and chemical traits in relation to Fremont cottonwood (*Populus fremontii*) stand age. Also shown are the surface elevation above the channel bed and the distance from the low-flow channel edge. Soils were collected from the San Pedro Riparian National Conservation Area (top 15 cm) (unpublished data of authors).

50 cm (McQueen and Miller 1972, Minckley and Clark 1981), but temperatures vary considerably seasonally and spatially. Values of 35°C at a depth of 0.05 m have been recorded in unshaded riparian soils, with much cooler temperatures under dense forest canopy (McQueen and Miller 1972, Andersen 1994).

The high potential evaporation rates that occur along the San Pedro River (see chap. 2), particularly in its lower reaches, influence soil chemistry. On terraces, low rainfall and high evapotranspiration rates permit carbonate and sulfate salts to accumulate in the soil profile or on the surface, producing soil pH values commonly in the alkaline range (median pH of 8.1 for San Pedro riparian soil samples). Other salts accumulate in floodplains underlain by a shallow water table, with values increasing near the soil surface (Gary 1965). During prolonged periods with no flooding, ions concentrate in the soils. However, with electrical conductivity (an indirect measure of salt content) values of less than 2 dS m^{-1} (fig. 14.2), floodplain surface sediments of the San

Pedro are an order of magnitude less saline than those of heavily regulated and flood-reduced rivers in the Southwest (Van Hylckama 1980, Shafroth et al. 1995).

TIME

Some properties of riparian soils, such as salinity or nutrient content, change rapidly in response to fluctuations in streamflow conditions, while others change over time in scales of weeks to decades. For example, potassium is a mobile nutrient and is often imported in solution via streams and groundwater and readily lost from the ecosystem by leaching (Boggs and Weaver 1994). Some forms of nitrogen also are highly soluble (e.g., nitrate) and can be rapidly flushed from soils during flood pulses; nitrogen levels also can increase abruptly after floods deposit fine, nutrient-rich sediments (Adair et al. 2004).

During periods when microbial organisms are highly active, organic matter content can change appreciably over periods of weeks. During the spring flood season along a cold desert river in Colorado, organic matter levels in leaf packs declined by about half, with decomposition rates increasing as a function of inundation time (Andersen et al. 2003). Decomposition of other types of organic matter can play out over longer time periods. Along a flow-regulated reach of the Rio Grande in New Mexico that was flooded on an experimental basis, the estimated decay rate for cottonwood (*Populus*) logs was 0.065 per year, with a predicted half-life of 10.6 years (Ellis et al. 1999). Multiple cycles of annual flooding are necessary to reduce the standing stock of forest floor litter that can accumulate during long periods of no flooding on dammed rivers (Molles et al. 1995).

Soils in riparian forests vary over successional time, as evidenced through studies of chronosequences. Along the San Pedro River, texture and chemistry of surface soils differed between stands of sapling, middle age, and old Fremont cottonwood (*Populus fremontii*), with silt, available phosphorus, organic matter, and electrical conductivity all increasing with forest age (Bagstad et al. 2006) (fig. 14.3). Along desert rivers in Colorado, clay, sodium, potassium, pH, and total soil nitrogen all increased with forest age (Adair et al. 2004). Over successional time, vertical accretion typically increases the elevation of the soil surfaces above the flood-prone elevation, while the channel thalweg continues to shift away; thus, in situ processes begin to dominate the soil-forming process. Nutrient-rich silts and clays are deposited during small flood pulses, while at the same time organic matter and nutrients accumulate due to autogenic soil-building processes in concert with reduced frequency of litter scour (Johnson et al. 1976, Bush and Van Auken 1986, Adair et al. 2004). The old forest stands produce a soil that is more characteristic of upland soils, i.e., organic matter accumulates and regularly decreases with

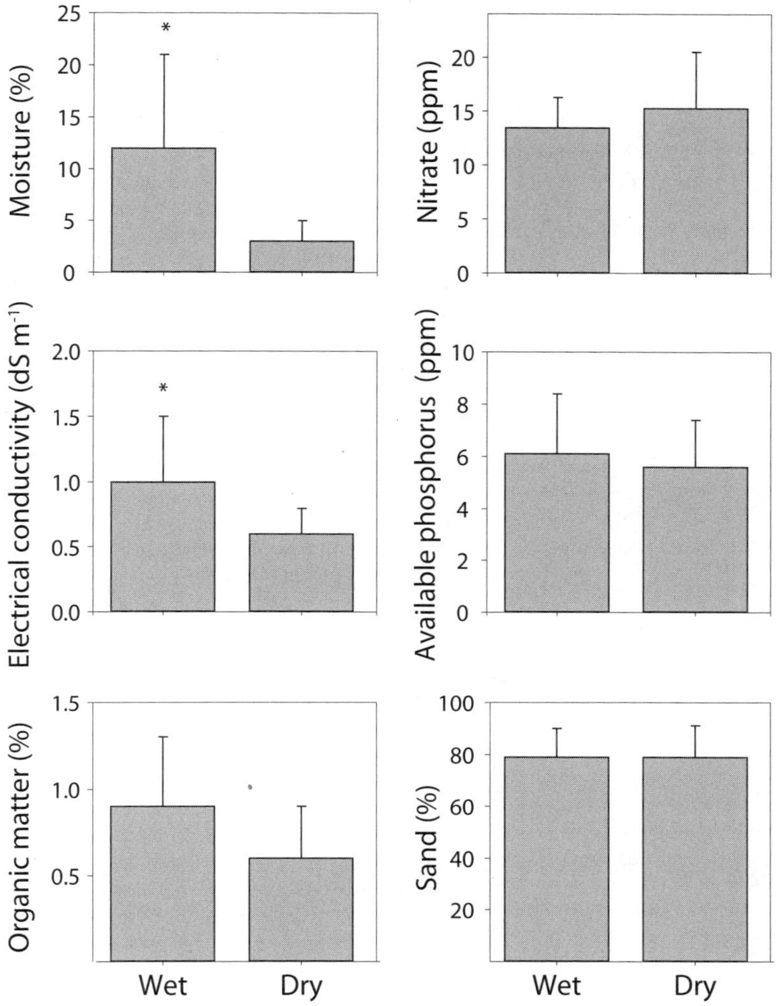

Fig. 14.4. Comparison of physical and chemical traits of stream bar soils (top 15 cm) between San Pedro River sites with high streamflow permanence (surface flow present more than 70 percent of the time during a two-year period; n=10) and sites with more intermittent flow (n=8). Data are averages for June and August 2000, from sites within the lower and upper basins. Asterisk indicates significant difference at $p<0.10$, based on Mann-Whitney U tests (unpublished data of authors).

depth, soil structure develops, and carbonates, clays, and other materials eluviate and illuviate.

BIOTIC INFLUENCES

Soil microorganisms are the principal drivers of biogeochemical cycles. Bacteria are responsible for many of the chemical transformations that occur in soils, and they drive processes such as nitrogen fixation, nitrification, and denitrification (see chap. 18). Some nitrogen-fixing bacteria live freely in riparian soils (Ota et al. 1991), while others form a symbiotic relationship with particular plant species or families. Nitrogen-fixing bacteria in the

Rhizobium and *Bradyrhizobium* genera inhabit roots of mesquite (*Prosopis* spp.) (Virginia and Jarrel 1983, Virginia 1986, Klemmedson and Tiedemann 1986, Jenkins 2003), a legume tree that is abundant in the San Pedro riparian zone; this symbiosis results in high levels of soil nitrogen under mesquite canopy (Schade and Hobbie 2005). Biological soil crusts composed of cyanobacteria, algae, lichens, and mosses can develop on stable riparian soil surfaces and play roles in nutrient cycling and soil stabilization (Dobkin et al. 1998, Belnap and Lange 2001).

Fungi together with bacteria play pivotal roles in decomposition and release of nutrients from organic residues to the soil and thus to plant roots. Humus, the by-product of microbial decomposition of organic matter, enhances water-holding capacity, nutrient exchange, and soil structure (Paul and Clark 1996). Soil fungi also influence riparian soils by producing exudates and binding soil particles with their hyphal strands (see chap. 4).

Riparian vegetation influences soil properties by providing inputs of organic matter, influencing sediment deposition, and modifying the chemistry of the rhizosphere (root zone) through uptake and release of chemical compounds. Vegetation types on the San Pedro floodplain that have high productivity, such as hydroriparian shrublands, have more organic matter in their soil than do less-productive patches such as xeroriparian shrublands (Bagstad et al. 2006). Similarly, stands of cottonwood trees can input more leaf litter to the soil surface than stands of saltcedar (Ellis et al. 1998). Vegetation types with high density of woody plant stems, such as saltcedar shrublands, can concentrate clay in the soils because the stems reduce flow velocities and allow fine particles to settle (Gary 1965, Stromberg 1998b, Bagstad et al. 2006). Exuded salt on stems of some halophytic plants also may flocculate clay and induce settling (Everitt 1980).

Riparian plant species differentially concentrate cations and anions, potentially affecting the chemistry of the soil. For example, although Fremont cottonwood and Goodding willow (*Salix gooddingii*) had generally similar foliar cation profiles, they differed somewhat in Ca and Mg content (Busch and Smith 1995). Saltcedar (*Tamarix*) had higher leaf nitrogen content than cottonwood (Tibbets and Molles 2005), and inorganic nitrogen has been found to be high in soils under saltcedar (Lesica and DeLuca 2004). Halophytic phreatophytes such as saltcedar can have higher Na:K ratios in leaf tissue than do glycophytes such as cottonwood and willow, and thus potentially salinize surface soils. However, amounts of chloride and other chemicals in plant leaves and in surface soil also vary considerably depending on the chemical composition of groundwater (Hem 1967, McQueen and Miller 1972). On the San Pedro River, electrical conductivity levels in saltcedar soils do not vary significantly from other forest types (Stromberg 1998b, Bagstad et al. 2006), likely because there are low levels of source salts to concentrate and because soils are frequently flushed by floodwaters.

Small animals influence riparian soils through their feeding and earth-moving activities. Earthworms and larger fossorial mammals such as Botta's pocket gopher (*Thomomys bottae*) create pathways for the movement of water and gases through the soil and exert significant influence on soil structure (Anderson 1987, Jones et al. 1994; and see chap. 6). Nymphs of cicadas (including Apache cicadas [*Diceroprocta apache*] and dog-day cicadas [*Tibicendealbatus*]) redistribute water in floodplain soils through their feeding activities and, as they emerge from the soil, form vertical tunnels that influence water infiltration (Anderson 1994). Surface-active arthropods such as carabid beetles shred plant litter and thus help to mineralize nutrients by physically breaking up organic matter (Ellis et al. 2001); however, insects and other macroinvertebrates played a minor role in the breakdown of cottonwood leaf litter along desert rivers in Colordao (Andersen and Nelson 2003, 2006).

Soils in areas with high use from livestock can become compacted, with little water or air penetration, high runoff, and reduced plant growth (Trimble and Mendel 1995); these changes can slowly reverse with time since livestock removal (Rogers 1994). Irrigated agriculture also alters soil properties. Tillage churns and mixes distinct soil horizons, altering the dynamics of water infiltration, storage, and availability. Cultivation leads to the depletion of soil organic matter and alters nutrient concentrations, while irrigation may increase the concentration of salts in the soil. Along the San Pedro, several attributes of the soils change with time since field abandonment (Oleksyszyn 2001). Soils of some abandoned fields on stream terraces have less available phosphorus, more sand, or similar texture, but similar pH, electrical conductivity, nitrates, and organic matter than soils of nearby sacaton grassland (Richter et al. 2002).

Soils of the San Pedro

Soils of valley systems such as the San Pedro River can be classified into four groups: (1) those of newly deposited stream bars, (2) soils of backwater wetlands, (3) floodplain soils, and (4) terrace soils (Platts et al. 1987, Phillips et al. 2001, Lewis et al. 2003).

STREAM BAR SOILS

Soils associated with new bars in the stream bed—including mid-channel bars, lateral bars, and point bars—are very young and lack pedogenic horizons. These laterally accreted, mixed-stream alluvial deposits are the initial state from which other soils develop in the system and differ from them in several ways. Stream bar soils along the San Pedro River have significantly more sand (mean of 79 percent), less clay (10 percent), less organic matter (<1 percent), and higher electrical conductivity than soils of the floodplain (Bagstad et al. 2006).

Stream bar soils vary depending on local hydrology and associated vegetation development. Over the length of the San Pedro River, soil moisture and electrical conductivity of stream bars are greater at sites with more permanent streamflows (fig. 14.4). Organic matter is somewhat elevated at the wetter sites, reflecting the denser cover of herbaceous wetland vegetation that develops where the stream flows year-round (see chap. 1).

Stream bar soils also vary locally depending on stream power. At regions of high unit stream power, such as the upstream end of stream bars and margins of point bars, the deposits tend to be coarse textured, with low water-holding capacity and nutrient status. Although groundwater level may be relatively close to the surface, the coarse texture of these soils limits capillary rise. In regions of lower unit stream power, bar deposits are finer textured, ranging from medium sands to silt, and have greater water-holding capacity. These local variations in texture may exert fine-scale influence on plant distribution and nutrient cycling.

BACKWATER/CIÉNEGA SOILS

Ciénegas are wetlands of warm areas of the American Southwest that have permanently saturated soils and are located along low-order streams with moderate flood intensities. In the remnant ciénegas that persist along the San Pedro River groundwater inflow creates seasonally to permanently ponded water, and thus hydric soils. Filling of pore space with water reduces oxygen diffusion rates, creating anoxic conditions and leading to development of redoximorphic features such as mottles.

Hydric soils are important sites of anaerobic activity. Anaerobic activity is favored by high organic-matter content, shallow water table, and warm temperatures of the surface layers (Mitsch and Gosselink 2000). Under anaerobic conditions, high microbial activity leads to denitrification, and significant amounts of nitrate can be reduced to nitrogen gas (C. A. Johnston et al. 2001). Anaerobic conditions slow decomposition and promote accumulation of organic matter, making wetland soils important sites of carbon storage. Globally, Histosols (organic-rich soils such as occur in ciénegas) account for approximately 25 percent of the terrestrial soil organic carbon pool while occupying only 1 percent of the land surface (Eswaran et al. 1995, Wilding 2000). Histosols are less prevalent in the desert Southwest than they were in the 1800s, when wetland conditions prevailed along the San Pedro and many other rivers.

Ciénega soils consist of layers of organic material interbedded with layers of silt, sands, or gravels (Rinne and Minckley 1991). The soils can build to

TABLE 14.1. Traits of surface soils (top 15 cm) within vegetation types and fluvial landforms of the upper San Pedro riparian corridor.

Also listed for each type are mean distance from the low-flow channel and elevation above the water table.

	Channel		Floodplain			Terrace	
	Streambar	2° channel	*Baccharis* shrubland	*Ericameria* shrubland	*Populus* forest	*Prosopis* forest	*Sporobolus* grassland
Distance (m)	00 ± 00	62 ± 45	47 ± 75	66 ± 65	100 ± 74	95 ± 99	105 ± 80
Elevation (m)	0.1 ± 0.3	1.0 ± 0.8	1.2 ± 0.9	2.6 ± 1.4	1.9 ± 0.7	5.2 ± 1.4	4.3 ± 1.2
Sand (%)	78 ± 25	38 ± 15	62 ± 21	71 ± 14	55 ± 27	50 ± 27	30 ± 12
Silt (%)	11 ± 15	36 ± 08	21 ± 10	13 ± 07	26 ± 16	27 ± 14	41 ± 11
Clay (%)	11 ± 11	26 ± 13	16 ± 12	11 ± 06	19 ± 13	23 ± 14	29 ± 07
Gravel (%)	27 ± 26	02 ± 02	14 ± 25	11 ± 07	5 ± 17	4 ± 8	00 ± 00
Organic matter(%)	0.7 ± 0.8	1.7 ± 0.4	1.0 ± 0.4	0.5 ± 0.5	1.3 ± 1.1	2.1 ± 1.1	2.4 ± 1.2
Nitrate (mg kg^{-1})	4.6 ± 2.6	2.7 ± 2.4	7.5 ± 7.9	5.6 ± 3.0	6.3 ± 5.2	13.6 ± 8.0	8.4 ± 5.7
Phosphorus (mg kg^{-1})	12 ± 04	17 ± 02	11 ± 07	16 ± 16	23 ± 18	17 ± 05	23 ± 11
EC (dS m^{-1})	2.7 ± 1.5	0.8 ± 0.1	0.6 ± 0.6	0.7 ± 0.4	2.3 ± 0.4	1.3 ± 0.9	1.0 ± 0.3

Note: Soils were collected in 1993 (see Stromberg et al. 1996).

depths of several meters as the vegetation grows and decomposes and as alluvial sediments are deposited during floods. The rate of soil accumulation can be rapid, as evidenced by the presence of a meter of organic peat overlying a mesquite stump felled within the last 150 years (Collins et al. 1981). Remnant organic deposits that are several meters deep are visible in the vertical arroyo walls of the San Pedro and other rivers in southern Arizona (Hendrickson and Minckley 1984).

FLOODPLAIN SOILS

Floodplains are topographically complex, dissected by ridges, swales, overflow channels, and abandoned channels. This complexity, along with heterogeneity of vegetation, leads to different flow velocities during over-bank flooding; this in turn creates deposits with a wide range of textures within a relatively small area (I. Moore and Burch 1986, Huggenberger et al. 1998, Johannes and Gurnell 2003). Textural content of soils from the upper San Pedro River reflects the spatial heterogeneity of the floodplain.

Sandy sediments predominate in surface soils of much of the San Pedro floodplain (table 14.1). Because of their coarse texture, floodplain soils have low water-holding capacity and little capillary rise (Brock 1985). The capillary fringe can extend about 1 to 2 m above the water table in clay soil, but only 15 to 30 cm in sandy soils (R. Zimmerman 1969, McQueen and Miller 1972). Nutrient retention and exchange in soils is usually associated with the negatively charged clay (<0.002 mm) particles, and thus the sandy floodplain soils can be relatively nutrient poor (see chap. 18). Even in areas of the floodplain with mesquite trees, soil nitrate can be less abundant than for terraces (table 14.1), owing in part to differences in soil textures and flood patterns between the two geomorphic zones.

Despite considerable topographic and vegetational complexity, there is a trend for sediment diameter to decrease across the floodplain with distance from, and elevation above, the channel (Alexander and Prior 1971, Johannes and Gurnell 2003). Clay content increases and sand content decreases, reflecting reduced stream power on the higher, distal surfaces (fig. 14.4). However, the conditions of the surface soils are not necessarily reflective of conditions at depth; the floodplain soils typically have lenses of sand or gravel within the profile, reflecting vagaries of past depositional events (Gary 1965, McQueen and Miller 1972).

Floodplain soils along the San Pedro River are young and poorly developed and classified as Entisols. The youngest lack an A horizon, while those on the highest floodplain positions do have an A horizon. At sites where litter fall is heavy and accumulates, such as below dense cottonwood/willow canopy on aggraded surfaces, a thin O horizon may develop. In the C horizons, where soils are little altered from the parent material, there is an irregular decrease of organic carbon, and redoximorphic features can be present in the lower depths. Soil structure, an indicator of profile development, is lacking in the lower horizons. Brookline and Quiburi soils (both Torrifluvents) are typical of soils in these floodplain positions and occupy significant areas along the San Pedro River (USDA, NRCS 2003a).

TERRACE SOILS

Terrace soils are located well above the flood-prone elevation and thus are similar to upland soils in having relatively advanced in situ pedogenesis, including a well-developed A horizon darkened by organic matter and development of one or more B horizons with illuvial accumulation of clays, carbonates, and other materials from upper horizons. Gulch, Major, and Kaboom soils are common on terraces. Gulch is a Calciargid, an Aridisol that has significant accumulation of clays and carbonates. Major and Kaboom soils are Calcigypsids, Aridisols with significant accumulation of carbonates and gypsum in the B horizons (USDA, NRCS 2003a).

Terrace soils are quite deep. In areas of the San Pedro where the river channel has downcut by several meters, multiple strata are visible in the arroyo walls. The surface soils of the San Pedro terrace sacaton (*Sporobolus*) grasslands are largely dominated by silts and clays (silty loam to silty clay loam), while coarser sediment layers of sandy loam to loam occur deeper in the profile. Flory (1937) observed multiple organic horizons beneath sacaton still connected to vestigial stems and roots over 2 m below.

Terrace soils receive alluvial material from adjacent uplands and furnish alluvium to floodplain soils. Alluvial sediments from adjacent uplands may be enriched in clay or silts, when derived as sheet flow from well-developed soils such as the desert grassland soils that border portions of the San Pedro. Alluvial sediments can also include larger particles, when arriving as sheet flow from overgrazed and eroded desert scrub soil, or where entering from ephemeral washes or alluvial fans. Sorting of sediment sizes, when it occurs, produces deposits with the long axis roughly parallel to the flow that deposited them, and perpendicular to flow of the stream in the valley.

Access to groundwater by the dominant plant species on terraces allows for high primary productivity, with cascading influences on soil traits. Along the Verde River, nitrate levels in mesquite soils were greater on terraces than uplands, likely reflecting the greater primary productivity of the former (Schade and Hobbie 2005). Along the upper San Pedro River, about 51 percent more organic carbon and 45 percent more total nitrogen were detected in soils of riparian mesquite woodlands than of upland desert scrub, and about 23 percent more organic carbon and 9 percent more total nitrogen were detected in soils of riparian grasslands than of upland desert scrub (Martens and McLain 2005). Soils of the terrace mesquite woodlands have greater available nitrogen (mean of 14.5 mg kg^{-1}) than the terrace grasslands (mean of 7.0 mg kg^{-1}) and also differ in texture with the soils of the grasslands having greater percentages of fine soil particles (table 14.1).

Conclusion

Riparian soils play a critical role in ecosystem function as they are regulators of fluxes of materials between the upland ecosystem and the aquatic ecosystem. Development of riparian soils, as for other soils in other contexts, is influenced by climate (inclusive of effects on stream hydrology), relief (topography), organisms, and time. Soils of the San Pedro River Valley are similar to those of other broad valley systems in that they are highly diverse, reflecting the complex of climatic conditions, fluvial processes, sediment sources, and vegetation types that occur along and within the riparian corridor.

Four suites of soils occupy the San Pedro River Valley: stream bar soils, backwater/ciénega soils, floodplain soils, and terrace soils. Stream bar soils are very young and poorly developed and generally sandier than other soils in the riparian zone. Stream bar soils have relatively high concentrations of salts, given the climatic setting of the San Pedro River, but values are much lower than on rivers that flood less frequently. Backwater/ciénega soils occupy low-lying floodplain positions with water at or near the soil surface seasonally or permanently. These soils can accumulate significant amounts of organic matter and show evidence of anaerobic conditions. Floodplains contain diverse soils, reflecting the topographic and vegetative complexity of their geomorphic surfaces. Although textural variation within patches can be high, silt and clay content of soils increases with distance from the active channel. Floodplain soils are highly dynamic, changing dramatically after large flood events and changing at slower rates over successional timescales. Terrace soils, because of their location above the flood-prone zone, are more stable and show evidence of long-term pedogenesis such as accumulation of clays and carbonates within the profile.

Humans have impacted riparian soils of the San Pedro for several hundred years, largely through cultivation and livestock grazing, but extent of change and rates of recovery from past uses are largely unknown. The high degree of fluvial activity on floodplains may significantly shorten recovery times relative to terrace sites. Important information on soil ecology also is lacking in other areas. For example, the soil biota of desert rivers undoubtedly is quite diverse and functionally important but has received little study. Likewise, the role of woody debris in riparian-zone soil formation has received limited attention. Additionally, while soil-plant relationships have been explored, gaps persist in our understanding of these relationships in riparian ecosystems.

SECTION 4

The Hydrologic Template

ROBERT MAC NISH,
KATHRYN J. BAIRD, AND
THOMAS MADDOCK III

Groundwater Hydrology of the San Pedro River Basin

FIFTEEN

Introduction

Though primarily a desert environment, some of the mountain blocks in the Basin and Range Physiographic Province of the southwestern United States have risen to elevations sufficient to capture annual precipitation in excess of 500 mm and support perennial streams in the mountain canyons. Streamflow from these higher mountain blocks, and groundwater seeping from the mountain blocks into the sediment-filled valleys, also supports perennial and intermittent flow in the streams that course down the valleys between the mountain blocks. The groundwater discharge to the stream is a small but important component of the annual streamflow, and is called baseflow.

In this chapter we explore the hydrologic infrastructure supporting the riparian ecosystem of one such stream, the San Pedro. We discuss some basic aspects of groundwater hydraulics and the interaction of the groundwater and surface water system in the riparian corridor in the present time and the foreseeable future. We use the extensively studied Sierra Vista subwatershed of the upper San Pedro basin as our specific example. The Benson subwatershed is similar to the Sierra Vista subwatershed, although that more northerly portion of the upper San Pedro basin does not contain as much high-elevation terrain. The lower San Pedro basin also differs in being less than half as wide between the mountain blocks as the upper basin.

Groundwater Overview

AQUIFERS

There are two types of aquifers in the San Pedro basin. One type is found in the mountains where water moves through fractures in the bedrock. The second type is found in the basins and is composed of unconsolidated sediments (clay, silt, sands, and gravels). There are two primary unconsolidated aquifers in the San Pedro basin—the regional aquifer and the floodplain (or stream) alluvial aquifer.

Regional aquifer. The regional aquifer extends throughout much of the basin and is comprised of materials eroded from the adjacent mountains as they rose in the geologic past. This material was transported by the canyon streams to the mountain front and deposited in coalescing fan-shaped deposits. The deposits tended to be coarsest, and thinnest, near the mountain front and finer and thicker (to 300 m and more) near the center of the basin. The mountains arose by a series of movements along the faults that separate the mountain blocks from the valley. These sporadic uplifts disrupted the earlier deposits of sediment and produced new depositional material. Geologists divide the regional aquifer into several distinct units based on their depositional relationships, while hydrologists usually treat them as a single unit, or as two units, with the lower unit being less permeable due to partial cementation and compaction of the sediments. In this chapter we treat the multiple horizons of varying sediment compositions and abilities to transmit water as a single aquifer.

The mountains rose over millions of years, and for most of that time there was no through-flowing stream. Because of this, lakes or playas formed, and deposits of clay and silt that do not transmit water readily were deposited near the basin centers. Eventually, the basin filled, and a through-flowing stream (the ancestral San Pedro River) became established, which transported sediment derived from upstream locales. So, while the regional aquifer contains rocks derived from the adjacent mountains, the floodplain aquifer deposited by the ancestral San Pedro contains rocks from Mexico as well as from local areas. Figure 15.1 shows the relationships of the various hydrogeologic units in the Sierra Vista subbasin.

Floodplain aquifer. The floodplain aquifer is a long, thin body of sediments of varying width and thickness. Both the San Pedro River and its tributaries have influenced the configuration of the floodplain aquifer. Historic fluctuations in climate have caused the river's behavior to alternate between ero-

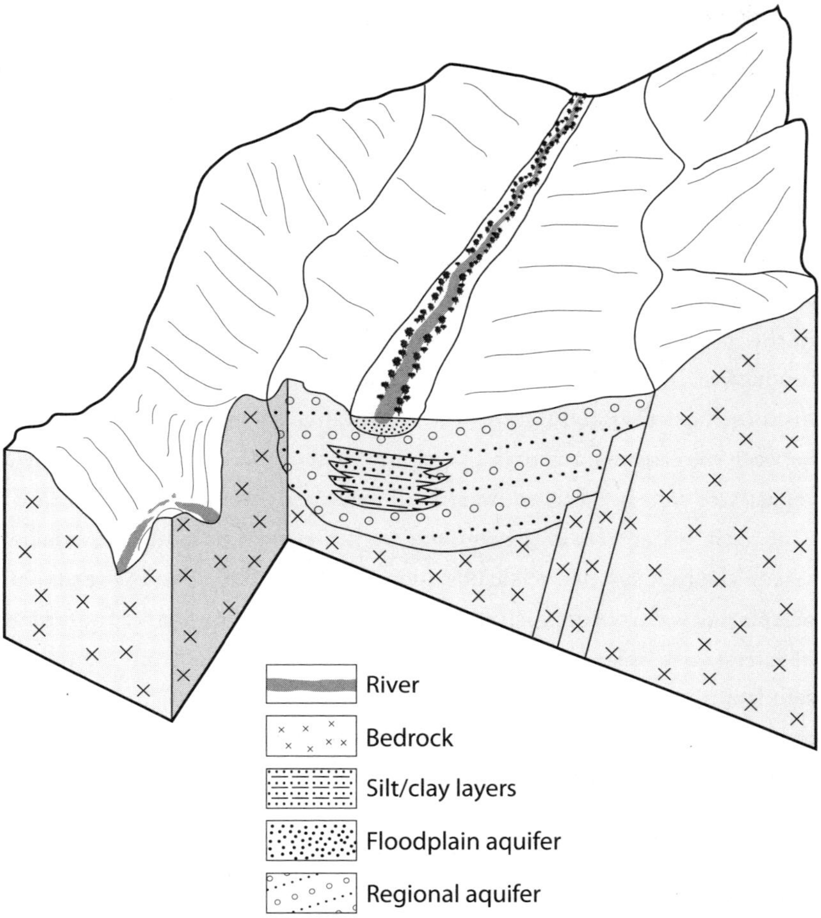

Fig. 15.1. Schematic block diagram showing the relationships and distribution of the hydrogeologic units in the Sierra Vista subbasin of the upper San Pedro River basin.

sional, where the river incises its channel more deeply, to depositional, where the channel fills with sediment and the river spreads out over the floodplain, depositing sediments and building up the floodplain. Local tributaries randomly contribute large sediment loads during the sometimes-spectacular summer thunderstorm season. These monsoon events may affect only one or two tributaries at a time, but their flood flows carry large sediment loads to the San Pedro and create temporary dams that cause aggradation upstream. The end result is a floodplain aquifer that thickens and thins (ranging from a meter or two to perhaps over ten meters in places) as well as widens and narrows in its course down the valley.

GROUNDWATER SOURCE AND FLOW RATES

Source of groundwater. Groundwater in the San Pedro basin, like all groundwater, is derived from rainfall. When rain falls on the ground, three things can happen to the water. It may percolate into the soil, with some used by plants and some reaching (and recharging) the water table (the surface below which all voids and interstices in the earth materials are fully saturated). It can concentrate into rivulets and run off on the land surface to the nearest stream channel. It can evaporate back into the atmosphere; in this desert environment, this is the fate of most of the precipitation.

Most groundwater recharge occurs at or near the mountain fronts, where precipitation is greatest. The recharge process is poorly understood, but there appear to be three main mechanisms. In the first mechanism, mountain precipitation percolates into the bedrock fractures and then discharges into the regional aquifer at the mountain front. While the bedrock aquifers in the mountains adjacent to the San Pedro are not thought to be highly productive, little is known of their hydraulic properties. A second mechanism includes overland flow off the mountainsides onto the basin soils in between the canyons, and direct precipitation onto these surfaces; if the field capacity of soils is exceeded, water will percolate through the unsaturated zone to the water table of the regional aquifer. However, during the growing season the root systems of shrubs and grasses are likely able to capture all percolating waters. And, such recharge to the regional aquifer may not occur in interstream areas lying more than a couple kilometers from the mountain front where sediments become finer (R. Scott et al. 2000b). The third mechanism is the infiltration of storm runoff from the mountain streams as they debouch from their canyons and flow in their gravel-lined channels over (and into) the regional aquifer. Up to 20 percent of the recharge to the regional aquifer is estimated to be by this mechanism (Coes and Pool 2005). The actual amount of groundwater recharge cannot be directly measured, so it historically has been estimated as a water budget residual, discussed later in this chapter.

Groundwater flow rates. Surface water flows in well-defined channels as streams and rivers, while groundwater flows in myriad tiny threads through the pores and interstices in the sediments or rock fractures. The rate at which groundwater flows in an aquifer is important to know, as it influences the design of groundwater-management systems. It depends on two factors. The first is the ability of the aquifer to transmit water, largely influenced by the size of the openings in fractures (for bedrock aquifers) and the size of the intergranular spaces (for unconsolidated aquifers). The second is the gradient (i.e., the difference in water levels at two points, divided by the distance between them), with higher (steeper) gradients causing water to flow faster. In unconsolidated aquifers, water flows more readily in coarse sands and gravels than in fine sands, silts, and sediments containing clay.

The flow of groundwater is very slow, with flow rates measured in units ranging from meters per day to millimeters per day. Considering the current estimates of 18.5 million m^3 annual recharge to the Sierra Vista subwater-

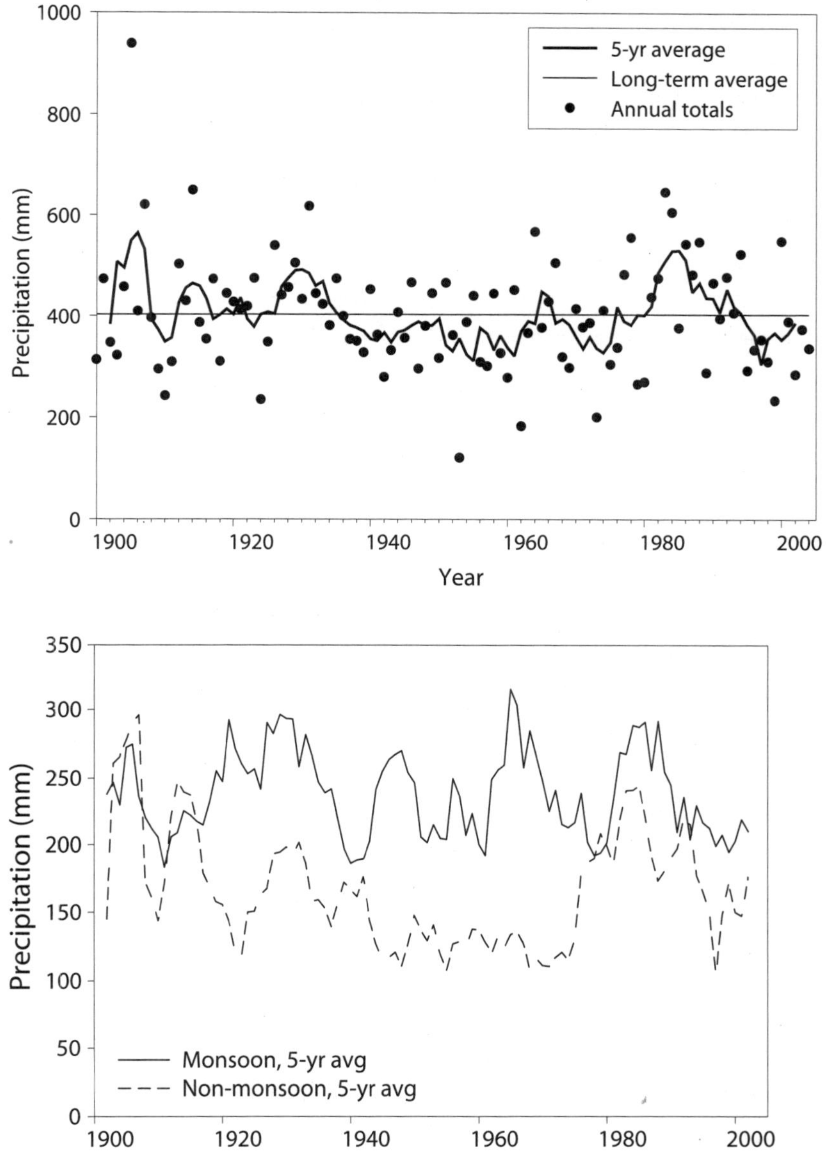

Fig. 15.2. (top; 15.2a): Annual precipitation at Fort Huachuca with a five-year moving average of the point data. (bottom; 15.2b): Monsoon and non-monsoon precipitation at Fort Huachuca, with annual point data smoothed with a five-year moving average.

shed (USDI 2005), and an estimated total volume of 20 km^3 (about half of the 38.24 km^3 of water in the regional aquifer of the upper San Pedro basin above St. David; ADWR 1991), the average time it takes a molecule of water to move from the recharge areas to the stream is about 1,100 years. Water moves streamward in the upper part of the regional aquifer more rapidly, perhaps in as little as 50 years, while water that circulates deeply in the partially consolidated lower units may take 10,000 years.

While there have been wet and dry periods over the past century, no long-term trend is apparent in the annual rainfall record at Fort Huachuca, which dates back to 1900 (fig. 15.2). (A small decline in annual rainfall is evident at Tombstone, which dates bask to the 1910s [B. Thomas and Pool 2006], although this decline may be an artifact of temporal changes in the location of this precipitation gaging station within the topographically diverse vicinity of Tombstone.) Near the mountain fronts, where most recharge seems to occur and where water tables are highest, these dry and wet periods can have significant effects on the elevation of the water table, with changes of 10 m or more possible (Pool and Coes 1999). These dramatic fluctuations in water levels become muted as water moves away from the recharge area, with water levels staying more constant during dry and wet periods, unless such periods extend for more than a decade. Thus, the amount of water reaching the stream or the floodplain aquifer from the regional aquifer is reasonably constant over time. In the lower San Pedro basin, the shorter distances from the recharge areas to the stream may reduce the dampening effects of distance described above.

GROUNDWATER FLOW PATHS

Like surface water, groundwater flows from higher elevations toward lower elevations due to the force of gravity. So, by ascertaining the slope or gradient of the water table, one can determine the direction of groundwater flow. This is accomplished by finding the elevation of the water table in a number of places, and then constructing a map with contours that show points of equal elevation of the water table.

Regional aquifer flow paths. Although the layers of clay present in some parts of the regional aquifer cause the water to follow a convoluted path, water in the regional aquifer flows toward, and ultimately reaches, the riparian corridor. This flow has both lateral and vertical components. Figure 15.3 shows the water levels in a pair of wells about 46 m west of the San Pedro River channel in the riparian corridor at Lewis Springs, east of Sierra Vista. Well #2, which penetrates to a depth of about 70 m in the regional aquifer, has a higher water level than well #3, which penetrates to about 7.3 m and taps the floodplain aquifer. Artesian pressure in the regional aquifer causes water in well #2 to rise above the levels of the water table in the floodplain aquifer. The difference in elevation of the two water levels divided by the 57.7 m that separate the opening of each well to the aquifer is the vertical gradient moving water upward from the regional aquifer into the floodplain aquifer.

Floodplain aquifer flow paths. Water in the floodplain aquifer predominantly flows in a downstream flow direction. Superimposed on this downstream gradient or slope on the water table are smaller gradients toward or away from the stream. The direction of these smaller lateral gradients changes with the carrying capacity of the floodplain aquifer. Where the floodplain aquifer is

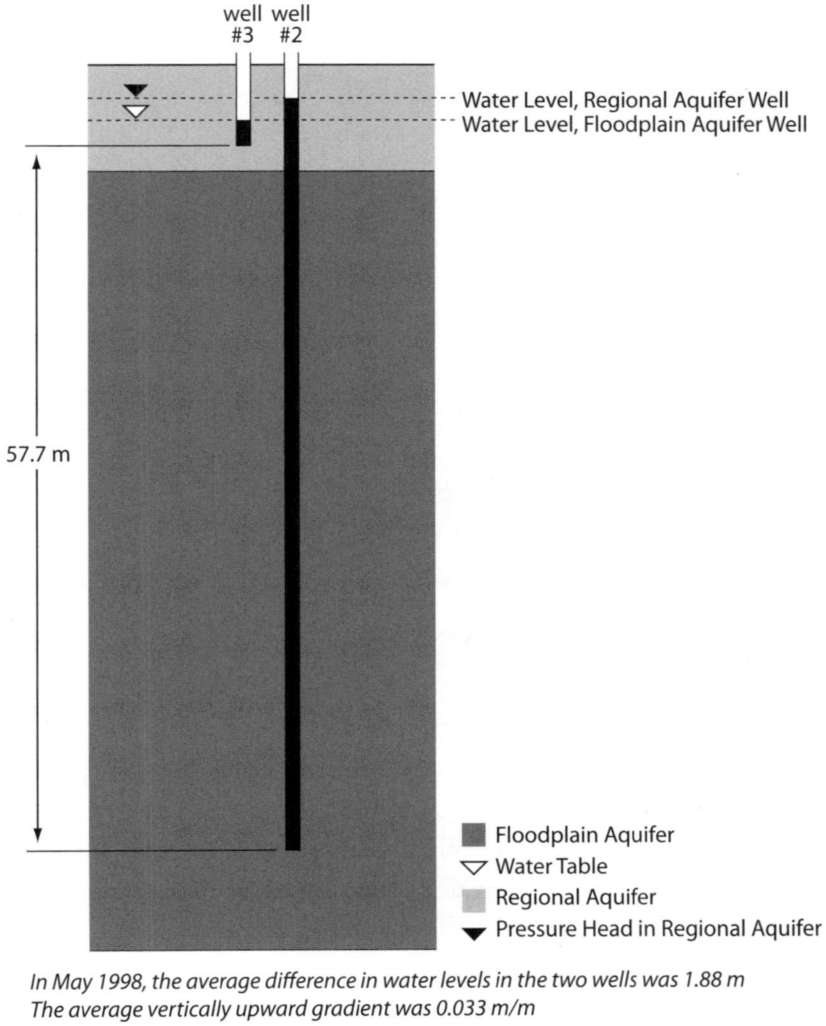

Fig. 15.3. Well pair at Lewis Springs site demonstrating the vertical gradient moving water from the regional aquifer to the overlying floodplain aquifer.

wide and/or thick, it can transport all the groundwater discharged from the regional aquifer, but where the aquifer narrows, or thins, it can't transport all this water, and so the stream transports what the aquifer cannot.

When the stream enters a reach where the floodplain aquifer widens or thickens, water moves into the aquifer from the stream; water-level contours show that water is moving laterally away from the stream as well as in a downstream direction (fig. 15.4). Such reaches are described as losing reaches as streamflow diminishes downstream. In reaches where the floodplain aquifer is getting thinner and/or narrower, water moves back toward the stream, as well as downstream. In these gaining reaches, streamflow increases downstream. In such reaches, the consistent presence of baseflow supports the designation of a perennial reach (see fig. 0.4 in Introduction). The presence of losing reaches creates complexities in trying to estimate the relative importance of various water sources for both the stream and the floodplain aquifer. The

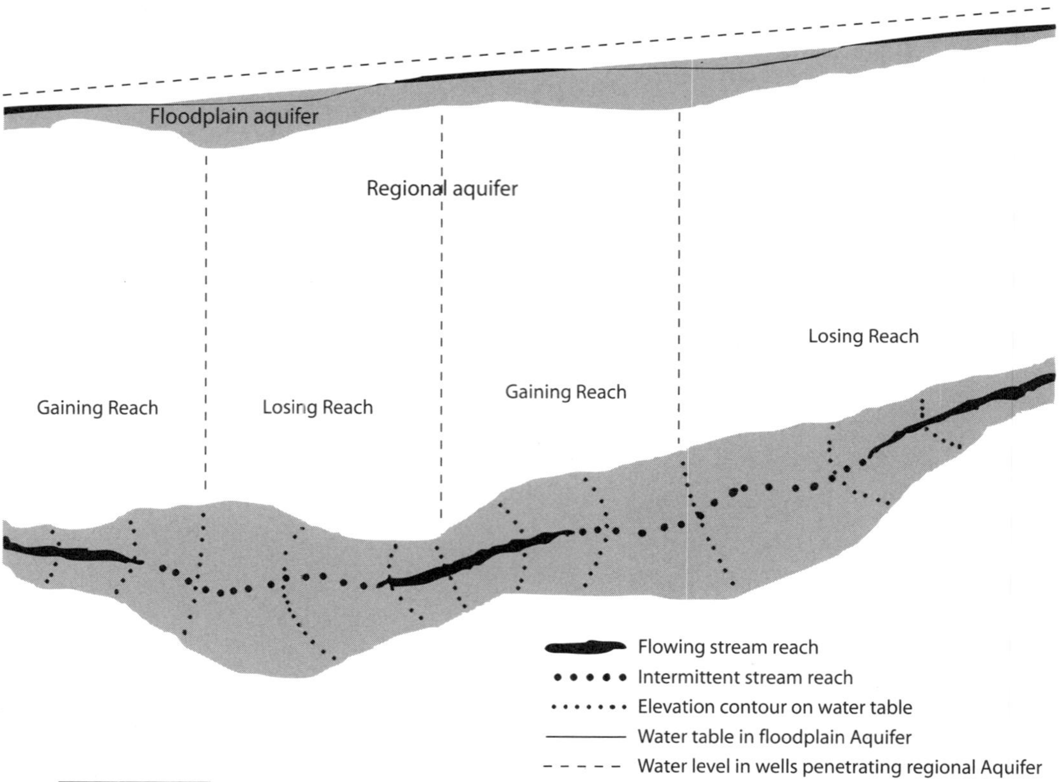

Fig. 15.4. Cross-sectional and plan views of the floodplain aquifer in gaining and losing reaches. Contours that are convex in a downstream direction show water is moving away from the stream. Contours that are concave in the downstream direction show that water is moving toward the stream.

complexities stem from the injection of storm runoff and baseflow contributions from an upstream gaining reach into the floodplain aquifer in losing reaches.

GROUNDWATER–SURFACE WATER INTERACTIONS IN THE FLOODPLAIN AQUIFER

When storm runoff arrives at a stream, the elevation of the stream surface is raised because of the additional flow. In the main channel, if the streamflow produced by this runoff exceeds a predefined threshold, it is considered a peak-above-base (flood peak) even if the flow stays confined within the channel banks. In a gaining reach, this increase in elevation of the stream surface serves to reverse the gradient of the water table, causing groundwater to cease discharging to the stream. The groundwater a little farther away that was still moving toward the stream suddenly can no longer move toward the stream, and so the water levels rise in the floodplain aquifer. This increase is described as bank storage or streamflow that has entered the aquifer. In gaining reaches, the amount of streamflow entering the aquifer is very small, with almost all the increase in water levels in the floodplain aquifer due to the backwater effect of the rising stream surface on the groundwater discharge. In contrast to this, in losing reaches, virtually all of any increase in water levels in the floodplain aquifer is due to stream water entering the aquifer.

As the length of perennial reaches in the San Pedro River has diminished since development of the basin's water resource began, the losing reaches have expanded, providing more aquifer space to capture surface waters during runoff events. In some losing reaches this may currently be the primary support of the riparian ecosystem. These recharged floodwaters also may be contributing to baseflow in a gaining reach downstream.

Effects of Wells Pumping Groundwater

CONES OF DEPRESSION

When a well is pumped, water moves into the well bore from the surrounding aquifer to replace the volume that was pumped out. A cone-shaped gradient or slope, called a cone of depression, is created as the water surface in the well is drawn down by the pumpage (Chow 1964) (fig. 15.5). The cone of depression continues to expand until the amount of water entering the cone balances the amount of water being pumped. Water can enter the cone of depression from various sources: laterally from the surrounding aquifer or stream channel; vertically downward from a usually dry, but sometimes wet, stream channel; or through human structures including leach fields, downspouts on buildings, and recharge basins. When there are multiple wells pumping water from the same aquifer, as in the Sierra Vista subbasin, the cones of depression of each well coalesce in such a way that the decline in the water level at a point is the sum of the incremental declines caused by each well affecting that point.

Although early data on water levels in the basin are sparse, contour maps of the groundwater table have been assembled depicting the water table configuration before significant pumping began (Goode and Maddock 2000). Prior to the development of the deep well turbine pump in the 1940s, the only large-capacity pumps were centrifugal pumps. So while they could pump large quantities of water, their suction lift was limited to about 5 m. Because the depth to water exceeded 5 m in most areas of the basin, even in much of the San Pedro riparian corridor, the water table was too deep for large-capacity pumps. Because wells pumping large amounts of water were non-existent prior to the 1940s, the water table that existed at that time is considered to be the natural or pre-development condition of the water table (also described as the steady-state condition). In some parts of the basin, development of water supplies from wells did not occur until the 1960s, and in these areas, 1960-era water levels are considered to be representative of steady-state conditions.

A recent (1997) map demonstrates the changes that more than 50 years of groundwater pumping have caused in the water table (Goode and Maddock 2000). When the differences in water table elevation from the steady state (1940) to 1997 are plotted, it is evident that the areas of the greatest pumping

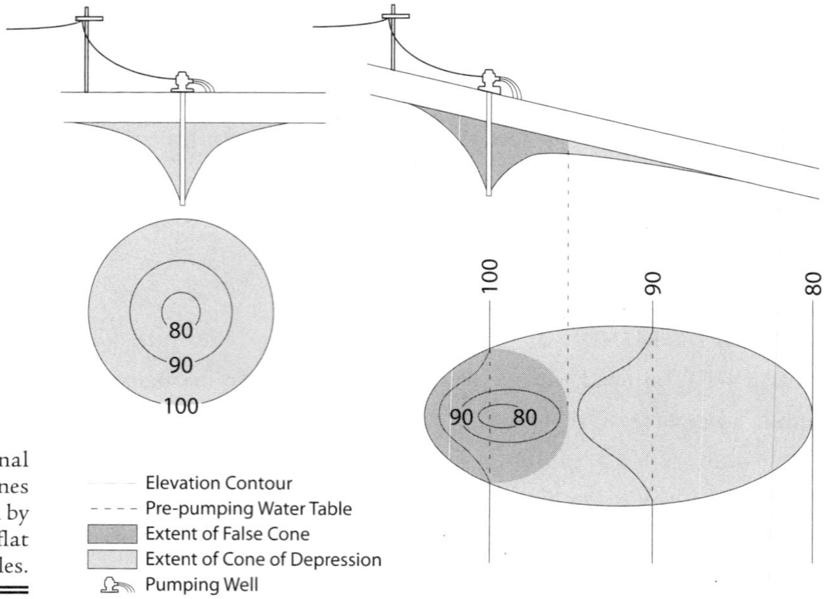

Fig. 15.5. Cross-sectional and plan views of cones of depression formed by pumping wells on flat and sloping water tables.

show the greatest declines (fig. 15.6). While discussions of cones of depression generally address major pumping centers, the composite cone of depression is literally basin wide in the regional aquifer. This map can also be used to estimate the reduction of the volume of water in the aquifer (also called groundwater mining) caused by pumping.

GROUNDWATER BUDGETS

Since 1974, groundwater models have been constructed to evaluate the groundwater resource and its management opportunities in the upper San Pedro basin, and in the Sierra Vista subwatershed in particular (Arizona Water Commission 1974, Freethey 1982, Villnow 1986, Putman et al. 1988, Rovey 1989, Vionnet and Maddock 1992, Corell et al. 1996, Goode and Maddock 2000). All groundwater models require the development of a conceptual water budget to test and evaluate. Mathematical models simulate the movement of water through a hydrologic system, and the modeler adjusts assumptions made for model values that could not be measured to achieve a best fit with values that could be measured, such as baseflow and water levels in wells.

Water budgets have several components, only some of which can be measured directly. The amount of groundwater discharged to the stream as baseflow, and the amount of groundwater pumped by water companies, municipalities, and military installations, are among those that can be measured directly. Some components—such as domestic well pumpage, riparian evapotranspiration (see chap. 2), and ephemeral stream recharge—can be inferred or estimated from partial measurement and extrapolation. The largest unmeasurable budget component is the groundwater recharge that occurs mainly at or near the mountain front. Coes and Pool (2005) estimated

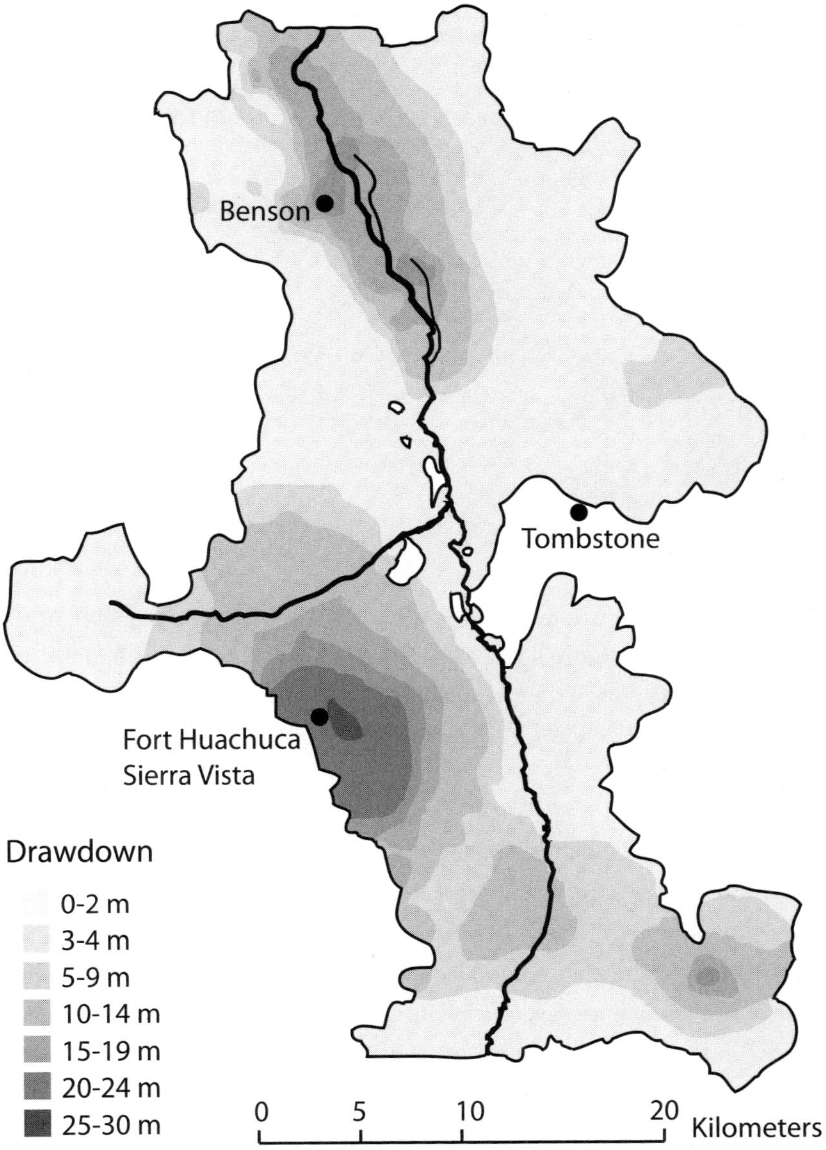

Fig. 15.6. Drawdown from 1940 to 1997 in the regional aquifer in the upper San Pedro basin (after Goode and Maddock 2000).

the relative volumes of recharge occurring under ephemeral stream channels crossing the regional aquifer, and in the future, isotopic and chemical signatures may help to further describe recharge distribution by source throughout the aquifer systems. Two smaller budget values that cannot be directly measured are inflows to, and outflows from, the basin as groundwater from adjacent basins upstream and downstream.

Measured and estimated budget values for the Sierra Vista subbasin are shown in table 15.1. It is important to note different authors use different methods and assumptions to prepare such water budgets, and even single authors may use different techniques when preparing multi-year budgets depending on the availability of data. Thus, great significance should not be placed on changes over time in budget values that are not directly measurable. Several

TABLE 15.1. Water budget values for the Sierra Vista subbasin.

Budget item	Vionnet and Maddock (1992)[1]	Corell et al. (1996)[2]	USDI (2006)[3]
Mountain front recharge	15,480,000m^3	19,360,000m^3	18,500,000m^3
Underflow from Mexico	4,600,000m^3	3,700,000m^3	3,700,000m^3
Riparian ET	9,740,000m^3	8,730,000m^3	9,500,000m^3
Baseflow out	3,080,850m^3	7,955,958m^3	4,000,000m^3
Underflow out	1,150,000m^3	543,000m^3	543,000m^3
Pumpage	16,874,030m^3	13,760,720m^3	20,350,000m^3
Groundwater "mined"*	7,929,672m^3	10,764,880m^3	12,193,000m^3

[1]Data estimates for 1988.
[2]Data estimates for 1991.
[3]Data estimates for 2002.
*Groundwater "mined" is a measure of the deficit between water entering the groundwater system and leaving it in that year.

budget values that are measurable (pumpage and baseflow) have changed considerably over time. In figure 15.7, the shaded area between total inflow and total outflow represents the extent of groundwater mining, or the groundwater budget deficit. It is noteworthy that the groundwater deficit is increasing on an annual basis.

EFFECTS OF THE COMPOSITE CONE OF DEPRESSION ON BASEFLOW

Even though the large pumping centers are several kilometers from the river, the near-basin-wide extent of the composite cone of depression is a powerful argument that groundwater pumping is creating the groundwater deficit and contributing to the declining baseflow of the upper San Pedro River. Temporally, the groundwater deficit tracks the amount of groundwater pumpage, and both are inversely related to the amount of baseflow (fig. 15.7). Patterns in a series of flow duration curves, which show the percentage of time, by decade, that streamflow equaled or exceeded a particular value, also are consistent with the idea that groundwater pumping is affecting San Pedro streamflow (see fig. 16.2). The stream is at or near baseflow about 40 percent of the time, and the portions of the curves to the right of the 60 percent gridline thus show the pattern of depletion. The first decade (1936–1945) represents pre-development flow conditions. Agricultural pumping for irrigation close to the stream in the Palominas-Hereford area is generally agreed to be primarily responsible for reducing baseflow prior to 1956 (Corell et al. 1996), and the 1946–1955 decade curve shows the effects of that near-stream pumpage. Depletion continued in subsequent decades but did not accelerate until the last two decades, when the effects of distant pumping as well as expanding domestic pumping throughout the subwatershed started impacting the stream. While the flow duration curves portray streamflow

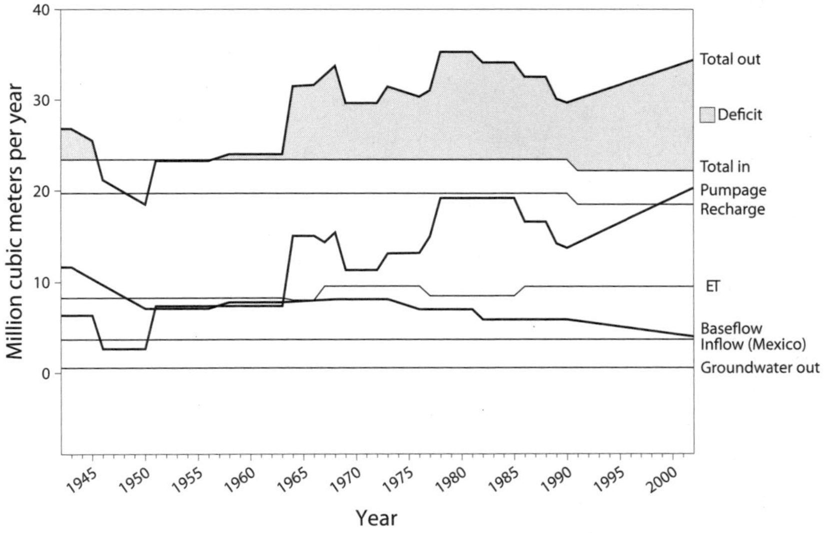

Fig. 15.7. Groundwater budget for the Sierra Vista subwatershed. The shaded area shows the deficit between supply and demand. Data from 1942 to 1990 from Corell et al. (1996); 2002 data from USDI (2006).

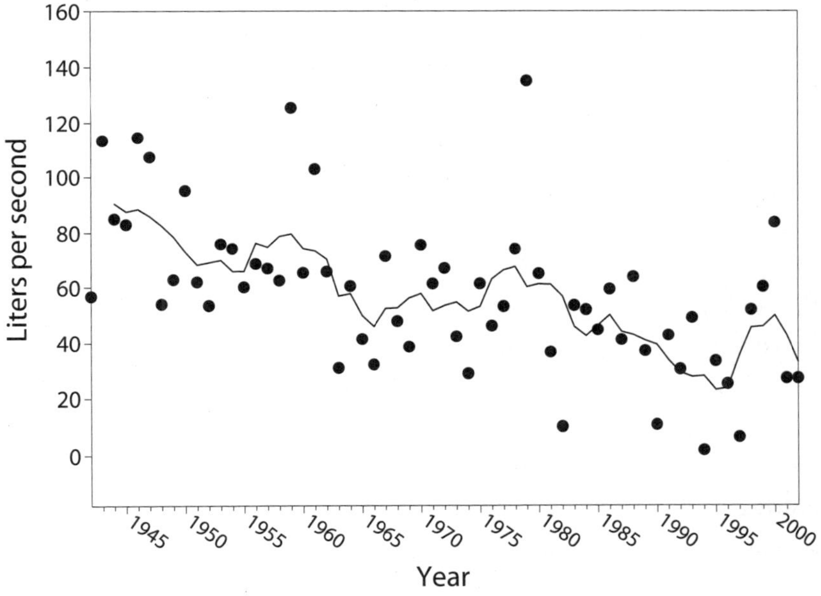

Fig. 15.8. Changes in baseflows in the upper San Pedro River. Annual seven-day low flows for the San Pedro River at Charleston, with a five-year moving average of the annual data to show trends. The seven-day low flow typically occurs in June, before the onset of the summer monsoon season.

depletion, the nature of the plot makes it difficult to decipher what is happening at the very lowest levels of streamflow. By using the average flow in the seven consecutive days of lowest flow each year (annual seven-day low flow) as an index of baseflow discharge, one sees that the baseflow has declined to about one-third of its amount in the early 1940s (fig. 15.8).

Empirical data from groundwater wells also support our contention that groundwater pumpage is a key cause of baseflow decline. Recent data from Lewis Springs, where there are two pairs of wells penetrating the floodplain and regional aquifers, show that since 1997, there has been a 15 to 20 percent reduction in the amount of water moving from the regional aquifer to the floodplain aquifer and thence to the stream (fig. 15.9). This figure demonstrates that the Lewis Springs area is being affected by the expanding cone

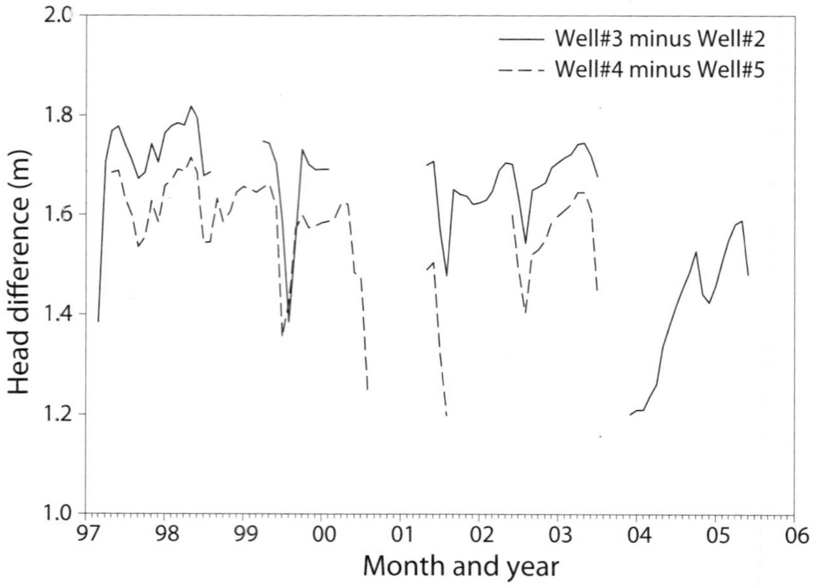

Fig. 15.9. Vertical head differences between the regional and floodplain aquifers in two well pairs at Lewis Springs: well #5 (~61 m deep) and well #4 (~7.3 m deep), and wells #2 and #3 (see fig. 15.3), from March 1997 through June 2005. Both well pairs show similar reductions in the upward vertical gradient from the deeper well. Wells #4 and #5 are within 10 m of the San Pedro river channel (on the west bank) and about 40 m east of wells #2 and #3.

of depression centered on Sierra Vista/Fort Huachuca, which is reducing the hydraulic gradient that moves water toward the river. As figure 15.5 shows, the cone of depression of a pumping well on a sloping water table may extend quite a distance in the down-gradient direction. (Sharma et al. 1997) noted that over the period 1985–1997, baseflow near Lewis Springs showed a slight increase attributed to a substantial reduction in agricultural pumping upstream, while farther downstream there was a reduction in baseflow, presumably due to groundwater pumping.

Some have hypothesized other causes for baseflow decline, including climatic and/or riparian vegetative changes (Thomas and Pool 2006). There have been significant declines in summer precipitation during recent decades, and such climatic changes may be influencing the stream baseflows, though no mechanism explaining such an effect has been postulated. Expansion of the riparian forest following historic channel entrenchment (see chap. 12) also may be a contributing factor, but the extent of this effect has not been definitively determined.

RECENT STREAMFLOW DECLINES

Starting on July 9, 2005, and continuing until the summer monsoon rains began on July 16, the U.S. Geological Survey stream gage at Charleston recorded zero flow for the first time since the gage was established at that site in 1942. A recent analysis by R. Koehler (written communication, 2005) based on the period of record for the Charleston gage from 1935 concludes that winter baseflows may reach a zero flow level at Charleston as early as 2010. As noted earlier in this chapter, there may be increased contribution to baseflow from floodwater infiltrating the floodplain aquifer in losing reaches. Though that mechanism provides a new source of water to sustain baseflow, it has not been sufficient to compensate for the diminishment of baseflow from groundwater discharge.

Should the present annual deficits continue their current trend (fig. 15.7), the composite cone of depression will continue to expand and deepen, and consequently, less and less water will reach the riparian corridor, baseflows will cease, and flow in the river will be driven entirely by occasional flood pulses. If tomorrow there were a catastrophic failure of the Western power grid and all groundwater pumping in the basin ceased, the cone of depression would continue to expand, even as it started to recover in the deepest parts of the cone, and, in the absence of effective mitigation, would create the same scenario for the river. If the power grid was never fixed, the hydrologic system would slowly recover to something approximating the conditions in 1940, but as a study of a groundwater mining operation planned in the Safford Valley demonstrates (USBLM 2003), this recovery process may take centuries.

Summary

The riparian ecosystem of the San Pedro basin, and the baseflows in the stream itself, are dependent on the discharge of groundwater from the regional aquifer, which in turn depends on rainfall. Though annual rainfall has fluctuated with decade-long wet and dry periods, the long-term average annual rainfall displays no obvious trend, and the great distance between the recharge areas near the mountains and the stream serves to modulate these short-term fluctuations; thus, the discharge from the regional aquifer to the riparian corridor under natural conditions would be nearly constant. Groundwater pumping in the basin has increased significantly over the last half century, and in the Sierra Vista subwatershed, groundwater pumpage now exceeds the rate of natural recharge. As most of the groundwater pumping is located between the recharge areas and the river, the pumps are intercepting water that otherwise would reach the riparian corridor. While the full impact of this pumping has not yet been felt at the river, the baseflow discharge of the river has diminished by 66 percent since 1942. As the full effects reach the river, baseflow will cease, as it did briefly in 2005 for the first time in the history of the Charleston stream gage. To preserve the river's flow, mitigation must involve elements designed to block the expansion of the basin-wide cone of depression in the riparian corridor.

KATHERINE K.
HIRSCHBOECK

Flood Flows of the San Pedro River

SIXTEEN

Introduction

This chapter addresses surface water flows of the San Pedro River with an emphasis on flood flows. In terms of total volume, the storm runoff or flood flow component of streamflow is much larger than the baseflow. Although individual flood flows can be very large, they are sporadic, with up to several months or more between such flows, and thus provide only an intermittent source of water to the stream. I discuss temporal patterns of flood regimes, from intra-annual to decadal, as well as spatial patterns over the length of the river.

Seasonal Patterns and Variability

The surface water hydrology of the San Pedro River basin has a distinctly seasonal regime. It is characterized by moderate and relatively consistent monthly streamflow during the winter wet season (primarily December through March), low streamflow in the late spring and early summer dry season (April through June), and greater and more variable flow during the summer wet season (July through September) (fig. 16.1). This seasonal pattern results from the spatial and temporal interplay of latitudinally shifting winter storm tracks (Aldridge and Hales 1984), and influxes of moist air

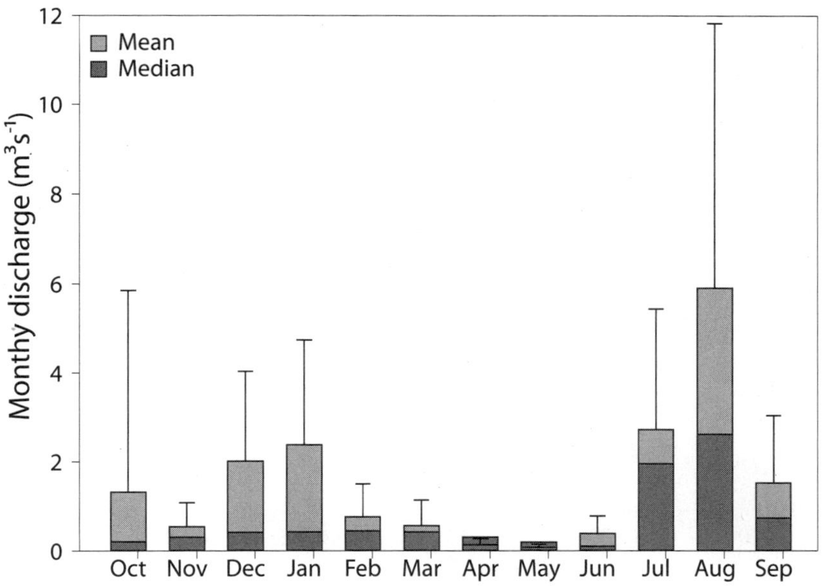

Fig. 16.1. Mean monthly streamflow variations for the San Pedro River at Charleston (#09471000), for 1943–2005. Values shown are the median and the mean plus one standard deviation.

during the summer monsoon thunderstorm season (Douglas et al. 1993, Adams and Comrie 1997). In addition, the occasional penetration of tropical storm-related moisture during summer or fall (especially in September and October) can trigger copious runoff-producing rains (W. Smith 1986, Roeske et al. 1989).

Desert rivers throughout the world have highly variable flow regimes (Tooth and Nanson 2000, Bunn et al. 2006). The high variability of the San Pedro River, even within the seasonal framework, is evident in flow duration curves (fig. 16.2). While discharges are much larger during the summer months, they also tend to be more variable. The steep flow duration curve calculated for the warm season months implies high flow variability, while the flatter slope for the cooler months suggests that streamflow is being sustained more consistently by either surface runoff or groundwater. Flow duration curves calculated by decade show that while the low-flow portion of the curves has steepened dramatically over the past several decades, the high-flow portion has varied more idiosyncratically (fig. 16.3). For example, the curve for 1976–1985 is distinctly flatter than those for other decades. This decade experienced an exceptionally large tropical storm-related flood (October 1977) in addition to multiple occurrences of extreme winter flooding (December 1978, January 1979, December 1984).

Floods

Flood events contribute significantly to the overall surface water regime of the San Pedro, and thus the seasonality of annual flood peak occurrence mirrors the seasonal pattern of mean monthly flows (fig. 16.4a). The large differences between mean and median discharge for nearly all months (fig. 16.1)

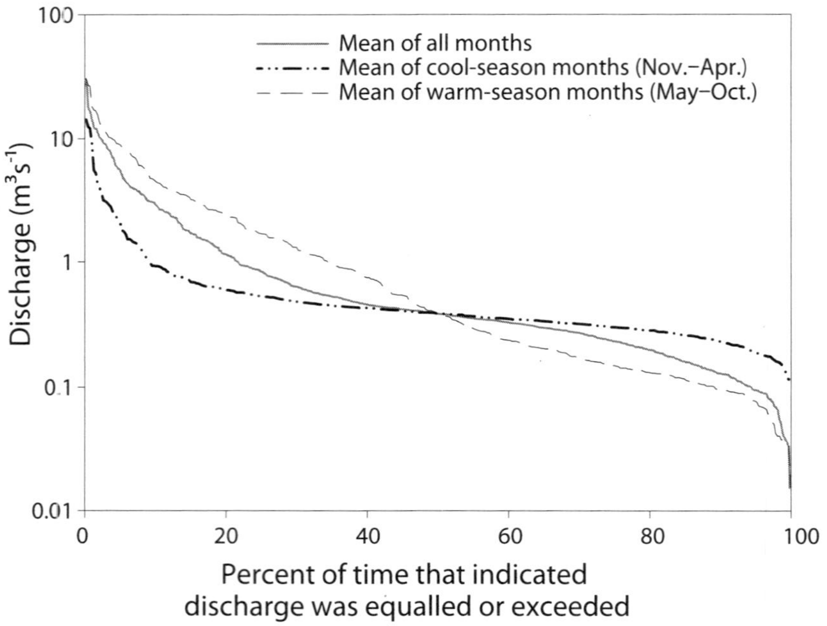

Fig. 16.2. Flow duration curves (based on 1943–2005 water year mean monthly streamflow) for cool-season months, warm-season months, and all months combined, for the San Pedro River at Charleston.

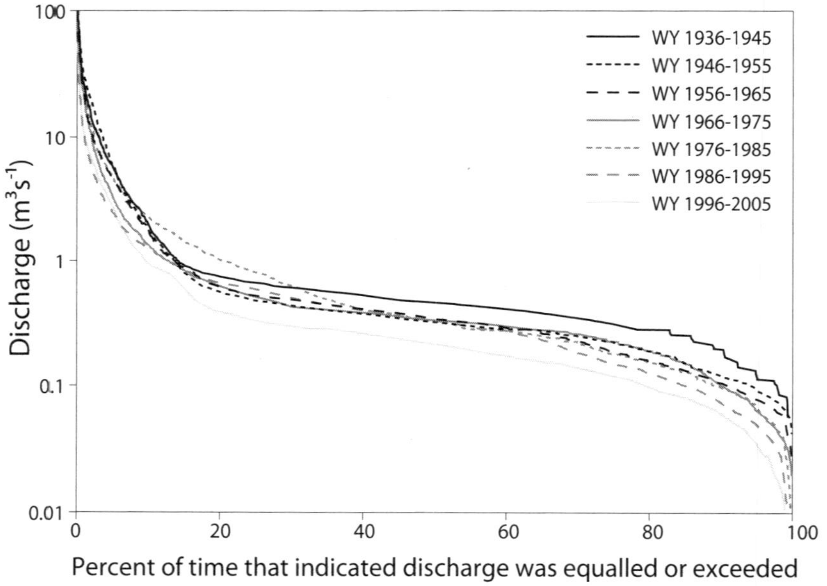

Fig. 16.3. Flow duration curves for seven decades of recorded daily mean streamflow of the San Pedro River at Charleston. Note: Prior to 1942, the Charleston gage was located at various positions along a six-mile reach of the river north of its present location, and thus the 1936–1945 decade may have included flow from the Babocomari River.

suggest that the flow from a relatively small number of extremely large events (floods) is an important component of the mean monthly discharge. Runoff-producing storms from three types of atmospheric patterns vary seasonally and interannually to control the frequency, magnitude, and timing of floods. Typically, the annual flow regime consists of a maximum peak in July or August, caused by intense summer convective rainfall, and one or more secondary winter flows in November through March, caused by synoptic storm precipitation (Hirschboeck 1991, House and Hirschboeck 1997).

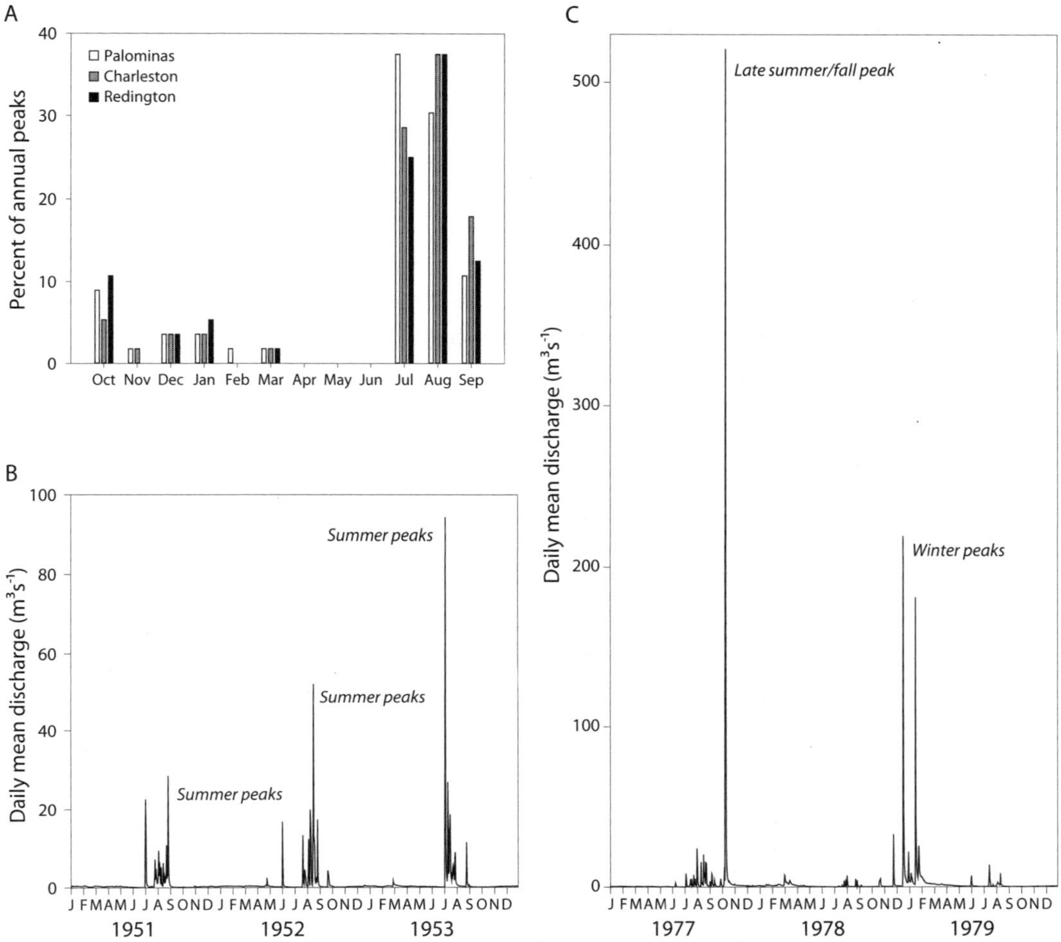

Fig. 16.4. Seasonal variations in San Pedro River flood peaks. (A) percent of annual floods occurring in each month at the Palominas, Charleston, and Redington gages for water years 1950 through 2005 (stations #9472000 and 9472050 combined for Redington), (B) three years of daily mean discharges at the Charleston gage during a period of dominance by summer flood peaks (maximum instantaneous peaks for the three years are 162 m^3s^{-1} on 7-2-1951, 222 m^3s^{-1} on 8-17-1952, and 243 m^3s^{-1} on 7-7-1953), and (C) three years of daily mean discharges at the Charleston gage during a period of dominance by fall and winter flood peaks (maximum instantaneous peaks for the three years are 672 m^3s^{-1} on 10-8-1977, 329 m^3s^{-1} on 12-18-1978, and 334 m^3s^{-1} on 1-8-1979).

SEASONAL FLOOD TYPES

Floods from convective storms. During 1950–2005, at the gaging stations of Palominas, Charleston, and Redington, almost 90 percent of the annual maximum flood peaks occurred during the months of July through October (fig. 16.4a). Two-thirds of the annual peaks occurred during July and August, the most active months of the summer monsoon thunderstorm season. Figure 16.4b illustrates the daily mean flow regime at the Charleston gage for three typical years during which summer flood peaks dominated. Several large peak-flow events are evident in the daily discharge data, and instantaneous flood peaks on a given day can be two to five times greater than the daily mean value. The flashiness of the flow regime and the occurrence of multiple peaks are indicative of frequent and variable convective storm activity.

Floods from tropical storms. September and October typically are much dryer than July and August, although susceptibility to intense, tropical storm–related precipitation increases during these late summer/early fall months. Figure 16.4c illustrates one such event, the severe flood of October 1977 linked to rainfall from Tropical Storm Heather in the eastern Pacific Ocean, which caused widespread flooding throughout southern Arizona (Aldridge and Eychaner 1984).

Floods from synoptic winter storms. A flood season occurs in winter in response to less-intense but more-sustained precipitation events, which usually yield moderate discharges. Only 10 percent of the annual maximum flood peaks in the San Pedro River were recorded in the months of December through March during 1950–2005. When these winter floods do occur, however, they tend to be of longer duration than the more flashy summer flood peaks; hence, the total flow volume of a winter event may exceed that of a summer event, even though the summer flood has a larger instantaneous discharge peak. Occasionally, a sequence of winter storms can lead to an exceptionally active winter flood season, such as the sequence of peaks in the winter of 1978–1979 (fig. 16.4c). Since 1965 there has been an increase in the frequency of maximum annual flood peaks occurring in winter in the San Pedro basin.

TEMPORAL VARIABILITY

Extreme events. All of the extreme discharges of record for each of the gaging stations in the river basin have occurred during May through October. All are associated with convective storms or with convective rainfall enhanced by moisture from dissipating tropical storms (table 16.1). The extremes of record in tributaries with small drainage areas are more likely to be caused by convective storms in July or August, in contrast to the record floods on the main trunk stream where all but two are associated with late-season eastern Pacific tropical storms.

No single storm event stands out as the cause of the largest flood of record at *all* the gaging stations, in part because of different periods of record at each station. Two notable tropical storms, however, produced the largest flood of record at several stations. The flood associated with the unnamed tropical storm of September 1926 produced exceptionally large flood stages from Palominas in the upper basin to Redington in the lower basin. The gage at Charleston recorded the greatest discharge in the basin for this event, and it dominates the annual flood time series for this station (fig. 16.5), as well as

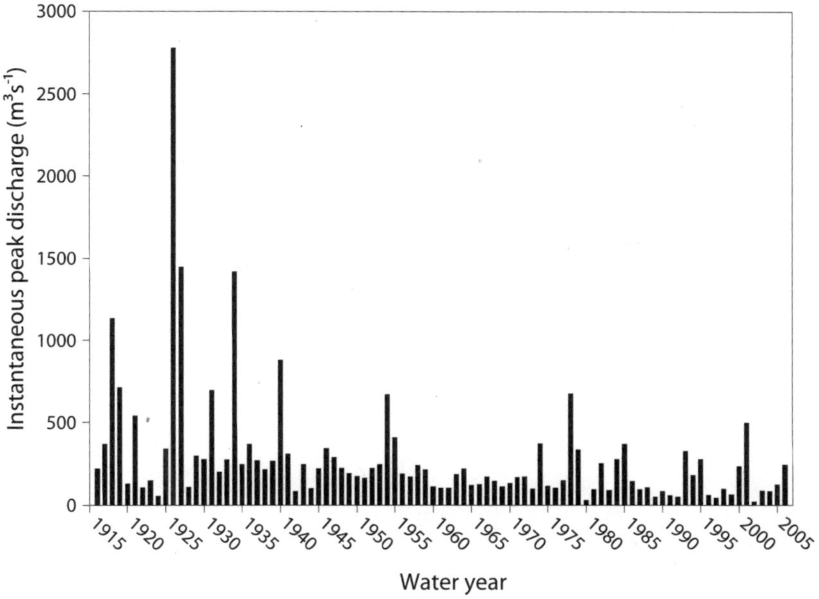

Fig. 16.5. Annual flood peak time series of the San Pedro River at Charleston (values are instantaneous peak flows). The maximum recorded peak (September 1926) was 2,775 m^3s^{-1}.

that of the Redington station downstream (not shown). Three gages in the lower basin experienced their greatest flood of record in response to intense precipitation linked to Tropical Storm Octave in October 1983 (water year 1984). This event also produced major flooding throughout central and southern Arizona, especially in the San Francisco, Gila, and Santa Cruz Rivers.

Intra-annual patterns. Limiting the discussion to only the largest floods in the record, or to the annual flood time series, neglects other peak flow events that play an important role in the flood hydroclimatology and riverine ecology of the basin. As suggested in daily mean discharge plots (figs. 16.4b, 16.4c), a given flow season in the San Pedro can be characterized by multiple flow events—some of substantial discharge—although only one will be recorded as the *annual* flood peak. Information on these lesser flood peaks can be obtained from a station's peaks-above-base record, which includes all discharges above a designated base discharge at a given recording station.

Figure 16.6 displays frequency histograms of the peaks-above-base record for the Charleston gage, separated by type of flood-causing storm for the period 1950–1985. The peaks were classified according to storm type based on a detailed analysis of weather charts and other sources (Hirschboeck 1987, 1988). (Histograms for the Palominas and Redington gages have patterns similar to that of the Charleston record.) This type of classification is more climatically accurate than a simple seasonal division of floods into winter, summer, and fall months (to represent synoptic, convective, and tropical storm–related floods, respectively) because some synoptic-scale circulation features (upper-level cutoff lows and short wave troughs) have contributed to enhanced flooding during the warm-season months, and some tropical storm–derived floods

TABLE 16.1. Maximum stage and discharge at San Pedro Basin stations for period of record.

	Station number 0947-	USGS station name	Drainage area (km^2)	Period of record	Water year	Date	Storm type linked to flooding	Stage (m)	Peak discharge (m^3s^{-1})	R.I.[3] (yrs)
Main Channel	0500	San Pedro at Palominas	1909	1926–05[1]	1926	09/28/26	Tropical storm (TS)	7.3	—	—
					1940	08/14/40		5.0	623	20
	1000	San Pedro at Charleston	3196	1916–05	1926	09/28/26	Tropical storm	6.7	2,775	>100
	1550	San Pedro near Tombstone	4481	1967–05	1978	10/09/77	TS Heather	3.5	685	50–100
	1800	San Pedro near Benson	6449	1966–76	1972	08/26/72	Convective	3.2	278	<10
	2000	San Pedro near Redington	7581	1926–97	1926	09/28/26	Tropical storm	6.6	2,549 [2]	>100
	2050	San Pedro at Redington Bridge near Redington	8019	1999–05	2003	07/22/03	Convective	4.6	167	—
	3100	San Pedro below Aravaipa Creek near Mammoth	11248	1980–84	1984	10/01/83	TS Octave	5.4	3,823	>100
	3500	San Pedro at Winkelman	11533	1919–83[1]	1984	10/01/83 [2]	TS Octave	—	3,823	>100

Tributaries	0900	San Pedro Tributary near Bisbee	13.6	1963–79	1965	09/04/65	TS Emily	3.7	41.3	—
	1087	Walnut Gulch 63.111 near Tombstone	0.6	1962–81	1976	07/27/76	Convective	—	15.3	>100
	1090	Walnut Gulch 63.009 near Tombstone	23.6	1967–81	1972	07/24/72	Convective	—	74.8	25–50
	1110	Walnut Gulch 63.015 near Tombstone	23.9	1955–81	1971	08/10/71	Convective	—	35.4	10–25
	1185	Walnut Gulch 63.103 near Tombstone	0.7	1963–81	1975	07/17/75	Convective	—	0.9	<10
	1195	Walnut Gulch 63.007 near Tombstone	13.5	1966–81	1972	08/12/72	Convective	—	73.3	50–100
	2100	Peck Canyon Tributary near Redington	20.8	1968–81	1972	08/12/72	Convective	2.8	123	>50
	3000	Aravaipa Creek near Mammoth	1391	1919–94[1]	1984	10/02/83	TS Octave	5.1	2,005	>100
	3200	Green Lantern Wash near Winkelman	9.4	1964–76[1]	1981	05/01/81	Convective	2.7	105	20–40

[1]Observations for some years in record are missing.
[2]Discharge or date is estimated.
[3]R.I. = Recurrence Interval (R.I. estimates based on Perry et al. 2001).

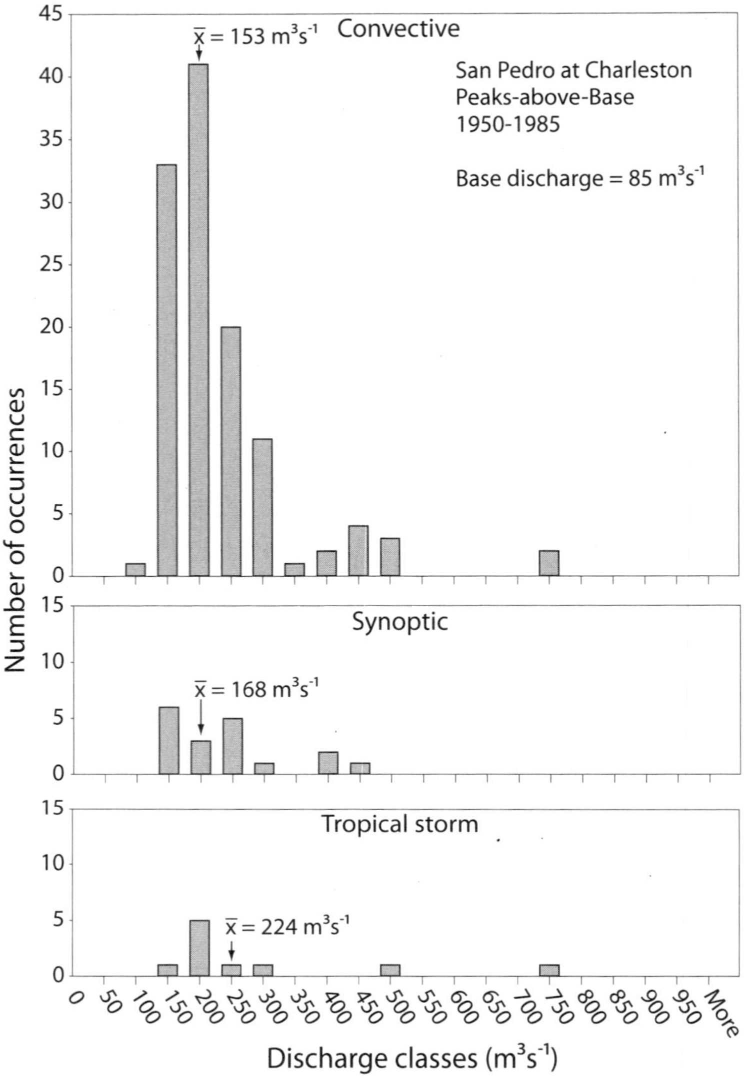

Fig. 16.6. Histograms of instantaneous peaks-above-base discharge for convective, synoptic, and tropical storm floods recorded at the Charleston gage (1950–1985).

have occurred in July and August, even though the tropical storm influence is most pronounced in September and October in southern Arizona.

The flood-type classification shows the dominance of frequent summer peaks that are produced by intense convective rainfall events but also emphasizes that most of these floods have moderate peak discharges (mean = 153 m^3s^{-1}), whereas the smaller populations of floods having synoptic or tropical storm influences have larger mean discharges (168 m^3s^{-1} for synoptic storms, 224 m^3s^{-1} for tropical storms). Hence, a shift to a climatic regime favoring more frequent synoptic or tropical storms could alter the magnitude and frequency of flooding in the San Pedro basin.

Decadal patterns. Figure 16.7, which shows the Charleston annual flood record classified by flood type, illustrates that abrupt shifts in flood type have occurred in the San Pedro basin over the twentieth century. This is seen most

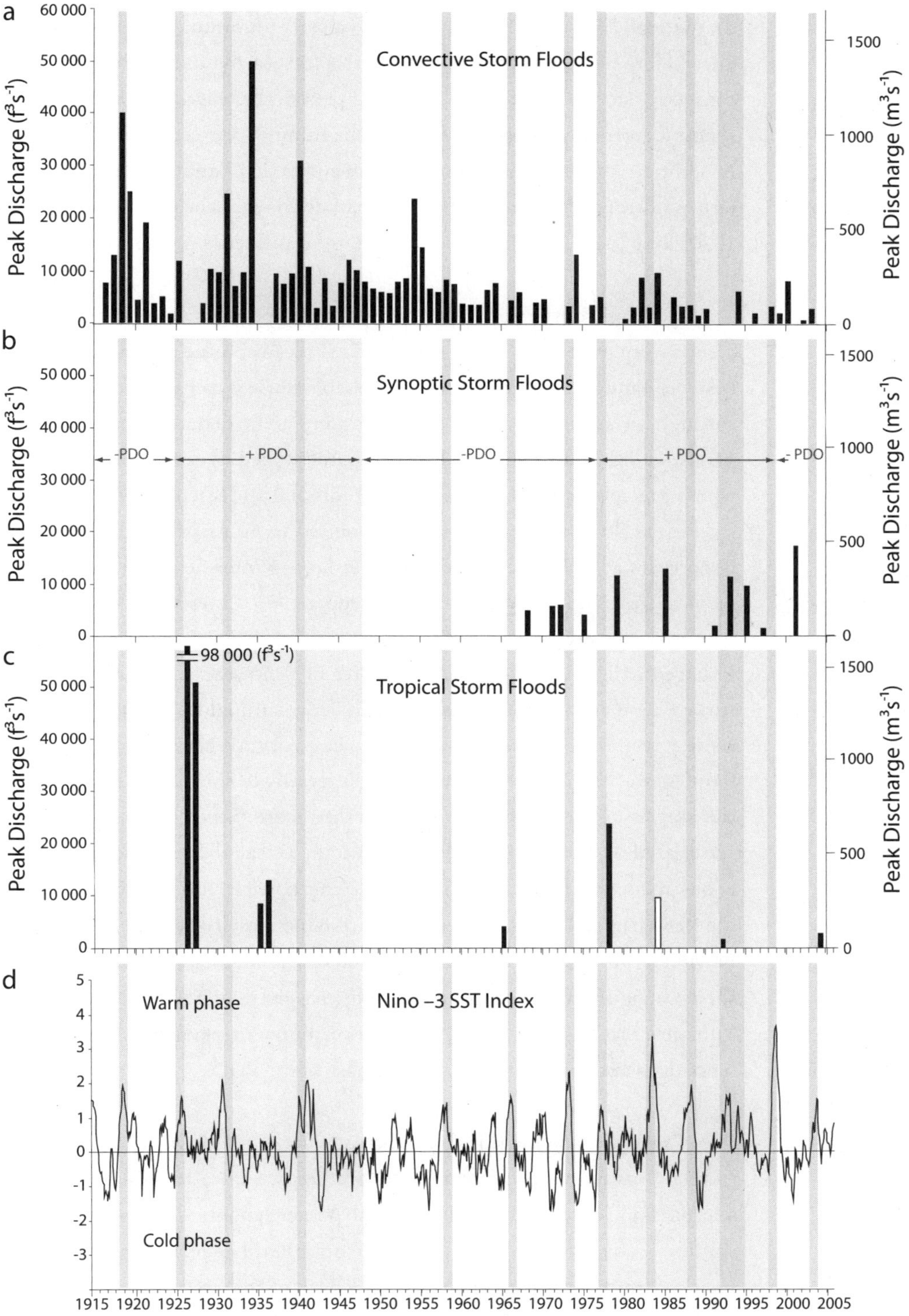

Fig. 16.7. Annual flood time series of the San Pedro River at Charleston, classified by flood-producing storm type: (a) convective, (b) synoptic, and (c) tropical, plotted with (d) the Niño-3 SST Index. Also shown are El Niño years (gray vertical lines) and positive (+) (shaded) and negative (-) phases of the PDO. (Flood values are instantaneous peak flows. An open bar is plotted for the October 1983 tropical storm flood which was the annual flood at downstream gages but not at Charleston.)

dramatically after 1965 when there was a shift for annual floods to be produced more frequently by synoptic and tropical storms and less frequently by convective storms. During the post-1965 period, the magnitudes of nearly all synoptic storm floods exceeded that of the summer convective annual floods. In addition, two very large tropical storm events (1977 and 1983) signaled the return of tropical storm floods which had dominated the early record in the 1920s and 1930s but then disappeared as an influence in the annual flood record.

The causes for the post-1965 shift in flood type, which has been observed on other regional rivers such as the Santa Cruz (Webb and Betancourt 1992), have been much debated. The most probable cause is a circulation response, through air-sea interactions, to changes in Pacific Ocean sea surface temperatures. El Niño years tend to be characterized by wetter-than-normal winter precipitation in the Southwest (Redmond and Koch 1991), and several studies on flooding in Arizona have pointed to an association between El Niño years and the occurrence of either large winter floods or late-season tropical storm floods (Webb and Betancourt 1992, Ely et al. 1994). It should be noted, however, that the Charleston record does not substantiate this linkage in all cases. Figure 16.7 compares El Niño years and the Niño-3 sea surface temperature index with the San Pedro annual flood series. There is some agreement between synoptic storm flood occurrences and the increased frequency of El Niño years since 1976 (especially in the 1990s), but the relationship is not consistent throughout the record. Several researchers have speculated on the possible role of the Pacific Decadal Oscillation (PDO) as a low-frequency influence on El Niño strength, such that positive phases of the PDO (shaded in fig. 16.7) enhance El Niño activity (see Gershunov and Barnett 1998, Gutzler et al. 2002). Seven of the nine tropical storm floods shown in figure 16.7 occurred during positive phases of the PDO, suggesting a possible low-frequency +PDO association, but winter synoptic flood occurrence does not appear to favor either phase of the PDO.

Spatial patterns. The magnitude and frequency of floods varies along longitudinal (upstream-downstream) river gradients. Figure 16.8 illustrates how selected San Pedro River floods linked to winter synoptic-scale precipitation, summer convective rainfall, and fall tropical storm–related precipitation each progressed downstream along the length of the river and were recorded sequentially at the main trunk stream gages in operation at the time of the flood. Overall, in the three types of floods, there is a tendency for peak discharges to increase with increasing drainage area in the upper basin above the Charleston gage, and then to decrease with increasing drainage area as the flood wave progresses downstream and into the lower basin. This is not surprising, because orographic effects cause more abundant precipitation in the

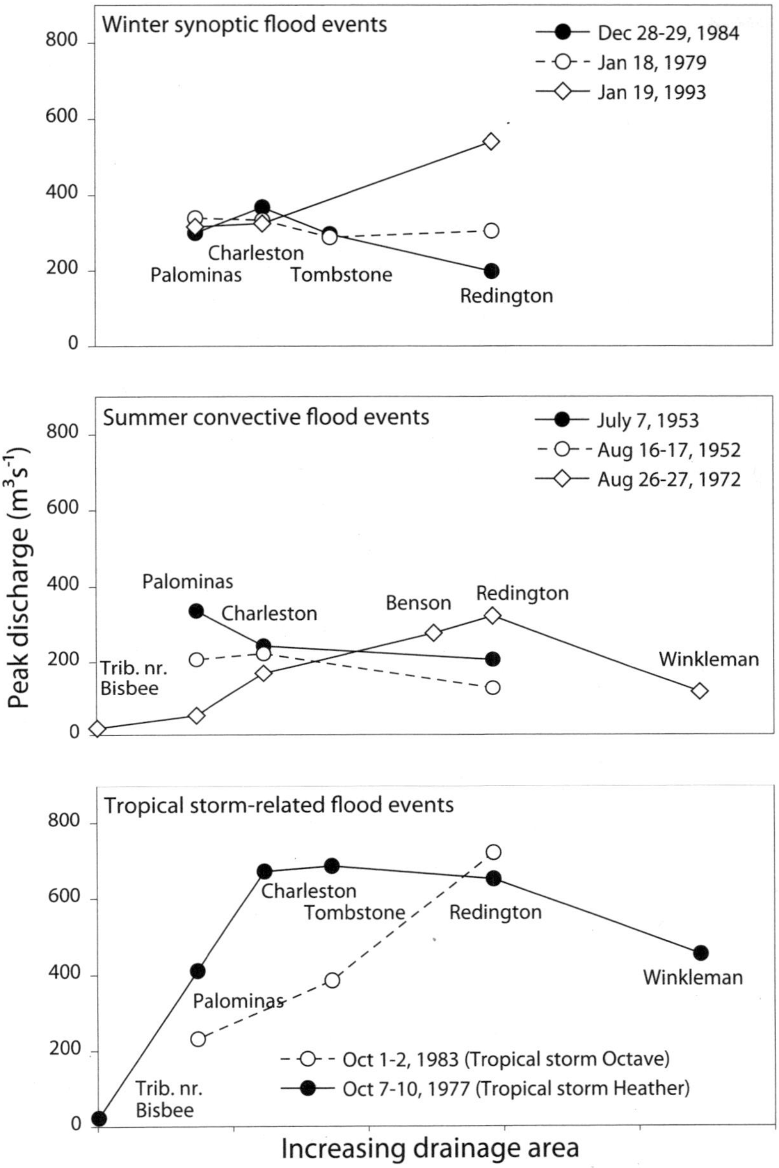

Fig. 16.8. Downstream progression of selected flood events of different types. (Values are instantaneous peak flows.)

mountains, and all of the large drainages from the high-elevation Huachuca Mountain range (except for the Babocomari) enter the San Pedro between the Palominas and Charleston gages. There have been exceptions to this trend, including the July 1953 convective event which *decreased* in size rapidly between Palominas and Charleston, suggesting a localized storm, and the 1983 flood associated with Tropical Storm Octave which *increased* dramatically as it traveled downstream and produced the flood of record at the lower basin stations near Mammoth and at Winkleman (table 16.1). Widespread heavy precipitation in the lower basin contributed to the downstream increase in 1983. Additionally, the higher drainage density in the lower basin can result in more rapid runoff and increased flood magnitude (see chap. 13).

Summary

The storm runoff or flood flow component of streamflow is much larger than the baseflow component, in terms of total volume, but is much more variable. Seasonally varying storm types cause the San Pedro streamflow and flood hydrographs to show a strong seasonal pattern. During the past century, about 90 percent of the annual maximum flood peaks occurred during July through October. Most of these were produced by summer monsoon convective thunderstorms, which dominate in July and August. Floods associated with eastern Pacific tropical storms occurred most often in late summer and fall. Less than 10 percent of the annual floods occurred in December through March. Though average annual rainfall has not changed over time, the fluctuations in the annual distribution of that rainfall have influenced the flooding experienced by the San Pedro River. During the 1920s and 1930s, the largest annual floods were dominated by tropical storm events, and this gave way to dominance of the convective storm-driven annual floods until 1965, when the synoptic winter storms became dominant in the production of annual floods. Twice in the last 30 years, annual floods caused by tropical storms were among the largest floods on record in this basin.

PAUL BROOKS AND
KATHLEEN LOHSE

Water Quality in the San Pedro River

SEVENTEEN

Introduction

In this chapter, we consider the temporal and spatial variation in San Pedro River water chemistry as it relates to local and regional land use patterns and to the hydrologic cycle. Because the chemistry of water integrates a wide range of physical and biological processes that occur as the water moves through the basin, monitoring water chemistry provides a mechanism to identify impacts from both natural and human-induced changes to the landscape. In semi-arid basins such as the San Pedro, the resultant changes in chemistry can limit the use of already scarce surface water.

Water quality, commonly defined as the physical, chemical, biological, and aesthetic characteristics of water, determines the suitability of water for purposes such as drinking, recreation, and commercial uses (USGS 2005). The chemical composition of water also influences the diversity and activity of aquatic organisms and their capacity to provide services such as nutrient filtering. In the United States, a growing concern and awareness of pollution impacts on water quality led to the enactment of the Clean Water Act of 1977. This act established a structure for regulating discharge of pollutants to waters and gave authority to the U.S. Environmental Protection Agency (EPA) to implement water pollution control programs (http://www.epa.gov/

regulations/laws/cwa.html). Since its enactment, federal programs have been implemented that enable local volunteer-based nongovernmental organizations (NGOs) to protect watersheds and quantitatively assess water quality. As populations grow and water use intensifies, water quality will become an increasingly important factor in political, social, and environmental decision making from local to national scales (National Research Council 1990). In the San Pedro basin, the combination of a rapidly growing population, expanding calls for protection of riparian ecosystems, and unknown consequences of climate change underscore the need to increase our understanding of interactions between land use within watersheds and water quality.

San Pedro Water Quality Overview

The quality of the groundwater that serves as the primary water source to the San Pedro River is generally good and is considered acceptable for most uses by EPA standards. Regular sampling of residential and municipal, and some sampling of agricultural groundwater wells, has revealed no widespread contaminant issues, although there have been isolated samples where fluoride concentrations have approached or exceeded state standards of 4.0 mg/L (Towne 2002). Total dissolved solids, often used as a rough guide of water quality, are typically below drinking water standards of 500 mg/L. The exceptions to generally good groundwater quality are a few areas with either metal or nutrient contamination associated with mining and industrial land uses described below.

The quality of surface water in the San Pedro River and tributaries is more variable than that of groundwater. Of the 331 miles of surface water in the San Pedro basin assessed by the Arizona Department of Environmental Quality (ADEQ), 70 miles are considered acceptable for all uses, 84 miles are impaired, and the conditions of the remaining reaches are either inconclusive or acceptable for some but not all uses (ADEQ 2005) (fig. 17.1). Of the impaired reaches, metals and low pH from historic mining operations, high nitrate concentrations from the Apache Powder Superfund cleanup site, and fecal coliform bacteria from diffuse sources in urbanizing and agricultural reaches represent threats to wildlife and human health. Furthermore, total dissolved solid concentrations are elevated in drier, losing reaches, a result of evapo-concentration of solutes (Brooks and Lemon 2007). It is possible that declines in baseflow over the last several decades (Pool and Coes 1999) have exacerbated this problem, but water quality records are not of sufficient duration to confirm or refute this hypothesis.

Temporal Patterns in Surface Water Chemistry

Surface water chemistry in the San Pedro River is related to both spatial and temporal variability in hydrologic flowpaths (Harms 2004, Lewis et al. 2006,

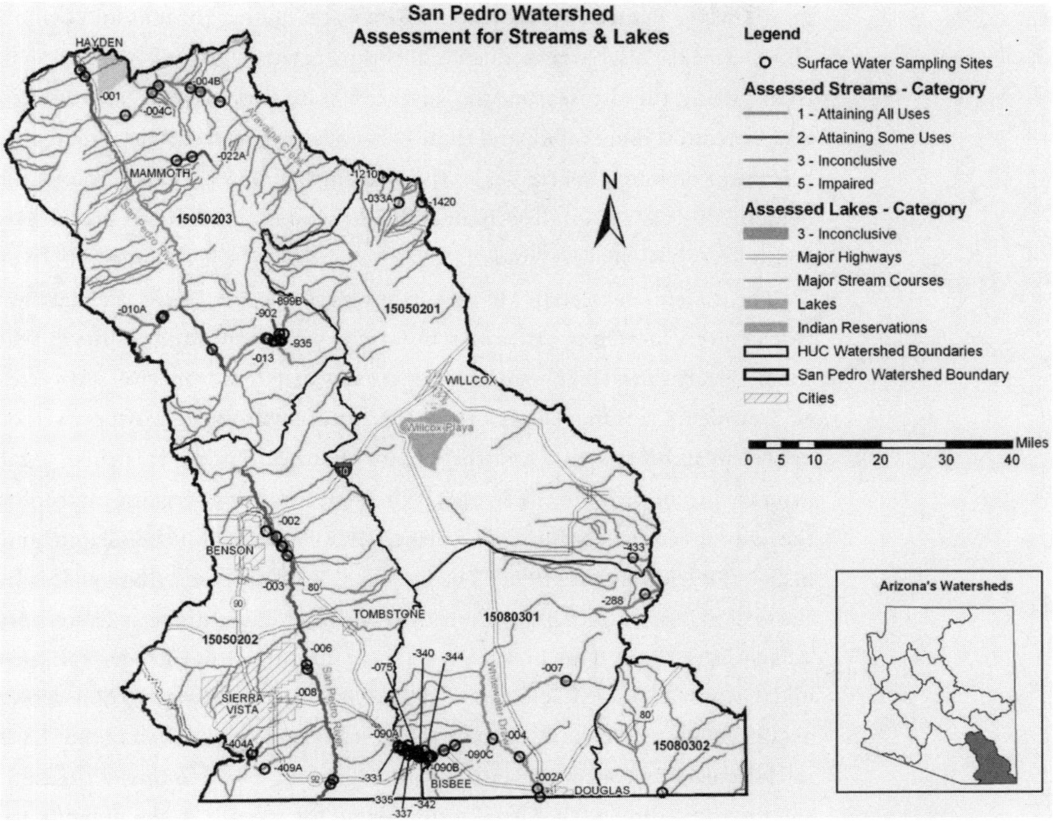

Fig. 17.1. Water quality status of assessed stream reaches in the San Pedro basin (after ADEQ 2005). Courtesy of Arizona Department of Environmental Quality.

Baillie et al. 2007, Brooks and Lemon 2007, Brooks et al. 2007, Meixner et al. 2007; and see chap. 18). During non-flood seasons, streamflow is supplied by a mixture of regional (basin-fill aquifer) and riparian (stream aquifer) groundwater sources (Baillie et al. 2007, Brooks and Lemon 2007; and see chap. 15). Extensive contact of the groundwater with subsurface minerals results in relatively consistent chemistry and quality of the water discharged to the stream as baseflow (Harms 2004, Lewis et al. 2006, Brooks and Lemon 2007). Nutrient concentrations are low, pH is neutral to slightly alkaline, and no major problems with organic or inorganic contaminants such as pesticides or metals have been identified. Some variation in stream water quality, primarily with respect to nutrients, results during the dry seasons due to interactions between water in the surface stream and water in the streambed and adjacent floodplain aquifer (the hyporheic zone).

During the monsoon season, episodic surface runoff and interflow from near-surface soil augment stream discharge. These flowpaths transport solutes and sediments from the terrestrial environment to the stream and result in higher and more variable solute concentrations (Harms 2004, Meixner et al. 2007). Multiple flood peaks typically occur during a single monsoon season with intervals of elevated baseflow discharge between peaks (see chap. 16). During these intervals, exchange between surface water and alluvial groundwater again exerts local control on water quality (Brooks and Lemon 2007). Because discharge is much higher during the intervals between floods than during the dry season, the advective transport of water and solutes downstream is more rapid, and these exchanges between surface and hyporheic occur on longer spatial scales than during low-flow periods (Brooks and Lemon 2007). Water quality changes during winter or fall floods on the San Pedro River have yet to be investigated.

The extreme seasonality in water sources to the San Pedro is characteristic of rivers located in arid and semi-arid environments and results in seasonal patterns in water chemistry that are very different from rivers in wetter environments. For months at a time, there is no hydrologic transport vector between catchment soils and the stream channel. During this time, algal primary productivity in the stream is the major source of organic matter for the aquatic ecosystem (Busch and Fisher 1981, Fisher et al. 1982, Grimm and Fisher 1989, Jones et al. 1996). The little dissolved organic carbon that is in the stream is produced in situ from algal and microbial exudates (Westerhoff and Anning 2000). This in-stream primary production is nitrogen (N) limited (Grimm et al. 1981, Jones et al. 1995a, Grimm and Petrone 1997, Marti et al. 1997, Valett 1993; and see chap. 18), resulting in surface waters that have low dissolved N concentrations. At the beginning of the monsoon season, solutes and sediment that have accumulated for months in the uplands are flushed to the riparian corridor, increasing carbon and nutrient concentrations in the stream water (Fisher et al. 1982, Jones et al. 1996, Jones et al. 1997, Parks and Baker 1997, Westerhoff and Anning 2000).

Brooks et al. (2007) documented this seasonality in an 18-month study of organic matter and inorganic nitrogen concentrations in the San Pedro River at Boquillas Ranch, approximately 55 km north of the Mexican border. Concentrations of dissolved organic carbon, dissolved organic nitrogen, nitrate, ammonium, particulate organic carbon (POC), and particulate organic nitrogen (PON) were low during the non-monsoon season (i.e., late fall, winter, spring, and early summer) and increased during the monsoon, with the largest changes observed in POC, PON, and nitrate. Spectrophotometric analyses confirmed that these increased concentrations were associated with solutes flushing from terrestrial sources. A major and still unaddressed question is the potential for these storms to transport contaminants to the riparian corridor along with carbon and nutrients. This may become an important issue in the future if changes in land use associated with a growing population increase potential contaminant sources within the basin.

The concurrent increases in discharge and concentration resulted in most of the annual organic carbon (96 percent) and nitrogen (97 percent) being transported in the stream during the two-month monsoon season. Of this,

92 percent of the carbon and 83 percent of the nitrogen were transported as particulates (Brooks et al. 2007). On other rivers, much of the organic matter similarly is transported to the stream as particulates (Parks and Baker 1997, Pusch et al. 1998, Jacobson et al. 2000). The fate of these compounds remains unknown, but their importance to the stream ecosystem may be greater than generally assumed. It is likely that this particulate organic matter is deposited with sediment on the streambed, providing the substrates for observed high rates of biogeochemical activity at the stream-hyporheic interface (Valett et al. 1994, Jones et al. 1995b, Morrice et al. 1997, Fisher, Grimm, Martí, Holmes and Jones 1998, Battin 2000).

Spatial Patterns in Surface Water Chemistry

INFLUENCE OF LAND COVER CHANGE ON SOLUTE DELIVERY TO THE STREAM CHANNEL

Nutrient and solute concentrations in stream ecosystems can exhibit considerable spatial variation related to the characteristics of the watershed (Aiken and Cotsaris 1995, Heiny and Tate 1997, Boyer et al. 2000, Snyder and Williams 2000, Findlay et al. 2001, Aryal et al. 2003). The substantial population growth and urban expansion that the San Pedro River basin is experiencing alters characteristics of the watershed (Parks and Baker 1997, Steiner et al. 2000). Much of the landscape has already been converted to agricultural and military land uses, and many of the undeveloped upland ecosystems now have increased abundance of nitrogen-fixing mesquite (see chap. 11). These transformations can cause excess loading of nutrients, solutes, and sediments to stream channels (Allan and Johnson 1997, Findlay et al. 2001, Martens 2003), a potential increase in trace metal concentrations associated with mining, and increases in enteric pathogens, bacteria, viruses, and parasites that pass from fecal matter from humans, cattle, and/or wildlife into surface waters.

Agriculture areas. Agriculture is a significant land use in the San Pedro River basin. The primary water quality issues associated with agriculture are sediment, pesticide, nutrient, and fecal matter delivery to the riparian corridor. A recent survey of stream water chemistry in the San Pedro suggests that agriculture has significantly increased nitrogen concentrations at several locations (Brooks and Lemon 2007). In repeat synoptic surveys along a 95-km section of the San Pedro, agricultural reaches (i.e., those with high amounts of agriculture within the subcatchment draining to the reach) had significantly higher nitrate-N concentrations during the dry season than either urban reaches (those with a large urban center in the subcatchment) or natural reaches (those with mostly undeveloped land in the adjacent uplands)

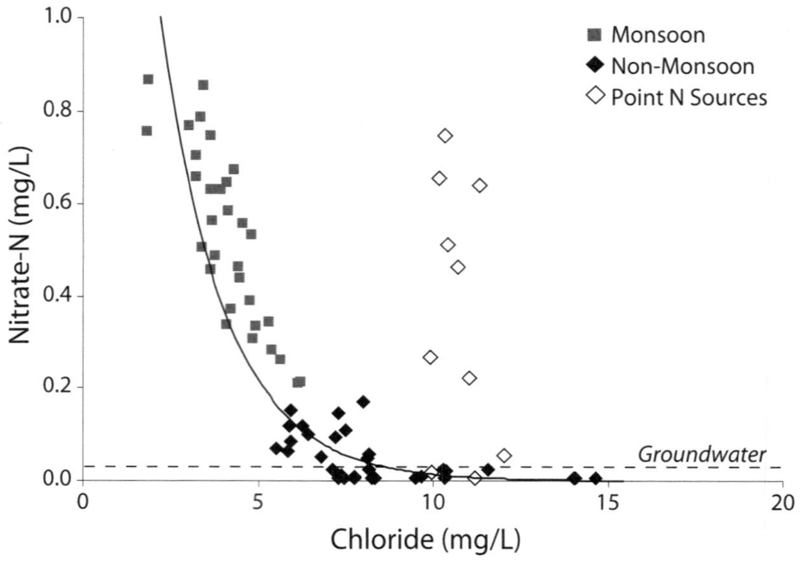

Fig. 17.2. Comparisons between chloride and nitrate-N concentrations in the San Pedro River. Higher chloride concentrations identify water that has been in the stream for a longer period with a higher likelihood that nitrogen would have been removed. The dashed line represents the average groundwater nitrate-N concentration in the San Pedro River basin (Pool and Coes 1999).

TABLE 17.1. **Mean (plus or minus standard deviation) nutrient concentrations in the San Pedro stream water associated with different land uses in the adjacent upland areas.**

	Land Uses		
	Undeveloped	Agriculture	Urban
Dissolved Organic Carbon			
Dry Season	1.82 ± 0.67[ab]	1.15 ± 0.86[b]	1.07 ± 0.44[a]
Wet Season*	3.70 ± 0.66	3.32 ± 0.52	3.24 ± 0.67
Dissolved Organic Nitrogen			
Dry Season	0.08 ± 0.04[b]	0.19 ± 0.22[b]	0.06 ± 0.03[b]
Wet Season*	0.13 ± 0.10	0.13 ± 0.08	0.12 ± 0.10
Nitrate Nitrogen			
Dry Season	0.01 ± 0.02[a]	0.23 ± 0.28[a]	0.06 ± 0.06[a]
Wet Season*	0.57 ± 0.19	0.53 ± 0.22	1.51 ± 2.27

Source: Data from Brooks and Lemon 2007.

[a]Difference between land uses is significant at $p<0.01$ level.

[b]Difference between land uses is significant at $p<0.05$ level.

*No significant differences were found between land uses during the wet season.

(table 17.1). Chronically elevated nitrogen concentrations in these agricultural reaches may indicate that N loads are exceeding biological demand (fig. 17.2).

Mining areas. The San Pedro River basin has a long history of mining, and many abandoned and active mine sites are distributed through the basin. Mining exposes primary minerals to both oxygen and water, resulting in accelerated mineral weathering, a decrease in pH, and elevated metal concentrations in surface water. Although a suite of minerals (e.g., P, S, Ca, Mg, Na) are released from the host rock of the catchment during this accelerated weathering, the major concern with mine drainage is the release of highly toxic metals (e.g., Cu, Zn, Cd, Ni). These metals can be toxic to both human and aquatic life at concentrations much lower than other dissolved minerals and require detailed analytical sampling to identify areas of impairment. In the San Pedro basin, sources of metals associated with mines present a potential for episodic metal transport to the riparian system in surface runoff as well as slow transport of mine wastes to the stream in groundwater. For example, Mule Gulch and its tributaries, which flow into the upper San Pedro from Bisbee, have copper and zinc concentrations and decreased pH values that exceed state standards. Dissolved copper is as high as 12,000 ug/L, or 185 times higher than wildlife and human health standards (ADEQ 2005). Groundwater in the headwaters of Greenbush Draw near Bisbee is also contaminated both from mining in the area and from Bisbee's wastewater treatment discharge, but not at the levels observed in Mule Gulch (ADWR 2005b). Sinks for these metals remain poorly characterized along the San Pedro, but metals are likely bound downstream of contaminant sources onto organic and Fe and Mn oxide mineral sediment fractions and possibly mobilized during floods (Bryan and Langston 1992).

The largest remaining active mine along the San Pedro is the Cananea mine, located in the headwaters of the river in northern Mexico. The Cananea facility is one of the largest open-pit copper mines in the world, and wastewater from the mine contains elevated iron, copper, manganese, and zinc concentrations with pH values as low as 3.1. An unintentional release of mine wastewater into the San Pedro in 1978 resulted in stricter regulation and improved wastewater control technologies, and this mine is no longer considered a major threat to the river's water quality in the United States (USBLM 1998b). However, a 1993 study by DICTUS (Departamento de Investigaciones Cientificas Tecnológicas, University of Sonora; DICTUS 1993) revealed that sulfates, cadmium, copper, chromium, lead, iron, manganese, and zinc remained elevated in surface water near the mine, with concentrations decreasing rapidly downstream towards the U.S.-Mexico border. Sulfates (6,000 mg/l) and total dissolved solids (9,000 mg/l), both indicative of mine drainage, in surface water near Cananea remained elevated in the late 1990s (Enlaçe Ecologico 1999); concentrations again decreased downstream. From

Ejido Jose Maria Morelos to the international border at San Pedro Palominas, slightly elevated concentrations of phosphorous, nickel, and sodium were also detected in surface water and wells. In a more recent study, elevated concentrations of cadmium, copper, chromium, lead, iron, manganese, and zinc were found in sediments near the city of Cananea, with concentrations decreasing downstream (Gomez-Alvarez et al. 2007). The most toxic metals (Cu, Zn) were predominately bound in Fe and Mn oxide fractions, while smaller amounts of Cu were found complexed with organics. Lead was found primarily in the residual phase, defined as inert metals held within the crystalline structure. Because metals held in Fe and Mn oxides or complexed with organics both can be released through abiotic and biotic redox reactions, these findings suggest that the mine will have a lingering effect on San Pedro water quality.

Industry and military. Routine monitoring by the Fort Huachuca military base has identified no water quality concerns associated with military land use (USDOD 2003). Concentrated industrial activity, however, results in point sources of pollutants to regional groundwater and the San Pedro River. Of note, groundwater in the area near St. David is contaminated with lead, sulfate, and nitrates from the former Apache Powder explosives manufacturing plant, designated a Superfund site by the EPA in 1990 (USEPA 2005). Recent groundwater remediation measures at the site include constructed wetlands to reduce nitrate in groundwater (>200 mg NO_3-N/L) to meet federal drinking water standards (10 mg NO_3-N/L) (Roudebush and Beilke 2006). Despite the remediation measures, elevated nitrogen concentrations along this reach of the San Pedro River suggest that contaminated groundwater may have entered the river (Brooks and Lemon 2007).

Cumulative impacts. Two emerging water quality concerns in the San Pedro River are enteric pathogens and fecal coliform bacteria. These represent a significant public health concern for human contact with surface water. Fecal coliform bacteria have been recorded at the border between the United States and Mexico (ADEQ 1992) and farther downstream. Recent monitoring revealed fecal coliform bacteria counts, in particular *Escherichia coli*, as high as 2636 CFU/100 ml near Benson (upper basin) and Hayden (lower basin), greatly exceeding state standards of 235 CFU/100 ml (ADEQ 2005). Potential sources of these bacteria include livestock and, in the upper reaches, untreated sewage from Cananea and other communities along the river in Sonora (Enlace Ecologico 1999).

The Role of Riparian Ecosystems on Water Chemistry

As discussed above, monsoon storms represent the major hydrologic transport vector from uplands to the stream channel. Analysis of both conserva-

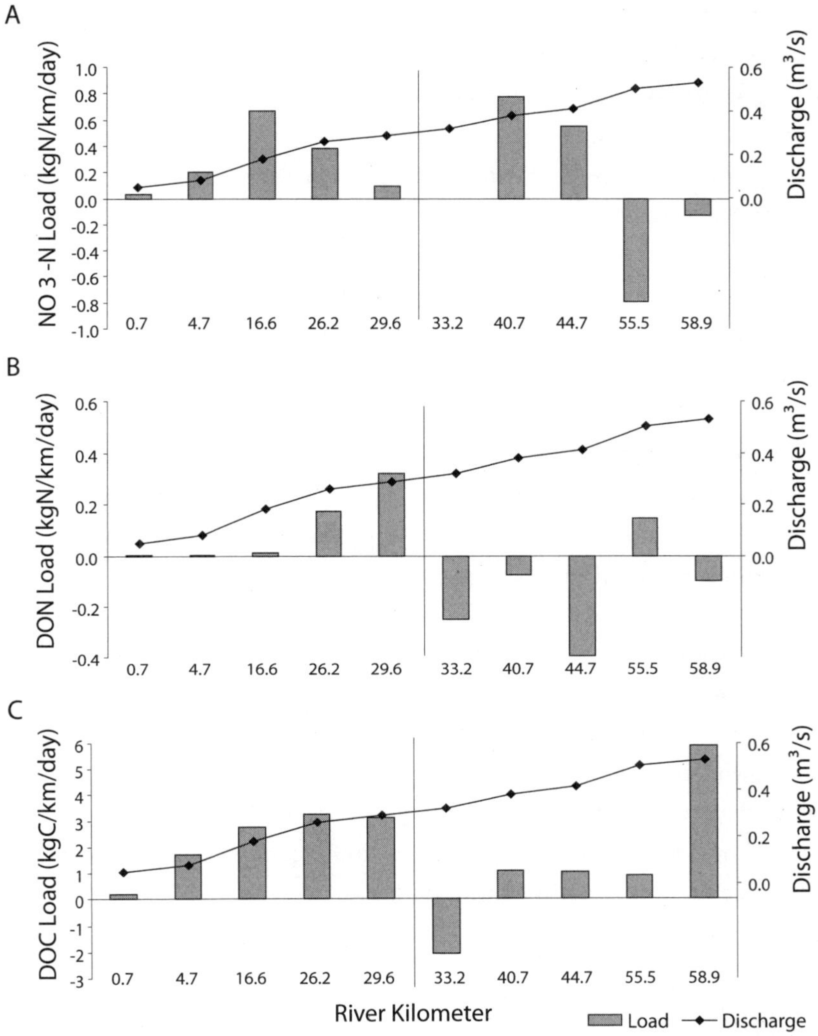

Fig. 17.3. Changes in the mass of Dissolved Organic Carbon (A), Dissolved Organic Nitrogen (B), and nitrate-N (NO_3-N) (C) transported between each sampling location during the synoptic sampling event on August 12–14, 2002; positive values indicate net input to the river, while negative values indicate net uptake. Values on the x-axis indicate distance from the Mexican border.

tive solutes and oxygen isotopes in the river indicate rapid mixing of water sources during and after monsoon storms (Brooks and Lemon 2007, Meixner et al. 2007). This mixing can mask local effects on water quality but allow larger-scale effects of land use and riparian structure to be seen. For example, Brooks and Lemon (2007) showed that both carbon and nitrogen concentrations increased in the 33 km of the river downstream from the Mexican border during monsoon-season baseflow (fig. 17.3). This is a region with high cover of both farm fields and mesquite shrublands, both of which have elevated soil nitrogen content (Martens 2003). The elevated stream nitrogen concentrations and load during the monsoon suggests that N inputs from upland systems may exceed biological demand for N along this reach of the river.

This pattern contrasts with the dry season when only reaches near N point sources were nitrogen-enriched, as shown in figure 17.2. Downstream from the agricultural reaches at km 33, the river transitions from a losing reach to a gaining reach, and the stream-riparian corridor becomes a sink for both nitrogen and carbon, indicating higher rates on biogeochemical cycling in the stream-riparian system. At km 58.9, loading of dissolved organic carbon increases due to two major ephemeral tributaries, Babocomari Wash and Walnut Gulch, that enter the San Pedro over a 3.5-km reach. Presumably, the patterns in chemistry below km 33 more closely reflect biogeochemical cycling in a stream setting not affected by agriculture.

Summary

Despite potential threats from mining, agriculture, and urbanization, water quality in the San Pedro River is generally good. However, there is evidence that biogeochemical processing in the stream-riparian system has been altered by human land and water use. Specifically, areas of the river appear to have switched from nitrogen sinks to nitrogen sources. This switch in N status appears to result both from reduction in streamflow and from the introduction of nutrients to the channel from upland agriculture. A major unknown in the San Pedro is how population growth will affect the transport of contaminants to the stream during annual monsoon floods. This may become an important issue in the future as changes in land use increase the potential contaminants within the basin.

DAVID BRUCE LEWIS,
TAMARA K. HARMS,
JOHN D. SCHADE,
AND NANCY B. GRIMM

Biogeochemical Function and Heterogeneity in Arid-region Riparian Zones

EIGHTEEN

Introduction

Rivers impress with their motion. Flowing water visibly sweeps materials by, in sharp contrast to the relative quiet of an observation point on an adjacent bank. The apparent stillness of these riparian zones, however, belies the dynamic flows occurring there. Riparian zones are open, thermodynamic, non-equilibrium systems. They exchange material with neighboring parcels of space while processing materials within their own boundaries.

Riparian zones showcase feedbacks among biological, chemical, and physical features. The age, size, density, and species composition of the riparian vegetation may dictate the demand for nutrients. Vegetation also affects the community of soil microbes, the stability of sediments, and the power of floods, thus indirectly regulating the release of nutrients and pollutants to downstream ecosystems and to the atmosphere as greenhouse gases. In turn, vegetation is influenced by microbes, nutrients, sediments, and floods. These feedbacks exhibit cycles as organisms grow, species composition changes, sediment sorting proceeds, and nutrient stores are altered. These cycles can be reset by floods, fires, and other disturbances. Multiple cycles are hierarchically related (Fisher, Grimm, Martí and Gomez 1998)—fast cycles occurring in small areas are nested in, and constrained by, slow cycles occurring in large

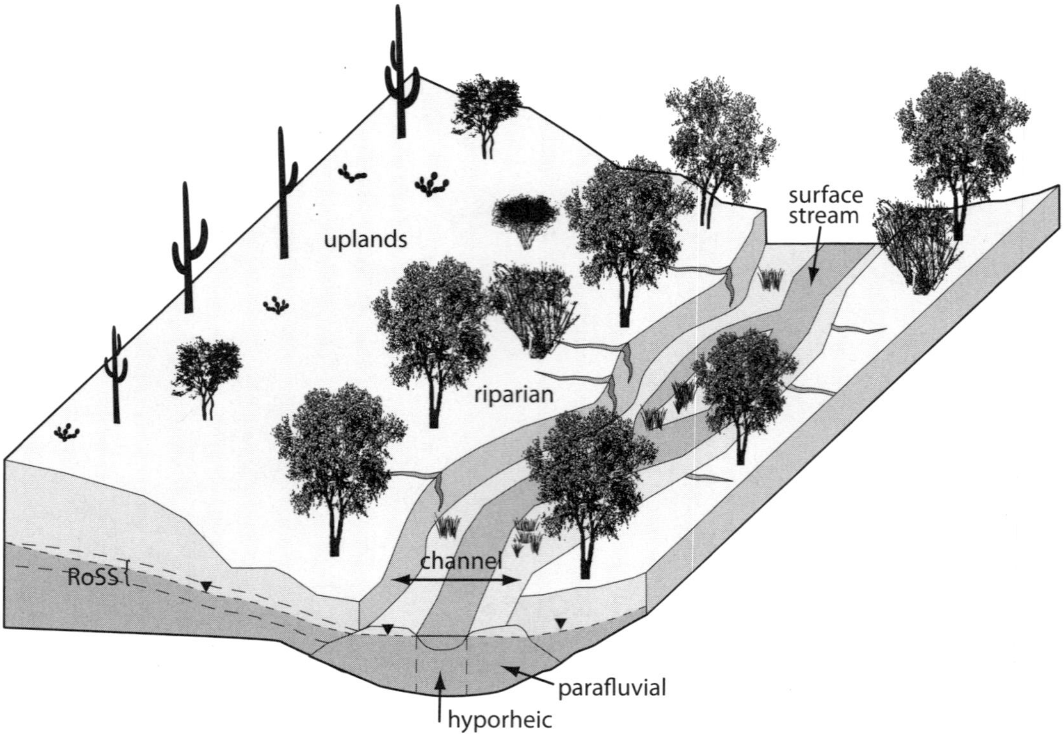

Fig. 18.1. Schematic drawing of upland, riparian, and channel landscape elements. Within the annually flooded channel are the surface stream, parafluvial zone, and hyporheic zone. The parafluvial and hyporheic zones are areas of saturated sediment within the channel. Their relative sizes change seasonally with fluctuations in the water table. Similarly, the relative sizes of the saturated and unsaturated layers of riparian soil change seasonally with water table fluctuations. RoSS stands for region of seasonal saturation. Illustration credit: David B. Lewis.

areas. This idea of cycles reset by disturbance shapes the conceptual model of arid-land riparian zones.

In this chapter, we provide structural and functional descriptions of riparian zones with respect to biogeochemical processes. Structurally, riparian zones are three-dimensional volumes (fig. 18.1). They are the interface between terrestrial and aquatic ecosystems (Grimm et al. 2003). They are also elements nested in a catchment, which further includes terrestrial uplands and the active channel (D. Lewis et al. 2006). Vertically, riparian zones extend up to the forest canopy (where they include air) and down through the rhizosphere and region of seasonal saturation (Baker et al. 2000, Huth 2003, Harms 2004). Functionally, riparian zones host a particular suite of feedbacks (fig. 18.2). Riparian zones are regions of successive and mutual influence between the routing of floodwaters and the location of alluvium deposits. Riparian zones are also regions where hydrology and geomorphology influence biota via regulation of water supply, and in turn, biota regulate streamflows and channel morphology. The biogeochemical features of any riparian zone depend

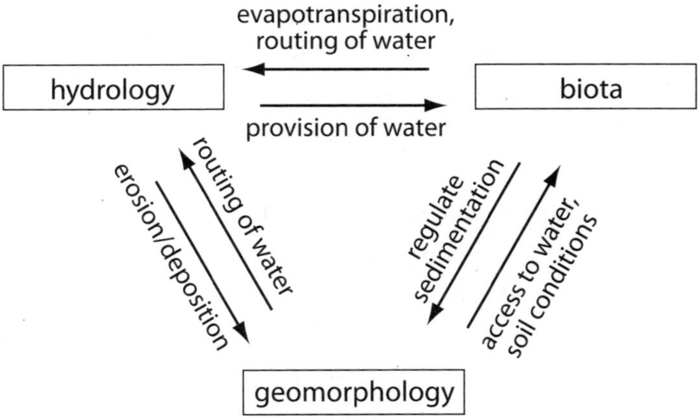

Fig. 18.2. Feedbacks occurring in the San Pedro River riparian zone among hydrological, biological, and geomorphic dynamics. Arrows represent directions of control, and arrow labels provide examples of controlling factors.

on the position of that riparian zone in the broader landscape, varying from ephemeral streams to large, low-gradient rivers. We discuss the biogeochemical functions riparian zones perform. Spatial and temporal variability are emphasized from the perspective of hierarchical patch dynamics. Finally, we point to some applied issues for riparian zones of the dry U.S. Southwest.

Riparian Zones and Big Questions in Biogeochemistry

Riparian zones have helped identify important issues in biogeochemical research (Grimm et al. 2003). These issues include nutrient limitation, material flux vs. recycling, and material retention. We use the term "material" to refer generally to atoms, molecules, and classes of molecules (e.g., organic carbon), any of which could be considered a nutrient or pollutant, depending on the context.

NUTRIENT LIMITATION

Nutrients are the elements required for life. The limiting nutrient is the least available nutrient relative to the needs of the organisms. Productivity in stream channels can be limited by either nitrogen (N) or phosphorus (P) (Elwood et al. 1981, Grimm and Fisher 1986). Which nutrient is actually limiting depends on the parent geology and length of time available for weathering, as P is mainly supplied from rock weathering and can become depleted over geologic timescales. Individual components of ecosystems can be limited by different elements at different times and in different places. Often, microbial processes are limited by the availability of organic carbon (organic C), an energy source for heterotrophic metabolism. Microbes, however, can also be limited by other nutrients (Howarth and Fisher 1976, Kaye and Hart 1997).

Riparian vegetation. In arid and semi-arid regions, riparian vegetation, once their water needs are met, are probably N limited (Lockaby and Conner 1999, Adair and Binkley 2002). Even if N supplies are ample, N limitation may occur if N is attached to complex organic molecules that are readily exported from the riparian zone in water flow. Nitrogen in this form is generally unavailable for direct uptake by plants, and some proportion of these molecules are recalcitrant to breakdown (which would liberate the N) via microbial respiration. No research, however, has documented whether N limitation broadly characterizes the San Pedro or southwestern riparian zones and if so, whether N limitation results from the coupled processes of rapid organic N formation followed by its dissolution and hydrologic export. Nevertheless, several sources of information lend credibility to these hypotheses. In terrestrial soils, N can quickly enter soil organic matter (Kaye et al. 2003), making it largely unavailable for plant uptake. In temperate forested catchments, hydrologic N losses can be high despite the N limitation status of the forest (Hedin et al. 1995). In the San Pedro, organic N loss from the riparian zone appears to be constant throughout the year, as the surface stream concentration of dissolved organic N does not fluctuate seasonally (Lewis et al. 2006).

Riparian microbes. Riparian zone microbes in the arid Southwest appear to be limited by organic C under certain conditions. In one Sonoran Desert stream, rates of denitrification along subsurface flowpaths are much higher within the organic C-rich rhizosphere of the riparian plant seepwillow (*Baccharis salicifolia*) than upstream of the plant (Schade et al. 2001). Along a Sonoran drainage network from upland rills to permanent channels, respiration pulses by soil microbes following rainfall events are limited by labile soil organic C, though labile C may be less important than water availability for annual respiration (Sponseller and Fisher 2006, Sponseller 2007). Microbial processes in riparian soils of the San Pedro River respond to pulses of water during the dry seasons, but to pulses of organic C during monsoons (Harms 2004). Given this importance of organic C as an energy source for heterotrophic microbes, two research frontiers loom large in dryland riparian zones. First, what is the source of organic C? Is it in situ breakdown of riparian zone organic matter, or material imported from uplands and the channel? Second, what is the molecular composition of the organic C? What proportion of it is recalcitrant (and thus a long-term sink for associated N atoms), and does this proportion differ between the uplands and channel, where moisture conditions and microbial species may differ?

There is little evidence that streams of the southwestern United States are P limited. This suggestion is surprising, given the age of this landscape. The faulting and crustal extension that caused downdrop of the basins in the Basin and Range Physiographic Province of southern Arizona initiated weathering 12–15 million years ago. Alluvium deposition into the basins has been going on ever since (McAuliffe 1994), but apparently P supplies to fluvial ecosystems remain sufficient.

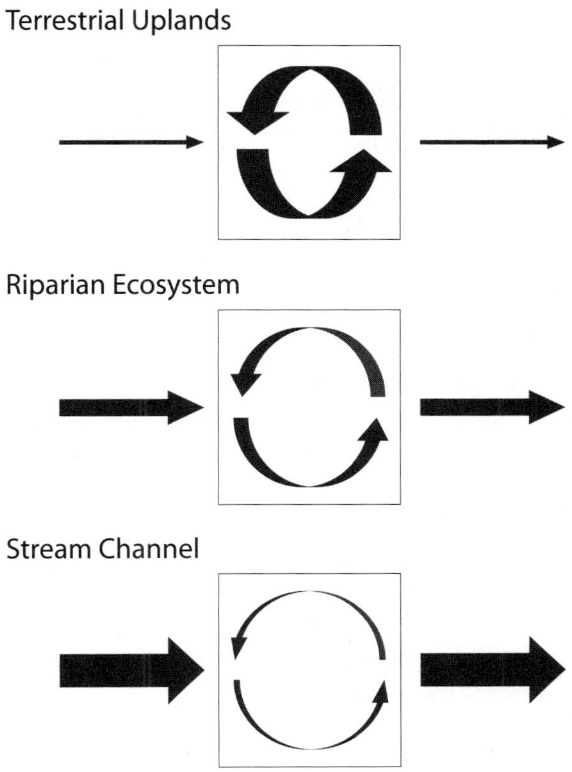

Fig. 18.3. Conceptualizations of mass balance budgets. Width of arrow represents amount of material. Box represents an ecosystem. Straight arrows are inputs to and outputs from the ecosystem. Curved arrows within boxes represent material cycling among pools within the ecosystem.

MATERIAL FLUX

One important problem in ecology is determining the rate of material recycling within an ecosystem relative to the rate of material flux through an ecosystem, and then comparing among systems the relative importance of recycling vs. throughput (Essington and Carpenter 2000, Lewis et al. 2006). At the San Pedro and other catchments of the Southwest, the riparian zone blends the recycling nature of the terrestrial uplands with the throughput nature of the active channel (fig. 18.3).

Terrestrial uplands. There is relatively little export of material from upland, terrestrial ecosystems along the San Pedro, nor from ecosystems on terraces situated above the active floodplain (e.g., phreatophytic *Prosopis* woodland and *Sporobolus* grassland). Aquatic vectors transport material from uplands to the riparian zone only in the event of large storms (see chap. 17), and the amount of material transported in such events remains an open question. Most storm runoff moves material only a short distance before infiltrating or evaporating, thus failing to reach lowland rivers (Welter et al. 2005). Gaseous exports generally are low from arid and semi-arid uplands because microbial activity is limited by a lack of water, but they do occur in pulsed events (Huxman et al. 2004, Sponseller 2007). In the uplands, most material remains in internal cycles driven by plant-soil exchanges and by short-distance hydrologic displacements. Episodic rainfall can, however, stimulate the export of carbon dioxide (CO_2), N_2, nitric oxide (NO), or nitrous oxide (N_2O) to the atmosphere from the uplands. Vertical leaching to groundwater is relatively minor (but see Walvoord et al. 2003).

Stream channel. In the surface stream and enveloping sediments of the active channel, horizontal discharge prevails. Inputs and outputs are relatively more important than internal cycles (fig. 18.3). The channel is characterized by more rapid flux and fewer opportunities for material to adsorb to sediments and be taken up by biota. Recycling within the stream does occur, however. In Sycamore Creek, channel nutrients are processed when channel water reenters the riparian zone (Schade et al. 2002, 2005), infiltrates within-channel sediments (Grimm and Fisher 1984, Holmes et al. 1994, Valett et al. 1994, Jones et al. 1995a), and evaporates and strands its material load (Stanley et al. 1997). Thus, the channel is not simply a pipe conveying material downstream (Bernhardt et al. 2003). Additionally, hydroclimatological conditions of this arid environment dictate that horizontal flux is not rapid year-round. During early to mid-summer, discharge is reduced, restricting channel flow to the hyporheic zone. This restriction in flow will increase opportunities for within-channel recycling and reduce rates of input and output. Nevertheless, unidirectional, horizontal flux is the principal feature of channel material budgets.

Riparian (floodplain) zone. Riparian zone material budgets are of an intermediate character (fig. 18.3). Advection is slower through the riparian zone than through the channel, affording greater opportunity for internal cycling of materials between soils and biota (Lewis et al. 2006). On the other hand, water and hydrologic flowpaths are more prevalent in the riparian zone than in the uplands, so fluxes of material in and out of the system are greater. This intermediate nature of riparian zones is evident when features of the riparian zone are averaged over long time periods and broad spatial extents. At any one time and place, the relative importance of input-output fluxes and internal cycles may more closely resemble that of uplands or channel.

In the San Pedro River riparian zone, the importance of inputs and outputs (table 18.1), relative to internal recycling, varies according to many factors. These include time of year; longitudinal position from headwaters in Sonora, Mexico, to confluence with the Gila River in south-central Arizona; valley geomorphology; and the material species or element in question. The throughput of more biologically usable forms of N (inorganic N) may fluctuate seasonally, with in situ recycling of inorganic N being important during seasons of rapid biological activity (Lewis et al. 2006). In contrast, the throughput (or processing length, after Fisher, Grimm, Martí, Holmes and

TABLE 18.1. Vectors by which atoms, molecules, and classes of material may enter or leave the riparian zone of the San Pedro River.

Inputs	Example
Dry atmospheric deposition	P bound to dust
Wet atmospheric deposition	N in rainwater
Atmospheric gaseous inputs	C uptake by plants
Sheet flow from uplands	Delivery of nutrients and organic matter of terrestrial origin
Floodwater from channel	Delivery of nutrients and organic matter of aquatic origin
Groundwater	Delivery of material from deep aquifer or from subsurface flow originating in the channel
Animals	Insects emerging from stream channel; terrestrial consumers of aquatic organisms; consumers of upland food sources
Outputs	**Example**
Gaseous	Denitrification, methanogenesis, aerobic respiration
Sheet flow to channel	Loss of nutrients and organic matter in dissolved and particulate (litter and sediment) form
Leaching	Loss of solutes to deep groundwater aquifer
Wind	Leaves and soil surface litter blown away
Litterfall	Leaves falling directly into channel
Animals	Riparian material exported by consumers

Jones 1998) of less usable forms (organic N) should be relatively great and constant year-round.

Floods facilitate both material flux through and material recycling within riparian zones. Floodwaters carry N from the channel to the riparian zone, both in overbank flow and in elevated discharge through subsurface flowpaths. Later, as overbank water and elevated groundwater recede, organic matter and N are exported back to the channel. Floodwaters also stimulate nutrient recycling. The rising riparian water table inundates previously dry soil, which contains a nutrient pool that had remained intact for lack of water. The newly available water dissolves and redistributes this pool, stimulating microbial activity and plant uptake (Baker et al. 2000, Schade et al. 2002, Heffernan and Sponseller 2004).

MATERIAL RETENTION AND WATER QUALITY PROTECTION

Riparian zones are noted for their retention potential. Inputs exceed outputs, and storage increases. This imbalance suggests that riparian zones can protect stream, lake, and estuarine water quality by retaining pollutants derived from uplands. At the San Pedro, most channel surface flow comes from precipitation and overland runoff. Some surface flow, however, does derive from the regional aquifer (see chap. 15), illustrating the connection of stream water with riparian and upland soils. This contact is variable in time and space.

Riparian zones have been characterized as filters because they remove nutrients from agricultural watersheds (e.g., Lowrance et al. 1984). In this filter model, water flows from upland to channel, and nutrients traverse the riparian zone before reaching the stream. In arid-region rivers and streams, a more complex array of flowpaths is possible, from generally longitudinal (e.g., flows in and out of riparian and hyporheic patches as water travels down the fluvial corridor) to lateral in either direction (riparian to stream or vice versa; Fisher, Grimm, Martí, Holmes and Jones 1998, Pool and Coes 1999, Schade et al. 2005). The capacity of the San Pedro and southwestern riparian zones to trap, incorporate, or otherwise retain many nutrients (P, S) and toxins (pesticides, trace metals) remains virtually unstudied. Progress has been made, however, on understanding NO_3^- retention by arid-land riparian plants. Riparian plant rooting zones host the anoxic conditions and supply the energy (as reduced organic C) necessary for denitrification (Schade et al. 2001, Schade and Lewis 2006).

Material budgets. Any discussion of material retention focuses on the budget model: change in storage (ΔS) = inputs (I) - outputs (O). The dynamics of the storage term in the riparian zone are probably intermediate to those in the uplands and channel (fig. 18.4). In the uplands, storage rates change slowly. Whether positive or negative, $\Delta S/\Delta t$ (storage/time) is relatively constant over decadal timescales. In the channel, storage-rates change varies rapidly (fig. 18.4). Diel changes in water demand by streambank vegetation cause hourly fluctuations in channel discharge, and thus in $\Delta S/\Delta t$. Floods sweep away sediments, algal mats, and other biota, causing $\Delta S/\Delta t$ rapidly to become negative. Following spates, nutrients begin accumulating within hours ($\Delta S/\Delta t$ quickly becomes positive), and nutrient pools fully recover in weeks (with positive $\Delta S/\Delta t$ declining), owing to sedimentation and incorporation into the biotic tissue of fast-growing, disturbance-adapted, pioneer organisms (Fisher et al. 1982, Grimm 1987).

In the riparian zone, we hypothesize that the storage term in the material budget model has more moderate dynamics, particularly when viewed over broad scales of space and time. At particular places over short time periods, $\Delta S/\Delta t$ may be very dynamic or relatively constant. Riparian zones can accumulate nutrients for months to years at a time as they add biomass. Then, they may quickly become net sources to the channel for brief periods (days

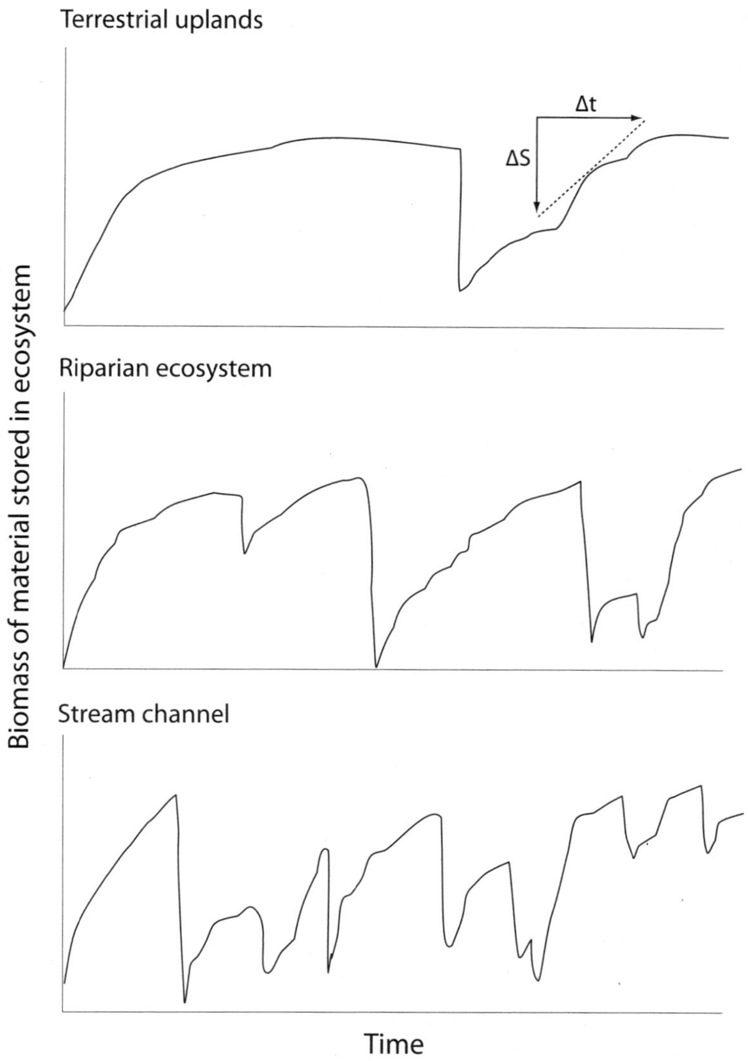

Fig. 18.4. Conceptualized time series for capital stored by an ecosystem, as compared among terrestrial uplands, riparian zones, and stream channels. An example of an instantaneous value for $\Delta S/\Delta t$ is shown.

to weeks) when this stored nutrient capital is scoured and swept downstream by floods. Floods, fires, and other disturbances cyclically destroy vegetation and create surfaces for colonization and deposition of soil organic matter. Riparian nutrient budgets generally exhibit cycles of accretion punctuated by catastrophic loss.

Spatial Heterogeneity

Biogeochemical heterogeneity in arid riparian zones is influenced by patchiness in water availability. Patchiness in water availability results from the uneven distributions of preferential flowpaths, pan evaporation and evapotranspiration rates (see chap. 2), and soil texture (see chap. 14). These distributions are influenced by the spatial arrangement of large physical structures, such as trees, which conduct, facilitate, and are influenced by a diverse array of biogeochemical processes.

INFLUENCES ON HETEROGENEITY

Water flow. Spatial heterogeneity in microbial processes and nutrient availability is partly governed by the distribution of water, the flow of which is regulated by slope, sediment size, woody debris, and live vegetation. These factors often interact to direct water along dendritic networks rather than in sheetflow, thereby restricting locations of biogeochemical activity (Sponseller and Fisher 2006).

Vegetation. Plants influence the heterogeneous distribution of nutrients. This influence is indirect, through plant control over water in both vertical and horizontal dimensions (see chap. 2), and is also direct, through nutrient uptake and release in litter fall and root turnover. Nutrient release can be particularly important for plants, such as mesquite (*Prosopis*), that host N-fixing symbiotic microorganisms in their root nodules (see chap. 14). Nitrogen concentrations are elevated in soils around mesquite trees, relative to soils in interplant spaces, perhaps owing to the provision of fixed N (Schade et al. 2003). While N fixers may supply nutrients to soil microorganisms, other plant species compete with soil microbes for nutrients (Kaye and Hart 1997).

Just as plant species identity influences water flux, it also influences nutrient dynamics. The quantity of nutrients and C contributed to soils varies among plant species. For example, sacaton grass (*Sporobolus wrightii*) produces senescent biomass that remains attached to the plant, while cottonwoods deposit significant quantities of biomass on the soil surface each fall.

Less well understood is the influence of plant C quality on soil microbial processes. In riparian zones, the chemical composition of plant tissue differs among plant species and among individual plants of the same species. The organic molecules that plants release to soil in root turnover and litter fall vary in size, structure, and N content. These features of organic molecules will determine whether they are relatively labile (subject to catabolism by soil microbes, whereupon C and N are mineralized) or recalcitrant (resistant to catabolism). The activity of varying types of mycorrhizal fungi also will influence the fate of these organic molecules (see chap. 4). Data from a San Pedro River terrace show how differences in organic compounds produced by different plant species generate heterogeneity in soil microbial activity. Soils of open, sacaton-dominated, and mesquite-dominated patches exhibit varying fluxes of CO_2 and N_2O gases to the atmosphere, indicating varying rates of microbial soil respiration and N processing (McLain and Martens 2005, R. Scott, Huxman et al. 2006).

Resource islands. Plants further regulate spatial heterogeneity in riparian zones by interacting with geomorphology (fig. 18.2; and see chap. 13). In dry regions, plants impart spatial heterogeneity by concentrating resources around themselves, creating islands of fertility. These islands of fertility are contingent on the lack of resources in interplant spaces. Islands of fertility will be eliminated (i.e., spatial heterogeneity will be suppressed) by factors that enhance productivity in the interplant spaces. Mesquite trees are particularly effective at generating islands of fertility. In addition to capturing wind- and water-borne particles, N-fixing mesquite also accumulate N within their vicinity. The magnitude of the fertility island, however, depends on the landscape position of the mesquite tree in question. Along the Verde River in arid central Arizona, mesquite-based islands of fertility are absent immediately alongside the active channel, where soil moisture is plentiful. Islands become more evident as soil moisture declines—and fertility of inter-mesquite spaces likewise declines—across the riparian zone away from the channel (Schade et al. 2003).

Another study, across a size gradient of southwestern riparian zones, similarly demonstrates how mesquite-based islands of fertility are mediated by landscape features. Islands of fertility around mesquite depend on the productivity of the tree. More productive trees have in their understory greater concentrations of soil organic matter, litter accumulation, and annual plant biomass. Because more productive trees are found in the riparian zone of higher-order streams, mesquite-based islands of fertility may increase with catchment drainage area. Flooding, however, also becomes more important with catchment size. In the largest catchments, despite the presence of the most productive mesquite trees, islands of fertility disappear as soil resources are redistributed and homogenized by floodwaters (Sponseller and Fisher 2006). Thus, riparian zone biogeochemistry is shaped by the interplay of tree-scale and catchment-scale processes.

Even after they have fallen, riparian trees of semi-arid regions continue to host resource islands. In the semi-arid riparian zone of the Sabie River, South Africa, a large flood (~100 yr recurrence interval) uprooted vegetation and deposited it as large woody debris. The woody debris piles act as resource nodes by accumulating fine sediments and by retaining C, N, P, and soil moisture. Resource accumulation was affected by pile characteristics such as size and species composition (Pettit and Naiman 2005).

Soils. Spatial heterogeneity in riparian nutrient content also arises due to variability in soil particle size (see chap. 14). Smaller particles have a higher affinity for charged constituents. In the riparian zone of the San Pedro, soil N concentrations are inversely correlated with sand content (Harms 2004). In two large, semi-arid basins in Colorado, high N content of freshly deposited sediments led to increased total N of sediments with site age on alluvial terraces (Adair et al. 2004). In a large temperate basin of similar size to the San Pedro, C, N, and P concentrations were positively correlated with alluvial sedimentation (Steiger and Gurnell 2002).

SCALE ISSUES

The degree of spatial heterogeneity depends on the scale at which the riparian zone is examined. Within two 0.5-km-long reaches of the upper San Pedro, there is distinct heterogeneity among subsystems (labeled regions in fig. 18.1) (Lewis et al. 2007). Dissolved inorganic N concentration in riparian groundwater is greater than in parafluvial subsurface water and surface stream water and is more spatially variable among points *within* the riparian zone than among points within the parafluvial zone or within the surface stream. Finally, riparian-zone dissolved inorganic N is dominated by the reduced form, ammonium (NH_4^+). When data are analyzed at finer scales, however, these generalities begin to break down. For example, at individual cross sections of the river there is no consistent pattern as to which subsystem (i.e., riparian zone, parafluvial zone, surface stream) has the greatest concentration of dissolved inorganic N or whether the N in the riparian zone is dominated by NH_4^+ or by the oxidized form, nitrate (NO_3^-).

Fine-scale (cm) heterogeneity can be generated by cascades of biologically mediated redox reactions (Dahm et al. 1998). This cascade results from a hallmark of riparian zones, the convergence of multiple flowpaths and the presence of disparate energy and element sources (McClain et al. 2003). Where elemental oxygen is available, microbial species and metabolic pathways adapted for aerobic respiration will dominate. When O_2 is depleted, reduction of NO_3^- to N gases (N_2, N_2O, NO_X) will ensue. The depletion of NO_3^- is then followed by the sequential reduction of manganese dioxide (MnO_2), iron(III) hydroxide ($Fe(OH)_3$), sulfate (SO_4^{2-}), and CO_2, with each electron acceptor assuming importance, in turn, as energetically more favorable ones are depleted. When water carrying a load of electron acceptors enters a slow-moving riparian flow path with ample opportunity to interact with the rhizosphere, bands in which electron acceptors are depleted will develop, with O_2 present at the upstream end of the flowpath and, perhaps, CO_2 reduction (methanogenesis) occurring <1 m farther along. These reactions are important, as they determine whether the element remains in the ecosystem as a solute or solid or is lost as a gas (e.g., $NO_3^-(aq) \rightarrow N_2(g)$), and whether the element is more or less toxic (e.g., various oxidation states of iron, Fe).

THE NITROGEN EXAMPLE

The N cycle demonstrates spatial variability in fluvial ecosystems. Different transformations of N (fig. 18.5) occur in different hotspots (fig. 18.6). Biological N fixation, the conversion of atmospheric N_2 to ammonia (NH_3) by the enzyme nitrogenase, supports all life on Earth because N is a component of proteins and nucleic acids, yet N in the predominant form (N_2, which is 78 percent of the atmosphere) cannot be used by most organisms. Because N fixation is energetically expensive, N-fixing microorganisms (diazotrophs) either

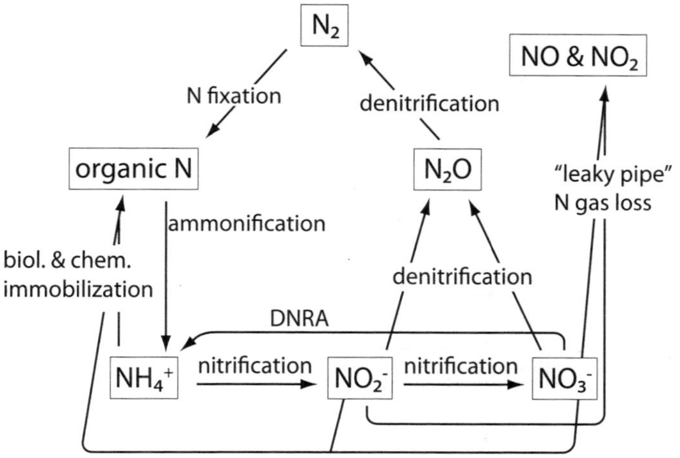

Fig. 18.5. Nitrogen-cycling processes probable in fluvial ecosystems. DNRA stands for dissimilatory nitrate (NO_3^-) reduction to ammonium (NH_4^+).

photosynthesize or live in an environment rich in organic C. For instance, in well-lit Sonoran Desert streams, N-fixing autotrophic cyanobacteria live on the parafluvial zone–surface stream interface, particularly at inwelling (upstream) edges of gravel bars (Henry and Fisher 2003). Additionally, heterotrophic bacteria fix nitrogen from within the root nodules of mesquite, from which they obtain energy as organic C. Nitrogen fixation rates are not perfectly correlated with mesquite distribution, however. Nitrogen fixers are facultative and often do not fix N when soil NO_3^- and NH_4^+ are available.

Nitrogen immobilization, ammonification, and nitrification are an interlinked suite of N transformations. In riparian zones, each is spatially heterogeneous, owing to heterogeneity in the others and in soil conditions and biological rates. Immobilization is the assimilation of N into organic matter. Immobilization rates vary according to rates of uptake by plants and microbes and according to the availability of labile ("usable") N (which is determined in large part by rates of ammonification and nitrification). Ammonification is the release of NH_4^+ during catabolism, by bacteria and fungi, of N-containing organic matter. Ammonification can occur under aerobic and anaerobic conditions. Ammonification rates are spatially heterogeneous in response to the distribution and C:N stoichiometry of the organic matter being catabolized. Under otherwise equal ammonification rates, the amount of NH_4^+ remaining in the soil will vary according to soil pH, as NH_4^+ loses an H^+ under high pH conditions (whereupon the resulting NH_3 can volatilize to the atmosphere and be lost from the fluvial ecosystem). Finally, nitrification is the oxidation of NH_4^+ to NO_3^- (fig. 18.7). Though heterotrophic nitrification occurs, autotrophic nitrification is generally thought to be more prevalent. Nitrification is carried out by taxa such as *Nitrosomonas* and *Nitrobacter*. Rates are spatially variable, being highest along the parafluvial zone–surface stream interface at the upstream edges of gravel bars—i.e., hotspots to which flowpaths deliver NH_4^+ and O_2 (Holmes et al. 1994).

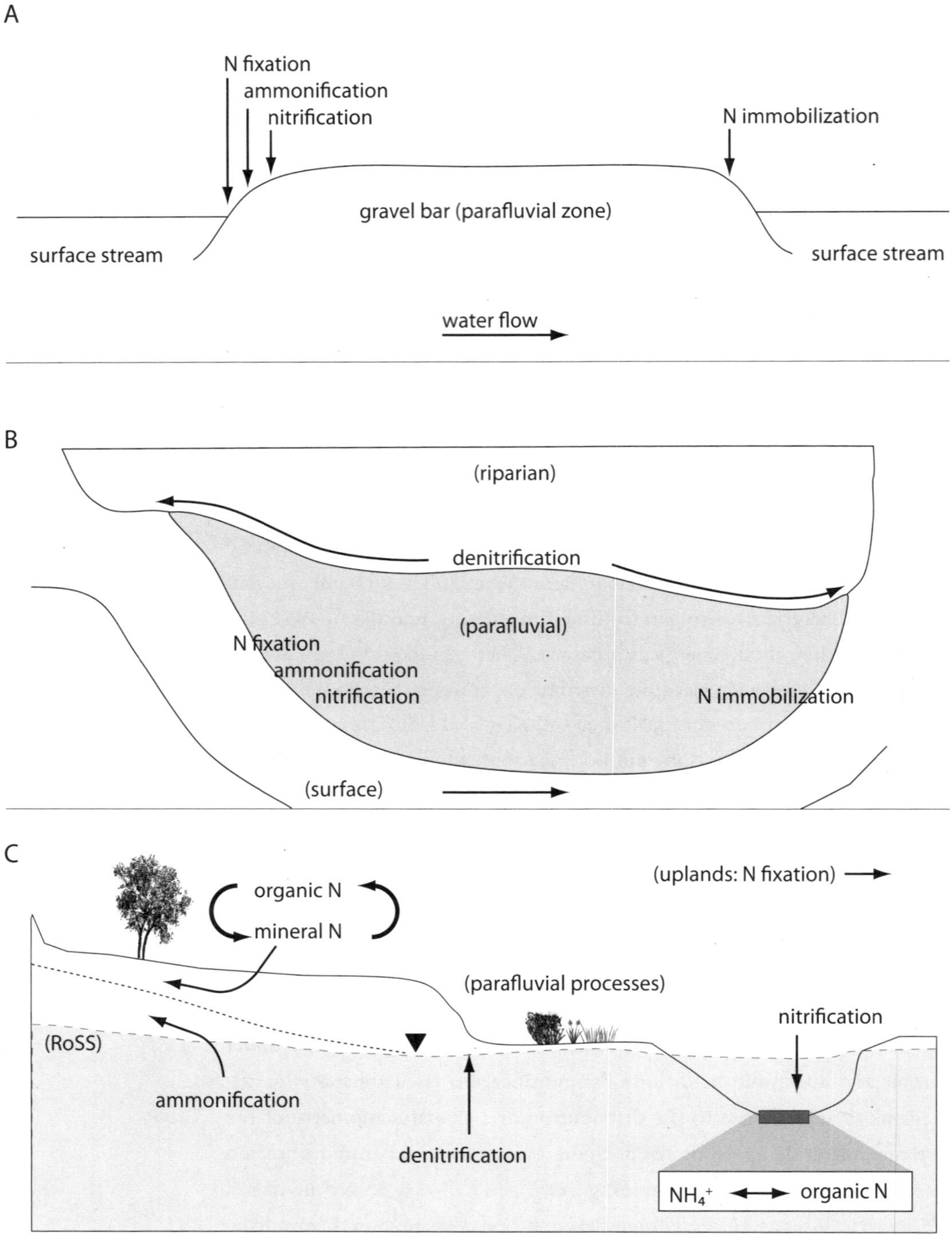

Fig. 18.6. Nitrogen cycling hotspots in the fluvial ecosystem. While important N transformations are shown, N transport vectors are not. (A) Elevation view of the surface stream–parafluvial zone interface. (B) Plan view of the parafluvial zone and the channel-riparian zone interface. (C) Elevation view of the fluvial ecosystem. The label "parafluvial processes" in panel (C) indicates the transformations described in panel (A).

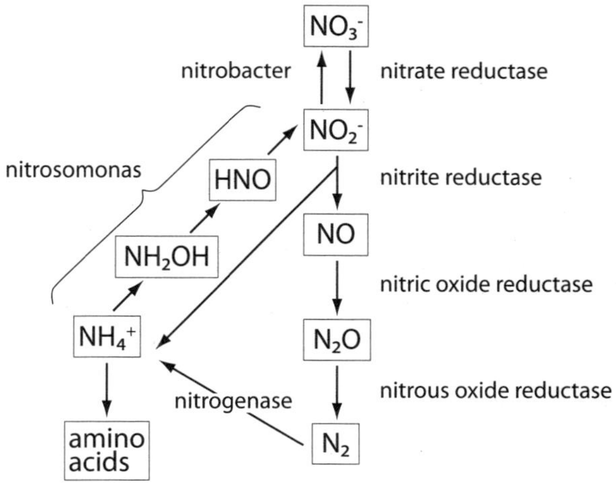

Fig. 18.7. Coupled nitrification-denitrification reaction showing intermediate species. Labels beside arrows indicate the microbial taxa or enzymes responsible for the conversion of N form. Adapted from Sprent (1987).

The product (NO_3^-) can then be swept through the gravel bar by subsurface flow paths, fueling denitrification and N immobilization at the middle and downstream edge of gravel bars (Schade et al. 2001, Henry and Fisher 2003).

Other N-transforming processes are equally variable in space (fig. 18.6). Denitrification is a heterotrophic respiratory pathway exhibited by some bacteria (e.g., *Pseudomonas*) that use NO_3^- as the terminal electron acceptor (fig. 18.7). Denitrification is dissimilatory with respect to N (the bacteria do not assimilate the N atom), and thus N gases are released to the atmosphere and N is lost from the ecosystem. Because it is a heterotrophic process that consumes NO_3^-, yet is energetically more costly than using O_2 as the electron acceptor, denitrification occurs only in anaerobic environments that host a supply of energy (organic C) and receive a steady supply of NO_3^-. Such places include riparian-channel interfaces, parafluvial rhizospheres, and waterlogged soils. Dissimilatory nitrate reduction to ammonium (DNRA) occurs in conditions similar to those favoring denitrification, with the possible difference that DNRA requires more alkaline soils and can prevail over denitrification as NO_3^- becomes limiting. Importantly, DNRA keeps N in the ecosystem, whereas denitrification does not. The importance of DNRA to the N cycle has been demonstrated for some wetlands and riparian zones, but its importance in southwestern riparian zones remains unstudied.

Further spatial heterogeneity is generated because multi-step processes such as nitrification and denitrification (fig. 18.7) proceed to completion in some places but not in others. The nitrification and denitrification "pipelines" may have leaks, through which N_2O, NO, and nitrogen dioxide (NO_2) gases escape (fig. 18.5). Owing to spatial variability in soil conditions, parafluvial and riparian zones may be heterogeneous in emissions of greenhouse (N_2O) and ozone-reactive (NO and NO_2) gases. In the San Pedro riparian zone, N_2O emissions are greater from moister soils and, given adequate moisture, from under mesquite than in the open or in sacaton grasslands (McLain and Martens 2006).

RESEARCH NEEDS

In dryland riparian zones like that of the San Pedro, opportunities for research on the importance of redox reactions fall into at least two domains. First, it would be interesting to know which redox reactions dominate, and how the various reactions are spatially distributed, in different seasons. Second, it would be interesting to know what abiotic redox reactions are characteristic of dryland riparian zones, and how these might be coupled with biologically mediated redox reactions.

Temporal Heterogeneity

Biogeochemical processes in arid and semi-arid riparian zones show temporal heterogeneity at multiple scales. Much of this variation results from episodic, seasonal, interannual, and decadal variation in hydrology (Welter et al. 2005; and see chap. 16). Episodic changes in hydrology occur when localized, intense storms produce overland flow from the uplands and nearly instantaneous rises in discharge known as flash floods. These flash floods regulate the structure and function of southwestern fluvial ecosystems (Fisher and Minckley 1978, Fisher et al. 1982, Fisher, Grimm, Martí, Holmes and Jones 1998, Grimm 1994, Lytle and Poff 2004). Periodic phenomena such as El Niño–Southern Oscillation (ENSO) influence the strength of winter rain seasons and possibly summer monsoons (Swetnam and Betancourt 1998) and hence stream discharge variation (Molles and Dahm 1990).

By exhibiting variable magnitudes and recurrence intervals, floods impart temporal heterogeneity to riparian zones. In the arid Southwest, in particular, both riparian and stream communities are usually successional (Fisher et al. 1982, Stromberg et al. 1991), meaning that vegetation is in an actively growing state and thus retains nutrients via assimilation (net uptake). This successional process can be reset by floods that scour the streambed and the floodplain, initiating a new phase of nutrient accumulation in biomass and organic matter (Fisher et al. 1982, Grimm 1987). In addition to disturbing vegetation and exporting nutrient capital from riparian zones, floods also import material in overland flow. The amount of material transported is determined, in part, by the storm timing, sequence, and total precipitation (Welter et al. 2005). Long-term chemistry data from Sycamore Creek suggest that an interannual pattern related to ENSO exists in nutrient transport. Grimm and Fisher (1992) hypothesized that this pattern reflects depletion of transportable upland soil nutrients during wet periods and accumulation during dry periods.

Not only do flood/non-flood cycles impart temporal heterogeneity, the biogeochemical consequences of floods themselves are temporally variable. When flooding occurs at the San Pedro, stream and soil water NO_3^- concentrations often increase owing to nutrient mobilization (rather than decrease owing to dilution by high discharge). The flood-generated increase in concentration, however, is particularly marked in dry years. Also during dry years, the flood-to-flood variability in NO_3^- and sulfate (SO_4^{2-}) concentrations declines from early to late in the flood season, as the system becomes more well mixed. Following periods of below-average precipitation, major runoff events are characterized by less export of soil water and solutes (Huth 2003). If water is routed to the channel not as soil water, but rather via overland flow, this finding suggests that the riparian zone may be relatively ineffective as a nutrient filter at the very time nutrient mobilization is most extreme, namely during post-dry-period runoff events.

Human Influences

CHANGES IN STREAMFLOW AND GROUNDWATER FLOW RATES

Human activities in the San Pedro basin are reducing water table depth and streamflow, eliciting changes in riparian biota. The biogeochemical consequences of these changes are important research frontiers (Pinay et al. 2002). Changes in the level of regional water tables have indirect effects on biogeochemistry by altering hydrologic flowpaths that link riparian zones to uplands and stream channels (see chap. 15). Switches in the direction of water flow from gaining (riparian zone toward stream) to losing (stream toward riparian zone) may affect biogeochemical cycling. In some cases, the riparian zone is a reducing environment, dominated by ammonification and denitrification, while the stream tends to an oxidizing environment, dominated by nitrification. Direction of flow between these two types of environments will determine the form of materials and the processes dominant at the riparian-stream interface.

LANDSCAPE CHANGE AND MATERIAL LOADING

Remotely sensed images of the San Pedro basin reveal, since the 1970s, increasing urban and agricultural areas, coupled with the invasion of woody plants at the expense of grasslands. Hydrologic models suggest that these land use and cover changes have produced greater annual and event runoff volumes, flashier flood response, and decreased water quality due to sediment loading (Miller et al. 2002). Agrarian practices such as cattle grazing and irrigated agriculture have several impacts on the structure and function of riparian zones, such as increased nutrient loading to the stream (see chap. 17). Cities and towns are additional sources of material, particularly owing to storm runoff (e.g., D. Lewis and Grimm 2007). The San Pedro receives septic leakage and has received industrial runoff. During the past several decades, output from a single nitrogen products company near the town of St. David has polluted drinking water wells along the river with toxic metals and inorganic and organic forms of N (USEPA 2005). In 1980, the EPA observed polluted wells to have NO_3^- concentrations of 470 mg N/L (USEPA 2005). To put this number in context, concentrations in riparian wells upstream of the polluted site, in 2001, exhibited a median value of 0.03 mg N/L and a maximum value of 4.9 mg N/L (Lewis et al. 2006, 2007).

Long-term addition of nutrients may lead to changes in plant community structure, which would alter nutrient demand and plant-microbe interactions. Plants exhibit physiological and morphological plasticity to changes in nutrient availability, thus altering the extent to which they retain nutrients and thereby mitigate loading (Schade and Lewis 2006). One research frontier vis-à-vis human impacts on riparian ecosystems is the response of plants, microbes, and ecosystem function to the interactive effects of nutrient loading and biocide loading, such as may occur in areas of the San Pedro receiving runoff from urban and managed conservation areas.

Summary

Riparian zones transform, transport, and store material at the interface between terrestrial uplands and aquatic habitats. In the desert southwestern United States, riparian zones are hierarchical systems of nested patches, creating a landscape of great spatial and temporal heterogeneity. Spatial heterogeneity derives from patchiness in water availability and soil organic matter, themselves variable owing to the uneven distributions of flowpaths, the plant community, and soil texture. Temporal heterogeneity is manifest at multiple scales owing to episodic, seasonal, interannual, and decadal variation in hydrology. Floods transport material to and from riparian zones, mobilize nutrients and organic matter, and regulate nutrient accumulation and limitation status.

Biogeochemical research in riparian zones can focus on material limitation, flux, and retention. Once water demands are met in arid and semi-arid riparian zones, primary producers are probably N limited and microbes organic C limited. These nutrients are imported and exported via several mechanisms of physical transport and biological transformation. Material can also be withheld from reaching the stream channel through incorporation into biological tissues, transformation into gaseous form, and stabilization within soil organic matter. In terms of internally recycling vs. transporting material, riparian zones are intermediate between uplands (recyclers) and active channels (transporters). Fundamental to riparian zone functions are redox reactions. In riparian zones especially, with their convergences of flowpaths and the presence of disparate sediments and element sources, a cascade of redox reactions can occur over centimeters.

The hydrology, geomorphology, and biotic community—and thus the consequent biogeochemical dynamics—of the San Pedro riparian zone have changed extensively through time and will continue to do so. Human activities with greatest influence on biogeochemical dynamics probably are those that cause loading of nutrients and pollutants directly into the fluvial ecosystem, coupled with those that alter the stream hydrology.

SECTION 5

Conservation and Management

GABRIELLE KATZ,
JEANMARIE A. HANEY,
CHARLES E. PARADZICK,
AND DAVID B. HARRIS

Mitigation, Restoration, and Endangered Species

HYDROLOGIC RESTORATION ON THE LOWER SAN PEDRO RIVER

NINETEEN

Introduction

Although the upper San Pedro River basin has received the lion's share of public and research attention, the lower San Pedro River basin (fig. 19.1) is an equally important terrain. Numerous Fremont cottonwood/Goodding willow (*Populus fremontii*/*Salix gooddingii*) habitats serve as stepping stones for migratory bird species using the river as a linkage between wintering habitat in Mexico and Central and South America and breeding habitat in North America (Skagen et al. 1998). The lower basin also supports breeding populations of the federally endangered southwestern willow flycatcher (*Empidonax traillii extimus*; Paradzick and Woodward 2003), which relies upon dense riparian tree and shrub communities associated with rivers and other water bodies (USFWS 2002c).

Although the San Pedro River is one of the few remaining un-dammed rivers in the southwestern United States, the river and its riparian ecosystems have not escaped other human impacts. Human activities in the lower basin that have influenced fluvial processes and ecosystem characteristics include surface water diversions, groundwater pumping, mining, recreation, livestock grazing, and clearing of terrace vegetation for agriculture (Manes 1968, Zimmerman 1969, Briggs 1996, Stromberg 1998a). Today, while the

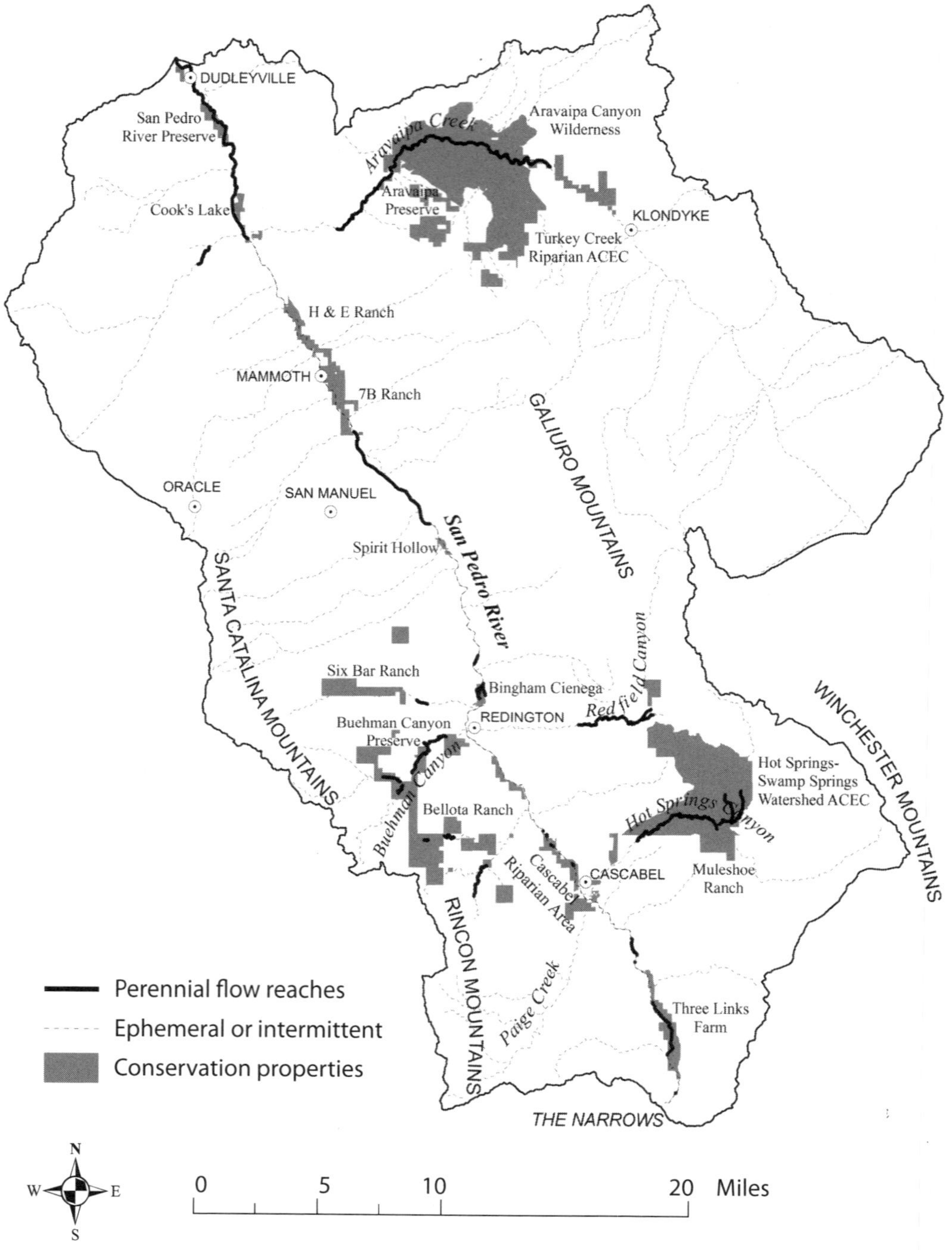

Fig. 19.1. Map of lower San Pedro River basin showing riparian conservation areas (public and private). Map prepared by Dale Turner, Arizona Nature Conservancy.

lower basin does contain many reaches with high-quality riparian habitat, other reaches are in a degraded condition, no longer supporting the hydrology, geomorphology, or ecosystem structure, composition, and function typical of non-impacted sites.

Numerous agencies and organizations actively promote conservation and restoration in the lower basin. These include The Nature Conservancy (TNC), Pima County, the U.S. Bureau of Reclamation (BOR), the U.S. Bureau of Land Management (BLM), the Saguaro Juniper Corporation, the Cascabel Hermitage Association, and the Salt River Project (SRP). This chapter emphasizes the work of TNC in part because TNC has pioneered much of the conservation and restoration effort in this area. The Nature Conservacy presence in the San Pedro basin dates to 1968, when the Canelo Hills Ciénega Preserve was established on a tributary of the river. Since then, TNC has focused considerable attention on conservation of biodiversity throughout the basin, with programs operating in Mexico, the upper basin, and the lower basin, and with ownership and leases now covering more than 40,400 ha. The Nature Conservacy also works in partnership with other agencies, nonprofit organizations, and private landowners to advance conservation objectives.

Conservation of dynamic ecosystems relies upon a sound understanding of the physical and ecological processes and conditions that are their foundation. For arid-region riparian and aquatic ecosystems such as those of the San Pedro River, the hydrologic regime is a primary determinant of ecosystem characteristics. The hydrologic regime includes two interrelated components: the river flow regime, characterized by patterns of variation in surface flow magnitude, duration, frequency, and timing (Poff et al. 1997); and the subsurface hydrologic regime, characterized by equivalent patterns of variation in depth to groundwater. The increased available moisture provided by streams results from both components, and their interaction. For example, surface streamflows (e.g., annual flow volume; Stromberg and Patten 1990a), overbank flows that replenish floodplain soil moisture (Reily and Johnson 1982, J. M. Friedman and Auble 2000), and the level of groundwater in the subsurface alluvial aquifer are all aspects of the hydrologic regime that influence riparian vegetation (see chap. 1).

Because of the importance of the hydrologic regime to the river and riparian ecosystem, effective conservation in the lower basin requires an understanding of basin hydrology and an ability to predict hydrologic response to management scenarios. Understanding and quantifying human-induced changes in the hydrologic system, and determining how to best mitigate those effects to protect and restore the riparian ecosystem, require substantial data and sophisticated tools. Over the past 10 years, TNC has utilized results from hydrologic modeling and data from hydrologic monitoring to inform conservation and management decisions in the lower basin. Initial modeling efforts culminated in a strategy to reduce ecosystem stresses by decreasing groundwater pumping in the stream alluvium. This strategy of irrigation curtailment is being applied in a collaborative manner by TNC (Haney 2005), the BLM, the BOR (USFWS 1996), and the SRP (SRP 2002). The strategy includes legal protection of water resources through issuance of new instream flow permits by the Arizona Department of Water Resources (ADWR), and a parallel strategy of conversion of existing surface water claims to instream flow (USBLM 1995).

Conservation efforts in the lower basin have taken on added significance because of the presence of an important population of the endangered southwestern willow flycatcher (Sferra et al. 1997). Listed as a federally endangered species in 1995 (USFWS 1995), the flycatcher relies on dense riparian habitat in the breeding season (Sogge and Marshall 2000). Entities required to mitigate for potential losses of flycatcher nesting habitat have been directing resources to the lower basin for habitat protection (USFWS 1996, SRP 2002). Because the flycatchers depend on habitat that combines specific vegetation and hydrologic conditions (Hatten and Paradzick 2003, Paradzick 2005), its protection necessitates a strong understanding of the linkages between hydrologic conditions, fluvial processes, and vegetation characteristics.

Regional Setting

Although the lower basin is close and accessible to the burgeoning Tucson and Phoenix metropolitan areas, it has so far not undergone extensive population growth and urban/suburban development. In 2000, the population in the central basin, which includes about 80 km of the river from the Narrows north to near San Manuel (but not including San Manuel), was reported to be 213 people (J. Haney, unpublished data). The towns of San Manuel and Mammoth had a combined population of 6,137 in 2000 (http://www.census.gov/). Land ownership in the lower basin consists chiefly of state (52 percent), private (18 percent), U.S. Forest Service (18 percent), and BLM (9 percent; J. Haney, unpublished data). Private ranch lands in valley bottoms typically have large state or federal upland grazing leases.

Important conservation management units on the lower San Pedro River include the Three Links Farm (TNC), riparian areas near Cascabel (BLM), Spirit Hollow Preserve (SRP), Bingham Ciénega (Pima County, TNC), H & E Farm (TNC), San Pedro River Preserve (TNC, BOR), Adobe Preserve (SRP), Stillinger Preserve (SRP), Black Farm Preserve (SRP), and Cook's Lake riparian mitigation area (BOR) (fig. 19.1). These units encompass a little over 42 km, or about one-third of the length of the river in the lower basin. Important conservation management units on tributaries include Muleshoe Ranch (TNC, BLM), Aravaipa Canyon Preserve (TNC, BLM), Buehman Canyon Preserve (TNC), and the Swamp Springs–Hot Springs Watershed (BLM, TNC)

TABLE 19.1. Pumping estimates by water use, lower San Pedro River basin

	Water use	
	(acre-ft/yr)	(m^3/dy)
Irrigation	16,910	57,109
Domestic	300	1,013
Municipal	920	3,107
Stockpond	940	3,175
Industrial	30	101
Mining	24,860	83,958
Total	43,960	148,463

Modified from J. D. Whittier and Maddock 2004; original data from ADWR 1991.

and Turkey Creek Riparian (BLM) Areas of Critical Environmental Concern (ACECs) (fig. 19.1).

Significant historic human impacts in the lower basin include mining, crop agriculture, livestock grazing, development of relatively small towns, and groundwater pumping in support of these activities (table 19.1). Several mines are present, with most extracting copper and related ores. The largest of these, the BHP San Manuel mine (formerly the Magma Copper Company mine), covers approximately 2,000 ha west of the San Pedro River, of which approximately 1,200 ha have been disturbed by mining. Mine tailings cover approximately 1,450 ha immediately adjacent to the river at the town of San Manuel (ADWR 1991). There are also several small flux mines in the lower basin that produce products used in the ore-extraction process, one located east of the river near San Manuel and one located west of the river near the Aravaipa confluence. In addition to the extraction of minerals, water is extracted by some mining companies in the lower basin. For example, although mining activities are not conducted on American Smelting and Refining Company (ASARCO) lands in the lower basin, the company extracts groundwater from land it owns downstream of Mammoth for use in its Hayden mine, located on the Gila River downstream from the San Pedro River confluence.

Agriculture has resulted in clearing and leveling of approximately 2,330 ha in the lower basin (ADWR 1991); substantial groundwater has been pumped to irrigate these fields (table 19.1). In addition, livestock grazing and off-road vehicle use in the riparian corridor occur in many reaches. Trespass in reaches that have been fenced for conservation purposes continues to be an ongoing management challenge.

Overall, groundwater pumping rates in the lower basin increased each decade between the 1940s and the 1990s (fig. 19.2). Historic stream diversions and groundwater pumping are associated with changes in the hydrologic

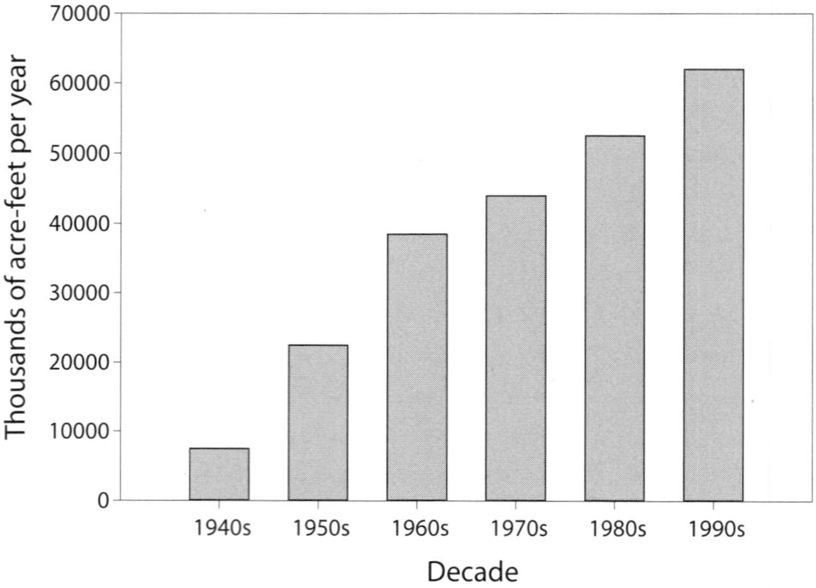

Fig. 19.2. Estimated average groundwater pumping by decade, lower San Pedro basin. Compiled from data summarized in J. D. Whittier and Maddock (2004) for growing season. Original data source is Arizona Department of Water Resources (ADWR 1991). Includes irrigation, mining, stock, public supply, domestic, and commercial/ industrial wells.

system; since the mid-twentieth century, groundwater levels have declined and streamflow has been reduced in some reaches (ADWR 1991). If streamflow were to become significantly smaller and less frequent, the San Pedro River would become an ephemeral stream in certain reaches where it was historically perennial or intermittent (ADWR 1991). This process appears to have already occurred in certain reaches of the river in the lower basin (Brown et al. 1981).

Conservation Importance and Opportunity

The lower basin is an important landscape, supporting significant biodiversity. Riparian habitat patches provide crucial breeding habitat for many bird species (USBLM 1990), and pairs of major tributaries linking the eastern and western ranges flanking the lower basin provide wildlife corridors that may allow for important genetic exchange between populations (Witham and Smith 1979, Ough and deVos Jr. 1984). For example, the Rincon-Catalina Mountains complex to the west is linked with the Galiuro-Winchester Mountains complex to the east by Buehman Canyon-Redfield Canyon, Hot Springs Canyon-Paige Canyon-Robles Canyon, and other tributary pairs. Farther downstream, Camp Grant Wash (Putnam Wash) and Aravaipa Creek provide a cross-basin link between the desert hill country to the west and the Sulphur Springs Valley and its flanking ranges to the east.

Under present conditions, many of the historic land uses and activities in the lower basin appear to have become economically unsustainable. For example, conditions for successful irrigated agriculture do not appear to be as favorable today as they were when these riparian lands were originally cleared and developed with irrigation infrastructure. This is evidenced by the number of farm operations for sale along the river corridor, the ongoing conversion of irrigated lands to rural residential uses, and the degree of subsidy required to support new agricultural uses. To quote one rancher-farmer speaking about his alfalfa farm, "Shoot, we can go to Oklahoma and truck in alfalfa cheaper than we can grow it here" (anonymous, personal communication to D. Harris, 2004). In addition to the decline in irrigated agriculture in the lower basin, the BHP San Manuel mine closed in 2002, contributing to the general loss of historic economic activities in the area.

The decline in traditional land uses dependent on groundwater extraction provides an unprecedented opportunity to protect and restore stream baseflows in the lower basin. As part of its conservation toolbox, TNC may purchase land in key areas to meet conservation goals. Other entities have also taken advantage of this opportunity, often motivated by federal mitigation requirements for loss of wetlands under the Clean Water Act and for impacts to listed species under the Endangered Species Act. For example, BOR and SRP have each acquired habitat as mitigation for impacts to southwestern willow flycatchers as a result of dam alteration and reservoir operation activities at Roosevelt Lake.

Conservation Approach

Initially, much of the land acquisition in the lower basin was directed toward the protection of existing wetland and riparian forest habitats. In 1989, the Pima County Regional Flood Control District acquired the Bingham Ciénega north of Cascabel and entered into a 25-year agreement with TNC to manage this significant wetland. In 1990, TNC and the BLM delineated five core conservation areas along the San Pedro River and identified approximately 8,500 ha for possible acquisition (USBLM 1990). They began a program of acquiring the highest-quality riparian habitat, including Fremont cottonwood/Goodding's willow floodplain forests and mesquite (*Prosopis*) bosques on the adjacent terraces. This acquisition work framed a program of conservation that has influenced numerous agencies and organizations with conservation objectives for the river.

Over time, a growing understanding of the relationship between hydrologic processes and riparian habitat characteristics led TNC to expand its perspective on conservation opportunities in the lower basin. In 1997, TNC initiated a planning effort for the central basin in which consideration of riparian potential, rather than existing condition alone, became an important criterion driving land conservation projects. Current conservation planning emphasizes the importance of hydrologic evaluation as a basis for land acquisition. An important aspect of this approach is assessment of the feasibility of improving hydrologic conditions in the river to benefit native fish and riparian habitat over the long term. As such, the conservation approach has expanded to include ecosystem restoration.

For TNC, the fundamental restoration approach that has evolved in the past 15 years invokes two overarching strategies: removal or reduction of system stresses, and restoration of landscape-level processes. The stresses at play in the lower basin are typical of many dryland riparian ecosystems—grazing pressure from cattle and horses, impacts from recreation and off-road vehicle access, lowered and unstable groundwater levels in the stream alluvium, and loss of baseflows (Briggs 1996, Ffolliott et al. 2004). Livestock grazing and off-road vehicle use in the stream corridor are management concerns because they can directly impact riparian vegetation, fluvial geomorphology, and water quality by depleting vegetative cover, reducing bank stability, and causing increased water turbidity (Catron et al. 2000). Processes and physical conditions of critical interest for restoration include those associated with the hydrologic regime (e.g., summer baseflows, groundwater levels) and fluvial geomorphic processes (e.g., erosion, sediment transport, deposition; see chap. 13). Floodplain vegetation is expected to recover in response to the dual strategy of stressor removal and process restoration, thereby providing riparian habitat as well as shade, bank stability, and organic matter inputs to the aquatic ecosystem.

One complicating aspect of river and riparian ecosystem conservation is the strong linkage between watersheds and the rivers that drain them. That is, watershed conditions influence important hydrologic and geomorphic processes such as the volume of surface runoff and the amount of sediment delivered to streams. For management efforts focused on a particular riparian zone, fluctuations in watershed conditions may impose a degree of uncertainty to management outcomes. For example, wildfire in the watershed may result in significant sediment inputs to the stream channel (see chap. 13), which may confound riparian management strategies aimed at protecting aquatic resources. Recognizing that watershed conditions influence river and riparian ecosystems, TNC and collaborating managers have employed practices such as grazing rest and prescribed burning at various properties in the lower basin (e.g., Muleshoe Ranch, Aravaipa Canyon Preserve, Bellota Ranch) to improve watershed conditions. This linkage was explicitly assessed in a management experiment at Muleshoe Ranch Cooperative Management Area, in which prescribed burning of shrub-invaded grasslands resulted in improved native grass cover on the uplands and improved aquatic conditions on Hot Springs Creek, a tributary of the lower San Pedro River (Gori and Backer 2005). However, to date, most of the conservation efforts on the mainstem of the lower San Pedro River have been focused in the riparian zone.

In response to the increasing recognition that successful conservation efforts must be informed by ongoing scientific study of the ecosystems being managed (e.g., Christensen et al. 1996, Richter et al. 2003, Palmer et al. 2005), managers in the lower basin have funded monitoring and scientific research. In the 1980s, TNC began a relationship with researchers from Arizona State University and began to support research into riparian vegetation processes (see chap. 1). Pima County also has supported research on biotic elements in the lower basin (e.g., Fonseca 1998). The Nature Conservancy and Pima County have also supported investigations and modeling of the surface and subsurface hydrology of the San Pedro River (Baird et al. 1997, Lombard 1998, Pima Association of Governments 2001, Whittier 2004). Other management agencies that have funded research in the lower basin include BOR, Arizona Game and Fish Department, and SRP (e.g., Munzer et al. 2005). The body of research from these and other investigators has grown through the years to support a paradigm of riparian conservation management that is currently employed by many of the conservation managers in the lower basin.

Since 1995, conservation in the lower basin has been strongly influenced by entities charged with management and mitigation of the endangered southwestern willow flycatcher. Mitigation conservation efforts have focused primarily on (1) acquisition of riparian habitat to maintain breeding areas, and (2) streamflow augmentation through acquisition of water rights and reduced diversion or groundwater pumping, with concomitant benefits to protected riparian habitat. In 1986, the BOR began a modification of Roosevelt Dam that added 23 m of dam height to increase reservoir storage. However, the Roosevelt Lake flycatcher population occurred within an area that would be repeatedly inundated as a result of this storage increase (Paradzick and Hatten 2004). Dam modification was considered a federal action; thus, the BOR consulted with the U.S. Fish and Wildlife Service (USFWS) under Section 7 of the Endangered Species Act, and a Biological Opinion was issued indicating jeopardy for flycatcher populations at Roosevelt Lake within the range of new storage space (USFWS 1996). On the other hand, because Roosevelt Lake stores a significant portion of the water for metropolitan Phoenix, it was also recognized that SRP, which operates the reservoir, must be able to utilize the reservoir to its maximum capacity and employ flexibility in operations (USFWS 1996, SRP 2002). The reasonable and prudent alternative included habitat acquisition and protection sufficient to support a minimum of 45 southwestern willow flycatcher territories. To meet these mitigation requirements, BOR purchased a farm and aquaculture operation near Dudleyville that became the San Pedro River Preserve and was turned over to TNC for management.

Riparian habitat along the San Pedro River is becoming increasingly important to flycatcher conservation as other known nesting locations within Arizona become degraded (SRP 2002, Munzer et al. 2005). In 2003, due to drought conditions, water levels in Roosevelt Lake fell to a level below that covered by the BOR Biological Opinion; willow flycatchers moved into vegetation that became established on the exposed delta sediments within this lower elevation. Because eventual inundation of the habitat would cause loss of nesting territories and possible bird mortality, colonization of the habitat required that the SRP consult with the USFWS to acquire an Incidental Take Permit under Section 10 of the Endangered Species Act (SRP 2002). As mitigation for the loss of flycatcher habitat at Roosevelt Lake, the SRP was charged with acquiring and managing 607 ha of riparian habitat and conducting additional habitat conservation measures equivalent to protecting 304 ha of habitat. SRP turned again to the lower basin to fulfill a part of this mitigation obligation (SRP 2002). Their San Pedro preserves now include the 57-ha Spirit Hollow Preserve near San Manuel, the 62-ha Adobe Preserve near Cook's Lake, the 16-ha Stillinger Preserve at the Aravaipa Creek confluence, and the 55-ha Black Farm on Aravaipa Creek (R. Valencia, personal communication, 2006). The main management strategy for conservation of flycatcher habitat on these properties is the protection of the natural ecosystem processes that support suitable breeding habitat. The primary strategies have been to exclude livestock grazing and reduce shallow groundwater withdrawal.

As of 2006, 190 ha of cottonwood/willow habitat and irrigated land had been acquired specifically for southwestern willow flycatcher mitigation on the San Pedro River (R. Valencia, personal communication, 2006), and more than 10 million dollars had been spent on research and monitoring of flycatcher populations on the lower San Pedro and at Roosevelt Lake (S. Sferra, personal communication, 2005). Thus, the flycatcher has become a driving force in riparian conservation along the lower San Pedro River.

Lower Basin Hydrology

The lower San Pedro River flows north in the San Pedro trough, an elongated structural depression bounded on the southwest by the Catalina core complex and associated uplands and bounded on the northeast by the faulted range fronts of the Galiuro and Dripping Springs Mountains (Dickinson 1991). The San Pedro trough contains basin-fill deposits (Quiburis Formation) that overlie a complex paleotopography (Dickinson 2003). Inset into these deposits is the recent floodplain alluvium, comprised of unconsolidated, chiefly coarse-grained sediments that immediately underlie, adjoin, and are in direct hydraulic communication with the stream.

The San Pedro River has two major flow components: direct surface runoff and baseflow (see chap. 15). Runoff occurs from precipitation events or snowmelt and can last a few hours to several weeks, until all flow is lost to either bank storage along the river, recharge of the underlying basin-fill deposits, evapotranspiration, or outflow from the basin. Baseflow results from the discharge of groundwater to the stream and sustains streamflow in dry seasons in perennial reaches. In the San Pedro basin, most of the groundwater that originates as mountain-front recharge moves through the basin-fill sediments and discharges to the floodplain alluvium aquifer (ADWR 1991). After reaching the floodplain alluvium, the groundwater moves as subsurface flow, is used by phreatophytic vegetation, discharges to the San Pedro River streambed as surface water flow, or is pumped for various cultural uses such as agricultural, municipal, and industrial uses (ADWR 1991).

Locations of perennial-flow reaches in the San Pedro River (see fig. 0.4) are controlled chiefly by the underlying geology. Where tilted blocks of the basin substratum have produced low-permeability bedrock at shallow depths below the river, the overlying basin-fill deposits are thinned, impeding movement of groundwater (Heindl 1963); groundwater in the alluvial aquifer is forced to discharge to the river in these areas. These conditions most likely exist in the perennial-flow reach about 10 km upstream from Redington and at Malpais Hill near Dudleyville. Also, presence of fine-grained layers in the basin-fill deposits can control locations of perennial flow (Pool and Coes 1999). Fine-grained layers may act as a confining unit or a perching unit, restricting the connection between the deeper regional aquifer and the shallow aquifer/river system. Such units likely exert control over the location of certain perennial reaches, such as those that occur near San Manuel and upstream of the Aravaipa Creek confluence.

The alluvial valley of the lower basin is narrower than that of the upper basin (Richard et al. 2000), resulting in less groundwater in storage. Thus, the aquifer-river hydrologic system is more responsive to removal of water from storage due to groundwater pumping, especially pumping from the recent alluvium aquifer. Groundwater pumping increased dramatically in the lower basin from 1940 to 1990, with average groundwater levels in the floodplain aquifer declining by 1.2 m between 1945 and 2004 (J. Haney, unpublished data). Because perennial flow (baseflow) is supported by discharge from the groundwater system, relatively small declines in water levels near the river may result in reduction or elimination of perennial flow. Conversely, comparable recovery of water levels near the river should result in increased perennial flow.

In recent decades, mining and agriculture have been the chief water uses in the lower basin (table 19.1). Prior to partial closure in June 1999, the BHP San Manuel mine diverted about 28,370,250 cubic m of water per year from the San Pedro River, about 24,000,000 cubic m of which was consumptively used (ADWR 1991). ASARCO owns approximately 100 ha 2 km south of Winkelman on the east side of the San Pedro River (ADWR 1991), where wells export water to the ASARCO smelting and refining facility at Hayden in the upper Gila watershed. The volume of water exported annually by ASARCO from the lower basin to the Hayden Unit from 1976 to 1989 ranged from 715,424 to 1,486,354 cubic m and averaged 974,456 cubic m (ADWR 1991). ASARCO also owns PZ Ranch, located near Dudleyville, 2.4 km north of the confluence of Aravaipa Creek with the San Pedro River. Water from the PZ Ranch is exported to the Ray Unit in the upper Gila watershed. The volume of water exported annually by ASARCO from the lower basin to the Ray Unit from 1978 to 1989 ranged from 0 to 3,547,514 cubic m and averaged 770,931 cubic m (ADWR 1991).

Agricultural water use has been comparable to that associated with mining (table 19.1). Because wells outside Active Management Areas in Arizona are not metered, agricultural water use cannot be directly measured. Instead, estimates of cropped area are made from aerial photography, types of crops are estimated based on available records, and approximate consumptive water use is then calculated. In 1990, 2,330 ha in the lower basin were under cultivation (ADWR 1991), resulting in a water use of approximately 26,987,500 cubic m per year (based on a consumptive water use of 11,582 cubic m per year per ha; ADWR 1991). Table 19.1, modified from J. D. Whittier and Maddock (2004) with original data from ADWR (1991), provides a slightly different irrigation water use than that given above. This illustrates one of the challenges of working on flow restoration through agricultural retirement in a setting where water use is not metered—estimates of water use and consumptive use may vary depending on assumptions used. Thus, water budgets are difficult to develop and must be based on stated assumptions.

Recently, significant reductions in groundwater pumping have occurred in the lower basin. Following full closure of the BHP San Manuel mine in 2002, pumping was reduced to its current level of 1,233,489 to 2,466,978 cubic m per year for municipal supply purposes (J. Parker, BHP Director of Community and Environmental Affairs, personal communication, February 24, 2005). The closure of this mine resulted in a reduction of approximately 24,000,000 cubic m per year of consumptive groundwater use (ADWR 1991), which formerly represented the largest extractive use in the lower basin. Retirement of irrigated agricultural land has resulted in reduction of approximately 6,167,445 cubic m per year of consumptive groundwater use (using 8,229 cubic m per ha as consumptive use). Thus, all else being equal, approximately 30,220,480 cubic m of additional water could be available annually to the river and riparian ecosystems compared to the 1990s. However, this water will be available into the future only if it is allocated and legally protected as instream flow.

Hydrologic conditions reflect climatic conditions for current and previous years, as well as geologic factors and land management. The extent of surface flow in the lower San Pedro River varies seasonally and interannually based on this complex interplay of factors. For example, during the extreme drought conditions of 2001–2003 (National Climatic Data Center 2004), only about 10 km of the approximately 137 river km exhibited surface water in the low-flow summer season (D. Harris, personal observation); during more normal rainfall conditions, about 26 km of perennial flow exist in the lower basin (D. Harris, personal observation). These climatic patterns are superimposed on flow patterns that are strongly influenced by historic geomorphic change, underlying geology, and groundwater pumping.

Hydrologic Analysis and Modeling as Basis for Land Acquisition

Consistent with the recognition of the need for scientific understanding of ecosystem processes to underpin conservation strategies, TNC has supported hydrologic investigations and monitoring efforts in the lower basin and used the study results to guide management strategies, purchases, and conservation plans. Several hydrologic models have been developed to investigate groundwater–surface water linkages in the lower basin and to evaluate the potential effects of cessation of irrigation pumping on groundwater conditions and surface flows (Jahnke 1994, Lombard 1998, Goode and Maddock 2000, Whittier 2004).

The first TNC project in the lower basin directed primarily toward improving riparian conditions through hydrologic enhancement was the 2001 purchase of the H&E Farm. This 214-ha farm located just north of Mammoth included 123 ha of irrigated alfalfa fields. Although hydrologic relationships are not fully understood in this area due to a lack of deep wells penetrating the regional aquifer, analyses of groundwater level information, groundwater pumping rates, and geologic data conducted prior to property acquisition suggested that termination of the approximately 2,713,676 cubic m per year of groundwater pumping at H&E Farm would result in increased extent of perennial flow in the reach between the farm and the Aravaipa Creek confluence. In June 2001, water level recovery on the order of 3 m was observed in irrigation wells at the farm, compared to water levels measured during farming operations in 1998 (J. Haney, unpublished data). Between June 2001 and July 2006, water levels in irrigation wells and shallow floodplain piezometers declined sharply. Then, following monsoon floods in July and August 2006, groundwater levels rose by 2–3 m and were sustained at relatively high levels through the summer of 2007. Despite drought conditions since 1998, biologists monitoring southwestern willow flycatcher populations report that nesting activity has increased since 2002 in the reach of the San Pedro River downstream of H&E Farm (Munzer et al. 2005). They credit this increased activity to improved riparian habitat condition, though there are no direct data to substantiate a connection between this improvement and reduced groundwater pumping at H&E Farm.

Following purchase of H&E Farm, TNC continued to pursue process-based ecosystem evaluation and restoration using more sophisticated hydrologic assessment techniques. Hydrogeologic assessment was conducted through modification and use of an existing groundwater model for the middle San Pedro basin (Jahnke 1994). This model was built on the MODFLOW platform, a three-dimensional, finite-difference groundwater flow model (McDonald and Harbaugh 1998). Stream-aquifer interactions were simulated using the USGS Streamflow-Routing Package (Prudic 1989). The purpose of the model was to better understand the streamflow and groundwater conditions and interactions in the riparian areas of the San Pedro River Valley, from Fairbanks to Redington. Lombard (1998) upgraded the model and used it to evaluate several pumping-elimination scenarios involving irrigation units located on the floodplain terraces in the reach from the Narrows to Redington. Model results allowed TNC to evaluate various irrigation-cessation options and to devise the most effective strategy to improve flow and riparian conditions on the river. TNC applied criteria that required a large magnitude response in the system before it would consider allocation of resources to a flow-management project.

Lombard (1998) found that substantial gains in surface flow frequency and groundwater levels could be obtained in several cases of irrigation retirement within the 32-km stretch of river evaluated. However, the greatest response in predicted groundwater conditions was to be obtained by the cessation of pumping on the Three Links Farm, located 3.2 km downstream from the Narrows. The farm produced alfalfa, small grains, and irrigated pasture from 1929 until 2002. At maximum production, 425 ha were under irrigation and cultivation, with the primary means of irrigation since the mid-1980s being pivot irrigation. Although actual pumping rates on the farm are not established by direct empirical data, for modeling purposes likely pumping rates were estimated at 3,947,165 cubic m per year, using published information on irrigation units, crop type, irrigation demand, well configuration, and location (ADWR 1991, Lombard 1998). Pumping elimination simulations predicted recovery of perennial flow in the river through much of the 10-km length of the farm and improved groundwater levels and flow conditions on an additional 13 km of stream downstream of the farm (Lombard 1998). Field observations were consistent with hydrogeologic modeling results in suggesting that groundwater pumping at the farm decreased surface flow in the river: visual monitoring of river conditions during a spring 1999 over-flight resulted in observations of a dry riverbed in the reach from the Narrows to Cascabel, except for an approximately 183-m reach downstream of Redrock Canyon (D. Harris, personal observation, April 1999). Aerial photography flown in summer 1995 documents a substantially reduced wetted area during periods of irrigation (D. Harris, unpublished data).

The Nature Conservancy acquired Three Links Farm in August 2002 and shut off the irrigation pumps. Groundwater level and streamflow data are not available for the farm prior to its acquisition by TNC. After the 2002 purchase, well locations and wellhead elevations were surveyed, and TNC commenced a program of quarterly groundwater level monitoring. Quarterly measurements of surface streamflow are also now conducted at four stations on the property, and precipitation at the farm is recorded daily. In 2003, TNC staff initiated a program of annual baseflow mapping for the 10-km river segment encompassed by the farm and, where accessible, downstream to Cascabel. This information is gathered in mid-June, typically the driest month of the year. The simple monitoring protocol involves walking the river and recording GPS coordinates at the start and end of each reach of surface water.

Groundwater levels and streamflow remained fairly stable at Three Links Farm over the period of TNC monitoring, October 2002 to the present. Groundwater levels exhibited small seasonal fluctuations through early summer 2006, and then increased by 0.61–1.8 m following floods in July and August (J. Haney, unpublished data). Streamflow volume was small, in the range of 0.015–0.06 cubic m per second, in perennial-flow locations. Winter streamflow increased over the monitoring period, whereas summer streamflow was constant (J. Haney, unpublished data). The spatial extent of mid-June surface flow fluctuated annually within the farm. Baseflow mapping indicated that 3.4 km, 4.2 km, 3.3 km, 3.0 km, and 5.5 km of surface flow occurred in 2003, 2004, 2005, 2006, and 2007, respectively (J. Haney, unpublished data).

Groundwater level and streamflow trends were small, but they did indicate some hydrologic improvement during the monitoring period. A variety of factors have likely influenced the rate and magnitude of hydrologic change. Such factors may include (1) severe regional drought since 1998, (2) slow response time of groundwater systems, and (3) complexity of the hydrogeologic system, which might result in distant effects of reduced or eliminated irrigation. Available monitoring data indicate that groundwater levels have declined throughout the lower basin since 1996 (J. Haney, unpublished data), suggesting that regional drought may be having a strong effect on observed conditions. On the other hand, considerable recharge to the groundwater system occurred during high flows in 2006, and we expect that irrigation termination will eventually result in groundwater recovery. Such improvement depends upon adequate precipitation to recharge the aquifer, ample time for water level recovery to occur, and protection from increased pumping in upgradient areas. Since groundwater decline occurred over several decades on the San Pedro River, it is likely that reversing this process will also require decades.

Riparian Vegetation: Current Conditions and Potential

A mosaic of riparian vegetation patch types occurs along the lower San Pedro River (see chap. 1). Vegetation patches vary from xeric, sandy open areas to more mesic, dense cottonwood/willow forests. The occurrence of riparian vegetation patch types reflects local surface water and groundwater hydrology, as well as local management practices. For the most part along the San Pedro River, pioneer cottonwood/willow forests occur where streamflow is perennial, or nearly so, and where the alluvial water table is close to the ground surface and does not fluctuate widely. On drier sites with intermittent or ephemeral flow and greater depth to groundwater, cottonwood/willow forest gives way to woodlands with less canopy volume, or to lower-stature shrublands dominated by saltcedar (*Tamarix* sp.).

As the number of riparian conservation and restoration projects has increased in recent years, there has been a heightened need for techniques to assess riparian ecosystems. Assessment protocols can be used to monitor changes in ecological condition, determine the need and type of restorative actions, assess restoration success, select high-quality sites during conservation planning, and facilitate determination of habitat acreages needed for compensatory mitigation. To assess riparian vegetation condition at restoration properties managed by TNC in the lower basin, researchers have collected field data and applied a stressor-response assessment model developed for the entire San Pedro River (Stromberg et al. 2006, G. Katz, unpublished data). This model was designed to assess riparian condition using a suite of vegetation traits (bioindicators) sensitive to site hydrologic condition. In the model, measures of vegetation structure, species composition, and population traits are used to determine the condition class of a site, using a calculated site score. The model provides a framework for interpreting baseline vegetation conditions in the context of site hydrology and enables prediction of future vegetation changes likely to result from hydrologic restoration.

CURRENT CONDITIONS

The three major TNC riparian conservation properties in the lower basin presently support the full range of riparian condition classes observed along

Fig. 19.3. (top; 19.3a): Riparian forest at Spirit Hollow Preserve reference site (condition class 3). (bottom; 19.3b): Channel and riparian vegetation at H&E Farm (condition class 1). Photo credits: Gabrielle Katz.

the lower San Pedro River as a whole (G. Katz, unpublished data). At the wet end of the spectrum, the San Pedro River Preserve supports tall, dense, multi-aged cottonwood/willow riparian forest stands, with some intermixed woodlands and grasslands. Saltcedar is present, but not dominant, and the streamside zone tends to be dominated by hydric herbs (fig. 19.3a). At the dry end of the moisture gradient, H&E Farm supports relatively little cottonwood/willow forest. Instead, the riparian zone is dominated by open sandy patches, burrobrush and seepwillow shrublands, and saltcedar woodlands (fig. 19.3b). Three Links Farm supports vegetation typical of wet to intermediate condition classes in the central part of the property and drier condition classes at the north end.

PREDICTED CHANGES

Riparian condition assessment models can be used to predict the ecological effects of stressors, such as stream dewatering, and also have great promise as predictors of ecological responses to process-based restoration. On the lower San Pedro River, we expect that hydrologic restoration (i.e., prolonged duration of surface flows and decreased depth to groundwater) will result in changes in riparian vegetation, including increased cottonwood/willow size class diversity and basal area, increased vegetation height, decreased abundance of open and shrubland patches (and increased abundance of forest patches), and higher streamside cover of hydric herbaceous vegetation. These riparian vegetation changes will occur if site hydrology is restored above specific threshold levels for water table depth and other key hydrologic factors (see chap. 1).

In addition to being dependent on rising groundwater tables, vegetational shifts from saltcedar to cottonwood/willow will require flood-related fluvial processes to scour existing vegetation and create new sites for seedling establishment. However, flood events of the appropriate size, timing, and duration to trigger cottonwood and willow seedling establishment occur relatively infrequently. Along the lower San Pedro River, cottonwood seedlings established, on average, about once every five years during the wet decades of the 1970s, 1980s, and 1990s, but only about once every 15 years in the three decades prior (Stromberg 1998a). Thus, riparian forest change in response to changing hydrologic conditions will take place over timescales of one or more decades.

TNC and collaborating researchers are committed to ongoing long-term monitoring of vegetation and hydrogeomorphic conditions at their restoration sites. The vegetation-hydrology relationships and thresholds delineated by riparian condition assessment models can help define management goals and can be used to predict shifts in vegetation community composition and structure. In addition to measurable hydrologic changes, the response of riparian vegetation to hydrologic restoration is considered to be a key measure of the success of TNC's restoration approach. By tracking changes over time, we will be able to evaluate the efficacy of stressor removal and process restoration as a conservation strategy in the lower basin.

Southwestern Willow Flycatcher: Conservation and Recovery

The southwestern willow flycatcher is a small migratory songbird that breeds in dense riparian forests along rivers and reservoir deltas in the southwestern United States and winters in southern Mexico, Central America, and northern South America (Sogge and Marshall 2000, USFWS 2002c). The subspecies was listed as federally endangered in 1995, with the causal agents listed as destruction, alteration and fragmentation of breeding-ground habitat, loss of wintering habitat, cowbird (*Molothrus ater*) nest parasitism, nest preda-

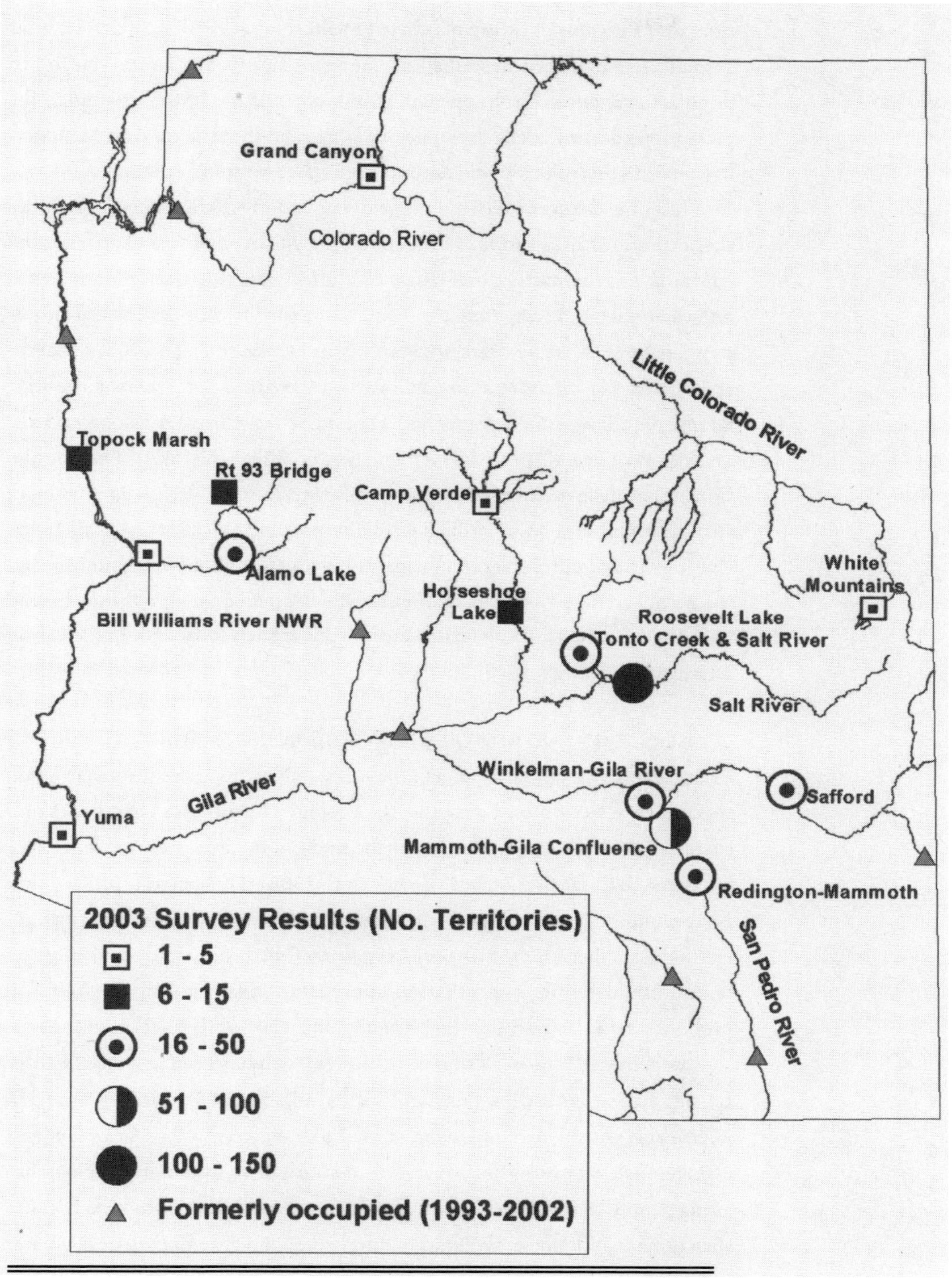

Fig. 19.4. Map of southwestern willow flycatcher abundance (number of territories detected) and distribution in Arizona in 2003 (A. Smith et al. 2004). Map also shows sites without detections in 2003 but which were occupied by flycatchers at least once during the breeding season between 1993 and 2002 (A. Smith et al. 2002, Paradzick and Woodward 2003, A. Smith et al. 2003).

tion, and invasion of exotic plants (e.g., saltcedar) (USFWS 2002c). Habitat degradation has been described as the most significant factor leading to population declines (Stoleson et al. 2000). When Unitt (1987) first noted the precipitous decline of the flycatcher, he suggested that the greatest reduction in the bird's distribution and abundance had occurred in Arizona.

Presently, the core breeding range of the endangered subspecies includes the four states of Arizona, California, New Mexico, and Nevada. Arizona contains approximately one-third of the known population rangewide, and the riparian forest along the lower San Pedro River sustains the largest population in Arizona (Paradzick and Woodward 2003). In 2003, researchers located 141 territories on the lower San Pedro River, whereas the next largest populations in Arizona had 111 and 32 territories, at Roosevelt Lake and Alamo Lake, respectively (A. Smith et al. 2004) (fig. 19.4). The population along the lower river has increased every year since 1997, when intensive surveys began and 43 territories were documented (McCarthey et al. 1998). The lower San Pedro River has probably been a stronghold for the subspecies for decades; in 1940, it was taxonomically described using a bird collected near the historic town of Feldman (now PZ Ranch located 5 km south of Dudleyville; Phillips 1948).

CONSERVATION AND RESTORATION OF BREEDING HABITAT

Conservation of willow flycatcher breeding habitat requires information on vegetation selection patterns and the ecological processes that create and sustain such habitats. Protection of the underlying processes is essential for long-term habitat persistence (Verner et al. 1986). This pattern-process link is especially critical for organisms, such as the flycatcher, that inhabit ephemeral habitats. Because southwestern riparian forests are highly spatially and temporally dynamic, conservation approaches must encompass their full range of natural variability (Finch et al. 2000, Graf et al. 2002).

Flycatcher habitat selection in the lower basin has been assessed at three spatial scales: landscape (Hatten and Paradzick 2003), patch (Paradzick 2005), and within-patch (Allison et al. 2003). Patterns of selection suggest a preference for young (<10-year-old) dense willow and saltcedar patches along perennial reaches (fig. 19.5). Birds may select these conditions because they provide more favorable microclimate (i.e., cooler temperatures), greater protection and cover from predators, higher food availability and foraging success, and greater amounts of dispersal habitat, which ultimately influence survival and fitness of the flycatcher. Long-term (>8 yrs) demographic monitoring of the San Pedro River population indicates that the population is stable or increasing (A. Tutor, personal communication). Thus, these habitat-selection patterns may serve as a model for restoration activities in other drainages across the flycatcher's range.

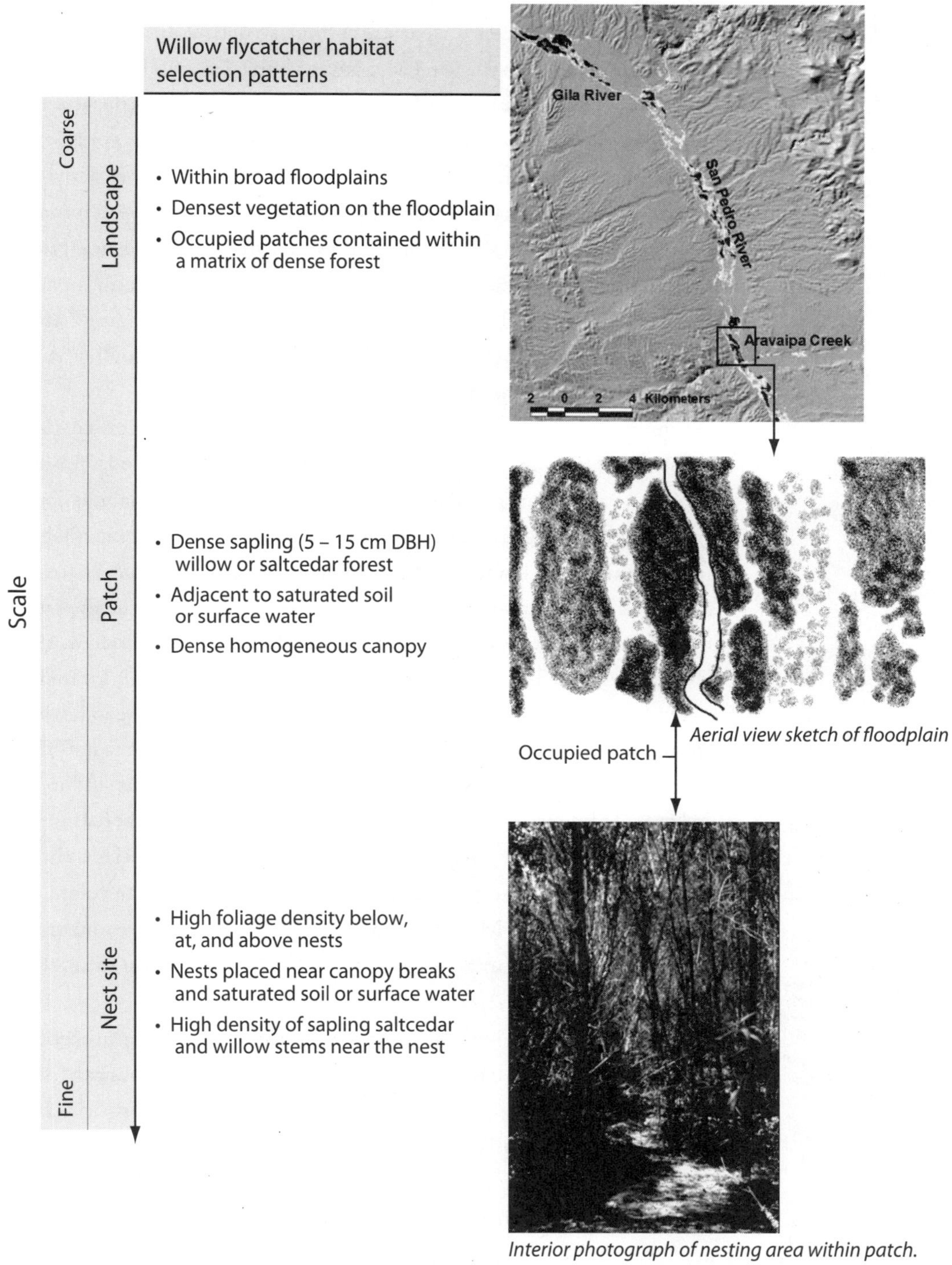

Fig. 19.5. Southwestern willow flycatcher habitat preferences at multiple spatial scales: landscape scale shows GIS-model results (black polygons) overlaid on digital elevation model of lower San Pedro River floodplain (modified from Hatten and Paradzick 2003, and Paradzick and Hatten 2004). Patch scale shows sketch of occupied flycatcher patch and local surrounding habitat (Paradzick, unpublished data). Nest scale shows photograph of nesting trees and within patch structure. Photo credit: Alexander Smith.

Several factors influence riparian habitat structure and composition across the landscape. On the San Pedro River, young dense willow and saltcedar patches develop after floods of suitable timing and intensity, on sites with shallow groundwater and high flow permanence. Preference for young trees suggests a critical temporal component to flycatcher restoration activities. Based on age-size class regressions (Stromberg 1998a), willow-dominated habitat may become suitable for nesting two to three years following germination and, under optimum growth, may remain suitable for only an additional six years. Thus, the abundance and distribution of breeding habitat can change quickly from forest succession alone, in addition to fluctuating in response to drought or flooding.

From the landscape perspective, the riparian corridor constitutes a dynamic or shifting mosaic of vegetation patch types, with floods, fire, and drought all producing spatial and temporal heterogeneity in habitat conditions. Remote sensing analyses highlight the dynamics of flycatcher habitat availability along the lower San Pedro River, while flycatcher abundance and distribution surveys, and movements of banded birds, reveal the dynamics of the flycatcher population (figs. 19.5, 19.6). Following a large flood in 1993, birds were restricted to patches, often dominated by saltcedar, located on high terraces that were not scoured by floodwaters. While these patches were far from the active stream channel, they were often flooded with runoff from irrigated agricultural fields during the breeding season. This additional water provided moist soils and dense foliage throughout the breeding season. In the ensuing years, trees reestablished in the river channel, patches of suitable breeding habitat developed, and birds found and began colonizing mid-channel patches. Flycatchers were documented to move to upstream sites that supported younger stands of vegetation (A. Smith et al. 2004, Munzer et al. 2005).

For managers, these dynamics pose conservation challenges. An occupied nesting site may provide habitat for only a few years. Without physical disturbance, at some point the habitat may become unsuitable for nesting as trees mature and vegetation structure changes. The parcel of land containing the occupied patch will have greater conservation value if the underlying environmental conditions that create young riparian forest in the long term are protected and restored. Periodic recruitment of riparian trees assures that habitat suitable for flycatcher nesting will persist over time. Such recruitment, in turn, depends upon fluvial processes and environmental conditions that create the physical foundation for the habitat (Graf et al. 2002). The lack of major dams on the San Pedro River, combined with maintenance of shallow groundwater tables and careful management of livestock grazing, supports these key ecological conditions and processes.

Southwestern willow flycatcher population growth and movement in

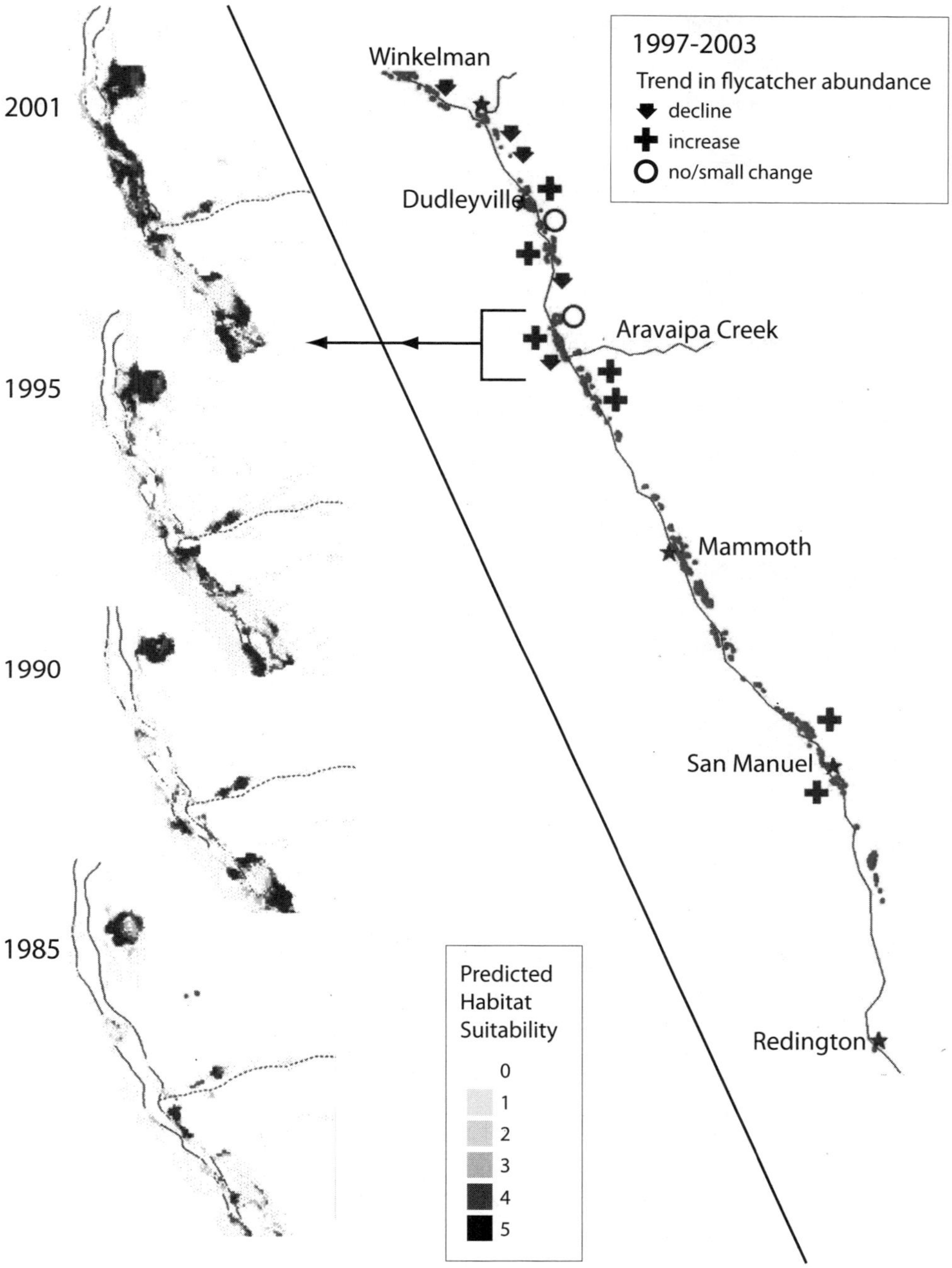

Fig. 19.6. Right side of figure shows trend in southwestern willow flycatcher abundance at sites on the lower San Pedro River, 1997–2004 (Arizona Game and Fish Department, unpublished data). Polygons shown are cells (30 x 30 m) of highest potential breeding habitat (Class 5) with a 100-m buffer produced from a GIS-based habitat model (modified from Hatten and Paradzick [2003], and Paradzick and Hatten [2004]). Left side of figure shows detailed model results near Aravaipa Creek inflow downstream to Cook's Lake for 1985–2001.

recent years, coupled with the conservation efforts of BOR, SRP, TNC and others to reduce ecosystem stressors, are hopeful signs that conditions along the lower San Pedro River will continue to support a dynamic riparian ecosystem that sustains the birds over the long term. The protection of ecological processes and the complex vegetation mosaic that they create and sustain will benefit other riparian obligate species in addition to the flycatcher. The willow flycatcher recovery plan lists 39 sensitive species (throughout the bird's range) that utilize similar habitat or would benefit from flycatcher habitat restoration actions (USFWS 2002c).

CONSERVATION OF THE FLYCATCHER ALONG OTHER DESERT RIVERS

While protection and restoration of flycatcher habitat along the lower San Pedro River has been a conservation emphasis, riparian habitat along some other streams has declined. As water demands increase with urban growth, and problems of water supply are exacerbated by drought conditions, less water is available for sustaining riparian ecosystems. For example, in 2001–2004, reduced flows released from Coolidge Dam, exacerbated by drought conditions, desiccated occupied flycatcher habitat along the Gila River downstream of the San Pedro River confluence (C. Paradzick, unpublished data). Flycatcher abundance at this location has declined as much as 80 percent since 1999, when abundance peaked at 69 territories (Paradzick et al. 2000, A. Smith et al. 2004). Upstream of Coolidge Dam, a large amount of habitat exists within the conservation pool of San Carlos Lake (Dockens et al. 2004). While a southwestern willow flycatcher management plan was developed between the San Carlos Apache tribe and USFWS to conserve habitat on tribal lands (USFWS 2005b), a broader view is needed to consider the downstream effects of reservoir operations on the riparian ecosystem. Similarly, farther upstream in the Safford Valley, habitat is available and some is occupied, but much is privately owned and only recently has land been acquired or put under conservation easement specifically to conserve riparian values (i.e., SRP mitigation for flycatcher impacts at Roosevelt Lake). This reach of the Gila River has been extensively developed for agriculture and could benefit from large-scale coordinated conservation actions similar to those used on the San Pedro River.

A successful model of coordinated river management in the flycatcher's range occurs on the Bill Williams River and Alamo Lake in west-central Arizona. Water and dam managers work closely with natural resource agencies as part of the Bill Williams River Corridor Steering Committee to incorporate Alamo Dam operational needs (flood control, water rights), recreation at the lake, and ecosystem needs to sustain riparian vegetation both above and below the dam. While conservation along the San Pedro River is needed and justified, managers must be careful not to place all the flycatcher eggs into one basket. More emphasis is needed to restore or protect the geomorphic and hydrological conditions that create and sustain suitable breeding habitat along other drainages in the flycatcher's range.

Conclusion

Desert riparian corridors provide crucial habitat for many species. The lower basin supports long stretches of Fremont cottonwood/Goodding's willow forest and saltcedar shrublands, which in turn provide habitat for many migrant Neotropical birds, bats, and insect pollinators, as well as for several threatened and endangered species, including the southwestern willow flycatcher. Arid-region riparian vegetation patterns strongly reflect hydrologic conditions and processes, which in turn reflect geologic and climatic conditions as well as land management practices. Thus, long-term conservation of species such as the flycatcher will depend upon landscape-scale protection of the processes that create and sustain suitable habitat. In this river setting, endangered species protection mandated by the Endangered Species Act hinges on protection of physical ecosystem processes.

The history of conservation in the lower basin highlights the utility of designing management approaches as scientific experiments. Such experiments should include (1) identification of conservation goals, (2) generation of hypotheses related to the effects of management activities on ecosystem conditions, (3) documentation of baseline conditions and monitoring of trends in response to conservation actions, (4) assessment of the trajectory of change in light of conservation objectives, and reevaluation of management approaches, and, possibly, (5) revision of management strategies in light of new information or if conservation goals are not being achieved. This adaptive management approach requires a commitment to monitoring and collaboration with scientific researchers, and also depends upon a willingness of managers to employ creative and flexible conservation measures.

In the lower basin, it became clear that a strategy focused solely on protection of existing high-quality riparian habitat was not sufficient to achieve long-term conservation objectives. Because of the dynamic nature of riparian ecosystems, and because of the significant influence of off-site processes and conditions on ecosystems within management units, managers expanded their perspective on conservation. This expanded view included explicit consideration of the physical processes that form the template for ecological systems in arid-region riparian zones. TNC, in particular, has employed analysis of riparian potential as a key conservation tool, focusing on the prospects for stressor removal and process restoration to achieve far-reaching restoration objectives in the lower basin.

This conservation approach may necessitate patience and a long-term perspective, however, since hydrology, geomorphology, and ecological conditions may not respond immediately to restoration actions. For example, recent drought conditions in southeastern Arizona have likely slowed river and riparian recovery in response to restoration actions at H&E and Three Links Farms. The occurrence of unexpected or delayed outcomes from management activities underscores the need for long-term commitment to rigorous documentation and ongoing monitoring of conservation projects. In the face of ongoing regional drought, altered watershed conditions, and the potential for new suburban developments to increase demand for groundwater, innovation and a flexible perspective will be required. Although the story is not over, we believe that the work accomplished to date in the lower basin provides a valuable case study in arid-region riparian conservation and management.

JULIET C. STROMBERG,
MARK D. DIXON,
RUSSELL L. SCOTT,
THOMAS MADDOCK III,
KATHRYN J. BAIRD,
AND BARBARA TELLMAN

Status of the Upper San Pedro River (United States) Riparian Ecosystem

TWENTY

Introduction

Because rivers are products of their watersheds, riparian preserves can be affected by off-site activities that alter the hydrologic cycle (Pringle 2000, 2001). We explore this issue of off-site effects for the San Pedro Riparian National Conservation Area (SPRNCA) (fig. 20.1). The U.S. Bureau of Land Management (BLM) strives to protect and enhance this desert riparian ecosystem and has filed for federal reserved water rights to maintain it (see chap. 22). Groundwater pumping, floodplain agriculture, off-road vehicle use, and mineral development have been halted within the SPRNCA, and a moratorium has been placed on livestock grazing (Yuncevich 1993). However, much of the upper San Pedro watershed is under state and private ownership and is steadily urbanizing (Steinitz et al. 2003). About 70,000 people live in the Sierra Vista subwatershed, with each using a portion of the basin's groundwater resources. Fewer people live in the Benson subwatershed, but considerable water is used in that area for agricultural purposes, and the population is growing rapidly (Kepner et al. 2004). There are concerns that the riparian ecosystem is being affected by human actions, notably groundwater pumping, occurring beyond the conservation area borders (Arias 2000; and see chap. 21).

Scientific studies play an important role in determining workable solu-

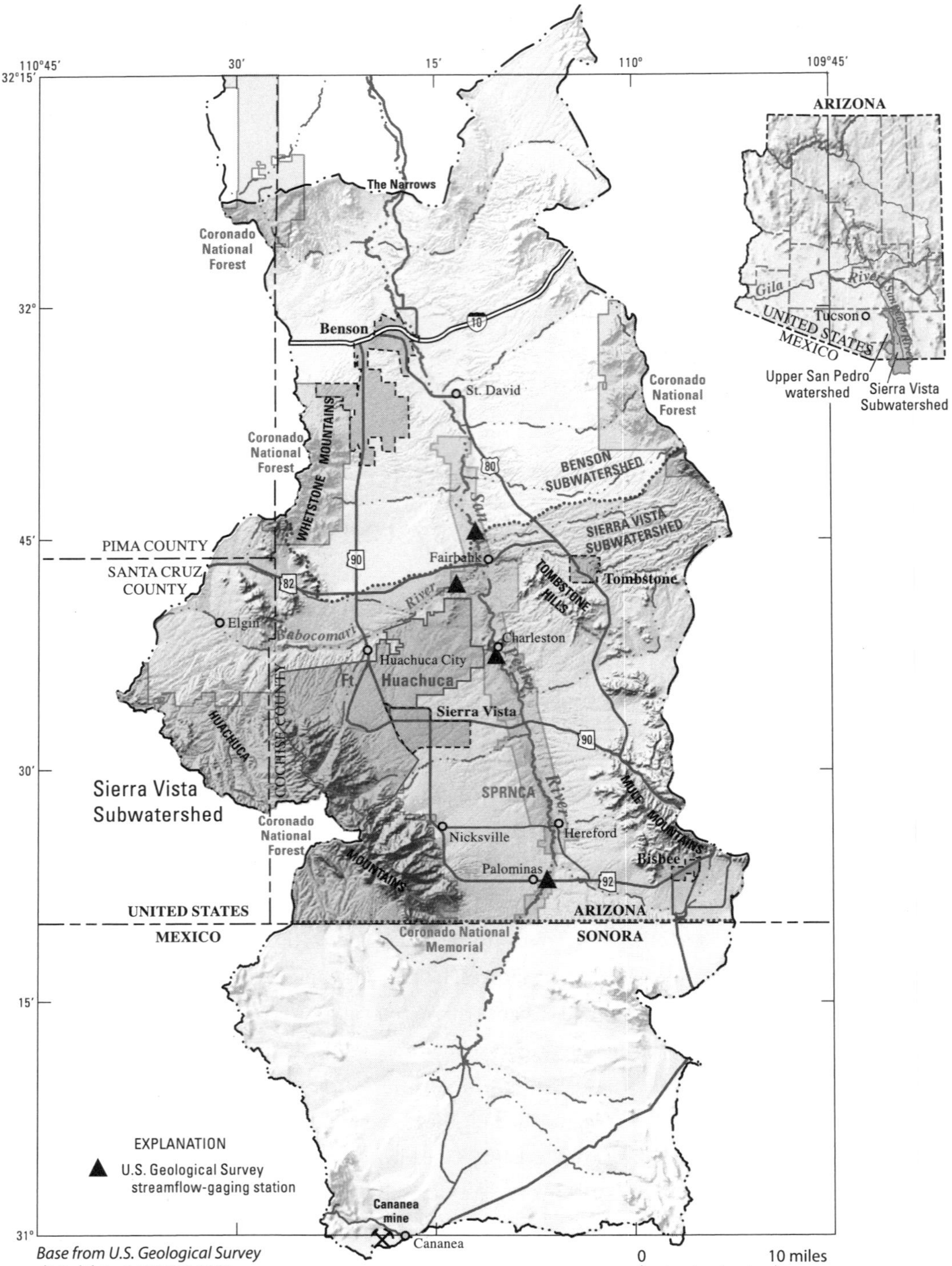

Base from U.S. Geological Survey digital data, 1:100,000, 1982 Universal Transverse Mercator projection, Zone 12

Fig. 20.1. Map of the upper San Pedro basin. Courtesy of James Leenhouts.

tions to water management challenges (Naiman et al. 2002, B. D. Richter et al. 2003) and are of value to stakeholders in the upper San Pedro watershed who are working towards the goal of sustainable groundwater management (see chap. 21). In this chapter we review some of the ecological issues that underpin the San Pedro conservation challenge. We discuss recent findings regarding water needs of the San Pedro riparian vegetation, explore the factors driving observed changes in San Pedro streamflows and riparian vegetation, and review ecological aspects of some management actions implemented to achieve sustainable water use.

How Much Water Does the San Pedro Riparian Ecosystem Need?

FLOW REGIMES

Various approaches can be taken to determine the amount of water, as well as the temporal and spatial patterns of water flow, that sustain aquatic and riparian biotic communities (Richter et al. 1997, Nilsson and Svedmark 2002, Postel and Richter 2003). A useful approach for determining environmental flow needs is the Building Block Methodology (King et al. 2003). For this, a recommended streamflow regime is constructed, on a month-by-month basis, by combining components of the flow regime that sustain different biotic elements or ecosystem functions. If the biotic components are selected judiciously, with each serving as a multi-species umbrella (Lambeck 1997), the composite flow regimes will encompass the processes and conditions that sustain a wide diversity of species and functional types.

Streamflow regimes can be characterized by their magnitude, timing, frequency, duration, and rate of change (Poff et al. 1997), with these applying both to the high-flow (floods) and low-flow (baseflow) components, and to surface (stream) and subsurface (groundwater) flow. The first building blocks of the flow regime are the low-flow conditions that provide water for survivorship of many organisms throughout seasonal dry periods. The second are small, annual floods for channel and habitat maintenance and for stimulating reproduction and growth of some aquatic and riparian organisms. The third are large, less-frequent floods that structure the channel, inundate the floodplain, recharge the floodplain (i.e., stream) aquifer, distribute nutrients and seeds across the floodplain, and create opportunities for establishment of riparian plants.

Table 20.1 portrays streamflow building blocks for the aquatic and riparian biota of the upper San Pedro River. Fish and other aquatic biota require surface flows in the channel year-round. Wetland plants along the channel edge depend upon saturated soils throughout the growing season, a condition associated with perennial streamflow (see chap. 1). Shallow groundwater (less than approximately 3 m below ground surface and with no more than

TABLE 20.1. Relationship of biota on the upper San Pedro River to components of the surface water/groundwater flow regime.

Seasons include the warm spring dry season (April through June), warm summer wet season (July through September), and the cool fall-winter (October through March) season.

Flow component	Frequency, magnitude, or fluctuation	Season	Function
Baseflows	Perennial (100%)	Cool and warm	Sustain riverine marsh vegetation; sustain fish and other aquatic organisms; provide drinking water for wildlife
Riparian groundwater	Depth of <3 m; inter-annual fluctuation <1 m	Cool and warm	Sustain dense, multi-age cottonwood/willow forests and seepwillow shrublands
Riparian groundwater	Depth of <3 m	Cool	Sustain high productivity of sacaton grasslands
Small floods	<5 yr return interval	Warm-wet	Increase productivity and diversity of warm-season plants
Small floods	<5 yr return interval	Cool	Increase productivity and diversity of cool-season plants
Large floods	>5 yr return interval	Warm-wet	Stimulate establishment and increase productivity of warm-season plants
Large floods	>5 yr return interval	Cool	Stimulate establishment of cottonwood/willow and other cool-season germinants; increase plant productivity; drive floodplain patch dynamics

1 m intra-annual fluctuation) maintains dense forests of Fremont cottonwood (*Populus fremontii*) and Goodding's willow (*Salix gooddingii*) and also maintains high productivity of big sacaton (*Sporobolus wrightii*) grass. Winter floods of suitable magnitude (sufficient to scour vegetation and mobilize sediment) and timing (with draw-down during the spring germination season) trigger the establishment of new generations of cottonwood and willow trees. Floods also play a role in the establishment of many other riparian and aquatic organisms, including certain fish. Summer floods provide pulses of productivity in riparian grasses, forbs, and trees, with cascading effects on insects, birds, and other animals. A range of small to large floods, in summer, fall, and winter, create spatially and temporally heterogeneous environments and maintain high diversity in the riparian zone. In concert, perennial streamflows, shallow groundwater, and seasonal rains and floods of varying intensity sustain a diverse and productive community of plants and animals.

GROUNDWATER QUANTITIES

Although many plant species in the San Pedro riparian corridor are sustained by floodwater or rain, the dominant overstory species, including cot-

TABLE 20.2. Groundwater use by major[1] vegetation types along the San Pedro River within the San Pedro Riparian National Conservation Area as measured in 2003.

Vegetation type	Total canopy area in SPRNCA (ha)	Groundwater use per unit area of canopy (mm yr^{-1})	Total estimated groundwater use in SPRNCA (1000 m^3 yr^{-1})	Percent of ET from groundwater
Cottonwood/willow, perennial reach	253	966[2]	2444[2]	100
Cottonwood/willow, intermittent reach	177	410	726	84
Mesquite	1154–1456[3]	689	7953–10035	75
Sacaton (<3 m to groundwater)	113–168	575	650–967	68
Saltcedar	72–108[3]	689	496–744	NA
Open water	73	1156	884	100
TOTAL IN SPRNCA	1842–2235		13113–15759	

Source: Leenhouts et al. 2006.

[1]Minor vegetation types not included in the table are seepwillow, and herbaceous types such as Johnson grass and Bermuda grass.

[2]Values are reported to the nearest millimeter, per convention in the water use literature, but this does not represent their level of certainty or accuracy.

[3]Range reflects uncertainty in percent canopy cover values in the vegetation map.

tonwood, willow, and mesquite (*Prosopis*), are sustained wholly or in part by groundwater. From a perspective of groundwater consumption, the question, "How much water is needed to support the current riparian vegetation?" can be reduced to a single number, with associated variance, for various desired river conditions (table 20.2). A suite of studies conducted along the San Pedro River has estimated the evapotranspiration (ET) rates for the common riparian plant associations. The area of each vegetation type has been mapped from aerial photographs, allowing for landscape-scale estimation of ET and the total consumptive water use of the riparian vegetation (Leenhouts et al. 2006; and see chap. 2).

These landscape-scale estimates of groundwater use have considerable variance. The variance arises because of measurement error (e.g., mapped patch area, ET rates per patch), but also because there is true spatial variance in ET rates among patches of a given type (due to differences in plant age, plant density, and groundwater levels) and true variance between years in water availability, temperature, and other factors that affect plant water use. The fraction of transpired water derived from groundwater (vs. soil water from rains or floods) also varies between vegetation types, as well as among sites and years, particularly for facultative phreatophytes such as mesquite.

Longer-term measurements, and better quantitative linkages of climate, water availability, and evapotranspiration, will be required to more precisely bracket this annual range of variance in total evapotranspiration and the fraction derived from groundwater.

The total water use by phreatophytes along the San Pedro varies spatially among river reaches, depending in part on local hydrogeomorphic conditions. Total riparian water use also undoubtedly has changed temporally over the past century, as dominance shifted from wetlands and grasslands to forests and woodlands, as cottonwood forests and floodplain area expanded in the decades following river entrenchment, and as forests and grasslands were replaced by irrigated and then abandoned fields. The water needs and water use of the riparian vegetation will continue to change in coming decades, as the forests, grasslands, and marshlands are influenced by ongoing successional processes, ever-changing flood and fire regimes, changing climate, and the activities of beaver (see chaps. 1 and 3).

Spatio-temporal Changes in Streamflow

Only a small percent of the rain that falls in the upper San Pedro watershed flows to the river; most evaporates from the soil or is transpired by upland vegetation (see chap. 15). Some of this rainwater recharges the regional aquifer (although rates and processes of such are still poorly understood; Hogan et al. 2004), and then slowly discharges to the San Pedro floodplain aquifer and then to the stream (fig. 20.2). However, most of the annual streamflow in the upper San Pedro River derives from summer stormwater runoff, with some of the stored floodwater contributing to streamflow during dry seasons. Although the portion of the streamflow that is derived from the regional aquifer is but a small percentage of the total streamflow, this baseflow discharge is ecologically important in that it sustains surface water in the channel and a high water table in the floodplain aquifer during dry seasons of non-flood years.

Since about the 1940s, the total annual flow volume of the upper San Pedro River has declined (reflecting significant reductions in summer flows) as has its baseflow discharge (Pool and Coes 1999, Thomas and Pool 2006; and see chap. 15). There are few long-term data sets for groundwater levels in the San Pedro floodplain alluvium, but water levels in wells near Palominas (southern end of SPRNCA) are known to have declined by about 1 m from the 1950s through 1980s, sufficient to convert this stream reach from perennial to intermittent (Pool and Coes 1999). This particular hydrologic shift is largely attributable to historic agricultural pumpage from the floodplain aquifer (see chap. 15).

Presently, about half of the river in the SPRNCA has perennial flow, although greater lengths of the river have year-round flow during wet years.

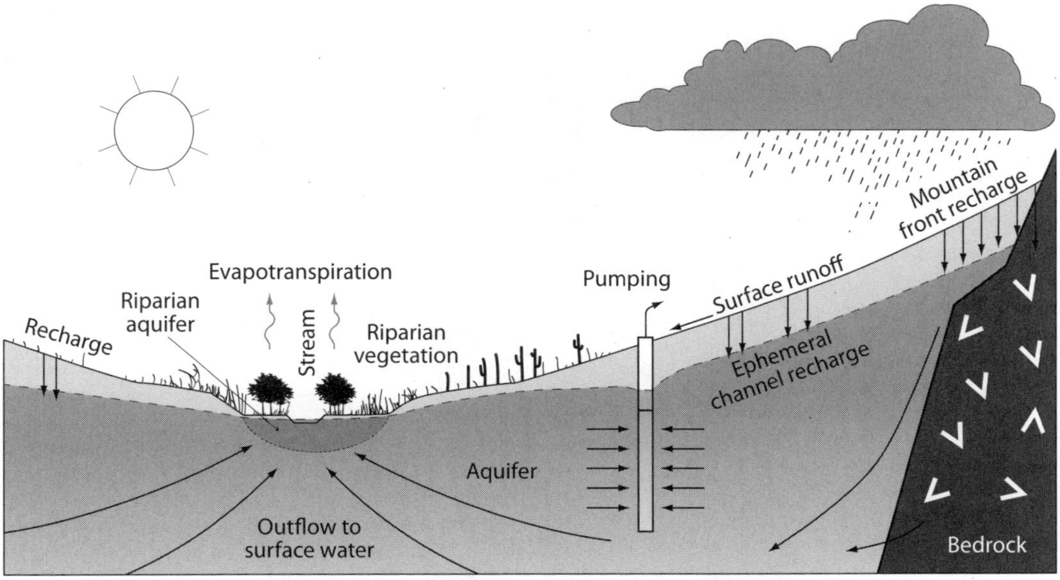

Fig. 20.2. Key hydrologic processes influencing streamflows in the San Pedro River. Illustration credit: Mike Buffington (as modified from U.S. Geological Survey Fact Sheet 086-00).

In 2001, for example, 75 percent (47 of 62 km) of the river in the SPRNCA had surface flow during the June dry season compared to 55 and 57 percent in 2002 and 2003 (fig. 20.3). The greater surface flow in 2001 was a result of discharge from the floodplain aquifer of recharge associated with a large flood in October 2000 (Leenhouts et al. 2006).

The consistently perennial reaches in the SPRNCA are concentrated in the south-central portion (Hereford to Fairbank). Here, year-round flow is maintained by shallow or exposed bedrock which forces groundwater up into the floodplain aquifer, discharging it into the stream. The late-1800s to early-1900s episodes of channel entrenchment may have contributed to perennial flows in this area by replacing alluvial sediments of moderate permeability with more permeable sands and gravels, thereby improving hydraulic connectivity between the regional aquifer and the river (Pool and Coes 1999).

Streamflows are intermittent in the northern tier of the SPRNCA. This is partly because of geologic features: the low permeability of the St. David Formation (a 300-m-thick layer of clays and silts) results in a poor hydraulic connection between the river and the regional aquifer near St. David and Benson (Goode and Maddock 2000). Additionally, San Pedro streamflow has been diverted seasonally into the St. David and Pomerene irrigation canals for over a century. These diversions, together with groundwater pumping, contribute to stream intermittency in the northern tier of the SPRNCA and in areas downstream (Vionnet and Maddock 1992; and see chap. 15).

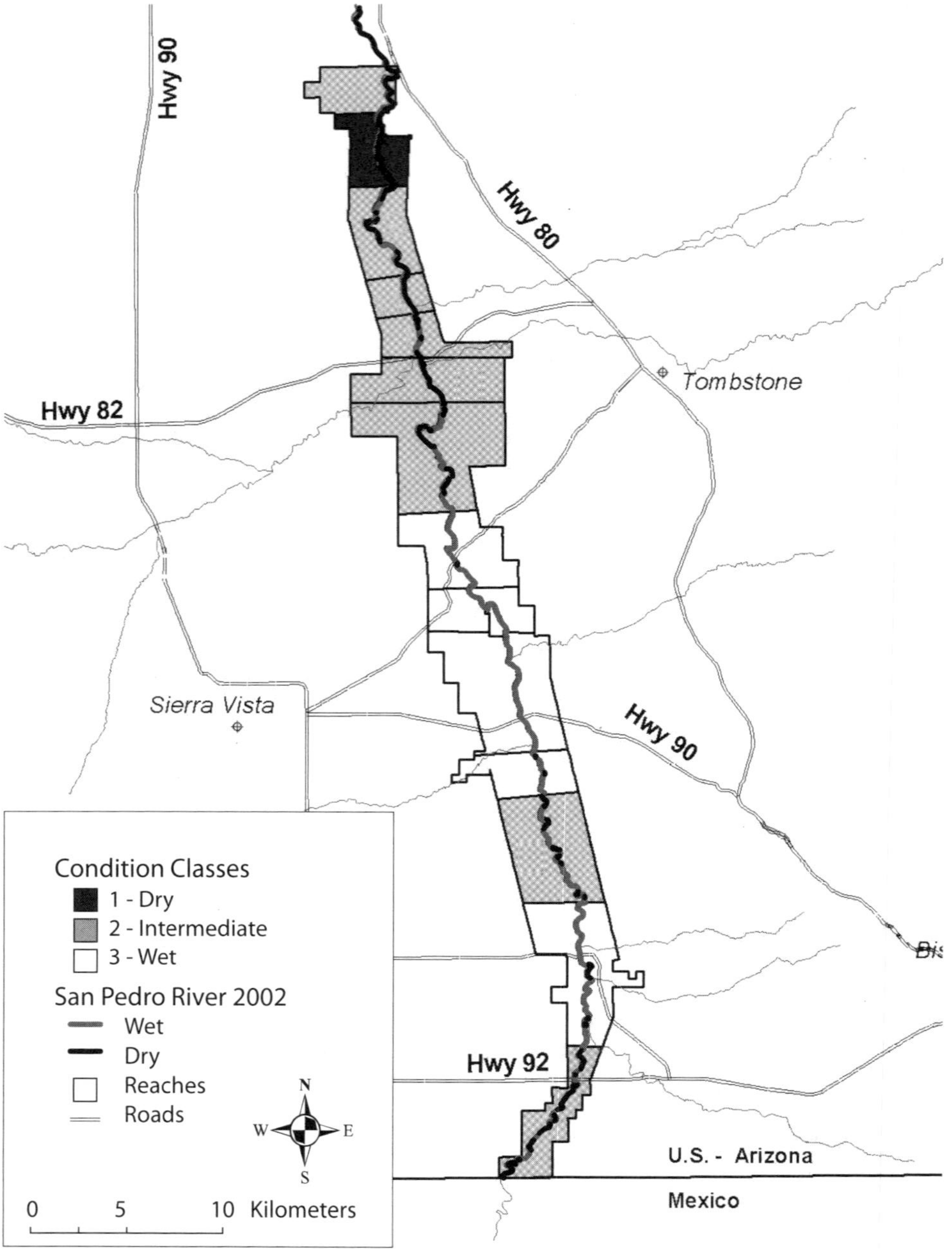

Fig. 20.3. Map indicating riparian condition class for 14 reaches within the San Pedro Riparian National Conservation Area (see Stromberg et al. 2006 for details). Also shown is the location of the stream channel in 2002. Illustration credit: Lainie Levick.

Drivers of Hydrologic Change

Riparian water tables and streamflow rates can be influenced by groundwater pumping from the floodplain aquifer (a more rapid and direct effect) and from the regional aquifer (a slower effect). Pumpage from the upper San Pedro floodplain aquifer mainly supplies irrigation water for row crops and pasture grasses, while pumpage from the regional aquifer supplies most of the urban water used throughout the basin. Both the amount of pumping and its location are influential.

Groundwater extraction in the upper basin, inclusive of that from the near-stream area, began to increase sharply in the 1940s (see chap. 15). However, pumpage rates from the near-stream area declined considerably in the Sierra Vista subwatershed after 1988 due to retirement of some irrigated agricultural land associated with designation of the SPRNCA. In recent decades, agricultural pumping from the floodplain aquifer in the upper basin has been limited mainly to areas near St. David and Benson, private inholdings near Palominas, and parts of Sonora, Mexico.

Urban population growth and municipal water use in the Sierra Vista area are increasing due to a variety of factors, including the nearby army base, attractiveness as a retirement community, and ecotourism. As of 2002, about 7,900,000 more cubic m of water were pumped annually from the Sierra Vista subbasin of the upper San Pedro River than were recharged from rainfall (see chap. 15). Most of this urban pumping has taken place between the mountain-front recharge zone and the river. Cones of groundwater depression have developed in the regional aquifer near Cananea (Sonora), near Sierra Vista (to depths in excess of 25 m), and near St. David and Benson (see chap. 15). Contour maps of regional groundwater levels suggest widespread declines over the past 50 years in many areas of the basin. This has reduced or reversed hydraulic gradients, thereby slowly reducing the amount of groundwater flowing from the regional to the stream aquifer. In addition, by converting stream reaches from perennial to intermittent or ephemeral, groundwater pumping (even when ceased) can continue to reduce streamflow rates by increasing rates of infiltration. Lag effects of past pumping on groundwater flow paths and streamflow rates can persist for centuries (Filippone and Leake 2005).

Streamflow rate and floodplain groundwater level are influenced by many factors other than extraction of water, including climate, riparian vegetation abundance and type, and watershed land cover and soil conditions. Determining the relative magnitudes of these various influences on historic streamflow declines is challenging. A decline in summer precipitation began about 1960 and contributes to declines in summer flood peaks and streamflow rates (Pool and Coes 1999, Thomas and Pool 2006). The recent drought (Gray et al. 2003) may be influencing low-flow conditions, but precipitation changes alone do not fully explain the long-term changes in streamflow (Thomas and Pool 2006).

Plants that use groundwater have a significant influence on daily and seasonal groundwater fluctuations, and as of 2002 accounted for about half of the total groundwater use in the Sierra Vista subwatershed. The post-entrenchment expansion of the high-water-use riparian cottonwood

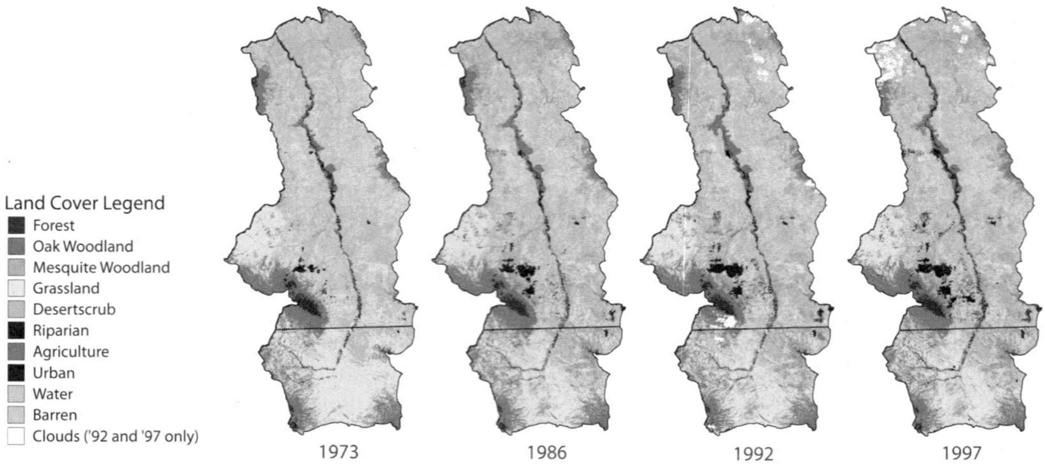

Fig. 20.4. Land cover of the San Pedro River watershed from headwaters in Mexico to Redington during four years, based on LANDSAT data. Figure courtesy of William G. Kepner, U.S. Environmental Protection Agency; report details are available at http://www.epa.gov/nerlesd1/land-sci/pdf/sw-watershed.pdf.

forests (see chaps. 1 and 12) may be contributing to the observed declines in streamflow rates (B. Thomas and Pool 2006), but the extent of this influence remains unknown in the absence of quantitative values for change in acreage of riparian cover types through time. Efforts are underway to reconstruct historic patterns of water use by riparian vegetation, and this information will be of value in refining understanding of feedbacks between vegetation and stream hydrology.

Range conditions throughout the watershed began to improve in about the 1930s and 1940s, following historic degradation of upland grasslands, and this may have reduced summer runoff amounts and flood size (Bahre and Shelton 1993; and see chaps. 11 and 12). Of the more recent land-cover changes in the upper San Pedro watershed (fig. 20.4), many—including the expansion of mesquite woodland, continued declines in grassland, and increases in urban and barren areas—are producing a more rapid or flashy watershed response to rainfall, with greater runoff rates, increased flood magnitude, and greater sediment erosion but less infiltration (Kepner et al. 2000, S. N. Miller et al. 2002). However, although the surface paving that accompanies urbanization is associated with increased surface runoff, it also may lead to increased aquifer recharge, by causing stormwater runoff to concentrate in ephemeral channels (Hernandez et al. 2000, Kepner et al. 2004). Effects of urbanization on hydrologic processes in the San Pedro watershed remains an area of active investigation.

Modeling Efforts

Researchers have been developing models of the relationship between regional groundwater, riparian zone water levels, and riparian vegetation abundance along the upper San Pedro for decades. These groundwater flow models show that the rate of flow between the regional aquifer and stream has changed over time because of groundwater pumping. They also suggest that the current levels of groundwater discharge into the San Pedro River may be only 30 percent of levels prior to 1940 (Goode and Maddock 2000; and see chap. 15). The groundwater models also have been used to predict groundwater level and streamflow changes under scenarios of population growth and water management in the Sierra Vista subbasin (Steinitz et al. 2003). The scenarios examined span a range from Open, which projects high population growth with no constraints on land development and no retirement of irrigated agriculture, to Constrained, with lower population growth than is anticipated, retirement of all irrigated agriculture, and growth concentrated in the existing urban areas. In the Open scenario, increased rates of pumping from the regional aquifer cause water levels to decline throughout the basin and cause streamflows in the San Pedro River to diminish. The Constrained scenario predicts a continued lowering of groundwater levels near the cities, but to a lesser degree than for the Open scenario, and also predicts rising stream water levels and riparian expansion in certain areas in response to retirement of irrigated agriculture. The reality likely will fall somewhere between these two extreme scenarios, depending on future management decisions.

Outputs from modeling scenarios can be used to guide planning (King et al. 2003, Tharme 2003). For the upper San Pedro River, a Decision Support System has been developed that allows policy makers and other stakeholders to use the model outputs to evaluate the potential economic and environmental impacts of different amounts and spatial distributions of groundwater pumping (McPhee and Yeh 2004; and see chap. 21). The output, of course, is only as good as the input. Development of a spatially explicit model that incorporates bi-directional linkages and feedbacks between stream hydrology and riparian vegetation (Baird et al. 2005, Loheide and Gorelick 2007), and that is calibrated using long-term data sets for hydrology and vegetation, will be necessary to provide the next iteration for more accurately predicting interactions between basin-fill groundwater levels, floodplain groundwater levels, and riparian vegetation over the length of the San Pedro River.

Effects of Streamflow Changes on Upper San Pedro Riparian Vegetation

Changes in San Pedro streamflow regimes over past decades have affected upper San Pedro riparian vegetation in many ways. Decadal variations in winter flood patterns have influenced population age structure of cottonwood and willow trees, with some decades being more favorable for establishment than others (see chap. 1). Effects of changing summer flood patterns on the riparian vegetation are less well understood. We know that small sum-

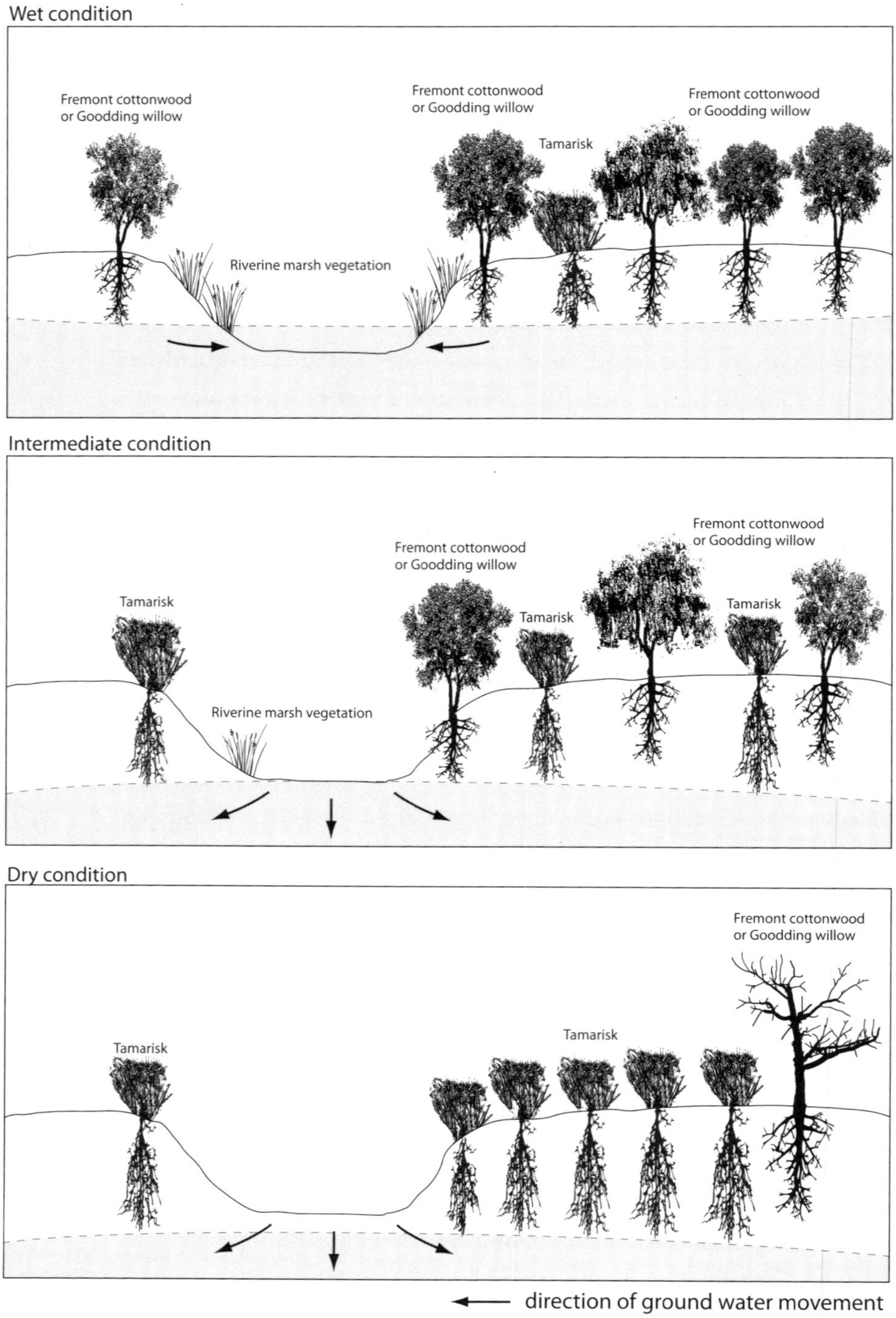

Fig. 20.5. Schematic of changes in riparian vegetation that occur as streams such as the San Pedro lose perennial flow and undergo declines in groundwater in the floodplain aquifer. In reaches with perennial flow and shallow groundwater (top panel), the vegetation is in a wet condition class (Stromberg et al. 2006), characterized by abundant cottonwood and willow trees in the floodplain and wetland plants along the low-flow channel. As the stream becomes increasingly intermittent, and as groundwater deepens and shows more inter-annual and intra-annual fluctuation, the moisture tolerance ranges for riparian plant species are sequentially exceeded and the vegetation shifts into intermediate and then dry condition classes.

mer floods create pulses of primary productivity and species diversity in the floodplain (see chap. 5), but vegetation response to the larger summer floods that typified some past decades has not been described.

Changes in low-flow conditions also undoubtedly have influenced vegetation. Shifts from perennial to intermittent flows, such as occurred in the Palominas area, likely were accompanied by declines in riverine marsh plants along the low-flow channel, but changes remain undocumented, given that focused monitoring of vegetation began only recently. Dendrochronology studies provide a small window to the past and provide evidence of some increase in saltcedar (*Tamarix*) in intermittent portions of the upper San Pedro over the past few decades (Stromberg 1998b, Leenhouts et al. 2006). These changes may be indicative of stream drying.

To provide a baseline for future monitoring, the riparian vegetation in the SPRNCA was assessed from 2001 to 2004 with a riparian condition model (Stromberg et al. 2006; figs. 20.3, 20.5). This analysis shows that about 40 percent of the San Pedro River in the SPRNCA is in a wet condition characterized by perennial flow, shallow groundwater, and an abundance of hydroriparian plant species. Most of these wet areas are in the south-central section. Floodplains in these areas support tall, dense, multi-aged cottonwood/willow forests, with intermixed areas of sacaton grassland and mesquite, seepwillow (*Baccharis salicifolia*), and rubber rabbitbrush (*Ericameria nauseosa*) shrublands. The stream channel is lined by wetland graminoids such as cattail (*Typha*) (fig. 20.6).

Another 55 percent of the SPRNCA, including reaches at the southern and northern ends, is in an intermediate-moisture condition, characterized by intermittent streamflows and low to no cover of wetland plants along the stream channel. Water stress effects, including leaf yellowing, reduced stem growth, and reduced evapotranspiration rates, are routinely observed on cottonwoods and willows in reaches of the San Pedro River with intermittent flow and summer declines in groundwater levels.

Six percent of the SPRNCA, in the reach downstream of the St. David diversion dam, is in a dry condition characterized by intermittent streamflows and floodplain groundwater conditions insufficient to sustain dense cottonwood/willow forests (figs. 20.3, 20.6). Saltcedar is the dominant pioneer plant. Survey data on vegetation condition are not yet available for portions of the upper San Pedro River outside of the SPRNCA boundary, but much of the riverbed downstream of the SPRNCA, near Benson, is dry.

Ecosystem monitoring will continue along the SPRNCA, mandated by federal legislation (see chap. 22). At varying intervals, there will be monitoring of stream surface flows, groundwater levels in the stream aquifer and regional aquifer, area of various vegetation types, and abundance of hydrologically sensitive biota. Downstream of the SPRNCA boundary, riverine

Fig. 20.6. Perennial-flow (above; 20.6a) and intermittent-flow (right; 20.6b) reaches of the upper San Pedro River. The photo of the perennial-flow site shows the low-flow channel, lined by *Populus/Salix* forests. The photo of the intermittent-flow site shows an unvegetated portion of the floodplain and *Prosopis* forests on the terrace. Photo credits: Elizabeth Makings.

ecosystem monitoring has been initiated by a local watershed group, the Community Watershed Alliance.

How Sensitive Is the Riparian Vegetation to Hydrologic Change?

Small changes in riparian groundwater levels and stream low-flows can have significant effects on the riparian community if values are near threshold levels for vegetation change. For example, in the northern tier of the SPRNCA,

water table levels in the stream aquifer are near the lower threshold for cottonwood and willow survivorship. Small declines in the water table (on the order of 0.5 m relative to levels measured in the floodplain in 2002) in this dry northern tier would drive shifts from the intermediate condition class to the dry condition class, with cascading effects across trophic levels (see chap. 8). In much of the central portion of the SPRNCA, in contrast, water tables are not approaching this particular threshold. Shifts between perennial and intermittent streamflow constitute another hydrologic threshold for riparian vegetation change. Restoration of perennial flows in reaches that are presently intermittent, such as those near Palominas, would restore riverine marsh communities and drive shifts from the intermediate condition class to the wet condition class.

Ecological Perspective on Management Options

To address the basin-wide groundwater overdraft, and balance the water budget, managers and policy makers in the Sierra Vista subwatershed of the upper San Pedro basin are taking actions both to increase the groundwater supply to the riparian corridor and to reduce demand by the riparian vegetation for groundwater (see chap. 21). Among these efforts are water conservation and re-use, recharge of municipal effluent, construction of stormwater retention basins, reduction of agricultural groundwater pumping, construction of watershed check dams, and burning of mesquite.

Scientific studies have been initiated to address the effects of many of these actions on water budgets. Studies also are needed to address their effects on ecological functions, to insure that potential consequences are understood. For example, stormwater retention basins have been constructed on washes in the upper San Pedro watershed to offset flow peaks resulting from upstream development and to increase percolation of water into the regional aquifer. While studies have been initiated to document the ensuing hydrologic changes, studies also are needed to determine how the retention basins will influence biotic processes, such as flows of plant seeds between the tributaries and main-stem river.

Another example of a project that could benefit from multidisciplinary research is the prescribed burning of mesquite and other shrubs by BLM personnel in the SPRNCA. The burning is intended to reduce riparian water use, reduce fuel loads, and restore riparian and desert grasslands. GIS-based models indicate that riparian groundwater use rates will decline in response to replacement of mesquite by sacaton, a species that uses less groundwater than mesquite, or by upland grasses, that use no groundwater (see chap. 21). However, many uncertainties remain. Burning of mesquite could create a shrubby growth form that no longer provides habitat for forest-affiliated birds, while allowing the plants to continue to transpire at high rates due to their well-established root system. Reduction in mesquite cover could reduce soil fertility (see chap. 14) or shallow soil moisture (see chap. 2), thereby affecting productivity and habitat quality of post-fire vegetation. Further study is warranted to determine how this management action would influence a range of riparian functions.

Summary and Conclusions

Do we know how much water is needed to sustain the San Pedro riparian ecosystem? The answer to this is a qualified "yes." Landscape-scale evapotranspiration rates, and the groundwater-derived component thereof, have been determined as one index of riparian vegetation water needs. Although there is high variance and although values will change given the dynamic nature of riparian ecosystems, it provides a measure of the amount of water that is needed to flow to the river to sustain the current levels of water use by the riparian vegetation. Further, hydrologic thresholds for plant-community maintenance have been quantified, thus determining groundwater levels and streamflow permanence needed to sustain various vegetation types.

Will these flows be maintained? This is a more difficult question to answer. The San Pedro case is different from some in that there is not a point source for management, such as a regulating dam for which the requisite flow releases can be prescribed (Richter et al. 2003). Rather, there are multiple groundwater wells distributed throughout the watershed and multiple land use changes taking place. Stream-groundwater interactions are complex and variable over time and space, and there remain unresolved issues regarding rates and extent of groundwater storage changes and the rate and extent of their impacts on San Pedro River flows and riparian biota in the basin. It does seem evident, however, that groundwater pumping beyond the SPRNCA borders has affected regional groundwater levels and groundwater flow paths, and thus streamflow and riparian condition. It also seems clear that there are unavoidable lag effects, with past human actions reverberating in the present landscape; just as vegetation trajectories have been set by past geomorphic events, groundwater flow paths will be affected for decades to come by past pumping. It also seems apparent that certain actions, such as retirement of irrigated agriculture, can be taken that will cause local increases in groundwater levels and streamflows. Results of the recently initiated ecosystem monitoring along the upper San Pedro will be a test of whether thriving urban-agricultural centers can co-exist with thriving riparian ecosystems, a challenge that can be difficult to meet (e.g., Logan 2002).

HOLLY RICHTER,
DAVID C. GOODRICH,
ANNE BROWNING-AIKEN,
AND ROBERT VARADY

Integrating Science and Policy for Water Management

TWENTY-ONE

The politicization of science is undoubtedly a slippery slope.
But so is the scientization of politics. The boundary organization
does not slide down either slope because it is tethered to both,
suspended by the coproduction of mutual interests.
(Guston 2001)

Introduction

Freshwater and the ecosystems from which it originates are indispensable to human health and survival. Yet, population growth, climatic variability, and land uses such as mining and agricultural practices along the U.S.-Mexico border challenge our ability to adequately manage this indispensable resource. As the Southwest grapples with ways to increase water supplies and ensure water quality for its burgeoning population, various institutional and political drivers of change, including government agencies at all levels and elected officials trying to serve their constituents' interests, directly affect water management policies in the region.

This chapter describes the collaborative efforts of scientists, agency representatives, non-governmental organizations, elected officials, and other stakeholders (table 21.1) to address water policy and management issues in the Sierra Vista subwatershed of the upper San Pedro River basin. The recent efforts of the Upper San Pedro Partnership (the Partnership) to manage groundwater resources through an adaptive management approach are

Participation by The Nature Conservancy in the development of this chapter was made possible through funding provided by the Department of Defense Legacy Program.

TABLE 21.1. The twenty-one member agencies and organizations of the Upper San Pedro Partnership, as of 2007.

Arizona Association of Conservation Districts
Arizona Department of Environmental Quality
Arizona Department of Water Resources
Arizona State Land Department
Audubon Arizona
Bella Vista Ranches
Cities of Sierra Vista, Tombstone, and Bisbee
Cochise County
Fort Huachuca (Department of Defense)
Hereford Natural Resource Conservation District
The Nature Conservancy
Town of Huachuca City
U.S. Agricultural Research Service
U.S. Bureau of Land Management
U.S. Bureau of Reclamation
U.S. Fish and Wildlife Service
U.S. Forest Service
U.S. Geological Survey
U.S. National Park Service

described. Adaptive management has been described as implementing projects or policies as experiments (Holling 1978, Walters 1986). This approach does not postpone action until "enough" is known, but instead acknowledges that time and resources are too short to defer *some* action, particularly actions to address urgent problems (Lee 1999).

The work of the Partnership has evolved the furthest of any watershed group in the San Pedro basin to date. This progress is at least partially attributable to the federal mandates and regulatory levers at play within this portion of the basin. However, there is a need for similar water management efforts in other portions of the basin, and the need for a more integrated basin-wide approach is apparent. The binational context of the upper San Pedro basin complicates integrated watershed management (Liverman et al. 1999), and the potential risks and uncertainty associated with ineffective basin-wide groundwater management remain high for all stakeholders. The Benson subwatershed is looking to Sierra Vista upstream for evidence of good management, and Sierra Vista is looking to the Sonoran portion of the basin for indications of progress in establishing watershed governance. Unfortunately, water within different portions of the basin have no effective venue for communication or coordination with one another. However, the Mexican Binational San Pedro Commission recently invited the Partnership to embark on a series of meetings for planning and information exchange.

Each of the three subwatersheds within the upper basin has responded in various ways to meet their respective groundwater-management challenges. Some efforts have made much more progress than others in terms of the implementation of technical studies and/or water management projects or policies. The ability to initiate studies or implement projects is largely dependent upon the resources available, which, in many cases, reflects the leverage provided from legal mandates and political context. Efforts in the Sierra Vista subwatershed have secured the most resources to date, where federal mandates associated with both Fort Huachuca and the San Pedro Riparian National Conservation Area (SPRNCA) strongly affect policy and decision making. However, even without this leverage, the motivation and initiative demonstrated by local residents in a given area can play a pivotal role in launching watershed initiatives, as has occurred in the Benson subwatershed.

In Mexico, the responsibility for water management has been decentralizing since the 1990s. The onus has shifted from the Mexican National Water Commission (CNA) to local water users such as state and municipal offices and agricultural water-use associations in irrigation districts (Browning-Aiken et al. 2004). The National Water Law in Mexico (Ley de Aguas Nacionales 1997) calls for the development of localized watershed councils and irrigation districts to serve water users, establish water infrastructure, and preserve water resources in the borderlands. However, in reality, Mexican municipalities along the U.S.-Mexico border remain very dependent upon the federal government or upon a mixture of national and international sources for investment in water infrastructure projects (Romero-Lankao 2002). Local initiatives along the northern Mexican border are linked to national policy demands. In addition, the CNA, despite its policy of decentralized water management, has been slow to support the development of a Mexican water council in the San Pedro, a necessary step in establishing good governance in the Mexican portion of the basin. Also complicating matters is the fact that Mexican environmental policy frequently runs counter to Mexican economic policy, especially in terms of mineral resources and maquiladoras along the northern border (Browning-Aiken et al. 2006). Considerable challenges to good governance remain with water policy implementation in Mexico, which hamper policy makers' abilities to address issues of equitable access to clean water, sustainable development of water resources, and ecosystem protection in the context of climate variability.

On the U.S. side of the border, the Sierra Vista subwatershed has been subject to considerable pressure from sociopolitical drivers with regard to water management over the past decade in the form of federal and/or international mandates, such as the Commission for Environmental Cooperation in 1999 and the Defense Authorization Act in 2004. Two significant national assets in that area, the SPRNCA and Fort Huachuca, home of the U.S. Army Intelligence Center, have resulted in a myriad of strong drivers and in the establishment of a highly developed, collaborative water management effort in this subwatershed.

In contrast, the Benson subwatershed, located downstream, lacks this level of federal participation. However, accelerated rates of residential development and large-scale master-planned communities are proposed for the area, likely a result of its proximity to Tucson. A newly established watershed coalition, the Community Watershed Alliance of the Middle San Pedro Valley, led by landowners, residents, and other organizations and government entities, was formed in 2005 to "promote collaboration and cooperation to advance research, education, and policies for the sustainable health of their watershed." They are assisting the U.S. Geological Survey and other agencies and organizations by helping to coordinate research and monitoring studies with private landowners in the area, and initiating watershed improvement projects, among other endeavors.

To a limited degree, monitoring and research data have been shared between individual scientists and agencies conducting studies throughout the upper San Pedro basin, yet regional water management planning or coordinated project implementation among the three subwatersheds has not begun. In addition, it remains unclear what entity might provide this type of integration in the future.

Drivers of Change in the Upper San Pedro: Federal and International Mandates and State Programs

The upper San Pedro River was designated as the first Riparian National Conservation Area in the United States in 1988, when the San Pedro Riparian National Conservation Area (SPRNCA) was established by the U.S. Congress as part of Public Law 100-696. The SPRNCA's enabling legislation established the existence of federally reserved water rights to maintain the riparian ecosystem (B. Childress, personal communication). Approximately a decade after the SPRNCA's designation by Congress, a team of experts was commissioned by the trinational (Canada, United States, Mexico) Secretariat of the Commission for Environmental Cooperation (CEC 1999a) to produce an interdisciplinary study of the upper San Pedro basin, intended to "serve as an example of how to protect a transboundary watershed" (Udall Center 1998, Varady et al. 2000). This study elevated the San Pedro basin issues to national attention (Christiansen 1999, Kingsolver 2000).

Following the CEC report, during the 1999 binational Divided Waters, Common Ground conference on the upper San Pedro River basin (Brady et al. 2000), then-Secretary of the Interior Bruce Babbitt provided further incentive for local decision makers to work together for effective watershed management. The secretary noted that if effective local management could not be brought to bear on the San Pedro's water issues, then external forces, such as the federal courts, would impose restrictions or management plans. Collectively, these federal drivers began to highlight the growing need for more collaborative approaches toward water management.

In 1999, Governor Jane Hull established the Rural Watershed Initiative to address rural watershed-management issues throughout Arizona. Administered by the Arizona Department of Water Resources, this program was intended to stimulate rural watersheds to develop their own water management plans with input from local citizens and stakeholders. Funding for water resource studies was provided to those rural areas that created watershed groups. The Partnership was one of seven initial watershed groups that formed in response to this initiative. To formalize and document the Partnership's commitment to work together, a total of 21 private, public, and non-governmental organizations eventually signed a Memorandum of Understanding (MOU) that defined the purpose of the group: "to coordinate and cooperate in the identification, prioritization, and implementation of comprehensive policies and projects to assist in meeting water needs in the Sierra Vista Subwatershed of the Upper San Pedro River Basin" (http://www.usppartnership.com). Nothing within the MOU limits the respective legal authorities or decision-making ability of any of the participants or requires expenditures of funds; membership in the group is strictly voluntary. This group serves as a "boundary organization," as described by Guston (2001), in that it blurs the boundaries between science and politics, directly linking scientists with decision makers and, in the process, not only serves the mutual interests of both but also leads to more productive decision making.

In 2002, the U.S. Fish and Wildlife Service issued Fort Huachuca a Biological Opinion on its compliance with Section 7 of the Endangered Species Act. The Biological Opinion addressed several endangered or threatened species potentially affected by Fort Huachuca's activities, including the southwestern willow flycatcher (*Empidonax traillii extimus*) and the Huachuca water umbel (*Lilaeopsis schaffneriana* var. *recurva*), both riparian species. Within this document the Sierra Vista subwatershed groundwater deficit was quantified as 5,144 acre-feet per year. Fort Huachuca committed to implement conservation measures resulting in water savings of 3,077 acre-feet per year toward this deficit. In addition, the Fort requested that the communities and agencies within the Partnership commit to reduce by 3,306 acre-feet per year

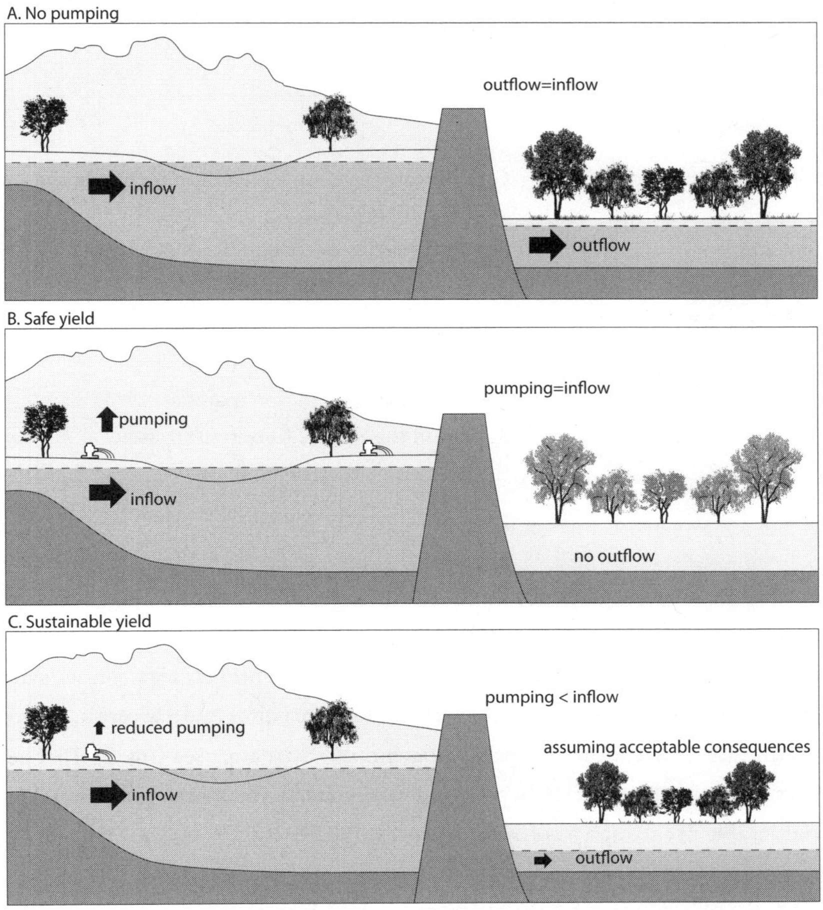

Fig. 21.1. Reservoir analogy to the response of a groundwater system to pumping comparing no pumping with safe yield and sustainable yield (USDI 2005).

the estimated groundwater pumping demands projected by 2011. Clearly, at this point, the role of the Partnership as a coordinating water management body had been recognized, even though the group had no formal regulatory authority.

One of the most recent legislative drivers within the Sierra Vista subwatershed is Section 321 of the National Defense Authorization Act of 2004, Public Law 108-136 (see chap. 22). The incentive, or "carrot," of this bill in terms of local interests is the potential for federal funding for conservation projects, scientific research, and monitoring needs. The cost, or "stick," is the requirement to report "measurable annual goals for the reduction of the overdrafts of the groundwater of the regional aquifer, identified specific water use management and conservation measures to facilitate the achievement of such goals, and identified impediments in current federal, state, and local laws that hinder efforts on the part of the Partnership to mitigate water usage in order to restore and maintain the sustainable yield of the regional aquifer by and after September 30, 2011" (Upper San Pedro Partnership 2004—Appendix 3). The legislation shifted the focus of

TABLE 22.2. Initial criteria established by the Upper San Pedro Partnership for sustainable yield of groundwater.

Social and economic	Environmental
· Sufficient water quantity for human needs · Fort Huachuca remains operational unless for reasons unrelated to water · Cost of living, specifically affordable housing and the cost of doing business, remains within the means of a diverse population · Maintain local participation in water management · Sustain water quality	· Groundwater levels in alluvial aquifer within the SPRNCA maintained · Stream baseflow and flood flows maintained · Accrete aquifer storage · Riparian habitat and ecologic diversity maintained · Water quality sustained in SPRNCA · Overall riparian condition maintained · Springs in the SPRNCA continue to flow

groundwater-management efforts from balancing a water budget, or safe yield, toward the concept of sustainable yield, a more comprehensive concept involving broader social, economic, and environmental factors. Under safe yield, wherein total groundwater withdrawals are equal to total recharge, no groundwater is allocated toward riparian or instream uses (fig. 26.1). The Partnership subsequently began to define criteria for meeting sustainable yield (table 2) through a consensus process (USDI 2005).

Constantly Rising Stakes: Increasing Demand and Prolonged Drought Conditions

Population growth and prolonged drought have also been key incentives for action. Increasing human demands on groundwater resources are widespread, as populations continue to grow on both sides of the international border. Much of the population growth within Cochise County on the U.S. side has occurred in unincorporated areas. The county as a whole experienced a growth rate of 10.6 percent from 2000 to mid-year 2004.

The recent multi-year drought, which began in many areas in 1999, pales in comparison to earlier periods of aridity in the western United States over the past 1,200 years (E. Cook et al. 2004), but nonetheless it is an important factor affecting water demand and supply in the region. Risks associated with drought are a product of both the severity of the event and the vulnerability of society (and the environment) to the event. At a minimum, recent climatic trends have increased the awareness by local decision makers regarding the amount of variability in precipitation patterns, and the uncertainty and vulnerability associated with managing for "average annual" rates of precipitation and groundwater recharge. Water and climate surveys assessed the quality and usability of climate and hydrologic information available to water managers and communities in the Mexican portion of the San Pedro River basin. The surveys indicated that the central concern for urban residents is the lack of reliable potable water due to frequent service breakdown—with climate change and variability, specifically drought and high temperatures, as contributing factors (Browning-Aiken et al. 2007).

Scientific Studies in the Sierra Vista Subwatershed

To meet the water management challenges presented by the federal mandates (i.e., federally reserved water rights, Biological Opinion for Fort Huachuca, Section 321 legislation), Partnership member agencies needed to establish sound research and monitoring programs that could be used to inform science-based decisions. The upper San Pedro basin has long served as an outdoor laboratory for scientific investigations aimed at increasing understanding of physical watershed processes. The U.S. Department of Agriculture's (USDA) Walnut Gulch Experimental Watershed has been researching hydrologic processes and soil erosion for more than 40 years, and other research efforts date back to the nineteenth century.

Since 1988, research investigations in the upper San Pedro basin have progressed so as to allow scientific information to be a much more integral part of water management decisions. Interagency and interdisciplinary efforts, such as the Semi-arid Land-Surface-Atmosphere (SALSA) Program and the University of Arizona's Center for the Sustainability of Semi-arid Hydrology and Riparian Areas (SAHRA), a National Science Foundation Science and Technology Center, have allowed investigators to address larger spatial scales and more complex ecosystem processes. SALSA identified the consequences of natural and human-induced change on basin-wide water balance and on ecological complexity of semi-arid basins at event, seasonal, interannual, and decadal timescales. SAHRA identified stakeholder-relevant questions on which to focus its scientific research, all of which bear on management of water resources in semi-arid regions.

The Partnership has funded or facilitated technical studies that have served three primary purposes: (1) to identify and acquire key data sets that can increase understanding of hydrologic systems and/or processes that have water management implications, (2) to develop predictive modeling tools and forecast specific outcomes in response to land/water management actions, and (3) to investigate and develop new water conservation, recharge, or augmentation strategies. One of the leading studies facilitated by the Partnership was the "Hydrologic requirements of and consumptive groundwater use by riparian vegetation along the San Pedro River, Arizona" (see chap. 20). This interagency research effort by the U.S. Geological Survey, the USDA-Agricultural Research Service, and Arizona State University served both to estimate evapotranspiration losses from the riparian corridor and define the hydrologic context needed to sustain that ecosystem over the long term. Other studies have been undertaken to better understand stormwater and groundwater recharge processes, develop a regional groundwater model, and develop a decision support system. Additionally, costs and benefits for dozens of potential water conservation and reclamation projects were analyzed, and several augmentation alternatives were assessed.

Collaborative Learning Processes: Bringing Scientists and Policy Makers to the Table

In addition to generating information, these technical investigations also served as a venue for initiating collaborative learning processes within and between groups of scientists and decision makers. This motivated them to jointly frame their information needs and initiate group processes toward building consensus on key issues. However, decision makers had to be patient enough to endure long technical discussions, and scientists had to engage in understanding fully the social, economic, and political sideboards and constraints. Discussion of the uncertainties involved and the potential risks associated with those uncertainties early on, even before actual results were produced, helped to manage expectations as to what science could actually deliver. In the long run, the process will hold participants more accountable to one another, increase confidence in results regardless of outcome, and help all participants attain a deeper understanding of complex information on which to base decisions.

Decision makers and scientists typically operate under different time frames and constraints, and often with misconceptions about one another. The world of decision makers revolves around societal values, beliefs, and perceptions; political considerations; definite deadlines; and limited budgets. The culture of scientists involves establishing facts, taking measurements, and making incremental progress over time (V. Baker 1998, Moran and Heilman 2000). Although scientists are often influenced, even unwittingly, by their own values and beliefs, their ability to work effectively with decision makers is greatly improved if they remain objective and unbiased by these factors as they design, implement, and ultimately communicate the results of scientific investigations. To work together effectively, scientists and decision makers must accommodate each other's needs.

Evaluating Success

In 2003, Partnership members and frequent attendees to Partnership meetings were surveyed to evaluate the midcourse effectiveness of the collaborative process (Browning-Aiken et al. in review). The survey's intent was to identify strengths and weaknesses in the group's collaborative process, and to pinpoint strategies in organizational structure and problem-solving processes from which other Arizona watershed organizations could benefit. Participants were asked about (1) the nature of basin water issues, (2) management goals and priorities, (3) organizational structure, (4) stakeholder identification and positions, (5) the method of selecting and interpreting scientific and technical information, (6) the nature of stakeholder collaboration within the watershed, (7) the processes of planning and decision making, (8) the method of leader or facilitator selection, including the qualities of effective leadership, and (9) the method of establishing authority within the regional community.

One clear finding was that success in accomplishing the group's mission was correlated with the extent to which scientific "research findings [have] been interpreted or used by the Partnership to make management decisions." Ninety percent of the Partnership members and participants considered scientific studies as one of the most important projects undertaken by the watershed initiative. This finding pointed to the key role of research and monitoring—from the viewpoint of participants in watershed group meetings—in achieving the group's mission of sustainably managing water. Another finding was that almost half (47 percent) rated both the group's accomplishment of its mission and its capacity to identify water problems as relatively high, while the Partnership's success in addressing basin water problems was rated either as very high (37 percent) or average (37 percent).

Adoption of an Adaptive Management Framework

Collaborative water-conservation planning efforts by the Partnership have continued to evolve since the establishment of the group in 1998. During their first two years, Partnership members focused on defining common ground: shared groundwater-management goals and objectives, and associated information needs. The original intent of these planning efforts was to develop one comprehensive, long-term conservation plan for the Sierra Vista subwatershed by 2005. By 2002, their initial goal of developing one definitive plan evolved into a more complex, yet flexible, ongoing adaptive management planning process. They began to recognize that certain key management concepts that became apparent during ongoing investigations could be applied to decision making before the studies were finalized, and they wanted to be able to assess the benefits of projects that were already underway so that they could adjust their objectives and strategies as appropriate. Finally, they continued to discover additional gaps in their knowledge as their understanding of key issues increased. Thus, they established an iterative planning process and produced their first annual plan in 2003.

To construct the plan, the Partnership inventoried and categorized all member agency water management projects, policies, and programs currently underway. From this baseline inventory, the Partnership prioritized additional collaborative efforts for the coming year and published this information in their "Working Water Conservation Plan" completed in 2003. This plan was subsequently updated and revised based on new information and developments, and an annual adaptive management cycle became established. The shift to an adaptive management approach, while representing perhaps the best chance for incorporating good science into decision making over the long term, also represented a huge increase in the commitment of the stakeholders. No single document now marks the end of their planning efforts, defines total funding needs for science, or even assures absolute certainty or success as a result of these efforts. The definition of success itself became even more of a moving target as a more comprehensive understanding of this complex hydro-ecological system was developed. The overall goal of the Partnership's adaptive management planning process is "to ensure an adequate long-term water supply is available to meet the reasonable needs of both the area's residents and property owners (current and future), and the San Pedro Riparian National Conservation Area" (http://www.usppartnership.com). This goal is interpreted in terms of the aforementioned legislative and regulatory drivers affecting the area.

STRENGTHS AND CHALLENGES OF THE ADAPTIVE, COLLABORATIVE MANAGEMENT FRAMEWORK

This adaptive management framework, like all others, involves active and focused learning on the part of both scientists and decision makers and an acknowledgement that their current understanding of the issues is not perfect. The process also has the potential to openly communicate the Partnership's increasingly detailed understanding of complex systems to external audiences, including the general public (Holling 1978, Walters 1986, 1997).

One great challenge in collaborative groundwater management relates to timing issues associated with planning, research, and the implementation of projects and policies. Research and conservation planning typically do not commence until there is a perceived problem that merits attention. Unfortunately, the acquisition of data of sufficient quality and credibility to enable decision making requires a substantial investment of time and resources. Therefore, decision making will be limited, at least at the onset of planning efforts, by the amount of information that is readily available and considered acceptable to all interests concerned. Additional data collection will likely be required, particularly to support decisions that involve high financial, political, or environmental risks, and/or high levels of uncertainty. Thus, certain decisions could be delayed due to lack of information.

One strength of an adaptive management approach is that it allows actions that have low risk or uncertainty to be taken early on. Partnership member agencies realized that the implementation of certain water management strategies requires substantial information through monitoring, research, and modeling efforts as well as political assessments, while other projects represented relatively low-risk strategies whose implementation could be more immediate. In 2003, the Partnership identified more than 100 projects for immediate implementation, ranging from the repair of leaky infrastructure, car-wash water recycling, voluntary retirement of agricultural pumping through conservation easements, recharge of treated effluent, and reintroduction of beavers. In their 2004 Water Management and Conservation Plan, the Partnership prioritized additional projects for implementation, including the development of model codes and ordinances, the establishment of water-conservation surcharges for excessive use, exploration of a transfer-of-development-rights (TDR) program, and other measures. Other projects with greater uncertainty, higher political risks, and/or significant costs were targeted for additional feasibility studies and/or evaluation through use of a decision-support system (see Case Study Example #2).

Common Ground through Informed Decisions: Case Study Examples

EXAMPLE #1

One major point of controversy along the upper San Pedro River in the past revolved around the need for all water users in the basin who are part of the water problem to also be part of the solution. To reach equity among water users, many stakeholders believed that the riparian forest, which transpires considerable amounts of groundwater, needed to reduce its total consumptive water use. However, the concept of cutting down cottonwoods to keep the river flowing was not considered an acceptable management alternative by the U.S. Bureau of Land Management (BLM). Environmental and conservation interests also strongly opposed this concept, since one of the primary drivers for groundwater management in the basin revolves around providing adequate water to sustain the SPRNCA's riparian forest.

The interagency study on groundwater use described previously (Leenhouts et al. 2006) resulted in information that created new management options and choices. Isotopic analysis of the water transpired by mesquite, including those plants that had established on floodplain grassland habitats, revealed that these shrubs were utilizing significant volumes of groundwater, in addition to opportunistically using precipitation. Removal of these recently established scrublands, in locations that historically had been sacaton grasslands, would significantly reduce per area water consumption rates, while also improving habitat conditions for some species and decreasing habitat for others (see chaps. 7 and 8). Vegetation mapping efforts revealed that the mesquite stands within the SPRNCA were far more

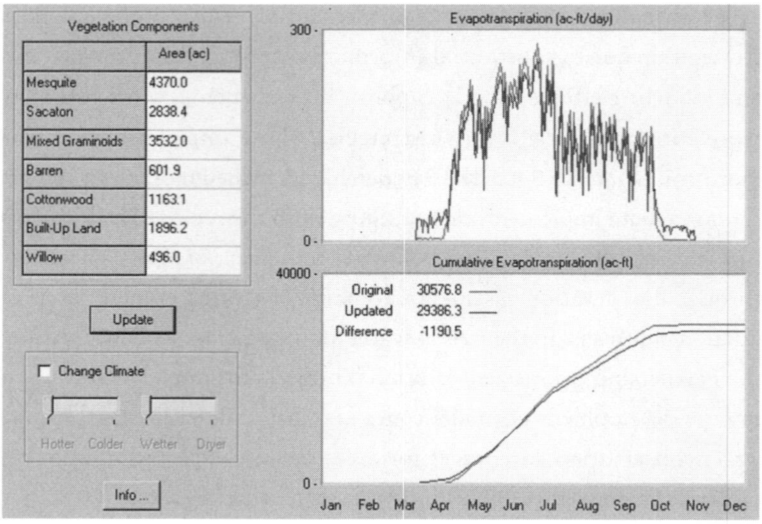

Fig. 21.2. "Screen capture" of a software program showing the change in SPRNCA evapotranspiration as a function of change in various vegetation types defined by the user to simulate potential land management actions (e.g., prescribed burn of mesquite and revegetation with desert grasses).

abundant than cottonwood forests, and that the total consumptive use of groundwater by mesquite was approximately triple that of cottonwood and willow forests (mesquite used between 6448–8135 acre-feet of groundwater per year, while cottonwood/willow forests used a total of approximately 2569 acre-feet per year) (see chap. 3). Moreover, BLM had already developed the capacity to implement a prescribed-fire program.

Another closely associated product of this same study is the Riparian Vegetation Evapotranspiration (ET) Tool (R. Scott, Watts et al. 2003). Although the ET Tool was developed based on extensive field work, large empirical data sets, and complex calculations and analyses, the result is a user-friendly GIS-based interface which can be employed by resource managers with basic training and support. This tool encapsulates many complex biophysical processes in a form that allows decision makers to easily evaluate what is important *to them*—various potential management scenarios and their resulting impacts on habitat conditions and riparian water use (fig. 21.2).

This new knowledge and tool enabled BLM management to prioritize burn plans with the incentive to maximize water savings, while still meeting the agency's ecological goals for the SPRNCA. It also enabled the BLM to contribute their part toward "equitable" groundwater management in the region. The effective use of such tools over the long term requires an ongoing commitment from scientists and programmers for training, support, and future modification as conditions and management issues change.

EXAMPLE #2

A decision support system (DSS), developed by SAHRA and linked to a regional groundwater model, is being used by the Partnership to evaluate different combinations of management options (scenarios), such as the possible relocation of municipal wells, construction of additional recharge facilities, and various water-augmentation strategies. Such a tool allows for the consideration of spatial and temporal groundwater-management concerns, as opposed to a simple annual "bottom line" water-budget approach. The former approach is essential for maintenance of the hydrologic context needed to sustain all 43 miles of the SPRNCA. Questions that can be asked of the DSS include: How will a proposed project, or group of projects collectively, influence groundwater levels at a specific point along the river? Or, where could municipal wells for a particular community be located to minimize the impact on the river?

Most importantly, development of the conceptual model upon which the DSS is based provided those Partnership members engaged in its development with a useful road map for understanding the complex and dynamic interaction of water management variables and functions (H. Richter 2006). A total of 74 individual water management options were initially identified for consideration within the DSS, and each was described in terms of estimated annual yield in acre-feet, cost per acre-foot, and other factors. As of 2006, the San Pedro DSS is now available online for decision-making purposes by Partnership members, but outreach applications have yet to be developed (Serrat-Capdevila and Browning-Aiken in review).

Meeting the Information Needs of Decision Makers

Very few elected officials and policy makers read scientific journals after spending a long day at city hall. Can we blame them? How often do scientists attend city council meetings in the evenings to improve their local political savvy? Given these realities, how then could we best ensure that local decision makers would have access to the best technical information currently available and that scientists would be aware of the complex political and social factors that might influence many aspects of their studies?

Most scientists typically design their studies in response to themes suggested by their sources of research funding. Once tools and knowledge are developed, results are published in peer-reviewed professional journals. The pace at which this information diffuses into the decision-making arena can be painfully slow, or non-existent. Efforts at technology transfer, in which research products or findings are transferred from their originators to users (E. Rogers 1995, Lai and Guynes 1997), represent a more proactive approach for scientists to get their knowledge and tools to decision makers. In this process, scientists identify a potential information user, or vice versa; thus, a scientist–decision maker relationship is established. In this type of knowledge transfer, the scientist develops his or her research protocol after direct communication with potential users or decision makers. In essence, the research and resulting information or products are conceived and designed by both the scientist and the decision maker.

Within the Upper San Pedro Partnership, scientists and decision makers have embraced this type of knowledge transfer. This represents a step beyond the traditional type of knowledge transfer resulting from strictly scientist-conceived research. This approach does not imply that scientists should eschew peer-reviewed journal publications. The journal publication process helps maintain currency of knowledge and is pivotal to the reward system of many research organizations and universities. The complexity of applied research issues typically allows ample opportunity for publishable research. However, those scientists who address applied management issues need to work more efficiently by continuing to publish while also making a commitment to work closely with policy and decision makers.

ESSENTIAL BUILDING BLOCKS OF THE SCIENTIST/DECISION MAKER INTERFACE

Making a long-term commitment to a partnership can be difficult for scientists if their research is largely supported by grants awarded on a typical three-year-or-less funding cycle. Many investigators have conducted research in the San Pedro basin under these circumstances. However, they are not the same scientists that decision and policy makers look toward as a reliable source of information over the long term. Decision makers recognize that these researchers will be providing only temporary assistance with their information needs, and therefore they do not invest the time to build strong working relationships. Several funding agencies have recognized this weakness and are now beginning to fund centers with substantially longer time horizons, such as SAHRA and the Climate Assessment for the Southwest project (CLIMAS; established with seed money from the National Oceanic and Atmospheric Administration).

A commitment by relatively senior scientists is necessary to effectively integrate science with policy. Elected officials and decision makers, who manage large staffs and significant budgets, expect to deal with scientific counterparts who can also make substantial programmatic commitments to their shared endeavors. While it is important that graduate students, postdoctoral researchers, and new researchers contribute to investigations, they are often not in a position to initiate or revise investigations in a relatively rapid manner. Several federal research agencies, and their senior scientists, have had a long history of conducting investigations in the upper San Pedro basin, and this has helped to develop mutual trust and respect between the agencies and local entities.

Senior technical-staff members are often expected to serve as facilitators in collaborative planning processes, in addition to their job responsibilities as scientists. While many technical specialists such as engineers or scientists can provide valuable experience and authority in this role (Wakeman 1997), the need for professional facilitation cannot be underestimated in complex collaborative decision-making processes if participants lack the time, neutrality, training, and/or experience to effectively facilitate a consensus-based process (Leach and Pelkey 2001, Browning-Aiken et al. 2004).

Effective communication is essential to building trust and to building successful watershed partnerships (Leach and Pelkey 2001). Successful communication requires openness, understanding, and listening. Openness and understanding imply patience to allow for the unimpeded sharing of everyone's ideas, information, and data. Scientists and decision makers both have an innate tendency to come to the table with their version of the answer already in mind. However, frequently the best answers in collaborative problem solving are generated from combining the input from multiple participants, or from an entirely new concept that emerges through well-facilitated group discussions (Imperial and Hennessey 2000, National Policy Consensus Center 2002, Pahl-Wostl 2002, Imperial and Kauneckis 2003).

Learning the nuances of the different vernacular spoken by scientists, decision makers, and laypersons takes commitment and patience, and may be no less difficult than committing to learning a foreign language. Scientists must make a conscious commitment not to communicate as if they were attending a scientific meeting. There is no quicker way to lose the attention of policy and decision makers than to speak in technical jargon and acronyms. This requires adopting a common language among partners that is comfortable for all. All partners also can benefit from developing an eye for glazed-over expressions and body language that suggest frustration or confusion.

Listening implies true two-way communication between scientists and partnership members. Sufficient discussion must take place during the design of research programs to ensure that researchers are properly focused on the real-life needs of the decision makers. On the other hand, the policy and decision makers must understand the limits and uncertainties associated with the scientific methods and research results. Because research is typically an iterative process, the principles of adaptive management and ongoing two-way communication between scientists and decision makers are essential.

Location is also an important factor in maintaining productive working relationships between partners. In most cases, it is essential to meet on the home turf of local managers and decision makers, within the basin or watershed of interest. These locations are typically not adjacent to research institutions, universities, or federal agency offices. Therefore, scientists must make an additional commitment to travel frequently and participate in partnership meetings in person, on-site.

Scientists should deliver results to decision makers that contain only the essential information needed to move forward. Many scientists want to study numerous facets of a problem in intricate detail and share all of that detail with others since it is fascinating to them. This tendency must be tempered by the time and resource constraints under which decision makers operate (V. Baker 1998). Data should be interpreted in straightforward terms that are relevant to the decisions at hand. If trust and transparency are developed early in investigative efforts, decision makers will be more confident that a strong foundation of data analyses underpins the bottom line results, allowing the conclusions to withstand close scrutiny by other interests.

The Importance of Community Engagement

Certain water management conservation strategies are more complex, costly, or politically difficult to implement than others. While some strategies, such as water conservation efforts in the home and the recharge of treated effluent, may be widely acceptable to most people, other strategies, such as the transfer of development rights, may require a much deeper level of understanding to attain support from residents. The process of engaging the general public in decision making can be just as important, if not more important, than the engagement of community leaders and decision makers. Elected officials in particular need to know that their constituency will support their decisions.

Recent efforts in California provide an example of the importance of community engagement to the successful implementation of water management efforts. In 1999, a $55-million water reclamation project that Los Angeles officials claimed would drought proof the city was derailed by public outrage over the prospect of drinking recycled toilet water. In contrast, Orange County took a similar plan to the community. The county held neighbor pizza parties, water treatment plant tours, and scores of public meetings where they explained how water treatment processes would work. As a result, Orange County broke ground on a $487 million "toilet to tap" project that opened in 2007 and serves 140,000 families. The difference between these two efforts primarily revolves around differences in community engagement. Support solely by the scientific and political leadership is simply not enough.

In the case of the San Pedro, complex and controversial strategies such as water importation, the transfer of development rights, and surcharges for excessive water use all have the capacity to divide the community, as did "toilet to tap." These particular issues must be carefully managed by direct engagement of the community early on in planning processes. Toward that end, the Partnership conducted a series of community connector meetings during 2004 to begin this type of strategy-specific dialog and to provide citizens with an opportunity to thoughtfully consider issues, help shape their own destiny, and provide meaningful input. A professional firm was hired to assist the Partnership with the process of facilitating public involvement and to ensure effective and equitable participation throughout the Sierra Vista subwatershed. The Partnership held workshops in public meeting rooms, as well as in the living rooms of interested volunteers within Sierra Vista and nearby communities. Residents clearly expressed a need for fairness between conflicting interests, and a need for clear and consistent information, as opposed to conflicting messages. There was a wide divergence between rural and urban perspectives, and between the need for mandatory versus voluntary measures. A distrust of public institutions about doing what they say they will do, especially in terms of the use of funds, was also evident.

Summary and Lessons Learned

The combination of a complex regulatory framework, coupled with a myriad of interacting environmental, social, and political factors, presents tremendous challenges related to groundwater management within the Sierra Vista subwatershed for local decision makers, elected officials, and the general public. The need to apply strong science and the best analytical tools is apparent. But, in addition, the role that collaborative learning processes play in effectively transforming this science into informed decisions will continue to be one of the most essential factors that determines the future of this river, and the fate of the waters that sustain it.

To transfer some of the lessons learned, we conclude with this checklist of questions that can serve as a guidepost for other watershed planning groups. To determine if science and policy are effectively being integrated into water-management planning efforts for any watershed, participants must pose the following questions:

- Is the geographic scale of collaborative planning/learning efforts appropriate? Does the planning area represent a relatively homogeneous area within the watershed, without too many vast differences in political, social, or economic sideboards, or regulatory constraints, to overly confound efforts?
- If multiple water-management and planning efforts are underway within different parts of the same watershed, are they well coordinated with one another?
- If a watershed crosses international boundaries, what opportunities are available for binational dialogue that bridges language barriers, and institutional and policy differences, to address basic management concerns between decision makers, water managers, and scientists?
- Are professional facilitators integrated into planning and decision-

making processes to ensure balanced participation and consensus building?

- Are the federal, state, and local drivers for policy change, and their associated implications, clear to scientists, decision makers, and the public?
- Has an adaptive management framework been adopted by water managers that allows for the immediate implementation of low-risk strategies while additional information is being collected for higher-risk projects, or for those with more uncertainty in outcome?
- Are the research, management tools, and monitoring results developed by scientists actually being used to make decisions by policy makers on an ongoing basis?
- Do face-to-face communications between scientists and decision makers occur regularly?
- Do these communications include openness, understanding, and good listening skills between participants, at on-site locations?
- Do political leaders at the federal and state levels acknowledge and support the efforts of locally based, collaborative decision-making processes? Are they advocates for transferring pertinent aspects of these efforts to other watersheds facing similar challenges?
- Are policy and decision makers eager to continue seeking out funding and resources for scientific support and monitoring efforts as needs continue to change over time?
- Are senior-level scientists easily accessible to local decision makers?
- Is the general public, in addition to elected officials and scientists, actively engaged in planning efforts for future conservation projects and policies, especially issues with a high potential for controversy?

ROBERT GLENNON

The Conflict between Law and Science in the San Pedro River

"If the law supposes that," said Mr. Bumble,
"the law is a ass, a idiot."

Charles Dickens, *Oliver Twist*

TWENTY-TWO

The Disconnect between Groundwater and Surface Water in Arizona Law

The San Pedro River faces an uncertain future in large part because the legal rules governing water use do not conform to the scientific reality (Glennon and Maddock 1994, Glennon 2002). In the American West, the prior appropriation doctrine governs rights to use surface water (Lamb and Lord 1992). The essence of this doctrine—"first-in-time is first-in-right"—rewarded entrepreneurs with the senior, most valuable rights. During times of shortage, senior appropriators receive their full allotments while juniors may go without. This bedrock principle stimulated economic growth and encouraged farmers to construct canals to divert water from rivers to their fields. Although this doctrine encouraged economic development, it ignored the relative economic value of activities, instead treating all activities as identical. It created a perverse incentive to undertake water-intensive but low-value activities because the government did not require the appropriator to pay for the resource. At bottom, the prior appropriation doctrine profoundly changed water from a shared common resource into private property. It had disastrous environmental consequences because it permitted the complete dewatering of rivers and streams. The doctrine arose in the nineteenth century when nature was something to be conquered, and if farmers and miners completely dried up a river, that seemed a reasonable price to pay for economic progress.

In Arizona, different legal rules govern the use of groundwater and surface water. In the nineteenth century, when American courts were struggling with articulating principles to govern groundwater use, the science of hydrology was primitive. Lacking guidance from scientists as to how water moved beneath the surface of the earth, nineteenth-century judges essentially shrugged their shoulders and announced that anyone who could get the water out of the ground could use it. This doctrine, which became known as the "reasonable use" doctrine, allows anyone to pump groundwater in limitless quantities so long as the water is used on the land for a beneficial purpose, a loosely defined concept that found most anything to be beneficial.

The reasonable use doctrine seemed appropriate during the nineteenth century when it was thought that groundwater existed in vast underground lakes and was as ubiquitous as the air that we breathe. American judges were not helped by late-nineteenth-century dabblers in hydrology, such as Clesson S. Kinney, who published *A Treatise on the Law of Irrigation* in 1894. He believed that water beneath the ground moved in known channels that contained "subflow" and in unknown channels. For water in the known channels containing subflow, the prior appropriation doctrine would determine rights to use this water. For underground water flowing in unknown channels, the reasonable use doctrine governed. His essential but flawed hydrologic concept was that *some* groundwater was connected to surface flows while other groundwater percolated, like water through a coffee pot, in unknown channels.

In 1931, the Arizona Supreme Court adopted Mr. Kinney's theory and introduced the concept of "subflow" to Arizona law (*Southwest Cotton Company* 1931). Arizona law has never recovered. Sixty-two years later, the Arizona Supreme Court revisited the question of subflow and, unfortunately, reaffirmed it (*Gila River II* 1993). These rulings created a dichotomy in Arizona law between surface water, which is governed by prior appropriation and includes "subflow," and groundwater, which is governed by the reasonable use doctrine.

Over the course of the twentieth century, hydrology has become a very sophisticated science, grounded in complicated mathematical equations and computer models. We now know that surface water and groundwater are part of a hydrologic cycle (see chap. 15). Groundwater sometimes flows into rivers and streams providing baseflow, and in other situations, surface water may flow into aquifers. The terms surface water and groundwater more accurately describe the location of particular molecules of water rather than indicate different species of water with different chemical compositions. But the law has not kept pace.

The Arizona Supreme Court had another chance to revisit the subflow issue in 2000 (*Gila River IV* 2000). The court recognized that "'subflow' is not a scientific, hydrologic term" but decided not to disturb the precedent. *Gila River IV* is important because the Court defined subflow to include wells that are located within the "saturated floodplain Holocene alluvium" and wells located outside of this area whose cone of depression has extended into the Holocene alluvium. As a result of this decision, wells located anywhere in the immediate vicinity of the San Pedro River are pumping subflow, that is, appropriable surface water and regulated under the jurisdiction of the Gila River General Stream Adjudication. However, *Gila River IV* is unlikely to include wells located in Sierra Vista or its vicinity, some 8 or 10 miles west of the San Pedro River. These wells are intercepting water that is moving subsurface downgradient from the foothills of the Huachuca Mountains toward the river. But for the pumping of the wells, this water would reach the riparian corridor to sustain riparian vegetation and provide baseflow to the river.

Arizona groundwater law epitomizes the tragedy of the commons by encouraging the limitless pumping of a finite resource. Think of an aquifer as a giant milk-shake glass and think of each well as a straw in the glass. What the doctrine of reasonable use permits is a limitless number of straws in a single glass. Excessive groundwater pumping has caused the groundwater table to plummet, in some areas of Arizona by hundreds of feet, creating a sense of urgency that led to the seminal 1980 Arizona Groundwater Management Act. This progressive legislation broke the relentless cycle of overuse and exploitation by placing sharp limits on the ability of people to drill new wells in an "active management area (AMA)" or "irrigation non-expansion area (INA)." In AMAs and INAs, the reasonable use doctrine no longer has the same force. The San Pedro River basin is not within either an AMA or INA, and therefore the reasonable use doctrine continues to govern. Moreover, the state of Arizona leaves entirely unregulated, even in AMAs and INAs, residential wells, so-called "exempt wells." This exemption is based on the theory that it is not necessary to regulate these wells because they pump modest amounts of water. While each well may pump only one or two acre-feet per year, an acre-foot is 325,000 gallons, so each well is pumping more than one-half-million gallons. There are tens of thousands of these wells located around the state, with more wells being drilled all the time. This is an immense loophole in Arizona water law that is having adverse consequences on many riparian areas, including that along the San Pedro River.

The Arizona Supreme Court's decisions in *Gila II* and *Gila IV* expressly invited the legislature to regulate the San Pedro area, an invitation that the Arizona Legislature has thus far declined. Nor has the Arizona Department of Water Resources (ADWR) been assertive in taking steps to protect the San Pedro River from the effects of groundwater pumping.

Neighboring States

Arizona is hardly alone in failing to integrate groundwater and surface water law (Glennon and Maddock 1997, DuMars and Minier 2004). For example, California borrowed an unfortunate concept from the work of Mr. Kinney called "underflow" that embraces the idea that there are underground rivers connected to the surface and therefore regulated by the prior appropriation doctrine. Texas has an even more extreme separation between groundwater and surface water. Under an amusing legal doctrine called "the right of capture," Texas treats groundwater much in the way that the legal system treats wild animals. Groundwater is something that moves in unknowable ways and can be claimed and owned by the first person to pump it, just as a hunter might shoot a deer. In 1997, the Texas legislature enacted a statute that may eventually lead to major changes in how Texas water resources are used, but this timid reform effort largely leaves unchanged the basic rule that permits unrestricted pumping of groundwater (Glennon and Maddock 1997).

Some western states integrate groundwater and surface water by applying the prior appropriation doctrine to both groundwater and surface water (Glennon and Maddock 1997). Under this system, any senior user may receive protection against a junior, and it matters not whether each is using groundwater or surface water. These systems, dubbed conjunctive use systems, are highly touted by some commentators, but if the devil is in the details, so too is it with conjunctive management systems. Consider the state of New Mexico, which protects surface water diverters from harm caused by groundwater pumping by allowing the surface water diverter to switch to groundwater. This satisfies the needs and interests of the senior surface water appropriators but does so at substantial cost to the river itself. The surface water diverter who drills a well and begins to pump groundwater is almost assuredly going to draw water that will reduce the flow in the river. Although this system protects the individual appropriator, the cumulative impact on the river reduces the flow. The state of Washington's law offers the greatest protection for the environment. Not only does Washington completely integrate the priority of both groundwater and surface water users, the legislature has explicitly recognized the hydrologic connection between ground and surface water and has authorized the state's Department of Ecology to establish minimum water levels for streams, rivers, and lakes. These minimum levels become appropriations within the prior appropriation system. But even the state of Washington has problems because it, like every other American state, continues to allow essentially unlimited pumping by domestic users under the category of exempt wells.

The Role of Federal Law

In the absence of effective leadership by the ADWR, the Arizona Legislature, or city and county officials, might federal law play a role in protecting the San Pedro River? In 1988, Congress, under the leadership of Mo Udall, created the San Pedro Riparian National Conservation Area (SPRNCA)(Arizona-Idaho Conservation Act, S. 2840, 16 U.S.C., Public Law No. 100-696). When the federal government, through legislative or executive action, sets aside land from the public domain for specific federal purposes, it impliedly reserves sufficient water for that land to accomplish the purposes of the reservation. Federal reserved water rights carry with them a priority date as of the creation of the set-aside. The U. S. Bureau of Land Management (BLM) manages the SPRNCA and has acted aggressively to protect, conserve, and enhance it.

The federal reserved rights for the SPRNCA are powerful water rights in part because Congress *expressly* reserved water to protect the conservation area. The enabling legislation proclaimed that Congress was setting aside this land to protect the riparian area and expressly reserved water to fulfill the purposes of the conservation area. Therefore, BLM is in a strong position to argue that post-1988 pumping that interferes with these water rights is inconsistent with the federal legislation. BLM has also received some support from the Arizona Supreme Court which, in 1999, held that (1) federal reserved rights extend to groundwater that is not subject to prior appropriation under Arizona law, and (2) federal reserved right holders are entitled to greater protection from groundwater pumping than are water users whose rights are based only on state law (*Gila River III* 1999). As early as 1999, BLM was urged to ask the Special Master in the Gila River General Stream Adjudication to enjoin irrigation pumping that commenced after 1988. However, BLM feared that such a request would undermine BLM's efforts to work with other agencies and landowners to achieve a collaborative solution. As the years have gone on and local officials have done precious little to protect the river, BLM's position is beginning to harden. By the end of 2006, BLM filed a statement of claim in the Gila River General Stream Adjudication court for specific quantities of water to fulfill the federal reserved rights for the conservation area. These rights will include water rights to the stream's baseflow and floodflows to sustain aquatic and riparian habitat, and to springs and seeps that are located within the conservation area.

Recent Federal Legislation and the Need for State Law Reform

A few plant and animal species that live in or along the San Pedro River are protected under the Endangered Species Act. Their habitat could be adversely affected by actions, such as groundwater pumping, that take place in the San Pedro watershed. In 2004, however, Congress passed a rider to the National

Defense Authorization Act that limited Fort Huachuca's liability for actions outside its boundaries. Pushed initially by U.S. Representative Rich Renzi and in an amended fashion by U.S. Senator John McCain, the idea was to recognize that Fort Huachuca had acted aggressively to encourage water conservation but that actions in the civilian sector were undermining the effectiveness of these efforts. Section 321 of the National Defense Authorization Act (H.R. 1588) provides that water consumption in the civilian sector that is not the direct or indirect effect of Fort Huachuca's actions shall not be considered in determining what actions are required by the Department of Defense under the Endangered Species Act.

While the Renzi rider received savage criticism by the environmental community, almost unnoticed was its amended version that requires a report on water use management and conservation. It charges the Secretary of the Interior, in consultation with the Secretary of Agriculture and the Secretary of Defense and in cooperation with the Upper San Pedro Partnership (the Partnership), to prepare a report on water use management and conservation measures that are needed "to restore and maintain the sustainable yield of the regional aquifer" by the year 2011. The Secretary of Interior was directed to submit the initial report by the end of 2004. The purpose of the report is threefold: first, to set forth measurable annual goals for reducing the overdraft; second, to identify specific water use management and conservation measures to facilitate the goals; and third, to identify impediments in current federal, state, and local laws that hinder efforts to mitigate water usage in order to maintain the sustainable yield of the aquifer by the year 2011.

The U.S. Geological Survey (USGS) took the lead in preparing the draft report, but the section that addressed legal impediments was left to the local members of the Partnership in Sierra Vista to write. They chose not to address any of the serious legal impediments. Although they identified two pages of minor changes and slight modifications of state law to allow for conservation easements and the like, there is no mention of the bifurcated nature of Arizona water law. This is the legal impediment that may ultimately doom the San Pedro River to go dry from groundwater pumping.

Recent Developments

Cochise County has moved aggressively to enact ordinances encouraging or mandating water conservation (see Sierra Vista Sub-watershed Water Conservation and Management Policy Plan, adopted by the board of supervisors, March 21, 2006). Cochise County's efforts are at the cutting edge of what any county in Arizona is attempting in order to encourage water conservation. It has adopted a number of innovative requirements for new residential development and placed sharp restrictions on rezonings to higher residential densities. Cochise County's efforts should be applauded, particularly as the Arizona legislature has not given counties authority to deny new subdivisions because they lack adequate water. That said, however, water savings through increased conservation programs, whether voluntary or mandated, are dwarfed by the increased amount of water necessary to serve new subdivisions that have already been approved by Cochise County and Sierra Vista.

In June 2007, the Arizona Legislature passed a bill, signed by the governor, that would permit the creation of the Upper San Pedro Water District (see HB 2300, State of Arizona Senate, 48th Legislature, 2007). This law is only a preliminary step; it does not create the water district, but rather allows the Cochise County Board of Supervisors to place the issue on the ballot. If a water district is formed, it would be able to acquire and sell water or water rights, charge user fees for services provided by the district, and issue revenue bonds. However, it would not be permitted to require water measuring devices for wells, impose mandatory conservation requirements, or regulate water use within the district. Without such regulatory authority, it is unclear how the creation of such a district would reduce the impact of groundwater pumping on the San Pedro River.

In February 2006, an interesting dispute in Kingman, Arizona, brought forth a response by the Arizona Corporation Commission (ACC) that may have bearing on the San Pedro River. There are a number of large-scale developments proposed for Kingman, in northwest Arizona. These developments would, if built, transform Kingman into the third-largest metropolitan area in the state. The Mohave County Board of Supervisors has basically rubberstamped most developments, and the ADWR has little leverage to deny permission to build. Because Kingman lies outside an Active Management Area, the only authority of the ADWR is to determine whether the development has an "adequate" supply of water. But even if ADWR makes such a finding, the development can go forward so long as the sales information brochures marketed to prospective buyers note that there is a problem with the adequate supply of water. Given this vacuum, members of ACC (the public utility commission for the state of Arizona) have suggested that perhaps the ACC will deny the water company's operating certificate without proof that the development would have an adequate supply. This is a welcome development that may provide a way to halt new development in an area where there truly is lacking a supply of water. But its relevance for the San Pedro is unclear because there is plenty of groundwater in Cochise County. Pumping this water will harm the San Pedro but not exhaust the groundwater supply.

Members of the Partnership and officials in Sierra Vista and Cochise County are beginning to realize that the federal requirement to reduce pumping to sustainable yield by 2011 is fast approaching. As a consequence, there have been proposals to import water from outside the San Pedro basin, but the options are few. One is the Central Arizona Project (CAP), which transports Colorado River water from the western part of the state over to Phoenix and then south to Tucson. The Partnership asked the U.S. Bureau of Reclamation to look into the prospects of extending the CAP canal to Sierra Vista as a potential additional supply of water. The Bureau released its final draft of its feasibility study in June 2007 (USBOR 2007). The study concluded that it would cost between $160 and $400 million to construct a pipeline from the CAP canal's terminus near Tucson, and this price tag does not include $18–37 million in annual operating costs.

It is unrealistic that the CAP will be extended, as the Partnership has offered no way of coming up with this tremendous amount of money. Even if the federal government could be persuaded to use CAP water to augment directly the supply of the San Pedro River, Colorado River water has a different water chemistry, with high salinity, and would be unacceptable from an environmental perspective. Other CAP water users in Arizona will be opposed to using the state's already stressed Colorado River allocation in this fashion, and it is unclear how the Partnership would secure water rights to CAP water. Another short-term augmentation alternative recommended by the USGS for a feasibility study is to recover and treat water from Bisbee's Copper Queen Mine, then convey it to the SPRNCA for recharge. The USGS rejected proposals to deliver it for municipal use in Sierra Vista, Bisbee, or Naco. The initial reaction of those in the regulatory and environmental communities was great suspicion about the quality of water that has been sitting in copper mines.

Conclusion

All water use is subject to reasonable regulations by the state. Arizona has set limits in other critical areas of the state, but pumping in the San Pedro area remains unregulated. State law should condition new residential development on the demonstration of a long-term renewable water supply that does not harm the river. And local government should not approve rezonings unless the developer demonstrates that there is an adequate supply for the development and that the new water use will not harm sensitive habitat, such as the San Pedro River. Until state law is reformed, no amount of tinkering with the deck chairs on the Titanic will save the San Pedro River.

BARBARA TELLMAN

Conservation Activities in the Sonoran Part of the Upper San Pedro Watershed

TWENTY-THREE

Introduction

The Sonoran part of the San Pedro basin includes areas of high ecological value but presents a sharp contrast to the other basins in being less well studied scientifically. The watershed here has been used primarily for ranching and some small-scale farming for about 300 years. Copper mining occurs at the fringe of the watershed in Cananea, which sits at a high point at the junction of the Rio Sonora and San Pedro watersheds. Although this remote portion of Sonora has received little attention from historians, hydrologists, or biologists, in recent years interest in the area has increased.

The Watershed

The San Pedro River originates in northeastern Sonora, with a major infusion of water from the Huachuca Mountains in southeastern Arizona. Water flows from the mountains primarily through Guadalupe Canyon and crosses the border to enter the Rio Nutrioso (Rio Los Fresnos). Numerous small tributaries and springs, including sources in the Cananea region and surrounding mountains, contribute to the flow as it moves east to meet the burgeoning Río San Pedro just a few miles south of the border. From this point the river flows northward into Arizona.

The region was noted as prime grassland by the first American explorers and travelers in the area, and high-quality grasslands persist today. Wildlife corridors remain intact, for the most part, between the Sierra Madre Mountains to the south and the Sky Islands of northern Sonora and southern Arizona and New Mexico (Anon. 2004). Although the area has not been fully hydrologically or ecologically mapped, an ongoing inventory has located nine ciénegas. A cottonwood gallery forest occurs along much of the river in Sonora, forming part of one of the major Neotropical bird flyways in western North America (Anderson 2005).

A streamflow gage has been in place for more than 100 years in the Arizona section of the river, but flow measurements in the Sonoran section are spotty at best (Anon. 2004). There are no long-term measurements of groundwater levels. Preliminary studies (unpublished) indicate that the alluvium is narrow and shallow through most of the Guadalupe-Nutrioso portion of the river, so there is little storage capacity (F. Gray, personal communication). A farming area (where forage crops are grown) uses water from a small dammed section of the river, but ranching is the primary land use in the area all the way to the international border. Little solid information is available about the predevelopment water conditions. Based on works by Esparza (2002) and Kepner et al. (2002), it is estimated that agricultural groundwater use in the area increased from 3.5^{3}hm/year in 1980 to more than 10^{3}hm/year during 1999–2002 (Pool and Dickinson 2007).

Historical Background

The area has been heavily grazed since the first major Spanish introduction of cattle in the eighteenth century. A Spanish presidio existed on the south side of the Nutrioso for about 40 years, protecting mining interests in the Cananea region to the south. The site of the presidio today can barely be discerned on a private ranch.

Col. William Greene, the major proprietor of the Cananea mines at the turn of the twentieth century, purchased thousands of acres of ranch land in Sonora and in the United States north of the border, including most of the old Mexican land grants in the area (Sonnichsen 1974). The ranches were a dominant land use for more than half a century until nationalized and turned over to communal use under Mexico's ejido system (a system whereby under government sanction the ranch or farm is owned jointly by many individuals). The newly formed ejidos followed the boundary lines of the old Greene ranches (Longwell 1974). Because of economic pressures under the ejido system, overgrazing was common, with resulting damage to the watershed. Fencing to limit livestock use of the riparian areas was rare.

In the 1990s, changes in Mexican law led to the privatization of many of the ejidos.

Conservation Programs

CONSERVATION OF RANCHO LOS FRESNOS

A few private ranches have existed in the vicinity, including the Rancho Los Fresnos in the Nutrioso basin owned by the Elias family for more than 100 years. The ranch itself has five ciénegas, 15 km of cottonwood/willow forest, and one of the largest contiguous and highest-quality grasslands in Mexico. The family has over the generations taken great care to preserve the ranch's environmental quality, and when it became necessary to sell, they chose a buyer that would maintain that tradition. In 2005, the Arizona Nature Conservancy (TNC) and its Mexican partners, Biodiversidad y Desarrollo Armónico, A.C. (BIDA) and Naturalia, acquired the ranch for conservation purposes. BIDA is an Hermosillo-based nongovernmental organization founded in 2004 to promote the conservation and management of Sonora's and Mexico's natural resources. Naturalia is a non-profit conservation organization formed in 1990 to protect Mexico's biodiversity. Naturalia will hold title to Los Fresnos and work with TNC and the Mexican Park Service to preserve a wildlife corridor between a jaguar sanctuary in the northern Sierra Madres and Rancho Los Fresnos (Anderson 2005).

Conservation strategies include land acquisition and management agreements, conservation easements, and other legally binding voluntary agreements that place conservation restrictions on lands. Lands will be managed to promote conservation and improve range conditions, and the ranch will serve as a demonstration site for landowners in the basin. Riparian restoration includes fencing to exclude cattle from wetlands. Erosion control structures are being built to reduce wetland and grassland degradation.

Grassland restoration and management programs include a ranching network to share practices, integrated fire management, erosion control, native grass restoration, and establishment of a native grass seed bank. Increased research and monitoring are integral parts of the project. Long-term funding will be assured though establishment of a conservation trust fund and application to Mexican and U.S. agencies as well as private foundations for assistance. Plans are to create a private reserve under Mexican law and have the springs, ciénegas, and river courses designated as critical aquatic habitat for endangered species under Mexican law. The efforts are intended to improve habitat for many species, including raptors such as the northern harrier and American bald eagle; numerous wading birds and waterfowl, including the mallard (Mexican variety), American coot, and American widgeon; and riparian forest species such as vermillion flycatcher, as well as native fish, frogs, and salamanders (Anderson 2005).

CONSERVATION AT RANCHO LAS PALMITAS

The Nature Conservancy and its partners work with other ranchers in the area to preserve and restore grasslands and riparian areas. The Miranda Quihui family, for example, is an active partner in preservation efforts of Rancho las Palmitas, in Ejido Jose Maria Morelos north of Cananea, Sonora. Restoration plans include fencing of the 1-ha ciénega and development of an alternate water source for cattle. Materials will be provided in part by Instituto de Medio Ambiente y el Desarrollo Sustentable de Estado de Sonora (IMADES). These efforts are intended to improve conditions for migratory bird species as well as for resident species such as the endangered Huachuca water umbel, Chiricahua leopard frog, and black-tailed prairie dog (Anon. 2004, Anderson 2005).

JULIET C. STROMBERG
AND BARBARA TELLMAN

Summary

WEAVING THE THREADS TOGETHER

One of our goals for this edited volume was to shed light on the processes that shape the San Pedro River's ecosystems, by gathering contributions from a variety of scientific disciplines; another was to explore how these scientific findings are being used to guide river and riparian conservation efforts. A related goal was to convey an understanding of historical riverine changes as a basis for providing guidance to those who are shaping the river's future. In this summary, we present some insights we have gained from reading the contributions, with respect to scientific findings, management implications, and policy considerations, and weave together common findings. We also highlight existing areas of scientific uncertainty and identify some constraints experienced by stakeholders in their quest for achieving desired riverine conditions.

Scientific Contributions and Management Implications

ECOSYSTEM CHANGE AND RESILIENCE

As evidenced by preceding chapters, the San Pedro riparian ecosystem has undergone extensive change through time. Major changes occurred during the late-nineteenth/early-twentieth century period of river entrenchment, as the river adjusted its form to changing inputs of water pulses and sediment from the watershed, triggered by the proximate cause of extreme flooding (see chap. 12). The organic-rich hydric soils that once prevailed along the San Pedro corridor prior to river entrenchment gave way to younger, mineral soils, with organic soils now much more spatially restricted (see chap. 14), and water tables declined. This radical restructuring of the stream's geomorphology and hydrology has had long-lasting effects on the biota. There have been major shifts in vegetation, with types associated with wet soils, moderate flooding, frequent fire, and/or fine-textured soils (such as ciénega wetlands and grasslands) declining in spatial extent and being replaced by types associated with drier conditions, greater flood intensity, less burning, and/or coarser soils (such as cottonwood/willow and mesquite forests and xeroriparian shrublands; see chap. 1). Within the riparian flora, some plant species have been extirpated, while many others have newly arrived—some are more prevalent than in the past, while others have declined (see chap. 5).

Changes also have occurred within the animal communities. Among fishes, species, including Colorado pikeminnow, were extirpated by the mid-1800s, while many new ones arrived in the 1900s (see chap. 10). Some wetland- and grassland-affiliated herpetofauna have declined, while other more opportunistic species have increased (see chap. 9). Long-term changes in the bird community are less well documented in our volume, although the present abundance of the riparian gallery forests clearly contribute to the diverse and abundant avian life (see chap. 8). There are known changes in the mammal community, with several large predators, including grizzly bear, jaguar, and grey wolf, no longer occurring along the river (see chap. 6). The historic keystone species (beaver) has been largely superceded by a new one (people), although beaver activity is once again on the rise. Some of these abiotic and biotic changes reflect human actions, including agriculture, hunting, mining, and species introductions, that occurred within the riparian corridor or throughout the watershed. Others reflect changing climates or a complex interplay of climate and human actions, with the precise roles of each still debated (see chap. 13). The changes are a product of human activities taken both directly within the riparian corridor and throughout the watershed.

A key finding that emerges from these collective reports is the resilience of desert rivers such as the San Pedro. Despite (and perhaps because of) changes in the particular species composition, a diverse and productive biotic community has persisted through major changes in stream geomorphology and hydrology. Dryland riparian ecosystems such as the San Pedro are resilient in part because they are structured primarily by abiotic processes (such as flows of water and sediment) rather than by biotic processes (such as competition), and in part because they are open systems, with high connectivity to the watershed (Holling 1973). Further, because stream flows fluctuate with dry and wet cycles, they harbor a biota that is adapted to disturbance and change. A bottom-line message for river management is that riparian biotic communities can persist through many types of environmental change, assuming that basic inputs of resources are not reduced below critical threshold levels.

ENVIRONMENTAL FLOWS

A key management concern on the San Pedro River today focuses on sustaining or restoring water resources and ensuring that basic inputs of water flow to the river as well as to the people who live near it. Localized water diversions and more widely dispersed groundwater pumping for farming, ore extraction, and urban use have influenced local water tables and surface flows, and today some once-perennial areas are intermittent or ephemeral, and water tables have dropped in places. In response, vegetation types characterized by more deeply rooted or more drought-tolerant vegetation types such as saltcedar shrublands have expanded in some parts of the river in the past several decades.

The riparian environmental flow studies that have been conducted on the San Pedro in response to concerns over stream water depletion can aid managers in their quest to attain or restore desired conditions. Hydrologic thresholds, with respect to minimum flows of surface water and groundwater, have been identified that regulate shifts from one riparian vegetation type to another (see chap. 20); these thresholds can serve as management targets. Valuable ecohydrological information also has been generated on rates of groundwater use by San Pedro riparian vegetation types, and on variations in these rates through time and space (see chap. 2). These studies provide critical input to the basin-wide water budgets that are fundamental to making decisions on sustainable water management practices.

Progress also has been made in understanding environmental flow needs for riparian animals, with some surprises. Avian studies show that linkages between streamflows and nesting birds are mediated in large part by linkages between stream hydrology and riparian vegetation structure (see chap. 8). Herpetological investigations reveal the importance of hydroperiod, with some frog taxa thriving along rivers with intermittent water flows and other more-recent arrivals such as bullfrogs thriving with more permanent water sources (see chap. 9). Arboreal riparian invertebrates appear to be more abundant at intermittent-flow sites than at perennial-flow sites and show a greater seasonal (monsoonal) pulse at the drier sites (see chap. 7). Dynamic effects of stream low-flows and high-flows on riparian taxa, interactions among trophic levels, and changes in biota along the hydro-climatic gradient from the upper to lower basin all are areas in need of further study; this information, collectively, can provide a more comprehensive model of biotic response to changing streamflow regimes.

SCIENTIFIC UNCERTAINTY

The many scientists who have been studying the San Pedro over the past decades have contributed to a deeper understanding of the processes that shape the patterns of the river's ecosystems, and this scientific input is being used to guide management decisions. But, despite the extensive research, considerable uncertainty remains in some areas regarding linkages between ecosystem components and causes of ecosystem change: to wit, uncertainty persists over the role of human actions in historic river entrenchment, an issue that may never be resolved (see chap. 13). Of more immediate concern, uncertainty also remains over the precise contributions of various factors to recent baseflow declines in the upper San Pedro (see chaps. 15 and 20); teasing out the effects of the many contributing factors will require yet more research, which is underway. Another area of hydrologic uncertainty is the timescales for ecosystem response to retirement of groundwater pumping, of interest to stakeholders over the entire river, including those in the lower basin where such efforts have been undertaken to restore habitat for endangered riparian bird species. Uncertainty also persists over the role of future climate change in exacerbating or ameliorating hydrologic conditions (see chap. 3).

HISTORIC LEGACIES

Another important finding that emerges from the chapters in this book is the legacies created by past events. Legacies of past land or water use affect many dryland ecosystems (Lake and Bond 2007) and can play out on timescales of decades to centuries. Knowledge of these legacies is critical for determining whether management intervention will be effective and sufficient or, in other cases, even necessary.

Hydrologic studies within the San Pedro basin elucidate the strong connection that the river has to its watershed and have increased understanding of the processes that link precipitation in the mountains to discharge of baseflow to the stream (see chap. 15). These water flow paths have been altered along some parts of the river by long-term groundwater pumping, and the cones of depression that have developed in the regional aquifer now intercept groundwater that might otherwise flow to the stream. These groundwater deficits generated during the mid- and late 1900s will shape groundwater flow paths and streamflows for decades to come. Despite the best efforts by stakeholders to implement changes to achieve desired conditions, the cones of depression represent historical debts to be paid; the groundwater changes may be irreversible on human timescales.

Other legacies have arisen in response to historic river entrenchment. Of note, this event set in motion a trajectory in the riparian forests that may influence future streamflows: after an initial expansion period, cottonwood/willow forests (major groundwater users) are expected to decline in the future and to give way to mesquite forests and/or sacaton grasslands, vegetation types that have lower consumptive water use (see chaps. 2 and 3). As these ongoing forest successional processes drive changes in riparian vegetation, and thus in evapotranspiration rates and plant water sources, stream water levels will change in turn.

Yet another legacy arises from long-term changes in land cover and soils in the watershed. Watershed land cover is strongly linked to stream hydrology and geomorphology (Graf 2001): each land cover type differentially influences surface runoff and infiltration processes, thereby determining whether water returns directly to the atmosphere, is transpired locally, or arrives at the river as a flood pulse or as groundwater. Historic topsoil loss, resulting from the historic confluence of climate extremes (drought and floods) and intensive land use (unregulated livestock grazing), led to irreversible state shifts in some parts of the watershed from grassland to scrubland; despite the many efforts to build check dams and halt tributary erosion, these landscape-scale changes will influence surface runoff and sediment erosion patterns for years to come (see chap. 11). The extensive mining of copper and other ores that has occurred over the length of the San Pedro River also will have long-lasting effects, by influencing stream water quality and aquatic biota (see chap. 17).

YET MORE DATA GAPS

In response to a changing economic climate, land use patterns are shifting along the San Pedro River, with lands once used for irrigated agriculture and/or livestock grazing now being used for ecosystem conservation purposes or for urban development. Land use also is changing in the watershed, with urban lands expanding and nitrogen-fixing mesquite trees increasing in some areas. Riparian ecosystem response to some of these changes needs further research. Of note, riparian ecosystem response to livestock removal has been poorly documented in the American West (Sarr 2002). While the studies that describe changes in vegetation and birds following removal of cattle from the San Pedro Riparian National Conservation Area (SPRNCA) are a key contribution (see chap. 8), there remains a need for longer-term, multi-disciplinary monitoring of linked changes in channel form, streamflows, soils, vegetation, and animals in response to livestock removal at various places along the river. Similarly, further study is needed on the many fallow fields that line the San Pedro River both to understand natural patterns of change and to inform ongoing restoration efforts (see chaps. 4 and 14).

From Science to Action: Some Legal, Economic and Policy Considerations

ECONOMIC ISSUES

Science is conducted within the larger framework of the human environment, including the political and economic conditions in the region. Policy decisions are only partially based on scientific information: although built on a scientific infrastructure, ecosystem management ultimately is driven by human values (N. L. Christensen et al. 1996). Because there are differences in value systems amongst stakeholders in a watershed, conflicts can arise over recommended management actions. How much public money should be spent on river preservation/restoration? What are the economic trade-offs between revenue from increased subdivision development and the tourism revenue from the attractions of the area if it is preserved as public land (Colby and Orr 2005)? Should the army base expand if increased water use may restrict water availability for the river? These and many other questions arise in public debate.

While mining and grazing were the most significant sources of upper basin change in the nineteenth century, in the past half-century, urbanization in the Sierra Vista area, accompanied by increased water demand, has predominated. Economic growth has become directly related to population growth, leading to controversy among advocates of economic growth, advocates of river preservation, and advocates of solutions to accommodate both. A variety of land management responsibilities complicates the picture, including those of private landowners and state and federal agencies with specific land management mandates (see chap. 11).

LEGAL ISSUES

Many laws influence the San Pedro River ecosystem, notably those related to water and to endangered species. As noted in chapter 22, policy decisions in the San Pedro basin are restricted by Arizona water law, because it does not direct or even allow groundwater and surface water to be managed conjunctively. The Groundwater Management Act is designed to protect groundwater supplies, sometimes at the expense of surface water for streams. Further, while it regulates groundwater pumping in the most populated areas of the state, it has little impact in the San Pedro basin. Litigation over Native American water rights also affects water use in the San Pedro basin, as do federal laws and court decisions on federal water rights. The legal framework is ever-changing, though, with laws, rules, and regulations modified in response to changing societal desires.

PARTNERSHIPS

Finding solutions to watershed problems requires comprehensive planning and participation and commitment of many stakeholders. A partnership has formed along the upper river that includes many viewpoints and representatives of government agencies involved in relevant decision making (see chap. 21). This Upper San Pedro Partnership (the Partnership) began as a local initiative and eventually was incorporated into federal law; they have successfully lobbied for funding of the hydrological and ecological studies that form the basis for decision making and have recommended policies on water conservation, reuse, and recharge. Members of the Partnership speak to the importance of establishing long-term relationships between senior scientists and policy makers, and of extensive community involvement, both to generate the needed scientific information and to use it effectively.

The Sonoran part of the watershed (see chap. 23) and the lower basin (see chap. 19) have different needs and solutions. Urban centers in both areas remain small (although they are increasing), and agriculture and ranching remain as important land uses. In both areas, non-profit organizations have been at the forefront of river management by purchasing land where ecological and hydrological conditions are conducive to preservation or restoration. Retirement of agricultural land to reduce water use, and fencing areas from cattle, have been major conservation efforts in these areas, as well as in the upper basin.

The San Pedro as a Reference River: Implications for Regional River Management

FLOOD PULSES

The scientific studies conducted along the San Pedro are being used not only to inform local conservation efforts but also to inform conservation efforts on other desert rivers that are undergoing rapid urbanization and water development, such as the Verde River in central Arizona. As one of the last un-dammed desert rivers in the American Southwest, the San Pedro also can serve as a reference river for the many regional drainages that are undergoing ecosystem restoration.

Ecosystem restoration can be defined as actions that result in the reestablishment of ecological processes, functions, and biotic/abiotic linkages and lead to a persistent, resilient system integrated with its landscape. For rivers, one of these key processes is flood pulses (see chap. 16). Floods are a fundamental force shaping ecosystem structure and function from biotic diversity (see "Scientific Contributions and Management Implications," p. 419, and "From Science to Action: Some Legal, Economic, and Policy Considerations," p. 424) to biogeochemical cycling (see chaps. 17 and 18), and the linkages described between San Pedro floods and various ecosystem parameters have implications for restoration of other rivers. For example, the relationships described between long-term flood patterns and establishment frequencies of cottonwood trees on the San Pedro River provide a model for managers who would like to see these pioneer trees regenerate on regulated rivers where flood regimes have been altered. The connections described between watershed nutrient pools, watershed runoff, and nutrient content of floodwater, and between monsoon floods and diversity and productivity of plants and terrestrial arthropods, add to the database relating river flooding to riparian ecosystem productivity and provide useful models for other rivers.

INTRODUCED SPECIES

Among certain taxa, including fish, herpetofauna, and plants, recent arrivals to the region (i.e., introduced species) are a major component of the San Pedro's biota. There are different schools of thought about the ecological value and roles of introduced species, and the range of viewpoints is reflected in the preceding chapters. One view is that introduced species are troublesome invaders that diminish native biodiversity and thereby degrade riparian and aquatic ecosystems. Another view is that introduced species add to overall diversity, sometimes fostering ecosystem resilience by carrying out functions no longer provided by other species under the prevailing environmental conditions.

The San Pedro River has served as a valuable laboratory for understanding ecological roles of introduced species, with saltcedar as a case in point. Where many environmental factors are changing in tandem, with complex synergies, it can be difficult to distinguish cause and effect and disentangle the direct effects of an introduced species from the indirect effects that led to its abundance (Gurevitch and Padilla 2004). The lack of major dams on the San Pedro has allowed for some level of this untangling. The finding that plant diversity remains high and soils remain non-saline in the San Pedro saltcedar patches has implications for approaches to introduced species management on flow-regulated rivers and points to manipulation of physical processes such as flooding as a fruitful approach to river restoration (see chap. 5).

The San Pedro studies also have revealed the important role of hydrology and land use in driving the vegetation changes that can lead to changes in bird habitat: replacement of saltcedar shrublands by cottonwood forest, with their more diverse bird life, hinges in large part on shifts in stream hydrology to perennial streamflows, in concert with periodic winter flood pulses (see chap. 8). This finding has implications for river restoration and similarly supports an approach that focuses on restoration of the physical template that shapes the biota.

When environmental conditions change, biotic interactions also can change, leading to unexpected outcomes. For example, beaver have been released into the upper San Pedro River, a historic part of their range. There is uncertainty, however, over which animal species will be favored by this action. Given the present environmental context, species once favored by ponds (such as certain native fish and frogs) may undergo declines as beaver activity increases, while some introduced species may undergo increases (see chaps. 9 and 10); some people favor the changes brought by the beaver, while others do not. Scientific study will continue to shed light on physical and biological changes occurring along the river and either confirm expectations or reveal unexpected twists. Such study, in turn, can guide local and regional river restoration efforts.

HETEROGENEITY

Another key lesson for regional restoration and conservation efforts that emerges from the San Pedro studies is the importance of habitat heterogeneity. Each of the different hydrogeomorphic zones along the river supports a different, but typically overlapping, suite of species. The low-flow channels of the perennial San Pedro River are lined by marsh plants and open channel bars that sustain animals such as white-footed mouse and killdeer. Off-channel ciénega wetlands sustain a different suite of plants and animals, including common yellowthroat and Mexican gartersnake. In the river floodplain, the cottonwood/willow forests provide habitat for animals, including raccoon, viceroy butterfly, and yellow warbler, and form mutualisms with hydric-affiliated fungi; the saltcedar shrublands sustain species, including southern grasshopper mouse and yellow-breasted chat, and affiliate with a different form of fungi. The biota on the river terraces remain connected to the floodplain aquifer via deep-rooted plants such as mesquite and sacaton. The mesquite woodlands provide habitat for animals including Bailey's pocket mouse, Lucy's warbler, and the bee *Dufourea boharti*; western harvest mouse, southern dogface butterfly, Botteri's sparrow, and others find homes in the sacaton grasslands. The desert upland habitats support their own set of species. In addition to this lateral zonation, species composition also changes along upstream-downstream gradients, highlighting the importance of a river-wide perspective to riparian conservation. Overall, the high diversity of species accrues from several factors: the location of the basin in a biotically and topographically diverse region, the existence of a spatially complex and temporally dynamic, flood-shaped riparian landscape, and abundant water resources provided by stream baseflows, shallow groundwater, and growing-season rains and floods.

Closing Thoughts: Envisioning and Monitoring the Future

Visualization exercises, informed by software such as decision support systems, are fruitful approaches for watershed management groups. Efforts to envision the future under various water and land use scenarios are underway for the San Pedro River and are of value in providing linkages between scientific output and stakeholder preferences. Such efforts hinge on models that relate stream hydrology to riparian ecology and to economic outcomes. Of course, models, whether conceptual or quantitative, carry their own set of caveats and depend on high-quality data input and understanding of processes; models of the San Pedro ecosystem continue to be refined by ongoing scientific inquiry.

Equally important to modeling future conditions is monitoring the present. The importance of ecosystem monitoring for understanding change and providing feedback on management actions is well recognized, although monitoring remains poorly funded on many rivers (Bernhardt et al. 2005). The coupled monitoring of hydrology and biota initiated in the San Pedro upper basin will yield key information on the status of the ecosystem and its response to ongoing climate change, forest succession, and geomorphic legacies, and to the many human efforts being undertaken in the watershed to balance the water budget. The ongoing monitoring of stream geomorphology, hydrology, and vegetation at sites along the lower San Pedro River that have been converted from farms to riparian conservation areas similarly will provide crucial information on rates and patterns of change and will validate or refute current understanding of ecosystem controls. Ongoing commitment to monitoring is essential, so that we may better anticipate our future.

Glossary

Accretion: The process of growth or enlargement by a gradual buildup; in riparian zones, applied to the vertical buildup of sediment.

Adsorption: The adhesion of a thin layer of molecules to the surfaces of solid bodies or liquids with which they are in contact, e.g., the adsorption of nutrients to soil particles.

Advection: The usually horizontal movement of a mass of fluid (gas or liquid), or transport of materials by such movement.

Aggradation: In riparian zones, the deposition of sediment on the channel, floodplain, or other surface in sufficient quantity to increase local elevation.

Allochthonous: Found in a place other than where they and their constituents were formed.

Alluvium: Sediment deposited by flowing water, as in a riverbed or floodplain, or delta, or by sheet flow in uplands.

Alpha diversity (of species): Species diversity within a particular area or habitat, measured at local scales.

Anoxic: Greatly deficient in oxygen.

Anurans: An order (Anura) of amphibians comprising the frogs, toads, and tree frogs.

Aquifer: A water-bearing stratum of permeable rock, sand, or gravel that can transmit significant quantities of water.

Arbuscular mycorrhizae: A type of mycorrhizae in which the fungus penetrates the protoplasm of plant root cells and develops structures called arbuscules.

Aridisol: A soil taxonomic order of dry climates characterized by low water content and low organic matter content and influenced by calcification and salinization; soils often have horizons of lime or gypsum.

Arroyo: A water-carved channel or gully in arid country, usually rather small with steep banks, that is dry most of the time.

Arthropods: Insects, arachnids, and other members of the phylum Arthropoda.

Assimilation: The incorporation or conversion of nutrients into the bodies of organisms.

Avulsion: A sudden change in form and course of stream or other body of water such as in a meander cutoff.

Azonal soils: Soils that are atypical of the surrounding common soils.

Bar (as in channel bar): An accumulation of sediment, which forms at the borders or in the channels of streams.

Baseflow: Portion of a stream's discharge derived from groundwater inflow.

Beta diversity (of species): A measure of the degree of species turnover along environmental gradients, or of the degree of difference in species between adjacent areas.

Bosque: Spanish word and regional name for riparian woodland or forest.

Braided channel: A channel pattern characterized by three or more intertwined channels.

Calcification: Concentration of calcium carbonate in a soil horizon.

Capillary fringe: The subsurface zone directly above the water table in which pores are saturated with water, but the fluid pressure is less than atmospheric. Water is held within pores by capillary forces.

Catchment (aka watershed, hydrologic basin, drainage basin): An area having a common outlet for its surface runoff.

Ciénega: Spanish word and regional name in areas of Hispanic culture for a type of spring-fed or groundwater-fed wetland (see Hendrickson and Minckley 1984).

Cohort: Individuals in a similar age class or group.

Coliform bacteria: Gram-negative, rod-shaped bacteria (such as *Escherichia coli*) present in aquatic environments and intestines of humans, cattle, and other animals.

Colluvium: Loose mass of soil or rock fragments that move downslope, largely by the force of gravity.

Competition: Active demand for a limited resource by two or more individuals.

Compound channel: A type of river channel common in arid regions, whereby a low-flow meandering channel is nested within a larger braided channel.

Cone of depression: A localized lowering of the water table due to groundwater pumping from one or more wells.

Denitrification: The chemical reduction of nitrate nitrogen to nitrogen gas by soil bacteria that occurs under anaerobic conditions.

Diel: A 24-hour period, including a day and the adjoining night.

Dioecious: Having the male and female reproductive organs on separate individuals of the same species.

Doctrine of Prior Appropriation: A doctrine that guides the legal use of water in many states in the western United States; the first person to use water (senior appropriator) acquires the right or priority to its future use against later users (junior appropriators).

Downcutting: Lowering of the streambed through erosion.

Ecotone: A transition area between two adjacent ecological communities or ecosystems.

Ectomycorrhizae: A type of mycorrhizae in which the fungal hyphae abut the cell walls of plant roots but do not penetrate the cell protoplasm.

Edaphic: Of or relating to the soil.

Ejido **system:** Process whereby the government promotes the use of communal land shared by the people of the community; common in Mexico.

El Niño: Significant increase in sea surface temperature over a portion of the Pacific Ocean that occurs at irregular intervals; produces weather patterns including wet winters in southwestern United States.

Endemic species: Species that evolved in, and are restricted to, limited geographic areas.

Endozoochory: Dispersal of seeds or other propagules via internal passage through an animal's digestive tract.

ENSO: Acronym for El Niño–Southern Oscillation; reflects the linkage between El Niño events and the Southern Oscillation.

Entisol: A soil taxonomic order characterized by lack of diagnostic pedogenic horizons and typically found on recent geomorphic surfaces.

Ephemeral stream: A stream drainage that is usually dry and that has surface water only as a result of runoff during precipitation events.

Erosion: Removal of materials from the surface through wind or water action.

Evapotranspiration: The combined moisture loss to the atmosphere from evaporation and transpiration.

Facultative, as in facultative riparian species or facultative phreatophyte: A condition that is not required. A facultative riparian species also may occur in non-riparian habitats; a facultative phreatophyte uses groundwater when available but also uses water from unsaturated soil layers.

Floodplain: Level land periodically submerged by floodwaters, or a plain built up by stream deposition.

Floodplain aquifer: The aquifer comprising the alluvium under and directly adjacent to a watercourse (aka stream aquifer).

Flora: Floristic inventory of a particular area.

Floristic elements: Groups of species with similar geographical ranges.

Fluvaquent: A soil taxonomic great group; an Entisol formed on river-deposited parent material and wet enough to display redoximorphic features.

Fluvial: Related to flowing water.

Fossorial mammal: Burrowing mammal having limbs adapted for digging.

Gaining stream: A stream or reach of a stream whose flow is being increased by inflow of groundwater.

Gamma diversity (of species): Total species diversity within the ecosystems of a region, sometimes referred to as geographic-scale species diversity.

Generalist species: A species with broad food or habitat preferences, or with broad tolerance ranges for environmental variables.

Glycophyte: A plant which cannot tolerate saline (salty) growing conditions.

Groundwater: Subsurface water in pores of geologic material; the top of the saturated zone is the water table.

Groundwater model: A numerical groundwater flow model is the mathematical representation of an aquifer in a computer.

Halophyte: A plant that can tolerate a high concentration of salt in the growing medium.

Herpetofauna: Collective term for the reptiles and amphibians of a given region.

Histosol: A soil taxonomic order characterized by accumulation of large amounts of organic material, i.e., a "peat soil."

Holocene: The most recent epoch of geologic time, extending from about 10,000 years ago to the present; the post-Pleistocene geologic epoch.

Host plant: The plant used for food by a specific herbivore, particularly butterflies. Butterflies typically lay their eggs on the host plant fed on by the caterpillar.

Hydraulic redistribution: The redistribution of water from one soil depth to another that results when the roots of deep-rooted plant species uptake water from one soil depth and subsequently leak it to another depth.

Hydrograph: Graph showing the water level (i.e., stage), discharge, or other property of a river with respect to time.

Hydroperiod: The frequency and duration of inundation or saturation; or the length of time during the year that a substrate is either saturated or covered with water.

Hyphae: One of the threads that make up the mycelium of a fungus.

Hyporheic zone: The region of saturated sediments beneath and beside the active channel.

Intermittent stream: Stream that ceases flowing during dry seasons but that remains connected to a shallow water table or subsurface flow.

Interrupted stream (aka spatially intermittent): Stream with reaches of perennial surface flow separated by reaches with intermittent flow.

Lateral: Of or relating to the side; in riparian ecosystems, often refers to gradients perpendicular to the stream channel.

Leaching: Removal of solutes from the soil by percolating waters.

Lentic: Living in swamps, ponds, lakes, or other standing or slow-moving water.

Longitudinal: Of or relating to length or the lengthwise dimension; in river courses, refers to the upstream-downstream dimension.

Losing (influent) stream: A stream or reach of a stream in which water flows from the streambed into the ground.

Lotic: Of, relating to, or living in moving water.

Meander: Broad, semicircular curves in a stream that develop as the stream erodes the outer bank of a curve and deposits sediment against the inner bank. A meander cutoff is a new, shorter channel across the narrow neck of a meander.

Meristem: A specialized section of plant tissue characterized by cell division and growth.

Morphospecies: Species defined by their anatomical features.

Mycelium: A network of fungal hyphae (filaments) forming the vegetative part of a fungus.

Mycorrhizae: Symbiotic associations between certain types of soil fungi and roots of plants.

Neotropical migratory birds: Birds that spend the winter months in Central and South America and summer months in North America.

Nitrification: The biochemical oxidation of ammonium to nitrate.

Nitrogen fixation: The metabolic assimilation of atmospheric nitrogen into ammonia by soil microorganisms and especially rhizobia.

Obligate, as in obligate riparian species or obligate phreatophyte: A condition that is required. For example, groundwater is essential to the survival of obligate phreatophytes and forms a primary component of their water supply.

Oxbow: A crescent-shaped piece of land occupying the abandoned channel of a stream meander that is isolated from the present channel by a meander cutoff and sedimentation.

Oxic: For soils, having a high redox potential.

Pacific Decadal Oscillation: Long-term (20–30 yr) pattern of Pacific Ocean climatic variability.

Parthenogenetic (aka parthenogenic and agamogenetic): Having unisexual reproduction in which young are produced by unfertilized females; reproduction not involving the fusion of male and female gametes.

Partners in Flight: A Western Hemisphere program to conserve Neotropical migratory birds; also known as Neotropical Migratory Bird Conservation Program.

Patch: Units or regions of the landscape that are relatively internally more-or-less homogeneous.

Patch dynamics: The change in the distribution of habitat patches in a landscape, generated by patterns of disturbance and subsequent patterns of vegetation succession.

Pedon: The smallest volume that can be called a soil. It extends to the limit of plant roots and has a surface area of 1 to 10 m^2 in size.

Perennial stream: A drainage that has permanent surface flow.

Phenology: Periodic biological phenomena (such as plant flowering or bird migration) that are correlated with climatic conditions, or the study of such events.

Phreatic: Of, relating to, or being groundwater.

Phreatophyte: A plant (often deep-rooted) that obtains its water from the water table or the layer of soil just above it.

Physiognomy: External aspect or outward appearance.

Piedmont: A sloping plane formed or lying at the foot of a mountain or mountain range.

Pioneer species: In ecology, the species that initially colonizes a site after a disturbance.

Point bar: Deposits of alluvium on the inside bank of a meander.

Recruitment: In population ecology, the addition of a new individual to a population through birth or immigration.

Redox potential: A measure (in volts) of the affinity of a chemical substance for electrons or electronegativity compared with hydrogen. A high positive redox potential (E_h) indicates an environment that favors oxidation reaction; a negative E_h indicates a reducing environment.

Redoximorphic features: Soil features resulting from prolonged saturation that causes reduction of iron and manganese compounds.

Regeneration niche: With respect to plants, the particular set of requirements for seedling establishment.

Regional aquifer: The aquifer extending out from the floodplain aquifer and hydrologically connected to it.

Relict vegetation: A vegetation type that developed under significantly different environmental conditions from those presently occurring.

Riffle: Fast-flowing, shallow segment of a stream where the surface of the water is broken over rocks or debris.

Riparian ecosystem: Ecosystem located along a river or other freshwater body that is ecotonal between the adjacent aquatic and terrestrial zones.

Riverwash: Deep stratified sand, gravel, cobbles, and stones from various sources.

Ruderal: Plants that are adapted to life in frequently disturbed environments by virtue of a suite of life-history traits including short lifespan and high allocation to reproduction.

Salination (aka salinization): Concentration of soluble salts (sulfates and chlorides) of calcium, magnesium, potassium, and sodium in a soil horizon by evaporation of water.

Seral: Refers to an intermediate stage in ecological succession.

Shifting mosaic: In ecology, a mosaic of patches, each in a different phase of successional development or reflecting different environmental conditions.

The mosaic of patches in a landscape changes through time in response to repeated disturbance events and to succession.

Species diversity: A measure of the number of species in a given area, sometimes expressed as species number weighted by their relative abundance.

Species evenness: The uniformity of abundance between species in a community.

Species richness: The total number of species in a community.

Sporocarp: A structure in or on which spores are produced.

Stream aquifer: The aquifer in alluvium, directly under and adjacent to the watercourse (aka floodplain aquifer).

Stream bank: The sloping land surface that connects a stream channel to its floodplain or adjacent terraces.

Stream power: Measure of energy of flowing water to do work (such as movement of sediment or any other particle in a stream channel). It is affected by discharge and slope.

Stream terrace: A level narrow plain usually with a steep front bordering a river, typically built from river sediments, and lying above the floodplain.

Streamflow regime: Characteristics of the regional streamflow patterns, including timing, magnitude, frequency, duration, and rate of change of flows.

Subbasin: A sub-unit of a water basin; also a legal term for an administrative portion of a water basin.

Succession: In community ecology, the temporal, and somewhat predictable, change in community structure through time following ecosystem disturbance.

Surface water: Water that occurs on the surface in lakes or watercourses.

Surface water right: Legal right to use or convey surface water, independent of the land on which the water originates or is used.

Swales: Elliptical depressions (often in the floodplain) formed by scour of sediment.

Symbiosis: The intimate living together of two dissimilar organisms in a mutually beneficial relationship.

Thalweg: Deepest portion of the stream channel, associated with highest velocities of the channel.

Torrifluvent: A soil taxonomic great group; an Entisol formed on river-deposited parent material with a torric (hot and dry) moisture regime.

Vadose zone (aka unsaturated zone): The subsurface zone above the water table and the capillary fringe in which pores within the geologic matrix are filled with both air and water.

Volant: Flying or capable of flying.

Water table (aka phreatic surface): The upper limit of the soil or underlying rock material that is wholly saturated with water.

Xeroriparian vegetation: Plants that grow in riparian habitats but that require little water; common in ephemeral streams.

Literature Cited

Abarca-Gonzales, F. J. 1989. Potential overlap on food habits between spikedace (*Meda fulgida*) and red shiner (*Cyprinella lutrensis*). Master's thesis, Arizona State Univ.

Adair, E. C. and D. Binkley. 2002. Co-limitation of first year Fremont cottonwood seedlings by nitrogen and water. *Wetlands* 22:425–429.

Adair, E. C., D. Binkley and D. C. Andersen. 2004. Patterns of nitrogen accumulation and cycling in riparian floodplain ecosystems along the Green and Yampa rivers. *Oecologia* 139:108–116.

Adams, D. K. and A. C. Comrie. 1997. The North American monsoon. *Bull. Amer. Meteor. Soc.* 78:2197–2213.

Addy, H. D., M. M. Piercey and R. S. Currah. 2005. Microfungal endophytes in roots. *Can. J. Bot.* 83:1–13.

ADEQ. 1992. *Occurrence of enteric pathogens in surface water of the San Pedro Basin; results from the ADEQ/ADHS surveillance parasite and virus monitoring program, Nov. 1991 through Sept. 1992.* Arizona Dept. of Environmental Quality.

——. 2005. *Status of water quality in Arizona-2004: Arizona's integrated 305(b) assessment and 303(d) listing report.* Arizona Dept. of Environmental Quality.

ADWR. 1991. *Hydrographic survey report for the San Pedro River watershed. Vol. 1. General assessment.* Arizona Dept. of Water Resources. Filed with the court, Nov. 20, 1991.

——. 2005a. *Upper San Pedro Basin Active Management Area review report.* Arizona Dept. of Water Resources.

——. 2005b. *Rural Arizona Watershed Alliance (RAWA) section: Upper San Pedro.* Arizona Dept. of Water Resources.

Agerer, R. 1991. Characterization of ectomycorrhiza. Pages 25–73 in *Techniques for the Study of Mycorrhiza.* J. R. Norris, D. J. Read and A. K. Varma, eds. Academic Press.

Aiken, G. and E. Cotsaris. 1995. Soil and hydrology—their effect on NOM. *J. Amer. Water Works Assoc.* 87:36–45.

Aldridge, B. N. and J. H. Eychaner. 1984. *Floods of Oct. 1977 in southern Arizona and March 1978 in central Arizona.* USGS Water-Supply Paper 2223.

Aldridge, B. N. and T. A. Hales. 1984. *Floods of Nov. 1978 to March 1979 in Arizona and west-central New Mexico.* USGS Water-Supply Paper 2241.

Alerstam, T. and A. Hedenström. 1998. The development of bird migration theory. *J. Avian Biol.* 29:343–369.

Alerstam, T. and Å. Lindström. 1990. Optimal bird migration: The relative importance of time, energy and safety. Pages 331–351 in *Bird Migration: Physiology and Ecophysiology.* E. Gwinner, ed. Springer-Verlag.

Alexander, C. S. and J. C. Prior. 1971. Holocene sedimentation rates in overbank deposits in the Black Bottom of the lower Ohio River, southern Illinois. *Amer. J. Sci.* 270:361–372.

Ali, M. M., G. Dickinson and K. J. Murphy. 2000. Predictors of plant diversity in a hyperarid desert wadi ecosystem. *J. Arid Environ.* 45:215–230.

Allan, J. D. and L. B. Johnson. 1997. Catchment-scale analysis of aquatic ecosystems. *Freshwater Biol.* 37:107–111.

Allen, C. D. 2002. Lots of lightning and plenty of people: An ecological history of fire in the upland Southwest. Pages 143–194 in *Fire, Native Peoples, and the Natural Landscape.* T. R. Vale, ed. Island Press.

Allen, J. A. 1895. On a collection of mammals from Arizona and Mexico, made by Mr. W. W. Price, with field notes by the collectors. *Bull. Amer. Museum of Natural History* 7:193–258.

Allen, M. F. 1987. Re-establishment of mycorrhizas on Mount St. Helens: Migration vectors. *Trans. British Mycology Soc.* 88:413–417.

——. 1991. *The Ecology of Mycorrhizae.* Cambridge University Press.

Allen, M. F., L. E. Hipps and G. L. Wooldridge. 1989. Wind dispersal and subsequent establishment of VA mycorrhizal fungi across a successional arid landscape. *Landscape Ecol.* 2:165–171.

Allison, L. J., C. E. Paradzick, J. W. Rourke and T. D. McCarthey. 2003. A characterization of nesting and non-nesting plots for southwestern willow flycatchers in central Arizona. *Studies in Avian Biol.* 26:81–89.

Allred, T. W. and J. C. Schmidt. 1999. Channel narrowing by vertical accretion along the Green River near Green River, Utah. *Geological Soc. of Amer. Bull.* 111:1757–1772.

American Bird Conservancy. 2003. *The American Bird Conservancy Guide to the 500 Most Important Bird Areas in the United States.* Random House Trade Paperbacks.

American Rivers. 1999. *America's Most Endangered Rivers of 1999.* Washington, D.C.

Ames, C. R. 1977. Wildlife conflicts in riparian management: Grazing. Pages 49–51 in R. R. Johnson and Jones 1977.

Amlin, N. M. and S. B. Rood. 2002. Comparative tolerances of riparian willows and cottonwoods to water-table decline. *Wetlands* 22:338–346.

Andersen, D. C. 1987. Below-ground herbivory in natural communities: A review emphasizing fossorial animals. *Quart. Rev. of Biol.* 62:261–286.

——. 1994. Are cicadas (*Diceroprocta apache*) both a keystone and a critical-link species in lower Colorado River riparian communities? *S.W. Nat.* 39:26–33.

Andersen, D. C. and D. J. Cooper. 2000. Plant-herbivore-hydroperiod interactions: Effects of native mammals on floodplain tree recruitment. *Ecol. Appl.* 10:1384–1399.

Andersen, D. C. and S. M. Nelson. 1999. Rodent use of anthropogenic and "natural" desert riparian habitat, lower Colorado River, Arizona. *Reg. Rivers: Res. and Manag.* 15:377–393.

——. 2003. Effects of river flow regime on cottonwood leaf litter dynamics in semi-arid northwestern Colorado. *S.W. Nat.* 48:188–201.

——. 2006. Flood pattern and weather determine *Populus* litter breakdown and nitrogen dynamics on a cold desert floodplain. *J. Arid Environ.* 64:626–650.

Andersen, D. C., S. M. Nelson and D. Binkley. 2003. Flood flows, leaf breakdown, and plant-available nitrogen on a dryland river floodplain. *Wetlands* 23:180–189.

Andersen, D. C., K. R. Wilson, M. S. Miller and M. Falck. 2000. Movement patterns of riparian small mammals during predictable floodplain inundation. *J. Mammology* 81:1087–1099.

Anderson, B. W., A. Higgins and R. D. Ohmart. 1977. Avian use of saltcedar communities in the lower Colorado River Valley. Pages 128–136 in R. R. Johnson and Jones 1977.

Anderson, B. W. and R. D. Ohmart. 1977. Vegetation structure and bird use in the lower Colorado River valley. Pages 23–34 in R. R. Johnson and Jones 1977.

Anderson, J. L. 1996. Floristic patterns on late Tertiary lacustrine deposits in the Arizona Sonoran Desert. *Madrono* 43:255–272.

Anderson, S. 2005. A sea of grass: Rancho los Fresnos. Unpublished proposal to the Wallace Research Foundation. The Nature Conservancy, Northwest Mexico Program.

Andrade, E. R., Jr. and W. D. Sellers. 1988. El Niño and its effect on precipitation in Arizona and western New Mexico. *J. Climatology* 8:403–410.

Anon. 2004. Application to the Sonoran Joint Venture Award Program. The Nature Conservancy and IMADES. Unpublished.

Antevs, E. 1955. Geologic-climatic dating in the West. *Amer. Antiquity* 20:317–335.

Anyon, R. 1999. *Hopi Traditional Uses of and Cultural Ties to the Coronado Nat. For.* USDA For. Serv., Coronado Nat. For., Tucson.

Apple, L. L. 1985. Riparian habitat restoration and beavers. Pages 489–490 in R. R. Johnson et al. 1985.

Archer, S., D. S. Schimel and E. A. Holland. 1995. Mechanisms of shrubland expansion—land-use, climate or CO2. *Climatic Change* 29:91–99.

Arias, H. M. 2000. International groundwaters: The Upper San Pedro River Basin case. *Nat. Res. J.* 40:199–221.

Arizona State University. 2002. *Lower Colorado River Basin fish database (SONFISHES).*

Arizona Water Commission. 1974. Status report of a study of the adequacy of the water supply of the Fort Huachuca Area, Arizona. In *Report on water supply, Fort Huachuca and vicinity, Arizona.* U.S. Army Corps of Engineers, Los Angeles District.

Arnell, N. W. 2003. Effects of IPCC SRES emissions scenarios on river runoff: A global perspective. *Hydr. and Earth Systems Sci.* 7:619–641.

Arnell, N. W. and C. Liu. 2001. Hydrology and water resources. Pages 191–234 in *Climate Change 2001: Impacts, Adaptation, and Vulnerability*. J. McCarthy, O. Canziana et al., eds. Cambridge University Press.

Arnold, J. G., J. R. Williams, R. Srinivasan, K. W. King and R. H. Griggs. 1994. *SWAT: Soil Water Assessment Tool.* USDA-Agricultural Research Service, Grassland, Soil and Water Research Laboratory, Temple, TX.

Arora, V. K. and G. J. Boer. 2001. Effects of simulated climate change on the hydrology of major river basins. *J. Geophysical Res.—Atmospheres* 106:3335–3348.

Aryal, S. K., R. G. Mein and E. M. O'Loughlin. 2003. The concept of effective length in hillslopes: Assessing the influence of climate and topography on the contributing areas of catchments. *Hydrological Processes* 17:131–151.

ASL. 1995. *Report on feasibility of groundwater recharge and sewage effluent reuse in the Sierra Vista Subwatershed.* ASL Hydrological Services. Phoenix.

Asplund, K. A. 1974. Body size and habitat utilization in whiptail lizards (*Cnemidophorus*). *Copeia* 3:695–703.

Auble, G. T., J. M. Friedman and M. L. Scott. 1994. Relating riparian vegetation to present and future streamflows. *Ecol. Appl.* 4:544–554.

Auble, G. T. and M. L. Scott. 1998. Fluvial disturbance patches and cottonwood recruitment along the upper Missouri River, Montana. *Wetlands* 18:546–556.

Augé, R. M. 2001. Water relations, drought and vesicular-arbuscular mycorrhizal symbiosis. *Mycorrhiza* 11:3–42.

Augé, R. M., J. L. Moore, K. Cho, J. C. Stutz, D. M. Sylvia, A. K. Al Agely and A. M. Saxton. 2003. Relating drought resistance of mycorrhizal *Phaseolus vulgaris* to soil and root colonization by hyphae. *J. Plant Physiology* 160:1147–1156.

Averill-Murray, A., S. Lynn and M. L. Morrison. 1999. Cowbird parasitism of Arizona Bell's vireos (*Vireo bellii arizonae*) in a desert riparian landscape: Implications for cowbird management and riparian restoration. *Studies in Avian Biol.* 18:109–120.

Bagstad, K. J., S. J. Lite and J. C. Stromberg. 2006. Vegetation, soils and hydrogeomorphology of riparian patch types of a dryland river. *Western North Amer. Nat.* 66:23–44.

Bagstad, K. J., J. C. Stromberg and S. J. Lite. 2005. Response of herbaceous riparian plants to rain and flooding on the San Pedro River, Arizona, USA. *Wetlands* 25:210–223.

Bahre, C. J. 1987. Wild hay harvesting in southern Arizona: A casualty of the march of progress. *J. AZ. Hist.* 28:69–78.

———. 1991. *A Legacy of Change: Historic Human Impact on Vegetation in the Arizona Borderlands.* University of Arizona Press.

Bahre, C. J. and C. F. Hutchinson. 1985. The impact of historic fuelwood cutting on the semidesert woodlands of southeastern Arizona. *J. For. Hist.* 29:176–186.

Bahre, C. J. and M. L. Shelton. 1993. Historic vegetation change, mesquite increase, and climate in southeastern Arizona. *J. Biogeog.* 20:489–504.

Baillie, M. N., J. F. Hogan, B. Ekwurzel, A. K. Wahi and C. J. Eastoe. 2007. Quantifying water sources to a semiarid riparian ecosystem, San Pedro River, Arizona using geochemical tracers. *J. Geophys. Res.* 112,G03S02, DOI:10.1029/2006JG000263.

Bailowitz, R. A. and J. P. Brock. 1991. *Butterflies of Southeastern Arizona*. Sonoran Arthropod Studies, Inc. Tucson.

Baird, K. J., M. Ronayne and T. Maddock III. 1997. *Preliminary vegetation and hydrologic analyses for Bingham Cienega*. University of Arizona. Dept. of Hydrology and Water Resources.

Baird, K. J., J. Stromberg and T. Maddock. 2005. Linking riparian dynamics and groundwater: An eco-hydrologic approach to modeling groundwater and riparian vegetation. *Environ. Manag.* 36:551–564.

Baird, S. F. and C. Girard. 1853. Descriptions of new species of fishes collected by Mr. John H. Clark, on the U.S. and Mexican Boundary Survey, under Lt. Col. Jas. D. Graham. *Proc. Acad. of Nat. Sci. of Philadelphia* 4:387–390.

——. 1854. Descriptions of new species of fishes collected in Texas, New Mexico, and Sonora, by Mr. John H. Clark, on the U.S. and Mexican Boundary Survey, and in Texas by Capt. Stewart van Vliet, U.S.A. *Proc. Soc. of Nat. Sci. of Philadelphia* 7:24–29.

Baker, J. M. and C. H. M. van Bavel. 1987. Measurement of mass flow of water in the stems of herbaceous plants. *Plant, Cell and Environ.* 10:777–782.

Baker, M. A., H. M. Valett and C. N. Dahm. 2000. Organic carbon supply and metabolism in a shallow groundwater ecosystem. *Ecol.* 81:3133–3148.

Baker, M. B., Jr., P. Ffolliott, L. DeBano and D. G. Neary, eds. 2004. *Riparian Areas in the Southwestern United States: Hydrology, Ecology, and Management.* Lewis Publishers.

Baker, V. R. 1977. Stream-channel response to floods, with examples from central Texas. *Geol. Soc. of Amer. Bull.* 88:1057–1071.

——. 1998. Hydrological understanding and societal action. *J. Amer. Water Res. Assoc.* 34:819–825.

Baker, W. L. 1990. Species richness of Colorado riparian vegetation. *J. Veg. Sci.* 1:19–124.

Baldocchi, D. D., E. Falge, L. Gu, R. Olson, D. Hollinger, S. Running, P. Anthoni et al. 2001. FLUXNET: A new tool to study the temporal and spatial variability of ecosystem-scale carbon dioxide, water vapor and energy flux densities. *Bull. Amer. Meteor. Soc.* 82:2415–2434.

Baldocchi, D. D., B. B. Hincks and T. P. Meyers. 1988. Measuring biosphere-atmosphere exchanges of biologically related gases with micrometeorological methods. *Ecol.* 69:1331–1340.

Banaszak, J. 1996. Ecological bases of conservation of wild bees. Pages 55–62 in *The Conservation of Bees.* A. Matheson, S. L. Buchmann, C. O'Toole, P. Westrich and I. H. Williams, eds. Academic Press.

Bandelier, A. F. 1892. *Final Report of Investigations among the Indians of the Southwestern United States.* Papers of the Archeological Institute of America. IV. Cambridge University Press.

Barber, W. E. and W. L. Minckley. 1966. Fishes of Aravaipa Creek, Graham and Pinal counties, Arizona. *S.W. Nat.* 11:313–324.

——. 1983. Feeding ecology of a southwestern cyprinid fish, the spikedace, *Meda fulgida* Girard. *S.W. Nat.* 28:33–40.

Barber, W. E., D. C. Williams and W. L. Minckley. 1970. Biology of the Gila spikedace, *Meda fulgida*, in Arizona. *Copeia* 1970:9–18.

Bardgett, R. D., J. M. Anderson, V. Behan-Pelletier, L. Brussaard, D. C. Coleman,

C. Ettema, A. Moldenke, J. P. Schimel and D. H. Wall. 2001. The influence of soil biodiversity on hydrological pathways and the transfer of materials between terrestrial and aquatic ecosystems. *Ecosystems* 4:421–429.

Barret, D. J., T. J. Hatton, J. E. Ash and M. C. Ball. 1995. Evaluation of the heat pulse velocity technique for measurement of sap flow in rainforest and eucalypt forest species of southeastern Australia. *Plant Cell and Environ.* 18:463–469.

Barrow, J. R. 2003. Atypical morphology of dark septate fungal root endophytes of *Bouteloua* in arid southwestern USA rangeland. *Mycorrhiza* 13:239–247.

Barrow, J. R. and R. E. Aaltonen. 2001. Evaluation of the internal colonization of *Atriplex canescens* (Pursh) Nutt. roots by dark septate fungi and the influence of host physiological activity. *Mycorrhiza* 11:199–205.

Barrow, J. R., K. M. Havstad, J. Hubstenberger and B. D. McCaslin. 1997. Seed-borne fungal endophytes on fourwing saltbush, *Atriplex canescens*. *Arid Soil Res. and Rehab.* 11:307–314.

Barrow, J. R. and P. Osuna. 2002. Phosphorus solubilization and uptake by dark septate fungi in fourwing saltbush, *Atriplex canescens* (Pursh) Nutt. *J. Arid Environ.* 51:449–459.

Bartlett, J. R. 1854. *Personal Narrative of Explorations and Incidents in Texas, New Mexico, California, Sonora, and Chihuahua Connected with the United States and Mexican Boundary Commission during the Years 1850, '51, '52, and '53.* Vols. 1–11. D. Appleton and Co.

Bastiaanssen, W. G. M., M. Menenti, R. A. Feddes and A. A. M. Holtslag. 1998. The Surface Energy Balance Algorithm for Land (SEBAL): Part 1 formulation. *J. Hydr.* 212:198–212.

Battin, J. 2004. When good animals love bad habitats: Ecological traps and the conservation of animal populations. *Conserv. Biol.* 18:1482–1491.

Battin, T. J. 2000. Hydrodynamics is a major determinant of streambed biofilm activity: From the sediment to the reach scale. *Limnology and Oceanography* 45:1308–1319.

Baum, C., U. Stetter and F. Makeschin. 2002. Growth response of *Populus trichocarpa* to inoculation by the ectomycorrhizal fungus *Laccaria laccata* in a pot and a field experiment. *For. Ecol. and Manag.* 163:1–8.

Baxter, J. W. and J. Dighton. 2001. Ectomycorrhizal diversity alters growth and nutrient acquisition of gray birch (*Betula populifolia*) seedlings in host-symbiont culture conditions. *New Phytologist* 152:139–149.

Beattie, A. J. and I. Oliver. 1999. Biodiversity buzzwords: Another reply to Goldstein. *Conserv. Biol.* 13:1514.

Beauchamp, V. B. 2004. Effects of flow regulation on a Sonoran riparian ecosystem, Verde River, Arizona. PhD diss., Arizona State Univ.

Beauchamp, V. B., J. C. Stromberg and J. C. Stutz. 2005. Interactions between *Tamarix ramosissima* (saltcedar), *Populus fremontii* (cottonwood) and mycorrhizal fungi: Effects on seedling growth and plant species coexistence. *Plant Soil* 275:221–231.

———. 2006. Arbuscular mycorrhizal fungi associated with *Populus-Salix* stands in a semiarid riparian ecosystem. *New Phytologist* 170:369–380.

———. 2007. Flow regulation has minimal influence on mycorrhizal fungi of a semiarid floodplain ecosystem despite changes in hydrology, soils, and vegetation. *J. Arid Env.* 68:188–205.

Beers, H. P. 1979. *Spanish and Mexican Records of the American Southwest.* University of Arizona Press.

Belnap, J. and O. L. Lange, eds. 2001. *Biological Soil Crusts: Structure, Function, and Management.* Springer-Verlag.

Belsky, A. J., A. Matzke and S. Uselman. 1999. Survey of livestock influences on stream and riparian ecosystems in the western United States. *J. Soil and Water Conserv.* 54:419–431.

Benda, L., D. Miller, P. Bigelow and K. Andras. 2003. Effects of post-wildfire erosion on channel environments, Boise River, Idaho. *For. Ecol. and Manag.* 178:105–119.

Bendix, J. 1997. Flood disturbance and the distribution of riparian species diversity. *Geog. Rev.* 87:468–483.

Bernhardt, E. S., G. E. Likens, D. C. Buso and C. T. Driscoll. 2003. In-stream uptake dampens effects of major forest disturbance on watershed nitrogen export. *Proc. Nat. Acad. of Sci.* 100:10304–10308.

Bernhardt, E. S., M. A. Palmer, J. D. Allan, G. Alexander, K. Barnas, S. Brooks, J. Carr et al. 2005. Synthesizing U.S. river restoration efforts. *Sci.* 308:636–637.

Bessey, C. E. 1904. The number and weight of cottonwood seeds. *Sci.* 20:118–119.

Bestgen, K. R. 1985. Distribution, biology and status of the roundtail chub, *Gila robusta*, in the Gila River basin, New Mexico. Master's thesis, Colorado State Univ.

Bestgen, K. R. and D. L. Propst. 1989. Distribution, status, and notes on the ecology of *Gila robusta* in the Gila River drainage, New Mexico. *S.W. Nat.* 34:402–412.

Betancourt, J. L. 1990. Tucson's Santa Cruz River and the arroyo legacy. PhD diss., University of Arizona.

Bever, J. D., P. A. Schultz, A. Pringle and J. B. Morton. 2001. Arbuscular mycorrhizal fungi: More diverse than meets the eye, and the ecological tale of why. *BioScience* 51:923–931.

Bezzerides, N. and K. R. Bestgen. 2002. *Status review of roundtail chub Gila robusta, flannelmouth sucker Catostomus latipinnis, and bluehead sucker Catostomus discobolus in the Colorado River basin.* Colorado State University Larval Fish Lab. 118:1–139.

Bhatia, N. P., A. Adholeya and A. Sharma. 1998. Biomass production and changes in soil productivity during longterm cultivation of *Prosopis juliflora* (Swartz) DC inoculated with VA mycorrhiza and *Rhizobium* spp. in a semi-arid wasteland. *Biol. and Fertility of Soils* 26:208–214.

Bibles, B. D. 1999. The relationship between productivity and habitat quality in gray hawks. PhD diss., University of Arizona.

Biggs, T. H., R. H. Webb and J. Quade. 1999. Fire frequency and spatial variability of soil biogeochemistry and plant chemistry in a southeastern Arizona desert grassland. Pages 77–89 in *Toward Integrated Research, Land Management, and Ecosystem Protection in the Malpai Borderlands: Conference Proceedings.* G. J. Gottfried, L. G. Eskew, C. G. Curtin and C. B. Edminster, eds. USDA For. Serv., Rocky Mountain Research Station RMRS-P-10.

Birkeland, G. H. 1996. Riparian vegetation and sandbar morphology along the lower Little Colorado River, Arizona. *Phys. Geog.* 17:534–553.

Blackburn, W. H., R. W. Knight and J. L. Schuster. 1982. Saltcedar influence on sedimentation in the Brazos River. *J. Soil and Water Conserv.* 37:298–301.

Blaney, H. F. and W. D. Criddle. 1950. *Determining water requirements in irrigated areas from climatological and irrigation data.* USDA (SCS) TP-96.

Blasius, H. B. 2002. Chemical removal of green sunfish (*Lepomis cyanellus*) from O'Donnell Creek, Arizona. *Proc. Desert Fishes Council* 34:41.

Bock, C. E. and J. H. Bock. 1978. Response of birds, small mammals, and vegetation to burning sacaton grasslands in southeastern Arizona. *J. Range Manag.* 31:296–300.

——. 1979. Relationship of collared peccary to sacaton grassland. *J. Wildllife Manag.* 43:813–816.

Bock, C. E. and Z. F. Jones. 2004. Avian habitat evaluation: Should counting birds count? *Frontiers in Ecol. and the Environ.* 2:403–410.

Bock, C. E., H. M. Smith and J. H. Bock. 1990. The effect of livestock grazing upon abundance of the lizard, *Sceloporus scalaris*, in southeastern Arizona. *J. Herp.* 24:445–446.

Bock, J. H. and C. E. Bock. 1986. Habitat relationships of some native perennial grasses in southeastern Arizona. *Desert Plants* 8:3–14.

Boeer, W. J. and D. J. Schmidly. 1977. Terrestrial mammals of the riparian corridor in Big Bend National Park. Pages 212–217 in R. R. Johnson and Jones 1977.

Boerner, R. E. J., B. G. DeMars and P. N. Leicht. 1996. Spatial patterns of mycorrhizal infectiveness of soils along a successional chronosequence. *Mycorrhiza* 6:79–90.

Bogeat-Triboulot, M. B., F. Bartoli, J. Garbaye, R. Marmeisse and D. Tagu. 2004. Fungal ectomycorrhizal community and drought affect root hydraulic properties and soil adherence to roots of *Pinus panaster* seedlings. *Plant and Soil* 267:213–223.

Boggs, K. and T. Weaver. 1994. Changes in vegetation and nutrient pools during riparian succession. *Wetlands* 14:98–109.

Bohrer, K. E., C. R. Friese and J. P. Amon. 2004. Seasonal dynamics of arbuscular mycorrhizal fungi in differing wetland habitats. *Mycorrhiza* 14:329–337.

Boudell, J. A. and J. C. Stromberg. 2008. Propagule banks: Potential contribution to restoration of an impounded and dewatered riparian ecosytem. *Wetlands.*

Bowers, J. E. and S. P. McLaughlin. 1996. Flora of the Huachuca Mountains, a botanically rich and historically significant sky island in Cochise County, Arizona. *J. AZ-Nev. Acad. of Sci.* 29:66–107.

Boyd, S. H. 1975. Inhibition of fish reproduction by *Rana catesbeiana* larvae. *Physiol. Zool.* 48:225–234.

Boyer, E. W., G. M. Hornberger, K. E. Bencala and D. M. McKnight. 2000. Effects of asynchronous snowmelt on flushing of dissolved organic carbon: A mixing model approach. *Hydrol. Proc.* 14:3291–3308.

Bradley, G. A., P. C. Rosen, M. J. Sredl, T. Jones and J. E. Longcore. 2002. Chytridiomycosis in native Arizona frogs. *J. Wildlife Diseases* 38:206–212.

Brady, W., S. McElroy, A. Chehbouni, D. C. Goodrich, D. Hadley, M. Hernandez, W. Kepner, B. McClure, A. Moote and D. Radtke, eds. 2000. *Proceedings of San Pedro Conference, Divided Waters—Common Ground.* Arizona State University.

Brakenridge, G. R. 1988. River flood regime and floodplain stratigraphy. Pages 139–156 in *Flood Geomorphology.* V. R. Baker, P. Patton and C. Kochel, eds. John Wiley & Sons, Inc.

Brand, L. A. 2004. Prediction and assessment of edge response and abundance for desert riparian birds in southeastern Arizona. PhD diss., Colorado State Univ.

Brand, L. A., J. C. Stromberg and B. R. Noon. In review. Avian density and nest success on the San Pedro River: Importance of vegetation type and hydrologic regime.

Brand, L. A., G. C. White and B. R. Noon. 2008. Factors influencing species richness and community composition of breeding birds in a desert riparian corridor. *Condor* 110:1–12.

Brandenburg, W. H. and K. B. Gido. 1999. Predation by nonnative fish on native fishes in the San Juan River, New Mexico and Utah. *S.W. Nat.* 44:392–394.

Brandt, H. 1951. *Arizona and Its Bird Life.* Bird Research Foundation.

Briggs, M. K. 1996. *Riparian Ecosystem Recovery in Arid Lands.* University of Arizona Press.

Brock, J. H. 1985. Physical characteristics and pedogenesis of soils in riparian habitats along the Upper Gila River basin. Pages 49–53 in R. R. Johnson et al. 1985.

——. 1986. Velvet mesquite seedling development in three southwestern soils. *J. Range Manag.* 39:331–334.

Brooks, P. D., P. A. Haas and A. K. Huth. 2007. Seasonal variability in the concentration and flux of organic matter and inorganic nitrogen in a semiarid catchment, San Pedro River, Arizona. *J. Geophys. Res.* 112, G03S04, DOI: 10.1029/2006JG000275.

Brooks, P. D. and M. M. Lemon. 2007. Spatial variability in dissolved organic matter and inorganic nitrogen concentrations in a semiarid stream, San Pedro River, Arizona. *J. Geophys. Res.* 112,G03S05, DOI: 10.1029/2006JG000262.

Brotherson, J. D., S. R. Rushforth, W. E. Evenson, J. R. Johansen and C. Morden. 1983. Population dynamics and age relationships of 8 tree species in Navajo National Monument, Arizona. *J. Range Manag.* 36:250–256.

Brouder, M. J. 2001. Effects of flooding on recruitment of roundtail chub, *Gila robusta*, in a southwestern river. *S.W. Nat.* 46:302–310.

Brower, A. 2004. *ET toolbox: Evapotranspiration toolbox for the Middle Rio Grande, a water resources decision support tool.* U.S. Bureau of Reclamation. Denver, CO.

Brown, A. G. 1997. Biogeomorphology and diversity in multiple-channel river systems. *Global Ecol. and Biogeog. Let.* 6:179–185.

Brown, A. M. and C. Bledsoe. 1996. Spatial and temporal dynamics of mycorrhizas in *Jaumea carnosa*, a tidal marsh halophyte. *J. Ecol.* 84:703–715.

Brown, B. T. 1993. Bell's Vireo (*Vireo bellii*). In *The Birds of North America, No. 35.* A. Poole and F. Gill, eds. The Academy of Natural Sciences, Philadelphia, PA, and The American Ornithologists' Union, Washington, D.C.

Brown, B. T. and R. R. Johnson. 1985. Glen Canyon Dam, fluctuating water levels, and riparian breeding birds: The need for management compromise on the Colorado River in Grand Canyon. Pages 76–80 in R. R. Johnson et al. 1985.

Brown, D. E., ed. 1994. *Biotic communities: Southwestern United States and Northwestern Mexico.* University of Utah Press.

Brown, D. E., N. B. Carmony and R. M. Turner. 1981. *Drainage map of Arizona showing perennial streams and some important wetlands.* Arizona Game and Fish Dept.

Brown, J. H. 1971. Mammals on mountaintops: Nonequilibrium insular biogeography. *Amer. Nat.* 105:467–478.

——. 1995. *Macroecology.* University of Chicago Press.

Brown, J. H. and E. J. Heske. 1990. Control of desert-grassland transition by a keystone rodent guild. *Sci.* 250:1705–1707.

Brown, J. H., T. J. Valone and C. G. Curtin. 1997. Reorganization of an arid ecosystem in response to recent climate change. *Proceedings of the National Academy of Sciences of the United States of America* 94:9729–9733.

Brown, L. R. and P. B. Moyle. 1991. Changes in habitat and microhabitat partitioning within an assemblage of stream fishes in response to predation by Sacramento squawfish (*Ptychocheilus grandis*). *Can. J. Fish. and Aquatic Sci.* 48:849–856.

Brown, R. L. and R. K. Peet. 2003. Diversity and invasibility of southern Appalachian plant communities. *Ecol.* 84:32–39.

Brown, T. J., B. L. Hall and A. L. Westerling. 2004. The impact of twenty-first century climate change on wildland fire danger in the western United States: An applications perspective. *Climatic Change* 62:365–388.

Browning-Aiken, A., B. Morehouse, A. Davis, M. Wilder, R. Varady, D. Goodrich, R. Carter, D. Moreno and E. D. McGovern. 2007. Climate, water management, and policy in the San Pedro basin: Results of a survey of Mexican stakeholders near the U.S.-Mexico border. *Climatic Change* 85:323–341.

Browning-Aiken, A., H. Richter, D. Goodrich, B. Strain and R. Varady. 2004. Upper San Pedro Basin: Fostering collaborative binational watershed management. *Water Res. Dev.* 20:353–367.

Browning-Aiken, A., T. Sprouse, G. Saliba and D. Garrick. In review. Collaborative natural resource management theory and practice: Lessons from three watershed organizations.

Browning-Aiken, A., R. G. Varady, D. Goodrich, H. Richter, T. Sprouse and W. J. Shuttleworth. 2006. Integrating science and policy for water management: A case study of the Upper San Pedro River Basin. Pages 1–39 in *Hydrology and Water Law—Bridging the Gap*. J. Wallace and P. Wouters, eds. IWA Publishing.

Bryan, G. W. and Langston, W. J. 1992. Bioavailability, accumulation and effects of heavy metals in sediments with special reference to United Kingdom estuaries: A review. *Environ. Poll.* 76:89–113.

Bryan, K. 1925. Date of channel trenching (arroyo cutting) in the arid Southwest. *Sci.* 62:338–344.

——. 1926. The San Pedro Valley, Arizona, and the geographical cycle. *Geol. Soc. of Amer. Bull.* 37:169–170.

——. 1928a. Historic evidence on changes in the channel of the Rio Puerco, a tributary of the Rio Grande in New Mexico. *J. Geol.* 36:265–282.

——. 1928b. Change in plant association by change in groundwater level. *Ecol.* 9:474–478.

Bryan, K., G. E. P. Smith and G. A. Waring. 1934. *Ground-water supplies and irrigation in the San Pedro Valley, Arizona.* USGS Open-File Report 67–31.

Bryan, S. D., A. T. Robinson and M. G. Sweetser. 2002. Behavioral responses of a small native fish to multiple introduced predators. *Environ. Biol. of Fishes* 63:49–56.

Bull, W. B. 1991. *Geomorphic Responses to Climatic Change.* Oxford University Press.

Bunn, S. E., M. C. Thoms, S. K. Hamilton and S. J. Capon. 2006. Flow variability in dryland rivers: Boom, bust and the bits in between. *River Res. and Appl.* 22:179–186.

Buol, S. W., R. J. Southhard, R. C. Graham and P. A. McDaniel. 2003. *Soil Genesis and Classification*, 5th ed. Iowa State University Press.

Burgess, S. S. O., M. A. Adams, N. C. Turner, D. A. White and C. K. Ong. 2001. Tree roots: Conduits for deep recharge of soil water. *Oecologia* 126:158–165.

Burgess, T. L. 1995. Desert grassland, mixed shrub savanna, shrub steppe, or semi-desert grassland? Pages 31–67 in *The Desert Grassland.* M. P. McClaran and T. R. Van Devender, eds. University of Arizona Press.

Burkham, D. E. 1972. *Channel changes of the Gila River in Safford Valley, Arizona, 1846–1970.* USGS Water-Supply Paper 655-G.

Busch, D. E. 1995. Effects of fire on southwestern riparian plant community structure. *S.W. Nat.* 40:259–267.

Busch, D. E. and S. G. Fisher. 1981. Metabolism of a desert stream. *Freshwater Biol.* 11:301–307.

Busch, D. E., N. L. Ingraham and S. D. Smith. 1992. Water uptake in woody riparian phreatophytes of the southwestern United States: A stable isotope study. *Ecol. Appl.* 2:450–459.

Busch, D. E. and S. D. Smith. 1995. Mechanisms associated with the decline of woody species in riparian ecosystems of the southwestern U.S. *Ecol. Monog.* 65:347–370.

Bush, J. K. and O. W. Van Auken. 1986. Changes in nitrogen, carbon, and other surface soil properties during secondary succession. *Soil Sci. Soc. of Amer. J.* 50:1597–1601.

Butler, D. 1995. *Zoogeomorphology: Animals as Geomorphic Agents.* Cambridge University Press.

Caldwell, M. M., T. E. Dawson and J. H. Richards. 1998. Hydraulic lift: Consequences of water efflux from the roots of plants. *Oecologia* 113:151–161.

Campoy-Favela, J. R. 1996. Evaluation of populations of *Cyprinodon macularius, Poeciliopsis occidentalis, Ictalurus pricei, Gila ditaenia* and *Gila intermedia* in shared watersheds between Arizona and Sonora. Instituto del Medio Ambiente y del Desarrollo Sustentable del Estado de Sonora (IMADES). Hermosillo, Sonora, Mexico.

Cane, J. H. 2001. Habitat fragmentation and native bees: A premature verdict? *Conserv. Ecol.* 5:3.

Cantrell, R. S., C. Cosner and W. F. Fagan. 2001. How predator incursions affect critical patch size: The role of functional response. *Amer. Nat.* 158:368–375.

Canty, J. M. 1991. *History of Mining in Arizona.* Amer. Assoc. of Mining Engineers. Tucson.

Cappuccino, N. and P. W. Price. 1997. *Population Dynamics: New Approaches and Synthesis.* Academic Press.

Carothers, S. W., R. R. Johnson and S. W. Aitchison. 1974. Population structure and social organization of southwestern riparian birds. *Amer. Zool.* 14:97–108.

Carpenter, J. 2005. Competition for food between an introduced crayfish and two fishes endemic to the Colorado River basin. *Environ. Biol. of Fishes* 72:335–342.

Carroll, A. B., S. G. Pallardy and C. Galen. 2001. Drought stress, plant water status, and floral trait expression in fireweed, *Epilobium angustifolium* (Onagraceae). *Amer. J. Bot.* 88:438–446.

Carveth, C. J., A. M. Widmer and S. A. Bonar. 2006. Comparison of upper thermal tolerances of native and nonnative fish species in Arizona. *Transactions of the American Fisheries Society* 135:1433–1440.

Castillo-Gamez, R. R., R. Arenas-Wong, G. Kruse-Llero and I. Parra-Salazar. 2005. Status of black-tailed prairie dog in Sonora, Mexico. Pages 511–514 in Gottfried et al. 2005.

Catron, J. L. E., S. H. Stoleson, P. L. L. Stoleson and D. W. Shaw. 2000. Riparian systems. Pages 281–327 in *Livestock Management in the American Southwest.* R. Jemison and C. Raish, eds. Elsevier.

Ceballos, G. 1985. The importance of riparian habitats for the conservation of endangered mammals in Mexico. Pages 96–100 in R. R. Johnson et al. 1985.

CEC. 1999a. *Ribbon of life: An agenda for preserving transboundary migratory bird habitat on the Upper San Pedro River.* Commission for Environmental Cooperation.

——. 1999b. *North American Important Bird Areas: A directory of 150 key conservation sites.* Commission for Environmental Cooperation.

——. 2004. *Commission for Environmental Cooperation Profile.*

Cerasale, D. J. 2004. Physiological assessment of avian refueling performance. Master's thesis, University of Montana.

Chaney, E., W. Elmore and W. S. Platts. 1990. *Livestock grazing on western riparian areas.* NW Resource Information Center, Eagle, ID.

Chen, Y. L., M. C. Brundrett and B. Dell. 2000. Effects of ectomycorrhizas and vesicular-arbuscular mycorrhizas, alone or in competition, on root colonization and growth of *Eucalyptus globulus* and *E. urophylla. New Phytologist* 146:545–556.

Childs, M. R. 2004. *Development of propagation techniques for loach minnow.* Arizona Game and Fish Dept. Phoenix.

——. 2006. Comparison of Gila topminnow and western mosquitofish as biological control agents of mosquitos. *West. North Amer. Nat.* 66:181–190.

Childs, M. R., R. W. Clarkson and A. T. Robinson. 1998. Resource use by larval and early juvenile native fishes in the Little Colorado River, Grand Canyon, Arizona. *Trans. Amer. Fish. Soc.* 127:620–629.

Chovanec, A. and J. Waringer. 2001. Ecological integrity of river-floodplain systems-assessment by dragonfly surveys (Insecta: Odonata). *Reg. Rivers—Res. and Manag.* 17:493–507.

Chow, V. T. 1964. *Handbook of Applied Hydrology.* McGraw-Hill.

Christensen, N. L., A. M. Bartuska, J. H. Brown et al. 1996. The report of the ecological society of America committee on the scientific basis for ecosystem management. *Ecol. Appl.* 6:665–691.

Christensen, N. S., A. W. Wood, N. Voisin, D. P. Lettenmaier and R. N. Palmer. 2004. The effects of climate change on the hydrology and water resources of the Colorado River basin. *Climatic Change* 62:337–363.

Christiansen, J. 1999. Arizona desert bird oasis in peril. *New York Times,* May 4, Final, Section F, Page 5, Column 2.

Christiansen, L. D. 1983. The Mormon Battalion in Cochise County and adjacent areas. *The Cochise Quarterly* 13:47.

Clark, S. 1987. Potential for use of cottonwoods as indicators of past floods. Pages 243–248 in *Proceedings of the International Symposium on Ecological Aspects of Tree-Ring Analysis.* Tarrytown, NY.

Clarkson, R. W., P. C. Marsh, J. A. Stefferud and S. E. Stefferud. 2005. Conflicts between native fish and nonnative sport fish management in the southwestern United States. *Fisheries* 30:20–27.

Clarkson, R. W. and W. L. Minckley. 1988. Morphology and foods of Arizona catostomid fishes: *Catostomus insignis, Pantosteus clarki,* and their putative hybrids. *Copeia* 1988:422–433.

Clarkson, R. W., A. T. Robinson and T. L. Hoffnagle. 1997. Asian tapeworm (*Both-*

riocephalus acheilognathi) in native fishes from the Little Colorado River, Grand Canyon, Arizona. *Great Basin Nat.* 57:66–69.

Clarkson, R. W. and J. C. Rorabaugh. 1989. Status of leopard frogs (*Rana pipiens* complex: Ranidae) in Arizona and southeastern California. *S.W. Nat.* 34:531–538.

Cleverly, J. R., C. N. Dahm, J. R. Thibault, D. J. Gilroy and J. E. Allred Coonrod. 2002. Seasonal estimates of actual evapo-transpiration from *Tamarix ramosissima* stands using three-dimensional eddy covariance. *J. Arid Environ.* 52:181–197.

Cleverly, J. R., S. D. Smith, A. Sala and D. A. Devitt. 1997. Invasive capacity of *Tamarix ramosissima* in a Mojave Desert floodplain: The role of drought. *Oecologia* 111:12–18.

Cochise County. 2002. *Cochise County comprehensive plan.* Adopted 1984 and last amended 2002. Bisbee, AZ.

Cochran, C. C. and M. L. Richardson. 2003. *Soil survey of Pima County, Arizona, eastern part.* Nat. Res. Cons. Serv. Washington, D.C.

Cockrum, E. L. 1960. *The Recent Mammals of Arizona: Their Taxonomy and Distribution.* University of Arizona Press.

Coes, A. L. and D. R. Pool. 2005. *Ephemeral-channel and basin-floor infiltration in the Sierra Vista Subwatershed, Arizona.* USGS Open-File Report 05-1023.

Cohan, D. R., B. W. Anderson and R. D. Ohmart. 1978. Avian population responses to salt cedar along the Lower Colorado River. Pages 371–382 in R. R. Johnson and McCormick 1978.

Colby, B. and P. Orr. 2005. Economic tradeoffs in preserving riparian habitat. *Nat. Res. J.* 45:15–31.

Coleman, D. C. and D. A. Crossley. 1996. *Fundamentals of Soil Ecology.* Academic Press.

Collen, P. and R. J. Gibson. 2001. The general ecology of beavers (*Castor* spp.), as related to their influence on stream ecosystems and riparian habitats, and the subsequent effects on fish—a review. Rev. in *Fish Biol. and Fish.* 10:439–461.

Collins, J. P. and H. M. Wilbur. 1979. *Breeding habits and habitats of the amphibians of the Edwin S. George Reserve, Michigan, with notes on the local distribution of fishes.* Occasional Papers of the Museum of Zoology, University of Michigan 686:1–33.

Collins, J. P., C. Young Jr., J. Howell and W. L. Minckley. 1981. Impact of flooding in a Sonoran Desert stream including elimination of an endangered fish population (*Peociliopsis o. occidentalis* Poeciliidae). *S.W. Nat.* 26:415–423.

Connell, J. H. 1978. Diversity in tropical rain forests and coral reefs. *Sci.* 199:1302–1310.

Constantz, G. D. 1980. Energetics of viviparity in the Gila topminnow (Pisces: Poeciliidae). *Copeia* 1980:876–878.

Cook, E. R., C. Woodhouse, C. M. Eakin, D. M. Meko and D. W. Stahle. 2004. Long-term aridity changes in the western United States. *Sci.* 306:1015–1018.

Cook, H. 1987. Too little–too much water and the Tombstone story. Pages 229–233 in *History of Mining in Arizona.* H. Cook, ed. Mining Club of the SW Foundation. Tucson.

Cooke, R. U. and R. W. Reeves. 1976. *Arroyos and Environmental Change in the American Southwest.* Clarendon Press.

Cooper, D. J., D. J. Andersen and R. A. Chimner. 2003. Multiple pathways for woody plant establishment on floodplains at local to regional scales. *J. Ecol.* 91:182–196.

Cooper, D. J., D. M. Merritt, D. C. Anderson and R. A. Chimner. 1999. Factors controlling the establishment of Fremont cottonwood seedlings on the Upper Green River, USA. *Reg. Rivers: Res. and Manag.* 15:419–440.

Corell, S. W., F. Corkhill, D. Lovvik and F. Putman. 1996. *A groundwater flow model of the Sierra Vista subwatershed of the Upper San Pedro Basin, southeastern Arizona.* Arizona Dept. of Water Resources Modeling Report 10.

Corman, T. E. 1988. Abundance, distribution, and habitat management of the reptiles and amphibians of the San Pedro Riparian National Conservation Area. Unpublished report to U.S. Bureau of Land Management.

Costa, J. E. and J. O'Connor. 1995. Geomorphically effective floods. Pages 45–56 in *Natural and Anthropogenic Influences in Fluvial Geomorphology* (The Wolman Volume). J. E. Costa, A. J. Miller, K. W. Potter and P. R. Wilcock, eds. Geophysical Monograph 89. Amer. Geophysical Union.

Cox, J. R. 1985. Above-ground biomass and nitrogen quantities in a big sacaton (*Sporobolus wrightii*) grassland. *J. Range Manag.* 38:273–276.

——. 1988. Seasonal burning and mowing impacts on Sporobolus-wrightii grasslands. *J. Range Manag.* 41:12–15.

Cox, J. R. and H. L. Morton. 1986. Big sacaton (*Sporobolus wrightii*) riparian grassland management: Annual winter burning, annual winter mowing, and spring-summer grazing. *Applied Ag. Res.* 1:105–111.

Cox, J. R., H. L. Morton, J. T. LaBaume and K. G. Renard. 1983. Reviving Arizona's rangelands. *J. Soil and Water Conserv.* 38:342–345.

Cristol, D. A. 1995. Early arrival, initiation of nesting, and social status: An experimental study of breeding female red-winged blackbirds. *Behavioral Ecol.* 6:87–93.

Crockford, R. H. and D. P. Richardson. 2000. Partitioning of rainfall into throughfall, stemflow and interception: Effect of forest type, groundcover and climate. *Hydr. Proc.* 14:2903–2920.

Crooks, K. R. and M. E. Soule. 1999. Mesopredator release and avifaunal extinctions in a fragmented system. *Nature* 400:563–566.

Cross, A. F. 1991. Vegetation of two southeastern Arizona desert marshes. *Madrono* 38:185–194.

Dafni, A. 2005. *Pollination Ecology: A Practical Approach.* Oxford University Press.

Dahlberg, A. 2001. Community ecology of ectomycorrhizal fungi: An advancing interdisciplinary field. *New Phytologist* 150:555–562.

Dahm, C. N., J. Cleverly, J. Coonrod, J. Thibault, D. McDonnell and D. Gilroy. 2002. Evapotranspiration at the land/water interface in a semi-arid drainage basin. *Freshwater Biol.* 47:831–843.

Dahm, C. N., N. Clifford, N. Grimm, P. Marmonier, M. Valett and P. Verview. 1998. Nutrient dynamics at the interface between surface waters and groundwaters. *Freshwater Biol.* 40:427–451.

Dahm, C. N. and M. C. Molles Jr. 1992. Streams in semi-arid regions as sensitive indicators of global climate change. Pages 250–260 in *Global Climate Change and Freshwater Ecosystems.* P. Firth and S. Fisher, eds. Springer-Verlag.

Dale, V. B. and L. A. Stewart. 1960. *Tungsten deposits of Cochise, Pima, and Santa Cruz Counties, Arizona.* U.S. Bureau of Mines. Washington, D.C.

Danzer, S. R., C. H. Baisan and T. W. Swetnam. 1996. The influence of fire and land use history on stand dynamics in the Huachuca Mountains of Southeastern Arizona. Pages 265–270 in *Effects of Fire on Madrean Province Ecosystems, A sym-*

posium proceedings. P. F. Ffolliott et al., eds. USDA For. Serv. Gen. Tech. Rep. RM-GTR-289.

Davis, F. W. and D. Roberts. 2000. Stand structure in terrestrial ecosystems. Pages 7–30 in *Methods in Ecosystem Science.* O. E. Sala, R. B. Jackson, H. A. Mooney and R. W. Howarth, eds. Springer.

Davis, G. P., Jr. 1982. *Man and Wildlife in Arizona—the American Exploration Period 1824–1865.* Arizona Game and Fish Dept. Phoenix.

Davis, M. A., J. P. Grime and K. Thompson. 2000. Fluctuating resources in plant communities: A general theory of invasibility. *J. Ecol.* 88:528–534.

Davis, O. K., T. Minckley, T. Moutoux, T. Jull and B. Kalin. 2002. The transformation of Sonoran Desert wetlands following the historic decrease of burning. *J. Arid Environ.* 50:393–412.

Dawson, T. E. and J. S. Pate. 1996. Seasonal water uptake and movement in root systems of Australian phreatophytic plants of dimorphic root morphology—A stable isotope investigation. *Oecologia* 107:13–20.

Deacon, J. E. and W. L. Minckley. 1991. Western fishes and the real world: The enigma of "endangered species" revisited. Pages 405–413 in W. Minckley and Deacon 1991.

DeBano, L. F., D. G. Neary and P. F. Ffolliot. 1998. *Fire's Effects on Ecosystems.* John Wiley & Sons, Inc.

DeLay, L., D. M. Finch, S. Brantley, R. Fagerlund, M. D. Means and J. F. Kelly. 1999. Arthropods of native and exotic vegetation and their association with willow flycatchers and Wilson's warblers. Pages 216–221 in *Rio Grande Ecosystems: Linking Land, Water, and People.* D. M. Finch, D. M., J. C. Whitney, J. F. Kelly and S. R. Loftin, eds. USDA For. Serv., Rocky Mountain Research Station RMRS P-7.

DeLoach, C. J., R. I. Carruthers, J. E. Lovich, T. L. Dudley and S. D. Smith. 2000. Ecological interactions in the biological control of saltcedar (*Tamarix* spp.) in the United States: Toward a new understanding. Pages 819–873 in *Proceedings of the X International Symposium on Biological Control of Weeds.* N. R. Spencer, ed. Montana State University.

DeMarais, B. D. 1986. Morphological variation in Gila (Pisces: Cyprinidae) and geologic history: Lower Colorado River basin. Master's thesis, Arizona State Univ.

Dennis, R. L. H., T. G. Shreeve and H. Van Dyck. 2003. Towards a functional resource-based concept for habitat: A butterfly biology viewpoint. *Oikos* 102:417–426.

Denno, R. F. and M. S. McClure, eds. 1983. *Variable Plants and Herbivores in Natural and Managed Systems.* Academic Press.

Desert Fishes Team. 2003. Status of federal and state listed warm water fishes of the Gila River basin, with recommendations for management. *Desert Fishes Team Report* 1:1–20.

——. 2004. Status of unlisted native fishes of the Gila River basin, with recommendations for management. *Desert Fishes Team Report* 2:1–25.

——. 2006. Analysis of recovery plan implementation for threatened and endangered warm water fishes of the Gila River basin. *Desert Fishes Team Report* 3:1–28 + appendix.

Dettinger, M. D., D. R. Cayan, M. Meyer and A. E. Jeton. 2004. Simulated hydrologic responses to climate variations and change in the Merced, Carson, and

American River basins, Sierra Nevada, California, 1900–2099. *Climatic Change* 62:283–317.

Devitt, D., A. Sala, K. Mace and S. Smith. 1997. The effect of applied water on the water use of saltcedar in a desert riparian environment. *J. Hydr.* 192:233–246.

Devitt, D., A. Sala, S. D. Smith, J. Cleverly, L. K. Shaulis and R. Hammett. 1998. Bowen ratio estimates of evapotranspiration for *Tamarix ramosissima* stands on the Virgin River in southern Nevada. *Water Res. Res.* 34:2407–2414.

Di Tomaso, J. M. 1998. Impact, biology, and ecology of saltcedar (*Tamarix* spp.) in the southwestern United States. *Weed Technology* 12:326–336.

Dickinson, W. R. 1991. *Tectonic setting of faulted Tertiary strata associated with the Catalina core complex in southern Arizona.* Geol. Soc. of Amer. Special Pub. 264.

——. 2003. Depositional facies of the Quiburis formation, basin fill of the San Pedro trough, southeastern Arizona Basin and Range Province. Pages 157–181 in *Cenozoic Systems of the Rocky Mountain Region.* R. G. Raynolds and R. M. Flores, eds. Rocky Mountain Section, Soc. for Sedimentary Geology.

Dick-Peddie, W. A., J. V. Hardesty, E. Muldavin and B. Sallach. 1987. *Soil-vegetation correlations of the riparian zones of the Gila and San Francisco Rivers in New Mexico.* U.S. Fish and Wildlife Service Biol. Report 87(9).

Dickson, L. L. and T. G. Whitham. 1996. Genetically based plant resistance traits affect arthropods, fungi, and birds. *Oecologia* 106:400–406.

DICTUS. 1993. *Estudio hidrologico del Estado de Sonora.* University of Sonora; Dept. de Investigaciones Cientificas y Tecnologicas. Hermosillo, Sonora, Mexico.

Diffenbaugh, N. S., J. S. Pal, R. J. Trapp and F. Giorgi. 2005. Fine-scale processes regulate the response of extreme events to global climate change. *Proc. of the Nat. Acad. of Sci.* 102:15774–15778.

Dixon, M. D. and W. C. Johnson. 1999. Riparian vegetation along the middle Snake River, Idaho: Zonation, geographical trends, and historical changes. *Great Basin Nat.* 59:18–34.

Dobkin, D. S., A. C. Rich and W. H. Pyle. 1998. Habitat and avifaunal recovery from livestock grazing in a riparian meadow system of the northwestern Great Basin. *Conserv. Biol.* 12:209–221.

Dobyns, H. F. 1981. *From Fire to Flood: Historic Human Destruction of Sonoran Desert Riverine Oases.* Ballena Press.

Dockens, P. E. T., C. E. Paradzick and J. R. Hatten. 2004. Application of a southwestern willow flycatcher GIS-based habitat model: An estimate of breeding habitat in Arizona, 2001. Pages 28–59 in *Mapping and monitoring southwestern willow flycatcher habitat in Arizona: A remote sensing approach.* P. E. T. Dockens and C. E. Paradzick, eds. Nongame and Endangered Wildlife Program Tech. Rep. 223. Arizona Game and Fish Dept. Phoenix.

Dohrenwend, J. C. 1987. Basin and Range. Pages 303–342 in *Geomorphic Systems of North America.* W. L. Graf, ed. Geol. Soc. of Amer. Centennial Special Vol. 2. Boulder, CO.

Donovan, L. A. and J. R. Ehleringer. 1992. Contrasting water-use patterns among size and life history classes of a semi-arid shrub. *Functional Ecol.* 6:482–488.

Douglas, M. E. and P. C. Marsh. 1998. Population and survival estimates of *Catostomus latipinnis* in northern Grand Canyon, with distribution and abundance of hybrids with *Xyrauchen texanus*. *Copeia* 1998:915–925.

Douglas, M. E., P. C. Marsh and W. L. Minckley. 1994. Indigenous fishes of western

North America and the hypothesis of competitive displacement: *Meda fulgida* (Cyprinidae) as a case study. *Copeia* 1994:9–19.

Douglas, M. W., R. A. Maddox, K. Howard and S. Reyes. 1993. The Mexican monsoon. *J. Climate* 6:1665–1677.

Drezner, T. D., P. L. Fall and J. C. Stromberg. 2001. Plant distribution and dispersal mechanisms at the Hassayampa River Preserve, Arizona, USA. *Global Ecol. and Biogeog.* 10:205–217.

DuBois, S. M. and A. W. Smith. 1980. *The 1887 earthquake in San Bernadino Valley, Sonora: Historic accounts and intensity patterns in Arizona.* Bureau of Geology and Mineral Technology. University of Arizona Special Paper 3.

Duffy, J. E., B. J. Carinale, K. E. France, P. B. McIntyre, E. Thebault and M. Loreau. 2007. The functional role of biodiversity in ecosystems: Incorporating trophic complexity. *Ecol. Let.* 10:522–538.

Dukes, J. S. and H. A. Mooney. 1999. Does global change increase the success of biological invaders? *Trends in Ecol. and Evol.* 14:135–139.

DuMars, C. T. and J. D. Minier. 2004. The evolution of groundwater rights and groundwater management in New Mexico and the western United States. *Hydrogeol. J.* 12:40–51.

Duncan, D. K. 1988. *Mammal inventory of the San Pedro Riparian National Conservation Area, Cochise County, Arizona.* U.S. Bureau of Land Management, San Pedro Project Office.

Duncan, D. K. and T. E. Corman. 1991. Relative abundance of desert shrews (*Notiosorex crawfordi*) within the San Pedro Riparian National Conservation Area, Arizona. *S.W. Nat.* 36:119–121.

Easterling, D. R., G. A. Meehl, C. Parmesan, S. A. Changnon, T. R. Karl and L. O. Mearns. 2000. Climate extremes: Observations, modeling, and impacts. *Sci.* 289:2068–2074.

Eberhardt, S. 1981. *San Pedro River basin water quality status report for period 1973–1979.* Prepared for Sierra Vista Economic Development Foundation. Arizona Dept. of Health Services.

Eby, L. A., W. F. Fagan and W. L. Minckley. 2003. Variability and dynamics of a desert stream community. *Ecol. Appl.* 13:1566–1579.

Eccleston, R. 1950. *Overland to California on the Southwestern Trail, 1849. Diary of Robert Eccleston.* G. P. Hammond and E. H. Howes, eds. University of California Press.

Egerton-Warburton, L. and M. F. Allen. 2001. Endo- and ectomycorrhizas in *Quercus agrifolia* Nee. (Fagaceae): Patterns of root colonization and effects on seedling growth. *Mycorrhiza* 11:283–290.

Ehleringer, J. R. and T. E. Dawson. 1992. Water uptake by plants: Perspectives from stable isotope composition. *Plant, Cell & Environ.* 15:1073–1082.

Ehleringer, J. R., J. Roden and T. E. Dawson. 2000. Assessing ecosystem-level water relations through stable isotope ratio analyses. Pages 181–198 in *Methods in Ecosystem Science.* O. E. Sala, R. B. Jackson, H. A. Mooney and R. W. Howarth, eds. Springer-Verlag.

Erhardt, A. 1985. Diurnal Lepidoptera: Sensitive indicators of cultivated and abandoned grassland. *J. Applied Ecol.* 22:849–861.

Ehrlich, P. R. 2003. Butterflies, test systems, and biodiversity. Pages 1–6 in *Butterflies: Ecology and Evolution Taking Flight.* C. L. Boggs, W. B. Watt and P. R. Ehrlich, eds. University of Chicago Press.

Ehrlich, P. R., D. S. Dobkin and D. Wheye. 1988. *The Birder's Handbook: A Field Guide to the Natural History of North American Birds.* Simon and Schuster, Inc.

Ehrlich, P. R. and P. H. Raven. 1965. Butterflies and plants: A study in coevolution. *Evolution* 18:586–608.

Ellis, L. M. 1995. Bird use of saltcedar and cottonwood vegetation in the Middle Rio Grande Valley of New Mexico, U.S.A. *J. Arid Environ.* 30:339–349.

——. 2001. Short-term response of woody plants to fire in a Rio Grande riparian forest, Central New Mexico, USA. *Biol. Conserv.* 97:159–170.

Ellis, L. M., C. S. Crawford and M. C. Molles. 2001. Influence of annual flooding on terrestrial arthropod assemblages of a Rio Grande riparian forest. *Reg. Rivers—Res. and Manag.* 17:1–20.

Ellis, L. M., C. S. Crawford and M. C. Molles Jr. 1997. Rodent communities in native and exotic riparian vegetation in the middle Rio Grande valley of central New Mexico. *S.W. Nat.* 42:13–19.

——. 1998. Comparison of litter dynamics in native and exotic riparian vegetation along the Middle Rio Grande of central New Mexico, USA. *J. Arid Environ.* 38:283–296.

Ellis, L. M., M. C. Molles and C. S. Crawford. 1999. Influence of experimental flooding on litter dynamics in a Rio Grande riparian forest, New Mexico. *Rest. Ecol.* 7:193–204.

Ellison, L. E. and C. Van Riper III. 1998. A comparison of small-mammal communities in a desert riparian floodplain. *J. Mammology* 79:972–985.

Elwood, J. W., J. D. Newbold, A. F. Trimble and R. W. Stark. 1981. The limiting role of phosphorus in a woodland stream ecosystem: Effects of P enrichment on leaf decomposition and primary producers. *Ecol.* 62:146–158.

Ely, L. L., Y. Enzel and D. Cayan. 1994. Anomalous North Pacific atmospheric circulation and large winter floods in the southwestern United States. *J. Climate* 7:977–987.

Emmerich, J. M. and P. A. Vohs. 1982. Comparative use of four woodland habitats by birds. *J. Wildlife Manag.* 46:43–49.

Emory, W. H. 1858–1859. *Report on the U.S. and Mexico Boundary Survey.* 2 vols. U.S. Dept. of Interior. Washington, D.C.

Engel-Wilson, R. W. and R. D. Ohmart. 1978. Floral and attendant faunal changes on the Lower Rio Grande between Fort Quitman and Presidio, Texas. Pages 139–147 in R. R. Johnson and McCormick 1978.

Enlace Ecologico, and Border Ecology Project. 1999. *Phase I Report of the Northeast Sonora Water Project: Water Quality for the Municipalities of Agua Prieta, Naco, and Cananea, Sonora, Mexico.* Agua Prieta, Mexico and Bisbee, AZ.

Eom, A. H., G. W. T. Wilson and D. C. Hartnett. 2001. Effects of ungulate grazers on arbuscular mycorrhizal symbiosis and fungal community structure in tallgrass prairie. *Mycologia* 93:233–242.

Epler, W. and G. Dillard. 1981. Phelps Dodge; A copper centennial, 1881–1981. *Arizona Pay Dirt* summer.

Esparza, J. G. 2002. Modelacion geohydrologica del aquifero del Rio San Pedro. Professional thesis in geology, Universidad de Sonora, Hermosillo, Sonora, Mexico.

Essington, T. E. and S. R. Carpenter. 2000. Nutrient cycling in lakes and streams: Insights from a comparative analysis. *Ecosystems* 3:131–143.

Eswaran, H., E. Van den Berg, P. Reich and J. Kimble. 1995. Global soil carbon resources. Pages 27–43 in *Soil and Global Change*. R. Lal, J. Kimle, E. Levine and B. A. Stewart, eds. Lewis Publishers.

Everitt, B. L. 1968. Use of the cottonwood in an investigation of the recent history of a floodplain. *Amer. J. Sci.* 266:417–439.

——. 1980. Ecology of saltcedar: A plea for research. *Environ. Geol.* 3:77–84.

——. 1995. Hydrologic factors in regeneration of Fremont cottonwood along the Fremont River, Utah. Natural and anthropogenic influences in fluvial geomorphology. *Geophys. Monog.* 89:197–208.

Fagan, W. F., C. M. Kennedy and P. J. Unmack. 2005. Quantifying rarity, losses, and risks for native fishes of the lower Colorado River basin: Implications for conservation listing. *Conserv. Biol.* 19:1872–1882.

Falck, M. J., K. R. Wilson and D. C. Andersen. 2003. Small mammals within riparian habitats of a regulated and unregulated aridland river. *West. North Amer. Nat.* 63:35–42.

Farley, G. H., L. M. Ellis, J. N. Stuart and N. J. Scott Jr. 1994. Avian species richness in different-aged stands of riparian forest along the middle Rio Grande, New Mexico. *Conserv. Biol.* 8:1098–1108.

Fenner, P., W. W. Brady and D. R. Patton. 1984. Observations on seeds and seedlings of Fremont cottonwood. *Desert Plants* 6:55–58.

——. 1985. Effects of regulated water flows on regeneration of Fremont cottonwood. *J. Range Manag.* 38:135–138.

Ferguson, T. J., C. Colwell-Chanthaphonh and R. Anyon. 2004. One valley, many histories: Tohono O'odham, Hopi, Zuni, and Western Apache history. *Archaeology Southwest* 18:1–15.

Fernandez, P. J. and P. C. Rosen. 1996. Effects of the introduced crayfish Orconectes virilis on native aquatic herpetofauna in Arizona. Unpublished report. Dept. of Biology, Grand Canyon University. Phoenix.

Ffolliott, P. F., L. F. DeBano, M. B. Baker Jr., D. G. Neary and K. N. Brooks. 2004. Hydrology and impacts of disturbances on hydrologic function. Pages 51–76 in M. B. Baker et al. 2004.

Filippone, C. and S. A. Leake. 2005. Time scales in the sustainable management of water resources. *Southwest Hydr.* Jan./Feb. 16–17.

Finch, D. M., J. Agyagos, T. McCarthey, R. M. Marshal, S. H. Stoleson and M. J. Whitfield. 2000. Management recommendation. Pages 107–118 in Finch and Stoleson 2000.

Finch, D. M. and S. H. Stoleson, eds. 2000. *Status, ecology, and conservation of the southwestern willow flycatcher.* USDA For. Serv. Gen. Tech. Rep. RMRS-GTR-60.

Findlay, S., J. M. Quinn, C. W. Hickey, G. Burrell and M. Downes. 2001. Effects of land use and riparian flowpath on delivery of dissolved organic carbon to streams. *Limnology and Oceanography* 46:345–355.

Fisher, S. G., D. E. Busch and N. B. Grimm. 1981. Diel feeding chronologies in two Sonoran desert stream fishes, *Agosia chrysogaster* (Cyprinidae) and *Pantosteus clarki* (Catostomidae). *S.W. Nat.* 26:31–36.

Fisher, S. G., L. J. Gray, N. B. Grimm and D. E. Busch. 1982. Temporal succession in a desert stream ecosystem following flash flooding. *Ecol. Monog.* 52:93–110.

Fisher, S. G., N. B. Grimm, E. Martí and R. Gomez. 1998. Hierarchy, spatial configuration, and nutrient cycling in a desert stream. *Australian J. Ecol.* 23:41–52.

Fisher, S. G., N. B. Grimm, E. Martí, R. M. Holmes and J. B. Jones. 1998. Material spiraling in stream corridors: A telescoping ecosystem model. *Ecosystems* 1:19–34.

Fisher, S. G. and W. L. Minckley. 1978. Chemical characteristics of a desert stream in flash flood. *J. Arid Environ.* 1:25–33.

Fleischner, T. L. 1994. Ecological costs of livestock grazing in western North America. *Conserv. Biol.* 8:629–644.

Fleishman, E., G. T. Austin, P. F. Brussard and D. D. Murphy. 1999. A comparison of butterfly communities in native and agricultural riparian habitats in the Great Basin, USA. *Biol. Conserv.* 89:209–218.

Floate, K. D. and T. G. Whitham. 1995. Insects as traits in plant systematics: Their use in discriminating between hybrid cottonwoods. *Can. J. Bot.* 73 1–13.

Flory, E. L. 1937. Silting did not discourage this plant. *Soil Conserv.* 2:193.

Fonseca, J. E. 1998. Vegetation changes at Bingham Cienega, the San Pedro River Valley, Pima County, Arizona, since 1879. *J. AZ-Nev. Acad. of Sci.* 31:103–116.

Fordyce, J. A. 2000. A model without a mimic: Aristolochic acids from the California pipevine swallowtail, *Battus philenor hirsuta*, and its host plant, *Aristolochia californica*. *J. Chem. Ecol.* 26:2567–2578.

Fort Huachuca. 1976. *History of a Frontier Post.* Fort Huachuca. Sierra Vista.

Fredlake, M., R. Gerhart and D. Krueper. 1993. *Habitat management plan: San Pedro Riparian National Conservation Area.* U.S. Bureau of Land Management and Arizona Game and Fish Dept. Region V (A4-WHA-T-4 San Pedro).

Fredrickson, E., K. M. Havstad and R. Estell. 1998. Perspectives on desertification: South-western United States. *J. Arid Environ.* 39:191–207.

Freethey, G. W. 1982. *Hydrologic analysis of the Upper San Pedro Basin from the Mexico U.S. boundary to Fairbank, Arizona.* USGS Water-Supply Paper 1819-D.

Friedman, J., G. Auble, P. Shafroth, M. Scott, M. Merigliano, M. Freehling and E. Griffin. 2005. Dominance of non-native riparian trees in western USA. *Biol. Invasions* 7:747–751.

Friedman, J. and V. J. Lee. 2002. Extreme floods, channel change, and riparian forests along ephemeral streams. *Ecol. Monog.* 72:409–425.

Friedman, J., W. R. Osterkamp and W. M. Lewis Jr. 1996a. Channel narrowing and vegetation development following a Great Plains flood. *Ecol.* 77:2167–2181.

———. 1996b. The role of vegetation and bed-level fluctuations in the process of channel narrowing. *Geomorph.* 14:341–351.

Friedman, J. M. and G. T. Auble. 2000. Floods, flood control, and bottomland vegetation. Pages 219–237 in *Inland Flood Hazards: Human, Riparian, and Aquatic communities.* E. E. Wohl, ed. Cambridge University Press.

Fuller, P. L., L. G. Nico and J. D. Williams. 1999. Nonindigenous fishes introduced into inland waters of the United States. *Amer. Fish. Soc. Special Pub.* 27:1–613.

Fulton, R. W. 1966. Milleville-Charleston, Cochise County, 1878—1889. *J. AZ. History* 7:9–22.

Fulton, R. W. and C. J. Bahre. 1967. Charleston, Arizona: A documentary reconstruction. *Arizona and the West* 9:41–64.

Galbraith, H., J. Price, M. Dixon and J. Stromberg. 2004. Development of HSI models to evaluate risks to riparian wildlife habitat from climate change and urban sprawl. Pages 148–168 in *Landscape Ecology and Wildlife Habitat Evaluation: Critical Information for Ecological Risk Assessment, Land-Use Management Activities, and Bio-*

diversity Enhancement Practices. L. A. Kapustka, H. Galbraith, M. Luxon and G. R. Biddinger, eds. ASTM STP 1458. ASTM International. West Conshohocken, PA.

Gange, A. C., V. K. Brown and G. S. Sinclair. 1993. Vesicular-arbuscular mycorrhizal fungi: A determinant of plant community structure in early succession. *Functional Ecol.* 7:616–622.

Gao, Q. and J. F. Reynolds. 2003. Historical shrub-grass transitions in the northern Chihuahuan Desert: Modeling the effects of shifting rainfall seasonality and event size over a landscape gradient. *Global Change Biol.* 9:1475–1493.

Gardiner, D. T. and R. W. Miller. 2004. *Soils in Our Environment.* Pearson Prentice Hall, Inc.

Garfin, G. and M. Lenart. 2007. Climate change effects of Southwest water resources. *Southwest Hydr.* Jan./Feb. 16–17, 34.

Gary, H. L. 1963. Root distribution of five-stamen tamarisk, seepwillow, and arrowweed. *For. Sci.* 9:311–314.

——. 1965. Some site relations in three flood-plain communities in central Arizona. *J. AZ. Acad. of Sci.* 3:209–212.

Gaskin, J. F. and B. A. Schaal. 2002. Hybrid *Tamarix* widespread in US invasion and undetected in native Asian range. *Proc. of the Nat. Acad. of Sci. of the USA* 99:11256–11259.

Gatewood, J. S., T. W. Robinson, B. R. Colby, J. D. Hem and L. C. Halpenny. 1950. *Use of water by bottom-land vegetation in the Lower Safford Valley, Arizona.* USGS Water-Supply Paper 1103.

Gazal, R., R. L. Scott, D. C. Goodrich and D. G. Williams. 2006. Controls on transpiration in a semiarid riparian cottonwood forest. *Ag. and For. Meteor.* 137:56–67.

Gershunov, A. and T. P. Barnett. 1998. Interdecadal modulation of ENSO teleconnections. *Bull. Amer. Meteor. Soc.* 79:2715–2725.

Gibson, C. A., J. L. Meyer, N. L. Poff, L. E. Hay and A. Georgakakos. 2005. Flow regime alterations under changing climate in two river basins: Implications for freshwater ecosystems. *River Res. and Appl.* 21:849–864.

Gido, K. B., J. F. Schaefer, K. Work, P. W. Lienesch, E. Marsh-Matthews and W. J. Matthews. 1999. Effects of red shiner (*Cyprinella lutrensis*) on Red River pupfish (*Cyprinodon rubrofluviatilis*). *S.W. Nat.* 44:287–295.

Gila River II. 1993. *In re the General Adjudication of All Rights to Use Water in the Gila River System and Source*, 175 Ariz. 382, 857 p. 2d 1236.

Gila River III. 1999. *In re the General Adjudication of All Rights to Use Water in the Gila River System and Source*, 195 Ariz. 411, ¶9, 989 p. 2d 739.

Gila River IV. 2000. *In re the General Adjudication of All Rights to Use Water in the Gila River System and Source*, 198 Ariz. 330, 9 p. 3d 1069.

Girmendonk, A. L. and K. L. Young. 1997. *Fish monitoring relative to impacts of the Central Arizona Project in the Gila River basin, Arizona: Results of the winter 1996–97 field season.* Arizona Game and Fish Dept. Tech. Rep. 119:1–24.

Givnish, T. J. 2002. Ecological constraints on the evolution of plasticity in plants. *Evolutionary Ecol.* 16:213–242.

Gleick, P. H. and E. L. Chalecki. 1999. The impacts of climatic changes for water resources of the Colorado and Sacramento-San Joaquin River Basins. *J. Amer. Water Res. Assoc.* 35:1429–1441.

Glendening, C. E. and H. A. Paulsen. 1955. *Reproduction and establishment of velvet mesquite as related to invasion of semidesert grasslands.* USDA Tech. Bull. 1127.

Glenn, E. and P. L. Nagler. 2005. Comparative ecophysiology of *Tamarix ramosissima* and native trees in western U.S. riparian zones. *J. Arid Environ.* 61:419–446.

Glenn, E., R. Tanner, S. Mendez, T. Kehret, D. Moore, J. Garcia and C. Valdes. 1998. Growth rates, salt tolerance and water use characteristics of native and invasive riparian plants from the delta of the Colorado River, Mexico. *J. Arid Environ.* 40:281–294.

Glennon, R. 2002. *Water Follies: Groundwater Pumping and the Fate of America's Fresh Waters.* Island Press.

Glennon, R. and T. Maddock III. 1994. In search of subflow: Arizona's futile effort to separate groundwater from surface water. *AZ Law Rev.* 36:567–610.

——. 1997. The concept of capture: The hydrology and law of stream/aquifer interactions. *Proc. of the Rocky Mountain Mineral Law Inst.* 43:22-1 to 22-89.

Glinski, R. L. and D. E. Brown. 1982. Mesquite (*Prosopis juliflora*) response to severe freezing in southeastern Arizona. *J. AZ-Nev. Acad. of Sci.* 17:15–18.

Gobl, F. 1996. Investigations on mycorrhizae and fine roots in an area used as forest pasture. *Phyton* 36:193–204.

Goldstein, P. Z. 1999. Functional ecosystems and biodiversity buzzwords. *Conserv. Biol.* 13:247–255.

Gomez-Alvarez, A., J. L. Valenzuela-García, S. Aguayo-Salinas, D. Meza-Figueroa, J. Ramírez-Hernández and G. Ochoa-Ortega. 2007. Chemical partitioning of sediment contamination by heavy metals in the San Pedro River, Sonora, Mexico. *Chemical Speciation and Bioavailability* 19:25–35.

Goode, T. C. and T. Maddock III. 2000. *Simulation of groundwater conditions in the Upper San Pedro Basin for the evaluation of alternative futures.* University of Arizona. Dept. of Hydrology and Water Resources HWR No. 00-030.

Goodrich, D. C., A. Chehbouni, B. Goff, B. MacNish, T. Maddock, S. Moran, W. J. Shuttleworth, D. G. W. Watts and L. H. Hipps. 2000. Preface paper to the Semi-arid Land-Surface-Atmosphere SALSA Program special issue. *Ag. and For. Meteor.* 105:3–20.

Goodrich, D. C., R. Scott, J. Qi, B. Goff, C. L. Unkrich, M. S. Moran et al. 2000. Seasonal estimates of riparian evapotranspiration using remote and in-situ measurements. *Ag. and For. Meterol.* 105:281–309.

Gordon, N. D., T. A. McMahon and B. L. Finlayson. 1992. *Stream Hydrology: An Introduction for Ecologists.* John Wiley & Sons, Inc.

Gori, D. F. 1993. *MonitoringNative Fish Populations on the Coronado Forest and Nature Conservancy Preserves in Arizona.* The Arizona Nature Conservancy, Tucson.

——. 1995. *Fish Monitoring Data (1991–1995) from The Nature Conservancy Preserves.* The Arizona Nature Conservancy, Tucson.

——. 1997. *Monitoring Results for Monitoring on Nature Conservancy Preserves, 1996.* The Arizona Nature Conservancy, Tucson.

Gori, D. and D. Backer. 2005. *Watershed improvement using prescribed burns as a way to restore aquatic habitat for native fish.* Pages 403–406 in Gottfried et al. 2005.

Gottfried, G. J., B. S. Gebow, L. G. Eskew and C. B. Edminster, eds. 2005. *Connecting mountain islands and desert seas: Biodiversity and management of the Madrean Archipelago II.* USDA For. Serv. Rocky Mountain Research Station Proceedings RMRS-P-36.

Govindarajulu, M., P. E. Pfeffer, H. R Jin, J. Abubaker, D. D. Douds et al. 2005. Nitrogen transfer in the arbuscular mycorrhizal symbiosis. *Nature* 435:819–823.

Graf, W. L. 1978. Fluvial adjustments to the spread of tamarisk in the Colorado Plateau region. *Geol. Soc. of Amer. Bull.* 89:1491–1501.

——. 1980. Riparian management: A flood control perspective. *J. Soil and Water Conserv.* 35:158–161.

——. 1982. Tamarisk and river-channel management. *Environ. Manag.* 6:283–296.

——. 1983. Flood-related change in an arid region river. *Earth Surface Processes and Landforms* 8:125–139.

——. 1985. *The Colorado River—instability and basin management.* Resource Publications in Geography, State College, PA.

——. 1988a. *Fluvial Processes in Dryland Rivers.* Springer-Verlag.

——. 1988b. Definition of flood plains along arid-region rivers. Pages 231–242 in *Flood Geomorphology.* V. R. Baker, P. C. Patton and C. Kochel, eds. John Wiley & Sons, Inc.

——. 1999. Dam nation: A geographic census of American dams and their large-scale hydrologic impacts. *Water Res. Research* 35:1305–1311.

——. 2001. Damage control: Restoring the physical integrity of America's rivers. *Ann. Assoc. of Amer. Geog.* 91:1–27.

Graf, W. L., J. Stromberg and B. Valentine. 2002. Rivers, dams, and willow flycatchers: A summary of their science and policy connections. *Geomorphology* 47:169–188.

Graham, J. D. 1852. *Report of Lieutenant Colonel Graham on the subject of the boundary line between the United States and Mexico.* U.S. 32nd Congress, 1st Session, Executive Document 121:1–250.

Granier, A. 1987. Evaluation of transpiration in a Douglas fir stand by means of sap flow measurements. *Tree Physiology* 3:309–320.

Gray, S. T., J. L. Betancourt, C. L. Fastie and S. T. Jackson. 2003. Patterns and sources of multidecadal oscillations in drought-sensitive tree-ring records from the central and southern Rocky Mountains. *Geophys. Res. Let.* 30:L12205.

Gregory, S. V., F. J. Swanson, W. A. McKee and K. W. Cummins. 1991. An ecosystem perspective of riparian zones. *BioScience* 41:540–551.

Gries, D., F. Zeng, A. Foetzki, S. K. Arndt, H. Bruelheide, F. M. Thomas, X. Zhang and M. Runge. 2003. Growth and water relations of *Tamarix ramosissima* and *Populus euphratica* on Taklamakan desert dunes in relation to depth to a permanent water table. *Plant Cell and Environ.* 26:725–736.

Griffen, G. F., D. M. Stafford Smith, S. R. Morton, G. E. Allan and K. A. Masters. 1989. Status and implications of the invasion of tamarisk (*Tamarix aphylla*) on the Finke River, Australia. *J. Environ. Manag.* 29:297–315.

Griffith, J. S. and T. R. Tiersch. 1989. Ecology of fishes in Redfield Canyon, Arizona, with emphasis on *Gila robusta intermedia. S.W. Nat.* 34:131–134.

Griffiths, D., G. L. Bidwell and C. E. Goodrich. 1915. *Native pasture grasses of the United States.* USDA Bull. 201:1–52.

Griggs Engineering. 1927. *San Pedro Municipal Water Conservation District report.* Phoenix. Manuscript. AZ Historical Soc. MS 0737.

Grime, J. P. 1973. Competitive exclusion in herbaceous vegetation. *Nature* 242:344–347.

——. 1977. Evidence for the existence of three primary strategies in plants and its relevance to ecological and evolutionary theory. *Amer. Nat.* 111:1169–1194.

Grimm, N. B. 1987. Nitrogen dynamics during succession in a desert stream. *Ecol.* 68:1157–1170.

——. 1994. Disturbance, succession, and ecosystem processes in streams: A case study from the desert. Pages 93–112 in *Aquatic Ecology: Scale, Pattern, and Process.* P. S. Giller, A. G. Hildrew and D. G. Faffaelli, eds. Oxford, Blackwell Scientific.

Grimm, N. B., A. Chacon, C. N. Dahm, S. W. Hostetler, O. T. Lind, P. L. Starkweather and W. W. Wurtsbaugh. 1997. Sensitivity of aquatic ecosystems to climatic and anthropogenic changes: The Basin and Range, American Southwest and Mexico. *Hydrol. Proc.* 11:1023–1041.

Grimm, N. B. and S. G. Fisher. 1984. Exchange between interstitial and surface water: Implications for stream metabolism and nutrient cycling. *Hydrobiologia* 111:219–228.

——. 1986. Nitrogen limitation in a Sonoran Desert stream. *J. North Amer. Benth. Soc.* 5:2–15.

——. 1989. Stability of periphyton and macroinvertebrates to disturbance by flash floods in a desert stream. *J. North Amer. Benth. Soc.* 8:293–304.

——. 1992. Responses of arid-land streams to changing climate. Pages 211–233 in *Global Climate Change and Freshwater Ecosystems.* P. Firth and S. G. Fisher, eds. Springer-Verlag.

Grimm, N. B., S. G. Fisher and W. L. Minckley. 1981. Nitrogen and phosphorus dynamics in hot desert streams of southwestern USA. *Hydrobiologia* 83:303–312.

Grimm, N. B., S. Gergel and W. McDowell. 2003. Merging aquatic and terrestrial perspectives of nutrient biogeochemistry. *Oecologia* 137:485–501.

Grimm, N. B. and K. C. Petrone. 1997. Nitrogen fixation in a desert stream ecosystem. *Biogeochem.* 37:33–61.

Griswold, T. L., F. D. Parker and V. J. Tepedino. 1998. Bees of the San Rafael Desert: Implications for the bee fauna of the Grand Staircase-Escalante National Monument. Pages 23–34 in *Learning from the Land: Grand Staircase-Escalante National Monument Science Symposium Proc.* L. M. Hill, ed. U.S. Bureau of Land Management. Salt Lake City, UT.

Groeneveld, D. P. and D. E. Crowley. 1988. Root system response to flooding in three desert shrub species. *Functional Ecol.* 2:491–497.

Grubb, P. J. 1977. The maintenance of species-richness in plant communities: The importance of the regeneration niche. *Biol. Rev.* 52:107–145.

Gurevitch, J. and D. K. Padilla. 2004. Are invasive species a major cause of extinctions? *Trends in Ecol. and Evol.* 19:470–474.

Guston, D. H. 2001. Boundary organizations in environmental policy and science: An introduction. *Sci., Technology & Human Values* 26:399–408.

Gutzler, D. S., D. M. Kann and C. Thornbrugh. 2002. Modulation of ENSO-based long-lead outlooks of southwestern U.S. winter precipitation by the Pacific Decadal Oscillation. *Weather and Forecasting* 17:1163–1172.

Haddad, N. M., D. Tilman, J. Haarstad, M. Ritchie and J. Knops. 2001. Contrasting effects of plant richness and composition on insect communities: A field experiment. *Amer. Nat.* 158:17–35.

Hadley, R. F. 1961. *Influence of riparian vegetation on channel shape, northeastern Arizona.* USGS Prof. Paper 424-C:C30–31.

Haig, S. L. 1996. Saltcedar (*Tamarix ramosissima*), an uncommon host for desert mistletoe (*Phoradendron californicum*). *Great Basin Nat.* 56:186–187.

Hair, J. D., G. T. Hepp, L. M. Luckett, K. P. Reese and D. K. Woodward. 1978. Beaver pond ecosystems and their relationship to multi-use natural resource management. Pages 80–92 in R. R. Johnson and McCormick 1978.

Hale, S. F., P. C. Rosen, J. L. Jarchow and G. A. Bradley. 2005. Effects of the chytrid fungus on the Tarahumara frog (*Rana tarahumarae*) in Arizona and Sonora, Mexico. Pages 407–411 in Gottfried et al. 2005.

Hall, E. R. 1981. *The Mammals of North America.* 2 vols. John Wiley & Sons, Inc.

Hall, L. S., M. A. Kasparian, D. Van Vuren and D. A. Kelt. 2000. Spatial organization and habitat use of feral cats (*Felis catus* L.) in Mediterranean California. *Mammalia* 64:19–28.

Halterman, M. D. 2002. Surveys and life-history studies of the yellow-billed cuckoo: Summer 2001. Unpublished report submitted to U.S. Bureau of Reclamation, Lower Colorado Regional Office.

Hamilton, T. H., I. Rubinoff, R. H. Barth Jr. and G. L. Bush. 1963. Species abundance: Natural regulation of insular variation. *Sci.* 142:1575–1577.

Hamlet, A. F. and D. P. Lettenmaier. 1999. Effects of climate change on hydrology and water resources in the Columbia River basin. *J. Amer. Water Res. Assoc.* 35:1597–1623.

Hancock, C. N., P. G. Ladd and R. H. Froend. 1996. Biodiversity and management of riparian vegetation in Western Australia. *For. Ecol. and Manag.* 85:239–250.

Haney, J. A. 2005. Hydrology and biodiversity conservation on the lower San Pedro River. Pages 311–315 in Gottfried et al. 2005.

Hannon, L. E. 2005. Native bee communities in an agricultural/wildland mosaic: Predictors of diversity and community composition. Master's thesis, Northern Arizona Univ.

Hanski, I. 1999. *Metapopulation Ecology.* Oxford University Press.

Hanson, J. D. 1982. Effect of light, temperature and water stress on net photosynthesis in two populations of honey mesquite. *J. Range Manag.* 35:455–458.

Hanson, R. B. 2001. *The San Pedro River—A Discovery Guide.* University of Arizona Press.

Harms, T. K. 2004. Impacts of plant community patchiness, vertical gradients, and temporal variability on microbial nitrogen transformations in a semi-arid riparian zone. Master's thesis, Arizona State Univ.

Harner, M. J., P. W. Ramsey and M. C. Rillig. 2004. Protein accumulation and distribution in floodplain soils and river foam. *Ecol. Let.* 7:829–836.

Harper, J. L. 1977. *Population Biology of Plants.* Academic Press.

Hartnett, D. C. and G. W. T. Wilson. 2002. The role of mycorrhizas in plant community structure and dynamics: Lessons from grasslands. *Plant and Soil* 244: 319–331.

Harvey, M. D. and R. A. Mussetter. 1993. A physical process–biological response model for spawning habitat formation for the endangered Colorado squawfish. *Rivers* 4:114–131.

Hass, C. C. 2001. *Landscape fragmentation and connectivity for carnivores in the Upper San Pedro Basin.* Fort Huachuca Wildlife Office.

Hastings, J. R. 1959. Vegetation change and arroyo cutting in southeastern Arizona. *J. AZ-Nev. Acad. of Sci.* 1:60–67.

Hastings, J. R. and R. M. Turner. 1965. *The Changing Mile.* University of Arizona Press.

Hatten, J. R. and C. E. Paradzick. 2003. A multiscaled model of southwestern willow flycatcher breeding habitat. *J. Wildlife Manag.* 67:774–788.

Haury, E., E. B. Sayles and W. W. Wasley. 1959. The Lehner Mammoth site, southeastern Arizona. *Amer. Antiquity* 25:2–30.

Haynes, C. V., Jr. 1968. Geochronology of late Quaternary alluvium. Pages 591–631 in *Means of Correlation of Quaternary Successions.* R. B. Morrison and H. E. Wright Jr., eds. University of Utah Press.

——. 1987. Curry Draw, Cochise County, Arizona: A late Quaternary stratigraphic record of Pleistocene extinction and paleo-Indian activities. Pages 23–38 in *Geol. Soc. of America Field Guide.* Boulder, CO.

Haynes, C. V., Jr. and B. B. Huckell, eds. 2007. *Murray Springs: A Clovis Site with Multiple Activity Areas in the San Pedro Valley, Arizona.* University of Arizona Press.

Hayward, B., J. E. Heske and C. W. Painter. 1997. Effects of livestock grazing on small mammals at a desert ciénega. *J. Wildlife Manag.* 61:123–129.

Hedin, L. O., J. J. Armesto and A. H. Johnson. 1995. Patterns of nutrient loss from unpolluted, old-growth temperate forests: Evaluation of biogeochemical theory. *Ecol.* 76:493–509.

Heffernan, J. B. and R. A. Sponseller. 2004. Nutrient mobilization and processing in Sonoran desert riparian soils following artificial re-wetting. *Biogeochem.* 70:117–137.

Hehnke, M. and C. P. Stone. 1979. Value of riparian vegetation to avian populations along the Sacramento River system. Pages 228–235 in R. R. Johnson and McCormick 1978.

Heindl, L. A. 1952a. Upper San Pedro basin (ground water). Pages 69–86 in *Ground water in the Gila River Basin and adjacent areas—A summary.* USGS Open-File Report 29.

——. 1952b. Lower San Pedro basin (ground water). Pages 87–100 in *Ground water in the Gila River Basin and adjacent areas—A summary.* USGS Open-File Report 29.

——. 1963. Cenozoic geology in the Mammoth area, Pinal County, Arizona. *USGS Bull.* 1141-E: E1–E41.

Heiny, J. S. and C. M. Tate. 1997. Concentration, distribution, and comparison of selected trace elements in bed sediment and fish tissue in the South Platte River Basin, USA, 1992–1993. *Arch. of Environ. Contamination and Toxicology* 32:246–259.

Hem, J. D. 1967. *Composition of saline residues on leaves and stems of saltcedar (Tamarix pentandra Pallas).* USGS Prof. Paper 491-C.

Hendricks, D. M. 1985. *Arizona Soils.* University of Arizona Press.

Hendrickson, D. A. and W. L. Minckley. 1984 (released in 1985). Ciénegas—vanishing climax communities of the American Southwest. *Desert Plants* 6:130–175.

Henry, J. C. and S. G. Fisher. 2003. Spatial segregation of periphyton communities in a desert stream: Causes and consequences for N cycling. *J. North Amer. Benth. Soc.* 22:511–527.

Hereford, R. 1984. Climate and ephemeral-stream processes: Twentieth-century geomorphology and alluvial stratigraphy of the Little Colorado River, Arizona. *Geol. Soc. Amer. Bull.* 95:654–668.

——. 1993. *Entrenchment and widening of the Upper San Pedro River, Arizona.* Geol. Soc. of Amer. Special Paper 282. Boulder, CO.

Hernandez, M., S. N. Miller, D. C. Goodrich, B. F. Goff, W. G. Kepner, C. M. Edmonds and K. B. Jones. 2000. Modeling runoff response to land cover and rainfall spatial variability in semi-arid watersheds. *Environ. Monitor. and Assess.* 64:285–298.

Hess, B. 2005. State says no AMA needed for basin. *Sierra Vista Herald*, March 10, 2005.

Hicken, E. J. 1984. Vegetation and river channel dynamics. *Can. Geog.* 18:111–126.

Hill, A. R. and M. Cardaci. 2004. Denitrification and organic carbon availability in riparian wetland soils and subsurface sediments. *Soil Sci. Soc. of Amer. J.* 68:320–325.

Hinchman, V. H. and K. W. Birkeland. 1995. Age prediction based on stem size for riparian cottonwood stands. *S.W. Nat.* 40:406–409.

Hinman, K. E. and T. K. Snow, eds. 2003. *Arizona bat conservation strategic plan.* Nongame and Endangered Wildlife Program Tech. Rep. 213. Arizona Game and Fish Dept. Phoenix.

Hipps, L. E., D. Cooper, W. Eichinger, D. Williams, S. M. Schaeffer et al. 1998. A summary of processes which are connected to evaporation of riparian and heterogeneous upland vegetation in arid regions. Pages 13–17 in *Proceedings from the Special Symposium on Hydrology*. E. F. Wood, A. G. Chehbouni, D. C. Goodrich, D. J. Seo and J. R. Zimmerman, eds. Amer. Meteor. Society. Boston.

Hirschboeck, K. K. 1987. Hydroclimatically defined mixed distributions in partial duration flood series. Pages 199–212 in *Hydrologic Frequency Modeling.* V. P. Singh, ed. D. Reidel Pub. Co.

——. 1988. Flood hydroclimatology. Pages 27-49 in *Flood Geomorphology.* V. R. Baker, R. C. Kochel and P. C. Patton, eds. John Wiley & Sons, Inc.

——. 1991. Climate and floods. In *National water summary 1988–1989—Hydrologic events and floods and droughts.* USGS Water-Supply Paper 2375:67–88.

Hoffmeister, D. F. 1986. *Mammals of Arizona.* University of Arizona Press and Arizona Game and Fish Dept.

Hoffmeister, D. F. and W. W. Goodpaster. 1954. The mammals of the Huachuca Mountains, southeastern Arizona. *Illinois Biol. Monog.* 24:1–152.

Hogan, J. F., F. M. Phillips and B. R. Scanlon. 2004. *Groundwater recharge in a desert environment: The southwestern United States.* Water science and applications series 9, Amer. Geophysical Union.

Holl, K. D. 1995. Nectar resources and their influence on butterfly communities on reclaimed coal surface mines. *Restor. Ecol.* 3:76–85.

——. 1996. The effect of coal surface mine reclamation on diurnal lepidopteran conservation. *J. Applied Ecol.* 33:225–236.

Holling, C. S. 1973. Resilience and stability of ecological systems. *Ann. Rev. of Ecol. and Systematics* 4:1–23.

——, ed. 1978. *Adaptive Environmental Assessment and Management.* John Wiley & Sons, Inc.

Holmes, R. M., S. G. Fisher and N. B. Grimm. 1994. Parafluvial nitrogen dynamics in a desert stream ecosystem. *J. North Amer. Benth. Soc.* 13:468–478.

Holmes, R. M., J. B. Jones, S. G. Fisher and N. B. Grimm. 1996. Denitrification in a nitrogen-limited stream ecosystem. *Biogeochem.* 33:125–146.

Hood, W. G. and R. J. Naiman. 2000. Vulnerability of riparian zones to invasion by exotic vascular plants. *Plant Ecol.* 148:105–114.

Hooke, J. M. 1996. River responses to decadal-scale changes in discharge regime: The Gila River, SE Arizona. Pages 191–204 in *Global Continental Changes: the Context of Palaeohydrology.* J. Branson, A. G. Brown and K. J. Gregory, eds. Geol. Soc. Special Pub. No. 115. London.

Horton, J. L. and J. L. Clark. 2001. Water table decline alters growth and survival of *Salix gooddingii* and *Tamarix chinensis* seedlings. *For. Ecol. and Manag.* 140:239–247.

Horton, J. L., T. E. Kolb and S. C. Hart. 2001a. Physiological response to groundwater depth varies among species and with river flow regulation. *Ecol. Appl.* 11:1046–1059.

——. 2001b. Responses of riparian trees to interannual variation in groundwater depth in a semi-arid river basin. *Plant, Cell and Environ.* 24:293–304.

Horton, T. R. and T. D. Bruns. 2001. The molecular revolution in ectomycorrhizal ecology: Peeking into the black-box. *Molecular Ecol.* 10:1855–1871.

House, P. K. and K. K. Hirschboeck. 1997. Hydroclimatological and paleohydrological context of extreme winter flooding in Arizona, 1993. Pages 1–24 in *Storm-Induced Geological Hazards: Case Histories from the 1992–1993 Winter Storm in Southern California and Arizona*, R. A. Larson and J. E. Slosson, eds. Geol. Soc. of Amer. Rev. in Eng. Geol. XI.

Howarth, R. W. and S. G. Fisher. 1976. Carbon, nitrogen, and phosphorus dynamics during leaf decay in nutrient-enriched stream ecosystems. *Freshwater Biol.* 6:221–228.

Howe, W. H. and F. L. Knopf. 1991. On the imminent decline of Rio Grande cottonwoods in central New Mexico. *S.W. Nat.* 36:218–224.

Huang, H. Q. and G. C. Nanson. 1997. Vegetation and channel variation; a case study of four small streams in southeastern Australia. *Geomorph.* 18:237–249.

Huckleberry, G. A. 1994a. Contrasting channel response to flooding on the middle Gila River. *Arizona. Geol.* 22:1083–1086.

——. 1994b. *Historical channel changes on the San Pedro River.* Arizona Geological Survey Open-File Report 96-15.

Hudson, J. W. 1967. Variations in the patterns of torpidity of small homeotherms. Pages 30–46 in *Mammalian Hibernation III*. K. C. Fisher et al., eds. Elsevier Publishing Company.

Huete, A., K. Didan, T. Miura, E. Rodriguez, X. Gao and L. Ferreira. 2002. Overview of the radiometric and biophysical performance of the MODIS vegetation indices. *Remote Sensing of Environ.* 83:195–213.

Huggenberger, P., E. Hoehn, R. Beschta and W. Woessner. 1998. Abiotic aspects of channels and floodplains in riparian ecology. *Freshwater Biol.* 40:407–425.

Hughes, F. M. R. 1997. Floodplain biogeomorphology. *Progress in Physical Geog.* 21:501–529.

Hughes, F. M. R. and S. B. Rood. 2003. Allocation of river flows for restoration of floodplain forest ecosystems: A review of approaches and their applicability in Europe. *Environ. Manag.* 32:12–33.

Hughes, J. M. 1999. Yellow-billed cuckoo (*Coccyzus americanus*). In *The Birds of North America*. A. Poole and F. Gill, eds. Philadelphia, PA. http://www.birds.cornell.edu/AllAboutBirds/BirdGuide/Yellow-billed_Cuckoo_dtl.html

Hultine, K. L., W. L. Cable, S. S. O. Burgess and D. G. Williams. 2003. Hydraulic redistribution by deep roots of a Chihuahuan desert phreatophyte. *Tree Physiology* 23:353–360.

Hultine, K. L., R. L. Scott, W. L. Cable and D. G. Williams. 2004. Hydraulic redistribution by a dominant, warm desert phreatophyte—seasonal patterns and response to precipitation pulses. *Functional Ecol.* 18:530–538.

Hultine, K. L., D. G. Williams, S. S. O. Burgess and T. O. Keefer. 2003. Hydraulic redistribution and diurnal storage capacitance in desert phreatophytes. *Oecologia* 135:167–175.

Humphrey, R. R. 1958. The desert grassland: A history of vegetational change and an analysis of causes. *Bot. Rev.* 24:193–253.

——. 1987. *90 Years and 535 Miles.* University of New Mexico Press.

Hunt, A. G. 1999. Understanding a possible correlation between El Niño occurrence frequency and global warming. *Bull. of the Amer. Meteor. Soc.* 80:297–300.

Hunt, C. B. 1967. *Physiography of the United States.* W. H. Freeman.

Hunter, W. C., B. W. Anderson and R. D. Ohmart. 1987. Avian community structure changes in a mature floodplain forest after extensive flooding. *J. Wildlife Manag.* 51:495–502.

Hunter, W. C., R. D. Ohmart and B. W. Anderson. 1987. Status of breeding riparian-obligate birds in southwestern riverine systems. *Western Birds* 18:10–18.

——. 1988. Use of exotic saltcedar (*Tamarix chinensis*) by birds in arid riparian systems. *Condor* 90:113–123.

Hupp, C. R. 1992. Riparian vegetation recovery patterns following stream channelization: A geomorphic perspective. *Ecol.* 73:1209–1226.

Hupp, C. R. and W. R. Osterkamp. 1996. Riparian vegetation and fluvial geomorphic processes. *Geomorph.* 14:277–295.

Huston, M. A. 1994. *Biological Diversity: The Coexistence of Species on Changing Landscapes.* Cambridge University Press.

——. 1999. Local processes and regional patterns: Appropriate scales for understanding variation in the diversity of plants and animals. *Oikos* 86:393–401.

Huth, A. K. 2003. Geochemical and isotopic mixing models: Two case studies in a snow-dominated and semi-arid environment. PhD diss., University of Arizona.

Hutto, R. L. 1989. The effect of habitat alteration on migratory land birds in a west Mexican tropical deciduous forest: A conservation perspective. *Conserv. Biol.* 3:138–148.

——. 1990. Measuring the availability of food resources. *Stud. in Avian Biol.* 13:20–28.

Hutton, N. H. 1859. Report on the El Paso and Fort Yuma Wagon Road. Pages 77–100 in *Report upon the Pacific Wagon Roads.* A. H. Campbell, ed. 35th Congress, 2nd Session, House Executive Document 108.

Huxman, T. E., K. A. Snyder, D. Tissue, A. J. Leffler, W. Pockman et al. 2004. Precipitation pulses and carbon balance in semi-arid and arid ecosystems. *Oecologia* 141:254–268.

Hyatt, M. W. 2004. *Assessment of Colorado pikeminnow and razorback sucker reintroduction programs in the Gila River basin.* Arizona Game and Fish Dept.

Idso, S. B. 1992. Shrubland expansion in the American Southwest. *Climatic Change* 22:85–86.

Imperial, M. T. and T. Hennessey. 2000. Environmental governance in watersheds: The role of collaboration. Prepared for 8th Biennial Conference of the International Association for the Study of Common Property, Bloomington, IN.

Imperial, M. T. and D. Kauneckis. 2003. Moving from conflict to collaboration: Watershed governance in Lake Tahoe. *Natural Res. J.* 43:1007–1055.

IPCC. 2007. Summary for Policymakers. Pages 1–18 in *Climate Change 2007: The Physical Science Basis. Contribution of Working Group I to the Fourth Assessment Report of the Intergovernmental Panel on Climate Change.* S. Solomon et al., eds. Cambridge University Press.

Irvine, J. R. and N. E. West. 1979. Riparian tree species distribution and succession along the lower Escalante River, Utah. *S.W. Nat.* 24:331–346.

Ives, R. L. 1942. The beaver-meadow complex. *J. Geomorph.* 5:191–203.

Jackson, R., S. Idso, R. Reginato and P. Pinter. 1981. Canopy temperature as a crop water stress indicator. *Water Res. Res.* 17:1133–1138.

Jackson, R. B., J. Canadell, J. R. Ehleringer, H. A. Mooney, O. E. Sala and E. D. Schulze. 1996. A global analysis of root distributions for terrestrial biomes. *Oecologia* 108:389–411.

Jackson, W., T. Martinez, P. Cuplin, W. Minckley, B. Shelby, P. Summers, D. McGlothin and B. Van Haveren. 1987. *Assessment of water conditions and management opportunities in support of riparian values.* U.S. Bureau of Land Management.

Jackson, W. T. 1952. *Wagon Roads West: A Study of Federal Road Surveys and Construction in the TransMississippi West 1846–1869.* University of California Press.

Jacobson, K. M. 2004. Mycorrhizal associations in dryland riparian forests of the southwestern United States. Pages 275–280 in *Fungi in Forest Ecosystems: Systematics, Diversity, and Ecology.* C. L. Cripps, ed. New York Botanical Garden Press.

Jacobson, P. J., K. M. Jacobson, P. L. Angermeier and D. S. Cherry. 2000. Variation in material transport and water chemistry along a large ephemeral river in the Namib Desert. *Freshwater Biol.* 44:481–491.

Jahnke, P. 1994. Modeling of groundwater flow and surface/groundwater interaction for the San Pedro River Basin from Fairbank to Redington. Master's thesis, University of Arizona.

Jakle, M. D. and T. A. Gatz. 1985. Herpetofaunal use of four habitats in the middle Gila River drainage, Arizona. Pages 355–358 in R. R. Johnson et al. 1985.

Janos, D. P. 1980. Mycorrhizae influence tropical succession. *Biotropica* 12:56–64.

——. 1983. Tropical mycorrhizas, nutrient cycles and plant growth. Pages 327–345 in *Tropical Rain Forests: Ecology and Management.* S. L. Sutton, T. C. Whitmore and A. C. Chadwick, eds. Blackwell Science.

Jastrow, J. D., R. M. Miller and J. Lussenhop. 1998. Contributions of interacting biological mechanisms to soil aggregate stabilization in restored prairie. *Soil Biol. Biochem.* 30:905–916.

Jenkins, M. B. 2003. Rhizobial and bradyrhizobial symbionts of mesquite from the Sonoran Desert: Salt tolerance, facultative halophily and nitrate respiration. *Soil Biol. and Biochem.* 35:1675–1682.

Jenny, H. 1941. *Factors of Soil Formation: A System of Quantitative Pedology.* McGraw-Hill Book Co.

Johannes, S. and A. M. Gurnell. 2003. Spatial hydrogeomorphological influences on sediment and nutrient deposition in riparian zones: Observations from the Garonne River, France. *Geomorph.* 49:1–23.

John, K. R. 1963. The effect of torrential rains on the reproductive cycle of *Rhinichthys osculus* in the Chiricahua Mountains, Arizona. *Copeia* 1963:286–291.

Johnson, C. N. 1996. Interactions between mammals and ectomycorrhizal fungi. *Trends in Ecol. and Evol.* 11:503–507.

Johnson, F. E. 1988. Taming flood water: The SCS effort in Bisbee. *Cochise Quarterly* summer 1988.

Johnson, J. D. 2005. Mycorrhizal fungal associations with *Populus fremontii* and

Tamarix ramosissima on the San Pedro River, Arizona. Master's thesis, Arizona State Univ.

Johnson, J. E. 2002. Predator recognition and avoidance by four endangered southwestern USA fishes. Pages 197–207 in *Libro jubilar en honor al Dr. Salvador Contreras Balderas.* M. L. Lozano-Vilano, ed. Universidad Autonoma de Nuevo Leon.

Johnson, J. E. and R. T. Hines. 1999. Effect of suspended sediment on vulnerability of young razorback suckers to predation. *Trans. Amer. Fish. Soc.* 128:648–655.

Johnson, N. C., J. H. Graham and F. A. Smith. 1997. Functioning of mycorrhizal associations along a mutualism-parasitism continuum. *New Phytologist* 135:575–585.

Johnson, R. 1977. *The Central Arizona Project 1918–1968.* University of Arizona Press.

Johnson, R. R. and L. T. Haight. 1985. Avian use of xeroriparian ecosystems in the North American warm deserts. Pages 156–160 in R. R. Johnson et al. 1985.

Johnson, R. R., L. T. Haight and J. M. Simpson. 1977. Endangered species vs. endangered habitats: A concept. Pages 68–79 in R. R. Johnson and Jones 1977.

Johnson, R. R. and D. A. Jones, eds. 1977. *Importance, preservation, and management of riparian habitat: A symposium.* USDA For. Serv. Gen. Tech. Rep. RM-43.

Johnson, R. R. and F. McCormick, eds. 1978. *Strategies for protection and management of floodplain wetlands and other riparian ecosystems.* USDA For. Serv. Gen. Tech. Rep. WO-12.

Johnson, R. R., C. D. Ziebell, D. R. Patton, P. F. Ffolliott and R. H. Hamre, eds. 1985. *Riparian ecosystems and their management: Reconciling conflicting uses.* USDA For. Serv. Gen. Tech. Rep. RM-120.

Johnson, W. C. 1992. Dams and riparian forests: case study from the upper Missouri River. *Rivers* 3: 229-242.

Johnson, W. C. 2000. Tree recruitment and survival in rivers: Influence of hydrological processes. *Hydr. Proc.* 14:3051–3074.

Johnson, W. C., R. L. Burgess and W. R. Keammerer. 1976. Forest overstory vegetation and environment on the Missouri River floodplain in North Dakota. *Ecol. Monog.* 46:59–84.

Johnston, C. A., S. D. Bridgham and J. P. Schubauer-Berigan. 2001. Nutrient dynamics in relation to geomorphology of riverine wetlands. *Soil Sci. Soc. of Amer. J.* 65:557–577.

Johnston, C. E. 1999. The relationship of spawning mode to conservation of North American minnows (Cyprinidae). *Environ. Biol. of Fishes* 55:21–30.

Jones, C. G., J. H. Lawton and M. Shackak. 1994. Organisms as ecosystem engineers. *Oikos* 69:373–386.

Jones, J. B., S. G. Fisher and N. B. Grimm. 1995a. Nitrification in the hyporheic zone of a desert stream ecosystem. *J. North Amer. Benth. Soc.* 14:249–258.

——. 1995b. Vertical hydrologic exchange and ecosystem metabolism in a Sonoran Desert stream. *Ecol.* 76:942–952.

——. 1996. A long-term perspective of dissolved organic carbon transport in Sycamore Creek, Arizona, USA. *Hydrobiologia* 317:183–188.

Jones, J. B., J. D. Schade, S. G. Fisher and N. B. Grimm. 1997. Organic matter dynamics in Sycamore Creek, a desert stream in Arizona, USA. *J. North Amer. Benth. Soc.* 16:78–82.

Jones, K. B. 1981a. *Distribution, ecology, and habitat management of the reptiles and amphibians of the Hualapai-Aquarius Planning Area, Mohave and Yavapai counties, Arizona.* U.S. Bureau of Land Management Technical Note 353. Phoenix.

———. 1981b. Effects of grazing on lizard abundance and diversity in western Arizona. *S.W. Nat.* 26:107–115.

———. 1988a. Distribution and habitat associations of amphibians and reptiles in Arizona: Comparisons by ecosystem. Pages 109–128 in *Management of amphibians, reptiles and small mammals in North America.* R. Szaro, K. Severson and D. Patton, eds. USDA For. Serv. Gen. Tech. Rep. RM-166.

———. 1988b. Herpetofauna of a natural vs. altered riparian ecosystem. Pages 222–227 in *Management of amphibians, reptiles and small mammals in North America.* R. Szaro, K. Severson and D. Patton, eds. USDA For. Serv. Gen. Tech. Rep. RM-166.

Jones, K. B. and P. C. Glinski. 1985. Microhabitats of lizards in a southwestern riparian community. Pages 342–346 in R. R. Johnson et al. 1985.

Jones, K. B., L. P. Kepner and T. E. Martin. 1985. Species of reptiles occupying habitat islands in western Arizona: A deterministic assemblage. *Oecologia* 66:595–601.

Jones, M. D., D. M. Durall and P. B. Tinker. 1990. Phosphorus relationships and production of extramatrical hyphae by two types of willow ectomycorrhizas at different soil phosphorus levels. *New Phytologist* 115:259–267.

Jones, M. D. and S. E. Smith. 2004. Exploring functional definitions of mycorrhizas: Are mycorrhizas always mutualisms? *Can. J. Bot.* 82:1089–1109.

Josephson, P. R. 2002. *Industrialized Nature.* Shearwater Books. Island Press.

Jumpponen, A. 2001. Dark septate endophytes—are they mycorrhizal? *Mycorrhiza* 11:207–211.

Jumpponen, A. and J. M. Trappe. 1998. Dark septate endophytes: A review of facultative biotrophic root-colonizing fungi. *New Phytologist* 140:295–310.

Junk, W. J., P. B. Bayley and R. E. Sparks. 1989. The flood pulse concept in river floodplain systems. Pages 110–127 in *Proceedings, International Large River Symposium.* D. P. Dodge, ed. Can. Special Pub. on Fish. and Aquatic Sci. 106.

Kalischuk, A. R., S. B. Rood and J. M. Mahoney. 2001. Environmental influences on seedling growth of cottonwood species following a major flood. *For. Ecol. and Manag.* 144:75–89.

Karns, H. J. 1954. *Unknown Arizona and Sonora, 1693–1721*; from the Francisco Fernández del Castillo version of Luz de Tierra Incognito, English translation of pt. 2. Silhouettes. Tucson.

Karrenberg, S., P. J. Edwards and J. Kollmann. 2002. The life history of Salicaceae living in the active zone of floodplains. *Freshwater Biol.* 47:733–748.

Kaye, J. P., D. Binkley and C. Rhoades. 2003. Stable soil nitrogen accumulation and flexible organic matter stoichiometry during primary floodplain succession. *Biogeochem.* 63:1–22.

Kaye, J. P. and S. C. Hart. 1997. Competition for nitrogen between plants and soil microorganisms. *Trends in Ecol. and Evol.* 12:139–143.

Kelly, J. F., L. S. DeLay and D. M. Finch. 2002. Density-dependent mass gain by Wilson's warblers during stopover. *Auk* 119:210–213.

Kelly, J. F., D. M. Finch and W. Yong. 2000. Vegetative associations of wood warblers migrating along the middle Rio Grande valley, New Mexico. *S.W. Nat.* 45:159–168.

Kennedy, L. J., J. C. Stutz and J. B. Morton. 1999. *Glomus eburneum* and *G. luteum*:

Two previously undescribed arbuscular mycorrhizal species from a semi-arid grassland. *Mycologia* 91:1083–1093.

Kennedy, L. J., R. L. Tiller and J. C. Stutz. 2002. Associations between arbuscular mycorrhizal fungi and *Sporobolus wrightii* in riparian habitats in arid southwestern North America. *J. Arid Environ.* 50:459–475.

Kepner, W. G. 1982. Reproductive biology of longfin dace (*Agosia chrysogaster*) in a Sonoran Desert stream, Arizona. Master's thesis, Arizona State Univ.

Kepner, W. G., C. M. Edmonds and C. J. Watts. 2002. *Remote sensing and geographic information systems for decision analysis in public resource administration: A case study of the 25 years of landscape change in a southwestern watershed.* EPA/600/R-02/039, US-EPA, Office of Research and Development, National Exposure Research Laboratory. Las Vegas, NV.

Kepner, W. G., D. J. Semmens, S. D. Bassett, D. A. Mouat and D. C. Goodrich. 2004. Scenario analysis for the San Pedro River, analyzing hydrological consequences of a future environment. *Environ. Monitoring and Assess.* 94:115–127.

Kepner, W. G., C. J. Watts, C. M. Edmonds, J. K. Maingi, S. E. Marsh and G. Luna. 2000. Landscape approach for detecting and evaluating change in a semiarid environment. *Environ. Monitoring and Assess.* 64:179–195.

Kessell, J. L. 1966. The puzzling presidio: San Phelipe de Guevavi, alias Terrenate. *New Mexico Hist. Rev.* 41:21–46.

King, J., C. Brown and H. Sabet. 2003. A scenario-based holistic approach to environmental flow assessments for rivers. *River Res. and Appl.* 19:619–639.

Kingsolver, B. 2000. The patience of a saint: San Pedro River. *Nat. Geog.* 197:80–97.

Kinney, C.S. 1894. *A Treatise on the Law of Irrigation.* WH Lowdermilk.

Klemmedson, J. O. and A. R. Tiedemann. 1986. Long-term effects of mesquite (*Prosopis juliflora*) removal on soil characteristics. II. Nutrient availability. *Soil Sci. Soc. of Amer. J.* 50:476–480.

Kline, N. and D. Swann. April 2001. Thousands of vertebrates die on Saguaro National Park roads. *Vermilion Flycatcher.* Tucson Audubon Soc.

Klironomos, J. N., J. McCune, M. Hart and J. Neville. 2000. The influence of arbuscular mycorrhizae on the relationship between plant diversity and productivity. *Ecol. Let.* 3:137–141.

Knighton, D. 1984. *Fluvial Forms and Processes.* Edward Arnold.

——. 1998. *Fluvial Forms and Processes: A New Perspective.* John Wiley & Sons, Inc.

Knighton, D. and G. Nanson. 1997. Distinctiveness, diversity, and uniqueness in arid zone river systems. Pages 185–203 in *Arid Zone Geomorphology: Process, Form, and Change in Drylands*, 2nd ed., D. S. G. Thomas, ed. John Wiley & Sons, Inc.

Knipe, O. D. and H. W. Springfield. 1972. Germinable alkali sacaton seed content of soils in the Rio Puerco Basin, west central New Mexico. *Ecol.* 53:965–968.

Knoerr, A. W. 1956. San Manuel—America's newest large copper producer. *Eng. and Mining J.* 157:75 ff.

Knopf, F. L., R. R. Johnson, T. Rich, F. B. Samson and R. C. Szaro. 1988. Conservation of riparian ecosystems in the United States. *Wilson Bull.* 100:272–284.

Kodric-Brown, A. 1977. Reproductive success and the evolution of breeding territories in pupfish (*Cyprinodon*). *Evolution* 31:750–766.

Koske, R. E. and J. N. Gemma. 1990. VA mycorrhizae in strand vegetation of Hawaii: Evidence for long-distance codispersal of plants and fungi. *Amer. J. Bot.* 77: 466–474.

Krebs, C. J. 1999. *Ecological Methodology*. Addison Wesley Longman.

Kremen, C. 1992. Assessing the indicator properties of species assemblages for natural areas monitoring. *Ecol. Appl.* 2:203–217.

Krombein, K. V., P. D. Hurd, D. R. Smith and B. D. Burks. 1979. *Catalog of Hymenoptera in America north of Mexico. Vol. 2. Apocrita (Aculeata)*. Smithsonian Institution Press.

Krueper, D. J. 1996a. Restoring riparian habitats raised avian numbers. *Bajada* 4:9.

——. 1996b. Effects of livestock management on southwestern riparian ecosystems. Pages 281–301 in Shaw and Finch 1996.

——. 1999. *Annotated checklist to the birds of the Upper San Pedro River Valley*. Report prepared for U.S. Bureau of Land Management, San Pedro Project Office.

Krueper, D. J., J. Bart and T. D. Rich. 2003. Response of vegetation and breeding birds to the removal of cattle on the San Pedro River, Arizona (U.S.A.). *Conserv. Biol.* 17:607–615.

Krueper, D. J., and T. E. Corman. 1987. *Avian inventory of the proposed San Pedro Riparian National Conservation Area interim report 1986–1987*. U.S. Bureau of Land Management, San Pedro Project Office.

Kurle, J. E. and F. L. Pfleger. 1994. The effects of cultural practices and pesticides on VAM fungi. Pages 101–130 in *Mycorrhizae and Plant Health*. F. L. Pleger and R. G. Linderman, eds. APS Press.

Kurschner, H. 2004. Phytosociological studies in the Alashan Gobi—A contribution to the flora and vegetation of Inner Mongolia (NW China). *Phytocoenologia* 34:169–224.

Kustas, W. P., M. S. Moran and J. M. Norman. 2003. Evaluating the spatial distribution of evaporation. Pages 461–492 in *Handbook of Weather, Climate and Water: Atmospheric Chemistry, Hydrology and Societal Impacts*. T. D. Potter and B. R. Colman, eds. John Wiley & Sons, Inc.

Lai, V. S. and J. L. Guynes. 1997. An assessment of the influence of organizational characteristics on information technology adoption decision: A discriminative approach. *IEEE Trans. on Eng. Manag.* 44:146–157.

Lake, P. S. and N. R. Bond. 2007. Australian futures: Freshwater ecosystems and human water usage. *Futures* 39:288–305.

Lamb, B. L. and E. Lord. 1992. Legal mechanisms for protecting riparian resource values. *Water Res. Res.* 28:965–977.

Lambeck, R. J. 1997. Focal species: A multi-species umbrella for nature conservation. *Conserv. Biol.* 11:849–856.

Langbein, W. and S. Schumm. 1958. Yield of sediment in relation to mean annual precipitation. *Trans. of the Amer. Geophys. Union* 39:1076–1084.

Langley, J. A. and B. A. Hungate. 2003. Mycorrhizal controls of belowground litter quality. *Ecol.* 84:2302–2312.

Lannoo, M. L., ed. 2005. *Amphibian Declines: The Conservation Status of United States Species*. University of California Press.

Larsen, E. W., E. H. Girvetz and A. K. Fremier. 2006. Assessing the effects of alternative setback channel constraint scenarios employing a river meander migration model. *Environ. Manag.* 37:880–897.

Larsen, E. W. and S. E. Greco. 2002. Modeling channel management impacts on river migration: A case study of Woodson Bridge State Recreation Area, Sacramento River, California. *Environ. Manag.* 30:209–224.

Larson, L. 1999. *Pomerene Arizona and the Valley of the San Pedro*. Self published. Benson, AZ.

Latham, R. E. and R. E. Ricklefs. 1993. Global patterns of tree species richness in moist forests: Energy-diversity theory does not account for variation in species richness. *Oikos* 67:325–333.

Lavigne, R. J. and M. K. Campion. 2001. The effect of ecosystem stress on the abundance and biomass of leafhoppers (Homoptera: Cicadellidae) on the shortgrass prairie. *J. Entomo. Res. Soc.* 3:5–16.

Leach, W. D. and N. W. Pelkey. 2001. Making watershed partnerships work: A review of the empirical literature. *J. Water Res. Planning and Manag.* 127:378–385.

Leake, J., D. Johnson, D. Donnelly, G. Muckle, L. Boddy and D. Read. 2004. Networks of power and influence: The role of mycorrhizal mycelium in controlling plant communities and agroecosystem functioning. *Can. J. Bot.* 82:1016–1045.

Leal, D. A., R. A. Meyer and B. C. Thompson. 1996. Avian community composition and habitat importance in the Rio Grande corridor of New Mexico. Pages 62–68 in Shaw and Finch 1996.

Lee, K. N. 1999. Appraising adaptive management. *Conserv. Ecol.* 3:3.

Leenhouts, J. M., J. C. Stromberg and R. L. Scott. 2006. *Hydrologic requirements of and consumptive ground-water use by riparian vegetation along the San Pedro River, Arizona*. USGS Scientific Investigations Report 2005-5163.

Leffler, A. J. and A. S. Evans. 1999. Variation in carbon isotope composition among years in the riparian tree *Populus fremontii*. *Oecologia* 119:311–319.

Leishman, M. R. and M. Westoby. 1994. The role of seed size in seedling establishment in dry soil conditions: Experimental evidence from semi-arid species. *J. Ecol.* 82:249–258.

Leliavsky, S. 1955. *An Introduction to Fluvial Hydraulics*. Constable.

Leopold, L. B., M. G. Wolman and J. P. Miller. 1964. *Fluvial Processes in Geomorphology*. W. H. Freeman and Co.

Lesica, P. and T. H. DeLuca. 2004. Is tamarisk allelopathic? *Plant and Soil* 267:357–365.

Leslie, L. L., C. E. Velez and S. A. Bonar. 2005. Diet and consumption rates of native fishes by nonnative fishes in the Verde River, Arizona. *Proc. of the Desert Fishes Council* 35:45.

Lettenmaier, D. P., A. W. Wood, R. N. Palmer, E. F. Wood and E. Z. Stakhiv. 1999. Water resources implications of global warming: A U.S. regional perspective. *Climatic Change* 43:537–579.

Levine, J. M. 2000a. Species diversity and biological invasions: Relating local process to community pattern. *Sci.* 288:852–854.

———. 2000b. Complex interactions in a streamside plant community. *Ecol.* 81:3431–3444.

Lewis, D. B. and N. B. Grimm. 2007. Hierarchical regulation of nitrogen export from urban catchments: Interactions of storms and landscapes. *Ecol. Appl.* 17:2347–2364.

Lewis, D. B., N. B. Grimm, T. K. Harms and J. D. Schade. 2007. Subsystems, flowpaths, and the spatial variability of nitrogen in a fluvial ecosystem. *Landscape Ecology* 22:911–924.

Lewis, D. B., J. D. Schade, A. K. Huth and N. B. Grimm. 2006. The spatial structure of variability in a semi-arid, fluvial ecosystem. *Ecosystems* 9:386–397.

Lewis, L., L. Clark, R. Krapf, M. Manning, J. Staats et al. 2003. *Riparian area management: Riparian wetland soils.* Tech. Reference 1737-19. U.S. Bureau of Land Management. Denver. BLM/ST/ST-03/001+1737.

Lindsay, E. H., G. A. Smith and C. V Haynes Jr. 1990. Late Cenozoic depositional history and geoarchaeology, San Pedro Valley, Arizona. Pages 9–19 in *Geologic Excursions through the Sonoran Desert Region, Arizona and Sonora.* G. E. Gehrels and J. E. Spencer, eds. Arizona G.S. Special Paper 7. Tucson.

Lite, S. J. 2003. San Pedro river riparian vegetation across water availability and flood disturbance gradients. PhD diss., Arizona State Univ.

Lite, S. J., K. J. Bagstad and J. C. Stromberg. 2005. Riparian plant species richness along lateral and longitudinal gradients of water stress and flood disturbance, San Pedro River, Arizona, USA. *J. Arid Environ.* 63:785–813.

Lite, S. J. and J. C. Stromberg. 2005. Surface water and ground-water thresholds for maintaining *Populus-Salix* forests, San Pedro River, Arizona. *Biol. Conserv.* 125:153–167.

Liverman, D. M., R. G. Varady, O. Chávez and R. Sánchez. 1999. Environmental issues along the United States-Mexico border: Drivers of change and responses of citizens and institutions. *Ann. Rev. of Energy and the Environ.* 24:607–643.

Livingston, M. F. and S. D. Schemnitz. 1996. Summer bird/vegetation associations in tamarisk and native habitat along the Pecos River, southeastern New Mexico. Pages 171–180 in Shaw and Finch 1996.

Lockaby, B. G. and W. H. Conner. 1999. N:P balance in wetland forests: Productivity across a biogeochemical continuum. *Bot. Rev.* 65:171–185.

Lodge, D. J. 1989. The influence of soil moisture and flooding on formation of VA-endo- and ectomycorrhizae in *Populus* and *Salix*. *Plant and Soil* 117:243–253.

——. 2000. Ecto- or arbuscular mycorrhizas—which are best? *New Phytologist* 146:352–354.

Lodge, D. J. and T. R. Wentworth. 1990. Negative associations among VA-mycorrhizal fungi and some ectomycorrhizal fungi inhabiting the same root system. *Oikos* 57:347–356.

Logan, M. 2002. *The Lessening Stream.* University of Arizona Press.

Loheide, S. P. and S. M. Gorelick. 2007. Riparian hydroecology: A coupled model of the observed interactions between groundwater flow and meadow vegetation patterning. *Water Res. Res.* 43:W07414.

Loiselle, P. V. 1994. Male spawning-partner preference in an arena-breeding teleost *Cyprinodon macularius californiensis* Girard (Atherinomorpha: Cyprinodontidae). *Amer. Nat.* 120:721–732.

Lombard, J. P. 1998. *Modeling report for San Pedro River stream flow simulations during the time period 1998–2028 using the University of Arizona Middle San Pedro River Basin MODFLOW model.* Draft report prepared for The Nature Conservancy, Arizona Chapter.

Longwell, A. R. 1974. The Cananea *ejidos*: From private ranch to collective in Sonora. PhD diss., University of Nebraska.

Louda, S. M. 1988. Insect pests and plant stress as consideration for revegetation of disturbed ecosystems. Pages 51–67 in *Rehabilitating Damaged Ecosystems*, Vol. 2. J. Cairns, ed. CRC Press.

Louda, S. M. and S. K. Collinge. 1992. Plant resistance to insect herbivores: A field test of the environmental stress hypothesis. *Ecol.* 73:153–169.

Lovich, J. E. and R. C. DeGouvenain. 1998. Saltcedar invasion in desert wetlands of the southwestern United States: Ecological and political implications. Pages 447–467 in *Ecology of Wetlands and Associated Systems.* S. K. Majumdar et al., eds. Penn. Acad. of Science.

Lowe, C. H. 1989. The riparianness of a desert herpetofauna. Pages 143–148 in *Proceedings of the California Riparian Systems Conference: Protection, Management, and Restoration for the 1990s.* D. Abell, ed. USDA For. Serv. Gen. Tech. Rep. PSW-110.

Lowe, C. H., D. S. Hinds and E. A. Halpern. 1967. Experimental catastrophic selection and tolerances to low oxygen concentration in native Arizona freshwater fishes. *Ecol.* 48:1013–1017.

Lowe, C. H., C. R. Schwalbe and T. B. Johnson. 1986. *The venomous reptiles of Arizona.* Nongame Branch Arizona Game and Fish Dept. Phoenix.

Lowrance, R., R. L. Todd, J. Fail, O. Hendrickson, R. Leonard and L. E. Asmussen. 1984. Riparian forests as nutrient filters in agricultural watersheds. *Bioscience* 34:374–377.

Lozano, G. A., S. Perrault and R. E. Lemon. 1996. Age, arrival date and reproductive success of male American Redstarts *Stephaga ruticilla. J. Avian Biol.* 27:164–170.

Lyon, J. and C. L. Sagers. 1998. Structure of herbaceous plant assemblages in a forested riparian landscape. *Plant Ecol.* 138:1–16.

Lytle, D. A. and N. L. Poff. 2004. Adaptation to natural flow regimes. *Trends in Ecology and Evolution* 19:94–100.

Lytle, D. A. and D. M. Merritt. 2004. Hydrologic regimes and riparian forests: a structured population model for cottonwood. *Ecology* 85: 2493–2503.

MacArthur, J. K. and J. W. MacArthur. 1961. On bird species diversity. *Ecol.* 42:594–598.

MacArthur, R. H. 1972. *Geographical Ecology.* Harper and Row.

MacArthur, R. H. and E. O. Wilson. 1967. *The Theory of Island Biogeography.* Princeton University Press.

Mack, R. N. 1981. Invasion of *Bromus tectorum* L. into western North America: An ecological chronicle. *Agro-Ecosystems* 7:145–165.

——. 1996. Predicting the identity and fate of plant invaders: Emergent and emerging approaches. *Biol. Conserv.* 78:107–121.

Mahoney, J. M. and S. B. Rood. 1998. Streamflow requirements for cottonwood seedling recruitment—an integrative model. *Wetlands* 18:634–645.

Makings, E. 2006. Flora of the San Pedro Riparian National Conservation Area. *Desert Plants* 22:3–103.

Malanson, G. P. 1993. *Riparian Landscapes.* Cambridge University Press.

Malo, J. E. and F. Suarez. 1996. New insights into pasture diversity: The consequences of seed dispersal in herbivore dung. *Biodiversity Let.* 3:54–57.

Manabe, S., P. C. D. Milly and R. Wetherald. 2004. Simulated long-term changes in river discharge and soil moisture due to global warming. *Hydrological Sci. J.* 49:625–642.

Manes, R. R. 1968. The fatal future. Part II: The San Pedro River. *Arizona Game and Fish Dept. Wildlife Views* 15:4–9.

Mantua, N. J., S. R. Hare, Y. Zhang, J. M. Wallace and R. C. Francis. 1997. A Pacific interdecadal climate oscillation with impacts on salmon production. *Bull. Amer. Meteor. Soc.* 78:1069–1079.

Marchetti, M. P. and P. B. Moyle. 2001. Effects of flow regime on fish assemblages in a regulated California stream. *Ecol. Appl.* 11:530–539.

Marchetti, M. P., P. B. Moyle and R. Levine. 2004. Alien fishes in California watersheds: Characteristics of successful and failed invaders. *Ecol. Appl.* 14:587–596.

Marks, J. B. 1950. Vegetation and soil relations in the lower Colorado desert. *Ecol.* 31:176–193.

Marler, M. J., C. A. Zabinski and R. M. Callaway. 1999. Mycorrhizae indirectly enhance competitive effects of an invasive forb on a native bunchgrass. *Ecol.* 80:1180–1186.

Marler, R. J., J. C. Stromberg and D. T. Patten. 2001. Growth response of *Populus fremontii*, *Salix gooddingii*, and *Tamarix ramosissima* seedlings under different nitrogen and phosphorus concentrations. *J. Arid Environ.* 49:133–146.

Marsh, P. C. 1987. Digestive tract contents of adult razorback suckers in Lake Mohave, AZ-NV. *Trans. of the Amer. Fish. Soc.* 116:117–119.

Marsh, P. C., F. J. Abarca, M. E. Douglas and W. L. Minckley. 1989. Spikedace (*Meda fulgida*) and loach minnow (*Tiaroga cobitis*) relative to introduced red shiner (*Cyprinella lutrensis*). Unpublished report to Arizona Game and Fish Dept. Arizona State University.

Marsh, P. C. and J. E. Brooks. 1989. Predation by ictalurid catfishes as a deterrent to re-establishment of hatchery-reared razorback suckers. *S.W. Nat.* 34:188–195.

Marsh, P. C., R. W. Clarkson and S. E. Stefferud. 2000. Native fish conservation and the Central Arizona Project. *Proc. Desert Fishes Council* 32:15–16.

Marsh, P. C. and B. R. Kesner. 2004. Analysis of fish population monitoring data for selected waters of the Gila River basin, Arizona, for the five-year period 1995–1999. Arizona State University.

Marsh, P. C. and C. A. Pacey. 2005. Immiscibility of native and non-native fishes. In *Proceedings of two symposia: Restoring Native Fish to the Lower Colorado River: Interactions of Native and Non-native Fishes, and Restoring Natural Function within a Modified Riverine Environment: The Lower Colorado River.* M. J. Brouder et al., eds. U.S. Fish and Wildlife Service, Albuquerque.

Marsh, P. C., C. A. Pacey and B. R. Kesner. 2003. Decline of the razorback sucker in Lake Mohave, Colorado River, Arizona and Nevada. *Trans. Amer. Fish. Soc.* 132:1251–1256.

Martens, D. A. 2003. Vegetation community impacts on soil carbon, nitrogen and trace gas fluxes. Pages 542–557 in *Interagency Conference on Research in the Watersheds.* USDA-Agricultural Research Service.

Martens, D. A. and J. E. T. McLain. 2005. Climate mitigation potential of the San Pedro River riparian zone. Pages 491–495 in Gottfried et al. 2005.

Martí, E., N. B. Grimm and S. G. Fisher. 1997. Pre- and post-flood retention efficiency of nitrogen in a Sonoran Desert stream. *J. North Amer. Benth. Soc.* 16:805–819.

Martin, P. S. 1963. *The Last 10,000 Years: A Fossil Pollen Record of the American Southwest.* University of Arizona Press.

Martinez, A. and J. Lopez-Portillo. 2003. Allometry of *Prosopis glandulosa* var. *torreyana* along a topographic gradient in the Chihuahuan desert. *J. Vegetation Sci.* 14:111–120.

Mattison, R. H. 1946. Early Spanish and Mexican settlements in Arizona. *New Mexico Historical Rev.* 21:282, 286, 288–289.

Mattson, W. J. and R. A. Haack. 1987. The role of drought in outbreaks of plant-eating insects. *BioScience* 27:110–119.

Mauchamp, A., S. Rambal and J. Lepart. 1994. Simulating the dynamics of a vegetation mosaic: A spatialized functional model. *Ecol. Modelling* 71:107–130.

Mayne, P. 1985. New county record possible for western hooknose snake *Gyalopion canum*. *Arizona Herpetological Association Newsletter* 14:2:7.

Mazzoni, R., A. A. Cunningham, P. Daszak, A. Apolo, E. Perdomo and G. Speranza. 2003. Emerging pathogen of wild amphibians in frogs (*Rana catesbeiana*) farmed for international trade. *Emerging Infectious Diseases* 9:995–998.

McAuliffe, J. R. 1994. Landscape evolution, soil formation, and ecological patterns and processes in Sonoran Desert bajadas. *Ecol. Monog.* 64:111–148.

———. 2003. The interface between precipitation and vegetation: The importance of soils in arid and semiarid environments. Pages 9–27 in *Changing Precipitation Regimes and Terrestrial Ecosystems: A North American Perspective*. J. F. Weltzin and G. R. McPherson, eds. University of Arizona Press.

McCabe, G. J., M. A. Palecki and J. L. Betancourt. 2004. Pacific and Atlantic Ocean influences on multidecadal drought frequency in the United States. *Proc. Nat. Acad. of Sci.* 101:4136–4141.

McCarthey, T. D., C. E. Paradzick, J. W. Rourke, M. W. Sumner and R. F. Davidson. 1998. *Partners in Flight southwestern willow flycatcher 1997 survey and nest monitoring report*. Nongame and Endangered Wildlife Program Tech. Rep. 130. Arizona Game and Fish Dept. Phoenix.

McCarthy, M. S. and W. L. Minckley. 1987. Age estimation for razorback sucker (Pisces: Catostomidae) from Lake Mohave, Arizona and Nevada. *J. AZ-Nev. Acad. Sci.* 21:87–97.

McClain, M. E., E. Boyer, C. Dent, S. Gergel, N. Grimm, P. Groffman, S. Hart, J. Harvey, C. Johnston, E. Mayorga, W. McDowell and G. Pinay 2003. Biogeochemical hot spots and hot moments at the interface of terrestrial and aquatic ecosystems. *Ecosystems* 6:301–312.

McClintock, J. H. 1921. *Mormon Settlement in Arizona*. Manufacturing Stationers. Phoenix.

McDonald, M. G. and A. W. Harbaugh. 1998. *A modular three-dimensional finite-difference ground-water flow model*. USGS TWI 6-A1.

McGrath, L. J. and C. van Riper III. 2005. *Influence of riparian tree phenology on lower Colorado River spring-migrating birds: Implications of flower cueing*. USGS Open-File Report 2005-1140.

McKinney, M. L. 2002. Urbanization, biodiversity, and conservation. *BioScience* 52:883–890.

McLain, J. E. T. and D. A. Martens. 2005. Nitrous oxide flux from soil amino acid mineralization. *Soil Biol. and Biochem.* 37:289–299.

———. 2006. Moisture controls on trace gas fluxes in semiarid riparian soils. *Soil Sci. Soc. of Amer. J.* 70:367–377.

McLaughlin, S. P. 1989. Natural floristic areas of the western United States *J. Biogeog.* 16:239–248.

———. 1995. *An overview of the flora of the sky islands, southeastern Arizona: Diversity, affinities and insularity*. USDA For. Serv. Gen. Tech. Rep. RM 264:66–70.

———. 2004. Riparian flora. 127–167 in M. B. Baker et al. 2004.

McNamee, G. 1994. *Gila: The Life and Death of an American River*. Orion Books.

McNatt, R. 1979. Discovery and loss of a new locality for the Gila topminnow, *Poeciliopsis occidentalis occidentalis* (Poeciliidae). *S.W. Nat.* 24:555–556.

McPhee, J. and W. W. G. Yeh. 2004. Multiobjective optimization for sustainable groundwater management in semiarid regions. *J. Water Res. Planning and Manag. ASCE* 130:490–497.

McPherson, G. R. 1995. The role of fire in the desert grasslands. Pages 130–151 in *The Desert Grassland.* M. P. McClaran and T. R. VanDevender, eds. University of Arizona Press.

McQueen, I. S. and R. F. Miller. 1972. *Soil-moisture and energy relationships associated with riparian vegetation near San Carlos, Arizona.* USGS Prof. Paper 655-E.

Mearns, E. A. 1907. *Mammals of the Mexican Boundary of the United States.* Government Printing Office. Washington, D.C.

Meffe, G. K. 1984. Effects of abiotic disturbance on co-existence of predator-prey fish species. *Ecol.* 65:1525-1534.

——. 1985. Predation and species replacement in American southwestern fishes: A case study. *S.W. Nat.* 30:173–187.

Meffe, G. K., D. A. Hendrickson, W. L. Minckley and J. N. Rinne. 1983. Factors resulting in decline of the endangered Sonoran topminnow *Poeciliopsis occidentalis* (Atheriniformes: Poeciliidae) in the United States. *Biol. Conserv.* 25:135–159.

Meffe, G. K. and W. L. Minckley. 1987. Persistence and stability of fish and invertebrate assemblages in a repeatedly disturbed Sonoran desert stream. *Amer. Midland Nat.* 117:177–191.

Meinzer, O. E. 1927. *Plants as indicators of groundwater.* USGS Water-Supply Paper 577.

Meixner, T., A. K. Huth, P. D. Brooks, M. H. Conklin, R. C. Bales, P. A. Haas and J. R. Petti. 2007. Influence of shifting flow paths on nitrogen concentrations during monsoon floods, San Pedro River, Arizona. *J. Geophysical Sci.-Biogeoscience* 112 (G3): Art. No. G03S03.

Melton, M. A. 1965. The geomorphic and paleoclimatic significance of alluvial deposits in southern Arizona. *J. Geol.* 73:1–38.

Menge, B. A. and J. P. Sutherland. 1987. Community regulation: Variation in disturbance, competition, and predation in relation to environmental stress and recruitment. *Amer. Nat.* 130:730–757.

Menges, C. M. and P. A. Pearthree. 1989. Late Cenozoic tectonism in Arizona and its impact on regional landscape evolution. *Arizona Geological Soc. Digest* 17:649–680.

Merigliano, M. F. 2005. Cottonwood understory zonation and its relation to floodplain stratigraphy. *Wetlands* 25:356–374.

Merrill, P. H. 1981. *Roosevelt's Forest Army: A History of the Civilian Conservation Corps 1933–1942.* Self published. Montpelier, VT.

Meyer, D. F. 1989. *The significance of sediment transport in Arroyo development.* USGS Water-Supply Paper 2349.

Michener, C. D. 2000. *The Bees of the World.* Johns Hopkins University Press.

Millar, R. G. 2000. Influence of bank vegetation on alluvial channel patterns. *Water Res. Res.* 36:1109–1118.

Miller, D., C. Luce and L. Benda. 2003. Time, space, and episodicity of physical disturbance in streams. *For. Ecol. and Manag.* 178:121–140.

Miller, R. M., D. R. Reinhardt and J. D. Jastrow. 1995. External hyphal production of vesicular-arbuscular mycorrhizal fungi in pasture and tallgrass prairie communities. *Oecologia* 103:17–23.

Miller, R. R. 1961. Man and the changing fish fauna of the American Southwest. *Papers Mich. Acad. Sci., Arts, and Let.* 46:365–404.

Miller, R. R., C. Hubbs and F. H. Miller. 1991. Ichthyological exploration of the American West: The Hubbs-Miller era, 1915–1950. Pages 19–40 in W. Minckley and Deacon 1991.

Miller, R. R., W. L. Minckley and S. M. Norris. 2005. *Freshwater Fishes of Mexico.* University of Chicago Press.

Miller, R. R. and H. E. Winn. 1951. Additions to the known fish fauna of Mexico: Three species and one subspecies from Sonora. *J. Wash. Acad. of Sci.* 41:83–84.

Miller, S. N., W. G. Kepner, M. H. Mehaffey, M. Hernandez, R. C. Miller et al. 2002. Integrating landscape assessment and hydrologic modeling for land cover change analysis. *J. Amer. Water Res. Assoc.* 38:915–929.

Miller, S. P. and J. D. Bever. 1999. Distribution of arbuscular mycorrhizal fungi in strands of the wetland grass *Panicum hemitomon* along a wide hydrological gradient. *Oecologia* 119:586–592.

Mills, G. S., J. B. Dunning Jr. and J. M. Bates. 1991. The relationship between breeding bird density and vegetation volume. *Wilson Bull.* 103:468–479.

Minckley, R. L., J. H. Cane, L. Kervin and T. H. Roulston. 1999. Spatial predictability and resource specialization of bees (Hymenoptera: Apoidea) at a superabundant, widespread resource. *Biol. J. Linnean Soc.* 67:119–147.

Minckley, W. L. 1965. Native fishes as a natural resource. Pages 48–60 in *Native Plants and Animals as Resources in Arid Lands of the Southwestern United States.* J. L. Gardner, ed. Committee on Desert and Arid Zones Research. Contribution 8. AAAS. Washington, D.C.

——. 1973. *Fishes of Arizona.* Arizona Game and Fish Dept. Phoenix.

——. 1981. Ecological studies of Aravaipa Creek, central Arizona, relative to past, present, and future uses. Final report to U.S. Bureau of Land Management District Office, Safford, Arizona. Arizona State University.

——. 1983. Status of the razorback sucker, *Xyrauchen texanus* (Abbott), in the lower Colorado River basin. *S.W. Nat.* 28:165–187.

——. 1985. Native fishes and natural aquatic habitats in U.S. Fish and Wildlife Service Region II west of the continental divide. Final report for U.S. Fish and Wildlife Service, Albuquerque, NM. Arizona State University.

——. 1987. Fishes and aquatic habitats. Pages 36–50 in *Assessment of water conditions and management opportunities in support of riparian values: BLM San Pedro River properties, Arizona. Project completion report.* W. L. Jackson, ed. U.S. Bureau of Land Management Service Center. BLM/YA/PT88/004+7200. Denver.

——. 1991. Native fishes of the Grand Canyon region: An obituary? Pages 124–177 in *Colorado River Ecology and Dam Management: Proceedings of a Symposium.* Santa Fe, NM. Nat. Acad. Press. Washington, D.C.

——. 1999a. Frederic Morton Chamberlain's 1904 survey of Arizona fishes, with annotations. *J. Southwest* 41:177–237.

——. 1999b. Ecological review and management recommendations for recovery of the endangered Gila topminnow. *Great Basin Nat.* 59:230–244.

Minckley, W. L. and N. T. Alger. 1968. Fish remains from an archaeological site along the Verde River, Yavapai County, Arizona. *Plateau* 40:91–97.

Minckley, W. L. and W. E. Barber. 1971. Some aspects of biology of the longfin dace, a cyprinid fish characteristic of streams in the Sonoran desert. *S.W. Nat.* 159:459–464.

Minckley, W. L. and T. O. Clark. 1981. Vegetation of the Gila River Resource Area, eastern Arizona. *Desert Plants* 3:124–140.

——. 1984. Formation and destruction of a Gila River mesquite bosque community. *Desert Plants* 6(1):23–30.

Minckley, W. L. and J. E. Deacon. 1968. Southwestern fishes and the enigma of "endangered species." *Sci.* 159:1424–1432.

——, eds. 1991. *Battle against Extinction: Native Fish Management in the American West.* University of Arizona Press.

Minckley, W. L. and M. E. Douglas. 1991. Discovery and extinction of western fishes: A blink of the eye in geologic time. Pages 7–17 in W. Minckley and Deacon 1991.

Minckley, W. L., D. A. Hendrickson and C. E. Bond. 1986. Geography of western North American freshwater fishes: Description and relationships to intracontinental tectonism. Pages 519–614 in *The Zoogeography of North American Freshwater Fishes.* C. H. Hocutt et al., eds. John Wiley & Sons, Inc.

Minckley, W. L. and P. C. Marsh. Forthcoming. *Inland Fishes of the Greater Southwest: Chronicle of a Vanishing Biota.* University of Arizona Press.

Minckley, W. L., P. C. Marsh, J. E. Brooks, J. E. Johnson and B. L. Jensen. 1991. Management toward recovery of the razorback sucker. Pages 283–357 in W. Minckley and Deacon 1991.

Minckley, W. L. and G. K. Meffe. 1987. Differential selection by flooding in stream fish communities of the arid American southwest. Pages 93–104 in *Community and Evolutionary Ecology of North American Stream Fishes.* W. J. Matthews et al., eds. University of Oklahoma Press.

Minckley, W. L. and J. N. Rinne. 1991. *Native fishes of arid lands: A dwindling resource of the desert southwest.* USDA For. Serv. Gen. Tech. Rep. RM-206:1–45.

Minshall, G. W., R. C. Paterson, K. W. Cummins, T. L. Bott, J. R. Sedell et al. 1985. Development in stream ecosystem theory. *Can. J. Fish. and Aquatic Sci.* 37:130–137.

Mitsch, W. J. and J. G. Gosselink. 2000. *Wetlands.* John Wiley & Sons, Inc.

Møller, A. P. 1994. Phenotype-dependent arrival time and its consequences in a migratory bird. *Behavioral Ecol. and Sociobiology* 35:115–122.

Molles, M. C., C. S. Crawford and L. M. Ellis. 1995. Effects of an experimental flood on litter dynamics in the middle Rio Grande riparian ecosystem. *Reg. Rivers—Res. and Manag.* 11:275–281.

Molles, M. C. and C. N. Dahm. 1990. A perspective on El Niño and La Niña: Global implications for stream ecology. *J. North Amer. Benth. Soc.* 9:68–76.

Monteith, J. L. 1965. Evaporation and the environment. *Symposium of the Soc. for Exper. Biol.* 19:205–234.

Montgomery, D. R. and T. B. Abbe. 2006. Influence of logjam-formed hard points on the formation of valley-bottom landforms in an old-growth forest valley, Queets River, Washington, USA. *Quaternary Res.* 65:147–155.

Mooney, H. A., B. B. Simpson and O. T. Solbrig. 1977. Phenology, morphology, physiology. Pages 26–43 in *Mesquite: Its Biology in Two Desert Ecosystems.* US/IBP Synthesis Series 4. B. B. Simpson, ed. Dowden, Hutchinson and Ross, Inc.

Moore, F. R., ed. 2000. Stopover ecology of Nearctic-Neotropical landbird migrants: Habitat relations and conservation implications. *Studies in Avian Biol.* No. 20.

Moore, F. R. and D. A. Aborn. 2000. Mechanisms of en route habitat selection: How do migrants make habitat decisions during stopover? *Studies in Avian Biol.* 20:34–42.

Moore, F. R., S. A. Gauthreaux Jr., P. Kerlinger and T. R. Simons. 1995. Habitat requirements during migration: Important link in conservation. Pages 121–144 in *Ecology and Management of Neotropical Migratory Birds*. T. E. Martin and D. M. Finch, eds. Oxford University Press.

Moore, F. R. and W. Yong. 1991. Evidence of food-based competition among passerine migrants during stopover. *Behavioral Ecol. and Sociobiology* 28:85–90.

Moore, I. D. and G. J. Burch. 1986. Modeling erosion and deposition: Topographic effects. *Trans. Amer. Soc. of Agricultural Engineers (ASAE)* 29:1624–1640.

Moran, M. S. and P. Heilman 2000. Foreword to the Semi-arid Land-Surface-Atmosphere (SALSA) Program Special Issue. *J. Ag. and For. Meterol.* 105:1–2.

Morisawa, M. 1968. *Streams: Their Dynamics and Morphology*. McGraw-Hill.

Morrice, J. A., H. M. Valett, C. N. Dahm and M. E. Campana. 1997. Alluvial characteristics, groundwater-surface water exchange and hydrological retention in headwater streams. *Hydro. Proc.* 11:253–267.

Morrison, M. L., C. J. Ralph, J. Verner and J. R. Jehl Jr., eds. 1990. Avian foraging: Theory, methodology, and applications. *Studies in Avian Biol.* 13.

Morton, J. B., S. P. Bentivenga and J. D. Bever. 1995. Discovery, measurement, and interpretation of diversity in arbuscular endomycorrhizal fungi (Glomales, Zygomycetes). *Can. J. Bot.* 73:S25–S32.

Mouw, J. E. B. and P. B. Alaback. 2003. Putting floodplain hyperdiversity in a regional context: An assessment of terrestrial-floodplain connectivity in a montane environment. *J. Biogeography* 30:87–103.

Moyle, P. B. and T. Light. 1996. Fish invasions in California: Do abiotic factors determine success? *Ecol.* 77:1666–1670.

Mpoame, M. and J. N. Rinne. 1983. Parasites of some fishes native to Arizona and New Mexico. *S.W. Nat.* 28:399–405.

Mueller, G. 1984. Spawning by *Rhinichthys osculus* (Cyprinidae) in the San Francisco River, New Mexico. *S.W. Nat.* 29:354–356.

Mueller, G. and P. C. Marsh. 2002. *Lost, a desert river and its native fishes: A historical perspective of the lower Colorado River.* 2002–0010 ed. Information and Technology Report USGS/BRD/ITR-2002-0010. U.S. Government Printing Office. Denver.

Mueller, G. and R. G. Wydoski. 2004. Reintroduction of the flannelmouth sucker in the lower Colorado River. *North Amer. J. Fish. Manag.* 24:41–46.

Muffey, B. W. 1938. The history of the Lower San Pedro Valley in Arizona. Master's thesis, University of Arizona.

Muldavin, E., P. Durkin, M. Bradley, M. Steuver and P. Mehlhop. 2000. *Handbook of wetland vegetation communities of New Mexico.* Vol. 1. *Classification and community description.* New Mexico Environment Dept., United States EPA, and New Mexico Natural Heritage Program.

Mund-Meyerson, M. J. 1999. Arthropod abundance and composition on native and exotic trees in the Middle Rio Grande riparian forest as related to avian foraging. Master's thesis, University of New Mexico.

Munzer, O. M., H. C. English, A. B. Smith and A. A. Tudor. 2005. *Southwestern willow flycatcher 2004 survey and nest monitoring report.* Nongame and Endangered Wildlife Program Tech. Rep. 244. Arizona Game and Fish Dept. Phoenix.

Murakami, M. and S. Nakano. 2002. Indirect effect of aquatic insect emergence on a terrestrial insect population through predation by birds. *Ecol. Let.* 5:333–337.

Murphy, D. D., A. E. Launer and P. R. Ehrlich. 1983. The role of adult feeding in egg production and population dynamics of the checkerspot butterfly *Euphydryas editha*. *Oecologia* 56:257–263.

Myers, T. J. and S. Swanson. 1995. Impact of deferred rotation grazing on stream characteristics in Central Nevada: A case study. *North Amer. J. Fish. Manag.* 15:428–439.

Myrick, D. F. 1975. *Railroads of Arizona*. Vol. 1. *The Southern Roads*. Howell-North Books. Berkeley, CA.

Nagler, P. L., J. Cleverly, E. Glenn, D. Lampkin, A. Huete and Z. Wan. 2005. Predicting riparian evapotranspiration from MODIS vegetation indices and meteorological data. *Remote Sensing of Environ.* 94:1–17.

Nagler, P. L., E. P. Glenn and T. L. Thompson. 2003. Comparison of transpiration rates among saltcedar, cottonwood and willow trees by sap flow and canopy temperature methods. *Ag. and For. Meterol.* 116:73–89.

Nagler, P. L., E. P. Glenn, T. L. Thompson and A. Huete. 2004. Leaf area index and normalized difference vegetation index as predictors of canopy characteristics and light interception by riparian species on the lower Colorado River. *Ag. and For. Meterol.* 125:1–17.

Nagler, P. L., R. L. Scott, C. Westenburg, J. R. Cleverly, E. P. Glenn and A. R. Huete. 2005. Evapotranspiration on western U.S. rivers estimated using the Enhanced Vegetation Index from MODIS and data from eddy covariance and Bowen ratio flux towers. *Remote Sensing and the Environ.* 97:337–351.

Naiman, R. J., S. E. Bunn, C. Nilsson, G. E. Petts, G. Pinay and L. C. Thompson. 2002. Legitimizing fluvial ecosystems as users of water: An overview. *Environ. Manag.* 30:455–467.

Naiman, R. J. and H. Décamps. 1997. The ecology of interfaces-riparian zones. *Ann. Rev. of Ecol. and System* 28:621–658.

Naiman, R. J., H. Décamps and M. E. McClain. 2005. *Riparia: Ecology, Conservation, and Management of Streamside Communities*. Academic Press.

Naiman, R. J., H. Décamps and M. Pollack. 1993. The role of riparian corridors in maintaining regional biodiversity. *Ecol. Appl.* 3:209–212.

Naiman, R. J., C. A. Johnston and J. C. Kelley. 1988. Alteration of North American streams by beaver. *BioScience* 38:753–762.

Nanson, G. C. 1986. Episodes of vertical accretion and catastrophic stripping: A model of disequilibrium floodplain development. *Geol. Soc. of Amer. Bull.* 97:1467–1475.

National Climatic Data Center. 2004. *Climate of 2003—Aug.: Arizona Drought*.

National Policy Consensus Center. 2002. *Watershed solutions: Collaborative Problem Solving for States and Communities*. Portland State University.

National Research Council. 1990. *A Review of the USGS National Water Quality Assessment*. National Research Council. National Academy Press.

——. 2002. *Riparian Areas: Functions and Strategies for Management*. National Academy Press.

The Nature Conservancy. 1991. Special issue, Last Great Places. *The Nature Conservancy* 41:May/June.

Nelson, S. M. 2003. The western viceroy butterfly (Nymphalidae: *Limenitis archippus obsoleta*): An indicator for riparian restoration in the arid southwestern United States? *Ecol. Indicators* 3:203–211.

Nelson, S. M. and D. C. Andersen. 1994. An assessment of riparian environmental

quality by using butterflies and disturbance susceptibility scores. *S.W. Nat.* 39:137–142.
——. 1999. Butterfly (Paplionoidea and Hesperioidea) assemblages associated with natural, exotic and restored riparian habitats along the Lower Colorado River, USA. *Reg. Rivers: Res. and Manag.* 15:485–504.
Neve, L. C. 1976. The life history of the roundtail chub, *Gila robusta grahami*, at Fossil Creek, Arizona. Master's thesis, Northern Arizona Univ.
Neville, J., J. L. Tessier, I. Morrison, J. Scarratt, B. Canning and J. N. Klironomos. 2002. Soil depth distribution of ecto- and arbuscular mycorrhizal fungi associated with *Populus tremuloides* within a 3-year-old boreal forest clear-cut. *Applied Soil Ecol.* 19:209–216.
Newkirk, W. 1966. Historical geography of Bisbee, Arizona. Master's thesis, University of Arizona.
Newman, R. A. 1989. Developmental plasticity of *Scaphiopus couchii* tadpoles in an unpredictable environment. *Ecol.* 70:1775–1787.
Nichols, W. F., K. T. Killingbeck and P. V. August. 1998. The influence of geomorphological heterogeneity on biodiversity. II. A landscape perspective. *Conserv. Biol.* 12:371–379.
Nijssen, B., G. M. O'Donnell, A. F. Hamlet and D. P. Lettenmaier. 2001. Hydrologic sensitivity of global rivers to climate change. *Climatic Change* 50:143–175.
Nilsson, C., G. Grelsson, M. Johansson and U. Sperens. 1989. Patterns of plant species richness along riverbanks. *Ecol.* 70:77–84.
Nilsson, C. and M. Svedmark. 2002. Basic principles and ecological consequences of changing water regimes: Riparian plant communities. *Environ. Manag.* 30: 468–480.
Nilsson, C., S. J. Xiong, M. E. Johanssonn and L. B. M. Vought. 1999. Effects of leaf-litter accumulation on riparian plant diversity across Europe. *Ecol.* 80:1770–1775.
Noss, R. F., E. T. LaRoe and J. M. Scott. 1995. *Endangered ecosystems of the United States: A preliminary assessment of loss and degradation.* Washington DC. U.S. Dept. of Interior, National Biological Service.
Odum, E. P. 1971. *Fundamentals of Ecology*, 3rd ed. Saunders.
Officer, J. E. 1987. *Hispanic Arizona.* University of Arizona Press.
Ohmart, R. D., B. W. Anderson and W. C. Hunter. 1985. Influence of agriculture on waterbird, wader, and shorebird use along the lower Colorado River. Pages 117–122 in R. R. Johnson et al. 1985.
Oksanen, T. 1990. Exploitation ecosystems in heterogenous habitat complexes. *Evol. Ecol.* 4:220–234.
Oleksyszyn, M. 2001. Vegetation and soil changes in secondary succession of abandoned fields along the San Pedro River. Master's thesis, Arizona State Univ.
Oliver, I. and A. J. Beattie. 1997. Future taxonomic partnerships: Reply. *Conserv. Biol.* 11:575–576.
Osmundson, D. B. and K. P. Burnham. 1998. Status and trends of the endangered Colorado squawfish in the upper Colorado River. *Trans. Amer. Fish. Soc.* 127:957–970.
Osmundson, D. B., R. J. Ryel and T. E. Mourning. 1997. Growth and survival of Colorado squawfish in the upper Colorado River. *Trans. Amer. Fish. Soc.* 126:687–698.
Ota, H, Y. Kurihara, S. Satoh and Y. Esashi. 1991. Development of acetylene-reduction

(nitrogen-fixation) activity on and around imbibed plant seeds. *Soil Biol. and Biogeochem.* 23:9–14.

O'Toole, C. and A. Raw. 1991. *Bees of the World.* Blandford Publishing.

Ough, W. D. and J. C. deVos Jr. 1984. Intermountain travel corridors and their management implications for bighorn sheep. *Desert Bighorn Council Trans.* 1984:32–37.

Pahl-Wostl, C. 2002. Participative and stakeholder-based policy design, evaluation and modeling process. *Integrated Assessment* 3:3–14.

Paine, R. T. 1966. Food web complexity and species diversity. *Amer. Nat.* 100:65–75.

Palmer, M. A., E. S. Bernhardt, J. D. Allan, P. S. Lake, G. Alexander et al. 2005. Standards for ecologically successful river restoration. *J. Appl. Ecol.* 42:208–217.

Paradzick, C. E. 2005. Southwestern willow flycatcher habitat selection along the Gila and Lower San Pedro Rivers, Arizona: Vegetation and hydrogeomorphic considerations. Master's thesis, Arizona State Univ.

Paradzick, C. E., R. F. Davidson, J. W. Rourke, M. W. Sumner, A. M. Wartell and T. D. McCarthey. 2000. *Southwestern willow flycatcher 1999 survey and monitoring report.* Nongame and Endangered Wildlife Program Tech. Rep. 175. Arizona Game and Fish Dept. Phoenix.

Paradzick, C. E. and J. R. Hatten. 2004. Southwestern willow flycatcher habitat change detection analysis: Roosevelt Lake, Arizona, 1985–2001. Pages 60–82 in *Mapping and monitoring southwestern willow flycatcher habitat in Arizona: A remote sensing approach.* P. E. T. Dockens and C. E. Paradzick, eds. Nongame and Endangered Wildlife Program Tech. Rep. 223. Arizona Game and Fish Dept. Phoenix.

Paradzick, C. E. and A. A. Woodward. 2003. Distribution, abundance, and habitat characteristics of southwestern willow flycatchers (*Empidonax traillii extimus*) in Arizona, 1993–2000. *Studies in Avian Biol.* 26:22–29.

Parke, J. G. 1857. Report of explorations for railroad routes. In *Explorations and surveys to ascertain the most practicable and economical route for a railroad from the Mississippi River to the Pacific Ocean.* U.S. 33rd Congress, 2nd Session, Senate Exec. Doc. 78. 7:1–469.

Parker, K. C. 2002. Fire in the pre-European lowlands of the American Southwest. Pages 101–142 in *Fire, Native Peoples, and the Natural Landscape.* T. R. Vale, ed. Island Press.

Parker, K. P. 1972. *An Illustrated Guide to Arizona Weeds.* University of Arizona Press.

Parks, S. J. and L. A. Baker. 1997. Sources and transport of organic carbon in an Arizona river-reservoir system. *Water Res.* 31:1751–1759.

Parmesan, C. and G. Yohe. 2003. A globally coherent fingerprint of climate change impacts across natural systems. *Nature* 421:37–42.

Pataki, D. E., S. E. Bush, P. Gardner, D. K. Solomon and J. R. Ehleringer. 2005. Ecohydrology in a Colorado River riparian forest: Implications for the decline of *Populus fremontii. Ecol. Appl.* 15:1009–1018.

Patrick, R. 2000. Rivers of the United States. Volume 5, Part A, *The Colorado River.* John Wiley and Sons.

Patten, D. T. 1998. Riparian ecosystems of semi-arid North America: Diversity and human impacts. *Wetlands* 18:498–512.

Pattie, J. O. 1905. The personal narrative of James O. Pattie. In Vol. 18, *Early Western Travels, 1748–1846.* R. G. Thwaites, ed. Arthur H. Clark Co.

Patton, P. C. 1988. Drainage basin morphometry and floods. Pages 51–64 in *Flood Geomorphology.* V. R. Baker, P. C. Patton and C. Kochel, eds. John Wiley & Sons, Inc.

Patton, P. C. and S. A. Schumm. 1981. Ephemeral stream processes: Implications for studies of Quaternary valley fills. *Quaternary Res.* 15:24–43.

Paul, E. A. and F. E. Clark. 1996. *Soil Microbiology and Biochemistry*. Academic Press.

Paxton, E., J. Owen and M. Sogge. 1996. Southwestern willow flycatcher response to catastrophic habitat loss. USGS Biological Resources Division unpublished document. Colorado Plateau Res. Station. Flagstaff. AZ.

Perry, C. A., B. N. Aldridge and H. C. Ross. 2001. *Summary of significant floods in the United States, Puerto Rico, and the Virgin Islands, 1970 through 1989.* USGS Water-Supply Paper 2502.

Petanidou, T. and W. N. Ellis. 1996. Interdependence of native bee faunas and floras in changing Mediterranean communities. Pages 210–226 in *The Conservation of Bees.* A. Matheson, S. L. Buchmann, C. O'Toole, P. Westrich and I. H. Williams, eds. Academic Press.

Peterjohn, W. T. and D. L. Correll. 1984. Nutrient dynamics in an agricultural watershed: Observations on the role of a riparian forest. *Ecol.* 65:1466–1475.

Pettit, N. E. and R. J. Naiman. 2005. Flood-deposited wood debris and its contribution to heterogeneity and regeneration in a semi-arid riparian landscape. *Oecologia* 145:434–444.

Pfister, L., J. Kwadijk, A. Musy, A. Bronstert and L. Hoffmann. 2004. Climate change, land use change and runoff prediction in the Rhine-Meuse basins. *River Res. and Appl.* 20:229–241.

Pflieger, W. L. 1975. *The fishes of Missouri.* Missouri Dept. of Conservation.

Phillips, A. 1968. The instability of the distribution of land birds in the Southwest. *Papers of the Archeological Soc. of New Mexico* 1:129–162.

Phillips, A. R. 1948. Geographic variation in *Empidonax traillii*. *Auk* 65:507–514.

Phillips, D. H., J. E. Foss, C. A. Stiles, C. C. Trettin and R. J. Luxmoore. 2001. Soil-landscape relationships at the lower reaches of a watershed at Bear Creek near Oak Ridge, Tennessee. *Catena* 44:205–222.

Phillips, J. V., D. McDoniel, M. P. Capesius and W. Asquith. 1998. *Method to estimate effects of flow-induced vegetation changes on channel conveyances of streams in central Arizona.* USGS Water-Resources Investigations Report 98-4040.

Phillips, W. S. 1963. Depth of roots in soil. *Ecol.* 44:424.

Pianka, E. R. 1967. On lizard species diversity: North American flatland deserts. *Ecol.* 48:333–351.

Pickett, S. T. A. and P. S. White. 1985. *The Ecology of Natural Disturbance and Patch Dynamics*. Academic Press.

Pima Association of Governments. 2001. *Bingham Cienega source water study: Final project report.* Tucson.

Pinay, G., J. C. Clement and R. J. Naiman. 2002. Basic principles and ecological consequences of changing water regimes on nitrogen cycling in fluvial systems. *Environ. Manag.* 30:481–491.

Platts, W. S., C. Armour, G. D. Booth, M. Bryant, J. L. Bufford, P. Cuplin et al. 1987. *Methods for evaluating riparian habitats with applications to management.* USDA For. Serv. Gen. Tech. Rep. INT-221.

Pockman, W. T. and J. S. Sperry. 2000. Vulnerability to xylem cavitation and the distribution of Sonoran Desert vegetation. *Amer. J. Bot.* 87:1287–1299.

Poff, N. L., J. D. Allan, M. B. Bain, J. R. Karr, K. L. Prestegaard, B. D. Richter and J. C.

Stromberg. 1997. The natural flow regime: A paradigm for river conservation and restoration. *BioScience* 47:769–784.

Polis, G. A., M. E. Power and G. R. Huxel. 2004. *Food Webs at the Landscape Level*. University of Chicago Press.

Pollock, M. M., R. J. Naiman and T. A. Hanley. 1998. Plant species richness in riparian wetlands—a test of biodiversity theory. *Ecol.* 79:94–105.

Pool, D. R. and A. L. Coes. 1999. *Hydrogeologic investigations of the Sierra Vista Subbasin of the Upper San Pedro River Basin, Cochise County, Southeast Arizona*. USGS Water-Resources Investigations Report 99-4197.

Pool, D. R. and J. E. Dickinson. 2007. *Ground-water flow model of the Sierra Vista Subwatershed and Sonoran portions of the upper San Pedro basin, southeastern Arizona, United States, and northern Sonora, Mexico*. USGS Scientific Investigations Report 2006-5228.

Postel, S. and B. Richter. 2003. *Rivers for Life: Managing Water for People and Nature*. Island Press.

Potts, S. G., B. Vulliamy, A. Dafni, G. Ne'eman and P. Willmer. 2003. Linking bees and flowers: How do floral communities structure pollinator communities? *Ecol.* 84:2628–2642.

Price, P. W. 1984. *Insect Ecology*. Wiley Publishers.

Pringle, C. M. 2000. Threats to U.S. public lands from cumulative hydrologic alterations outside of their boundaries. *Ecol. Appl.* 10:971–989.

——. 2001. Hydrologic connectivity and the management of biological reserves: A global perspective. *Ecol. Appl.* 11:981–998.

Propst, D. L. 1999. *Threatened and endangered fishes of New Mexico*. New Mexico Dept. of Game and Fish Tech. Rep. 1.

Propst, D. L. and K. R. Bestgen. 1991. Habitat and biology of the loach minnow, *Tiaroga cobitis*, in New Mexico. *Copeia* 1991:29–38.

Propst, D. L., K. R. Bestgen and C. W. Painter. 1986. *Distribution, status, biology, and conservation of the Spikedace (Meda fulgida) in New Mexico*. U.S. Fish and Wildlife Service Endangered Species Report 15. Albuquerque, New Mexico.

Prudic, E. D. 1989. *Documentation of a computer program to simulate stream-aquifer relations using a modular, finite-difference, groundwater flow model*. USGS Open-File Report 88-729.

Pusch, M., D. Fiebig, I. Brettar, H. Eisenmann, B. K. Ellis, L. A. Kaplan and M. A. Lock. 1998. The role of micro-organisms in the ecological connectivity of running waters. *Freshwater Biol.* 40:453–495.

Putman, F., K. Mitchell and G. Bushner. 1988. *Water resources of the Upper San Pedro Basin, Arizona*. Arizona Dept. of Water Resources, Hydrology Division.

Qi, J., M. S. Moran, D. C. Goodrich, R. Marsett, R. Scott et al. 1998. Estimation of evapotranspiration over the San Pedro riparian area with remote and in situ measurements. Pages 55–60 in *Proceedings from the Special Symposium on Hydrology*. E. F. Wood et al., eds. American Meteor. Soc.

Ragab, R. and C. Prudhomme. 2002. Climate change and water resources management in arid and semi-arid regions: Prospective and challenges for the 21st century. *Biosystems Eng.* 81:3–34.

Ramenofsky, E. L. 1984. *From Charcoal to Banking*. Westernlore Press.

Rea, A. M. 1983. *Once a River: Bird Life and Habitat Changes on the Middle Gila*. University of Arizona Press.

Read, D. J. and J. Perez-Moreno. 2003. Mycorrhizas and nutrient cycling in ecosystems—a journey towards relevance. *New Phytologist* 157:475–492.

Redecker, D., I. Hijri and A. Wiemken. 2003. Molecular identification of arbuscular mycorrhizal fungi in roots: Perspectives and problems. *Folia Geobotanica* 38:113–124.

Redecker, D., R. Kodner and L. E. Graham. 2000. Glomalean fungi from the Ordovician. *Sci.* 289:1920–1921.

Redmond, K. T. and R. W. Koch. 1991. Surface climate and streamflow variability in the western United States and their relationship to large-scale circulation indices. *Water Res. Res.* 27:2381–2399.

Reichman, O. J. and E. W. Seabloom. 2002. The role of pocket gophers as subterranean ecosystem engineers. *Trends in Ecol. and Evol.* 17:44–49.

Reighard, J. 1920. The breeding behavior of the suckers and minnows. *Biol. Bull.* 38:1–32.

Reily, P. W. and W. C. Johnson. 1982. The effects of altered hydrologic regime on tree growth along the Missouri River in North Dakota. *Can. J. of Bot.* 60:2410–2423.

Relyea, R. A. and E. E. Werner. 2000. Morphological plasticity in four larval anurans distributed along an environmental gradient. *Copeia* 2000:178–190.

Remington, R. K. 2002. Larval fish drift of Aravaipa Creek. Master's thesis, Arizona State Univ.

Renker, C., M. Zobel, M. Öpik, M. F. Allen, E. B. Allen, M. Vosátka, J. Rydlová and F. Buscot. 2004. Structure, dynamics and restoration of plant communities: Does arbuscular mycorrhiza matter? Pages 189–229 in *Assembly Rules and Restoration Ecology*. V. M. Temperton, R. J. Hobbs, T. Nuttle and S. Halle, eds. Island Press.

Reynolds, H. G. and J. W. Bohning. 1956. Effects of burning on a desert grass-shrub range in southern Arizona. *Ecol.* 37:769–776.

Rice, J., R. D. Ohmart and B. W. Anderson. 1983. Habitat selection attributes of an avian community: A discriminant analysis investigation. *Ecol. Monog.* 53:263–290.

Rich, T. D. 2002. Using breeding land birds in the assessment of western riparian systems. *Wildlife Soc. Bull.* 30:1128–1139.

Rich, T. D., C. J. Beardmore, H. Berlanga, P. J. Blancher, M. S. W. Bradstreet, G. S. Butcher et al. 2004. *Partners in Flight North American landbird conservation plan.* Cornell Lab of Ornithology. Ithaca, NY.

Richard, S. M., S. J. Reynolds, J. E. Spencer and P. A. Pearthree. 2000. *Geologic Map of Arizona*. Arizona Geological Survey Map 35.

Richerson, P. and K. Lum. 1980. Patterns of plant species diversity in California: Relation to weather and topography. *Amer. Nat.* 116:504–536.

Richter, B. D., J. V. Baumgartner, R. Wigington and D. P. Braun. 1997. How much water does a river need? *Freshwater Biol.* 37:231–249.

Richter, B. D., R. Mathews, D. L. Harrison and R. Wigington. 2003. Ecologically sustainable water management: Managing river flows for ecological integrity. *Ecol. Appl.* 13:206–224.

Richter, B. S. and J. C. Stutz. 2002. Mycorrhizal inoculation of big sacaton: Implications for grassland restoration of abandoned agricultural fields. *Rest. Ecol.* 10:607–616.

Richter, B. S., R. L. Tiller and J. C. Stutz. 2002. Assessment of arbuscular mycorrhizal fungal propagules and colonization from abandoned agricultural fields and semi-arid grasslands in riparian floodplains. *Applied Soil Ecol.* 20:227–238.

Richter, H. 2006. Participatory learning on the San Pedro: Designing the crystal ball together. *Southwest Hydr.* 5:24–25.

Richter, R. and J. C. Stromberg. 2005. Seed banks of two montane riparian areas: Implications for restoration. *Biodiversity and Conserv.* 14:993–1016.

Rickerl, D. H., F. O. Sancho and S. Anath. 1994. Vesicular-arbuscular endomycorrhizal colonization of wetland plants. *J. of Environ. Qual.* 23:913–916.

Ries, L. 2003. Placing edge responses into a predictive framework. PhD diss., Northern Arizona Univ.

Ries, L., R. J. Fletcher Jr., J. Battin and T. D. Sisk. 2004. Ecological responses to habitat edges: Mechanisms, models, and variability explained. *Ann. Rev. of Ecol., Evol., and Systematics* 35:491–522.

Rillig, M. C., S. F. Wright and V. T. Eviner. 2002. The role of arbuscular mycorrhizal fungi and glomalin in soil aggregation: Comparing effects of five plant species. *Plant and Soil* 233:325–333.

Rinne, J. N. 1989. Physical habitat use by loach minnow, *Tiaroga cobitis* (Pisces: Cyprinidae), in southwestern desert streams. *S.W. Nat.* 34:109–117.

——. 1991. Habitat use by spikedace, *Meda fulgida* (Pisces: Cyprinidae) in southwestern streams with reference to probable habitat competition by red shiner, *Notropis lutrensis* (Pisces: Cyprinidae). *S.W. Nat.* 36:7–13.

——. 1992. Physical habitat utilization of fish in a Sonoran desert stream, Arizona, southwestern United States. *Ecol. of Freshwater Fish* 1:35–41.

——. 1993. A wildlife viewpoint. Southwestern riparian-stream areas: Habitats for fish. In *Riparian management: common threads and shared interests.* B. Tellman et al., ed. USDA For. Serv. Gen. Tech. Rep. RM-226:46–51.

Rinne, J. N. and W. L. Minckley. 1991. *Native fishes of arid lands: A dwindling resource of the desert Southwest.* USDA For. Serv. Gen. Tech. Rep. RM-206:1–42.

Robertson, B. A. and R. L. Hutto. 2006. A framework for understanding ecological traps and an evaluation of existing evidence. *Ecol.* 87:1075–1085.

Robertson, F. N. 1991. *Geochemistry of ground water in alluvial basins of Arizona and adjacent parts of Nevada, New Mexico, and California.* USGS Prof. Paper 1406-C.

Robinson, A. T., P. P. Hines, J. A. Sorensen and S. D. Bryan. 1998. Parasites and fish health in a desert stream, and management implications for two endangered fishes. *North Amer. J. Fish. Manag.* 18:599–608.

Robinson, S. K., S. I. Rothstein, M. C. Brittingham, L. J. Petit and J. A. Gryzbowski. 1995. Ecology and behavior of cowbirds and their impact on host populations. Pages 428–460 in *Ecology and Management of Neotropical Migrant Birds.* T. Martin and D. Finch, eds. Oxford University Press.

Robinson, T. W. 1965. *Introduction, spread, and areal extent of salt cedar (Tamarix) in the western states.* USGS Prof. Paper 491-A:1–13.

——. 1970. *Evapotranspiration by woody phreatophytes in the Humboldt River Valley near Winnemucca, Nevada.* USGS Prof. Paper 491-D.

Rodgers, W. M. 1965. Historical land occupance of the upper San Pedro Valley since 1870. Master's thesis, University of Arizona.

Roeske, R. A., J. M. Garrett and J. H. Eychaner. 1989. *Floods of Oct. 1983 in southeastern Arizona.* USGS Water-Resources Investigations Report 85-4225-C.

Rogers, B. F. 1994. An ecological survey of the understory of southwestern mesquite bosques and the impacts of grazing on this habitat. Master's thesis, Arizona State Univ.

Rogers, E. M. 1995. *Diffusion of Innovations*, 4th ed. Free Press.

Romero-Lankao, P. 2002. Evaluación social de la transferencia del Distrito de Riego 011 "Alto Lerma." Pages 181–202 in *Primer encuentro de investigadores del agua en la Cuenca Lerma-Chapala*. B. Broehm et al., eds. Colegio Michoacán-Universidad de Guadalajara, Mexico.

Rood, S. B., J. H. Braatne and F. M. R. Hughes. 2003. Ecophysiology of riparian cottonwoods: Stream flow dependency, water relations and restoration. *Tree Physiology* 23:1113–1124.

Rood, S. B., A. R. Kalischuk and J. M. Mahoney. 1998. Initial cottonwood seedling recruitment following the flood of the century of the Oldman River, Alberta, Canada. *Wetlands* 18:557–570.

Rood, S. B. and J. H. Mahoney. 1990. Collapse of riparian poplar forest downstream from dams in western prairies: Probable causes and prospects for mitigation. *Environ. Manag.* 14:451–464.

Root, T. L., J. T. Price, K. R. Hall, S. H. Schneider, C. Rosenzweig and J. A. Pounds. 2003. Fingerprints of global warming on wild animals and plants. *Nature* 421:57–60.

Rose, M. D. and G. A. Polis. 1998. The distribution and abundance of coyotes: The effects of allochthonous food subsidies from the sea. *Ecol.* 79:998–1007.

Rosen, P. C. 1991. Comparative field study of body temperatures in garter snakes. *J. Herp.* 25:301–312.

——. 2005. *Lowland riparian herpetofauna: The San Pedro River in Arizona*. Pages 106–111 in Gottfried et al. 2005.

Rosen, P. C., R. B. Duncan, P. A. Holm, T. B. Persons, S. S. Sartorius and C. R. Schwalbe. 2002. Status and ecology of the giant spotted whiptail (Cnemidophorus burti stictogrammus) in Arizona. Unpublished report to Arizona Game and Fish Dept. Heritage Grant (IIPAM program) I99018.

Rosen, P. C., W. R. Radke and D. J. Caldwell. 2005. Herpetofauna of lowland bottomlands of southeastern Arizona: A comparison of sites. Pages 112–117 in Gottfried et al. 2005.

Rosen, P. C., S. S. Sartorius, C. R. Schwalbe, P. A. Holm and C. H. Lowe. 1998. Annotated checklist of the amphibians and reptiles of the Sulphur Springs Valley, Cochise County, Arizona. Pages 65–80 in *The future of arid grasslands*. B. Tellman, D. M. Finch, C. Edminster and R. Hamre, eds. USDA For. Serv. Proc. RMRS-P-3.

Rosen, P. C. and C. R. Schwalbe. 1988. Status and ecology of the Mexican and narrow-headed gartersnakes (Thamnophis eques megalops and Thamnophis rufipunctatus rufipunctatus) in Arizona. Unpublished report to U.S. Fish and Wildlife Service, Albuquerque, NM.

——. 1998. Status of native and introduced species of amphibians and reptiles at the San Bernardino National Wildlife Refuge. Unpublished report to Arizona Game and Fish Dept. Heritage Program and USFWS.

——. 2001. Bullfrog impacts on native wetland herpetofauna in southern Arizona. Unpublished report to Arizona Game and Fish Dept. Heritage Program (IIPAM I97041) and USFWS.

——. 2002. Widespread effects of introduced species on aquatic reptiles and amphibians in the Sonoran Desert region. Pages 220–240 in Tellman 2002.

Rosenberg, K. V., R. C. Ohmart, W. C. Hunter and B. W. Anderson. 1991. *Birds of the Lower Colorado River Valley*. University of Arizona Press.

Rosenzweig, M. L. 1995. *Species Diversity in Space and Time*. Cambridge University Press.

Ross, S. T. 2001. *Inland fishes of Mississippi*. Mississippi Dept. of Wildlife, Fisheries and Parks.

Rotenberry, J. T., R. E. Fitzner, and W. H. Rickard. 1979. Seasonal variation in avian community structure: Differences in mechanisms regulating diversity. *Auk* 96:499–505.

Rothermel, R. C. 1983. *How to predict the spread and intensity of forest and range fires*. USDA For. Serv. Gen. Tech. Rep. INT-143.

Roudebush, E. M. and P. J. Beilke. 2006. The Apache nitrogen wetland: Groundwater denitrification using constructed wetlands. *Southwest Hydr.* 5:22–23.

Rovey, C. 1989. *Hydrologic investigation of the Upper San Pedro Basin, Southeastern Arizona*. Preliminary report.

Rowley, W. D. 1985. *U.S. Forest Service Grazing and Rangelands: A History*. Texas A&M Press.

Rundel, P. W. and S. B. Sturmer 1998. Native plant diversity in riparian communities of the Santa Monica Mountains, California. *Madrono* 45:93–100.

Ruppert, J. B., R. T. Muth and T. P. Nesler. 1993. Predation on fish larvae by adult red shiner, Yampa and Green rivers, Colorado. *S.W. Nat.* 38:397–399.

Rychener, T. J. 2006. Pattern and process: Effects of riparian fire along the San Pedro River, Arizona. Master's thesis, Arizona State Univ.

Ryel, R. J., A. J. Leffler, M. S. Peek, C. Y. Ivans and M. M. Caldwell. 2004. Water conservation in *Artemisia tridentata* through redistribution of precipitation. *Oecologia* 141:335–345.

Sabo, J. L. 2003. Hot rocks or no hot rocks: Overnight retreat availability and selection by a diurnal lizard. *Oecologia* 136:329–335.

Sabo, J. L. and M. E. Power. 2002. River-watershed exchange: Effects of riverine subsidies on riparian lizards and their terrestrial prey. *Ecol.* 83:1860–1869.

Sabo, J. L., R. Sponseller, M. Dixon, K. Gade, T. Harms, J. Heffernan, A. Jani, G. Katz, C. Soykan, J. Watts and J. Welter. 2005. Riparian zones increase regional species richness by harboring different, not more, species. *Ecol.* 86:56–62.

Sala, A., S. D. Smith and D. A. Devitt. 1996. Water use by *Tamarix ramosissima* and associated phreatophytes in a Mojave Desert floodplain. *Ecol. Appl.* 6:888–898.

Sarr, D. A. 2002. Riparian livestock exclosure research in the western United States: A critique and some recommendations. *Environ. Manag.* 30:516–526.

Sarr, D. A., D. E. Hibbs and M. A. Huston. 2005. A hierarchical perspective of plant diversity. *Quarterly Rev. of Biol.* 80:187–212.

Sartorius, S. S. and P. C. Rosen. 2000. Reproductive and population phenology of the lowland leopard frog in a semi-desert canyon. *S.W. Nat.* 45:267–273.

Sasaki, A., M. Fujiyoshi, S. Shidara and T. Nakatsubo. 2001. Effects of nutrients and arbuscular mycorrhizal colonization on the growth of *Salix gracilistyla* seedlings in a nutrient-poor fluvial bar. *Ecol. Res.* 16:165–172.

Saville, N. M., W. E. Dramstad, G. L. A. Fry and S. A. Corbet. 1997. Bumblebee movement in a fragmented agricultural landscape. *Agriculture, Ecosystems and Environ.* 61:145–154.

Schade, J. D., S. G. Fisher, N. B. Grimm and J. A. Seddon. 2001. The influence of a riparian shrub on nitrogen cycling in a Sonoran Desert stream. *Ecol.* 82:3363–3376.

Schade, J. D. and S. E. Hobbie. 2005. Spatial and temporal variation in islands of fertility in the Sonoran Desert. *Biogeochem.* 73:541–553.

Schade, J. D. and D. B. Lewis. 2006. Plasticity in resource allocation and nitrogen use efficiency in riparian vegetation: Implications for nitrogen retention. *Ecosystems* 9:740–755.

Schade, J. D., E. Martí, J. R. Welter, S. G. Fisher and N. B. Grimm. 2002. Sources of nitrogen to the riparian zone of a desert stream: Implications for riparian vegetation and nitrogen retention. *Ecosystems* 5:68–79.

Schade, J. D., R. Sponseller, S. L. Collins and A. Stiles. 2003. The influence of *Prosopis* canopies on understory vegetation: Effects of landscape position. *J. Veg. Sci.* 14:743–750.

Schade, J. D., J. R. Welter, E. Martí and N. B. Grimm. 2005. Hydrologic exchange and N uptake by riparian vegetation in an arid-land stream. *J. North Amer. Benth. Soc.* 24:19–28.

Schaeffer, S. M., D. G. Williams and D. C. Goodrich. 2000. Transpiration of cottonwood/willow forest estimated from sap flux. *Ag. and For. Meterol.* 105:257–270.

Schluter, D. and R. E. Ricklefs. 1993. Species diversity: An introduction to the problem. Pages 1–10 in *Species Diversity in Ecological Communities: Historical and Geographical Perspectives.* R. E. Ricklefs and D. Schluter, eds. University of Chicago Press.

Schmid, H. P. 1997. Experimental design for flux measurements: Matching scales of observations and fluxes. *Ag. and For. Meterol.* 87:179–200.

Schmidt-Nielsen, K. 1964. *Desert Animals: Physiological Problems of Heat and Water.* Clarendon Press.

Schoenherr, A. A. 1974. Life history of the topminnow *Poeciliopsis occidentalis* (Baird and Girard) in Arizona and an analysis of its interaction with the mosquitofish *Gambusia affinis* (Baird and Girard). PhD diss., Arizona State Univ.

———. 1977. Density dependent and density independent regulation of reproduction in the Gila topminnow, *Poeciliopsis occidentalis* (Baird and Girard). *Ecol.* 58:438–444.

———. 1988. A review of the life history and status of the desert pupfish, *Cyprinodon macularius. Bull. Southern California Acad. Sci.* 87:104–134.

Scholes, R. T. and S. R. Archer. 1997. Tree-grass interactions in savannas. *Annual Rev. of Ecol. and Systematics* 28:517–544.

Schreiber, D. C. and W. L. Minckley. 1981. Feeding interrelations of native fishes in a Sonoran desert stream. *Great Basin Nat.* 41:409–426.

Schultz, C. B. and K. M. Dlugosch. 1999. Nectar and hostplant scarcity limit populations of an endangered Oregon butterfly. *Oecologia* 119:231–238.

Schultz, R. J. 1961. Reproductive mechanism of unisexual and bisexual strains of the viviparous fish Poeciliopsis. *Evolution* 15:302–325.

Schumm, S. 1977. *The Fluvial System.* John Wiley & Sons, Inc.

Schussler, A., D. Schwarzott and C. Walker. 2001. A new fungal phylum, the Glomeromycota, phylogeny and evolution. *Mycological Res.* 105:1413–1421.

Schwemm, M. R. and P. J. Unmack. 2001. Native catostomids as prey of introduced largemouth bass in a central Arizona stream. *Proc. Desert Fishes Council* 33:54.

Scoppettone, G. G. 1988. Growth and longevity of the cui-ui and longevity of other catostomids and cyprinids in western North America. *Trans. Amer. Fish. Soc.* 117:301–307.

Scott, J. A. 1986. *Butterflies of North America.* Stanford University Press.

Scott, M. L., J. M. Friedman and G. T. Auble. 1996. Fluvial process and the establishment of bottomland trees. *Geomorph.* 14:327–339.

Scott, M. L., G. C. Lines and G. T. Auble. 2000. Channel incision and patterns of cottonwood stress and mortality along the Mojave River, California. *J. Arid Environ.* 44:399–414.

Scott, M. L., P. B. Shafroth and G. T. Auble. 1999. Responses of riparian cottonwoods to alluvial water table declines. *Environ. Manag.* 23:347–358.

Scott, M. L., S. K. Skagen, and M. F. Merigliano. 2003. Relating geomorphic change and grazing to avian communities in riparian forests. *Conserv. Biol.* 17:284–296.

Scott, R. L., E. A. Edwards, W. J. Shuttleworth, T. E. Huxman, C. Watts and D. C. Goodrich. 2004. Interannual and seasonal variation in fluxes of water and carbon dioxide from a riparian woodland ecosystem. *Ag. and For. Meterol.* 122:65–84.

Scott, R. L., D. C. Goodrich and L. R. Levick. 2003. A GIS-based management tool to quantify riparian vegetation groundwater use. Pages 222–227 in *Proceedings: 1st Interagency Conference on Research in the Watersheds.* K.G. Renard, S. McElroy, W. Gburek, E. Canfield and R.L. Scott, eds. Benson, AZ.

Scott, R. L., D. Goodrich, L. Levick, R. McGuire, W. Cable, D. Williams et al. 2006. *Determining the riparian groundwater use within the San Pedro Riparian National Conservation Area and the Sierra Vista Sub-Basin, Arizona.* USGS Scientific Investigations Report 2005-5163:107–142.

Scott, R. L., T. E. Huxman, D. G. Williams and D. C. Goodrich. 2006. Ecohydrological impacts of woody-plant encroachment: Seasonal patterns of water and carbon dioxide exchange within a semiarid riparian environment. *Global Change Biol.* 12:311–324.

Scott, R. L., W. J. Shuttleworth, D. C. Goodrich and T. Maddock III. 2000a. The water use of two dominant vegetation communities in a semiarid riparian ecosystem. *Ag. and For. Meterol.* 105:241–256.

Scott, R. L., W. J. Shuttleworth, T. O. Keefer, and A. W. Warrick. 2000b. Modeling multiyear observations of soil moisture recharge in the semi-arid American Southwest. Water Resources Research 36: 2233–2247.

Scott, R. L., C. Watts, J. Garatuza, E. Edwards, D. C. Goodrich, D. G. Williams and W. J. Shuttleworth. 2003. The understory and overstory partitioning of energy and water fluxes in an open canopy, semiarid woodland. *Ag. and For. Meterol.* 114:127–139.

Seager, R., M. Ting, I. Held, Y. Kushnir, J. Lu, G. Vecchi, H.-P. Huang, N. Harnik, A. Leetmaa, N.-C. Lau, C. Li, J. Velez and N. Naik. 2007. Model projections of a more arid climate in southwestern North America. *Sci.* 316:1181–1184.

Segelquist, C. A., M. L. Scott and G. T. Auble. 1993. Establishment of *Populus deltoides* under simulated alluvial groundwater declines. *Amer. Midland Nat.* 130:275–285.

Serrat-Capdevilla, A. and A. Browning-Aiken. In review. Increasing socio-ecological resilience by placing science in the decision table: The role of the San Pedro Basin Decision Support System Model (Arizona).

Serrat-Capdevilla, A., J. B. Valdes, J. Gonzales Perez, K. Baird, L. J. Mata and T. Maddock III. 2007. Modeling climate change impacts—and uncertainty—on the hydrology of a riparian system: The San Pedro Basin (Arizona/Sonora). *J. of Hydrology* 347:48–66.

Sexton, O. J. and C. Phillips. 1986. A qualitative study of fish-amphibian interactions in three Missouri ponds. *Trans. Missouri Acad. of Sci.* 20:25–35.

Seymour, D. J. 2001. Seventeenth-century Sobaipuri political and economic relations in Northwestern New Spain. Unpublished manuscript.

Sferra, S. J., T. E. Corman, C. E. Paradzick, J. W. Rourke, J. A. Spencer and M. W. Sumner. 1997. *Arizona Partners in Flight southwestern willow flycatcher survey: 1993–1996 summary report.* Arizona Game and Fish Dept. Tech. Rep. 113.

Shafiqullah, M., P. E. Damon, D. J. Lynch, S. J. Reynolds, W. A. Rehrig and R. H. Raymond. 1980. K-Ar geochronology and geologic history of southwestern Arizona and adjacent areas. *Arizona Geol. Soc. Digest* 12:201–242.

Shafroth, P. B., G. T. Auble, J. C. Stromberg and D. T. Patten. 1998. Establishment of woody riparian vegetation in relation to annual patterns of streamflow, Bill Williams River, Arizona. *Wetlands* 18:577–590.

Shafroth, P. B., J. M. Friedman and L. S. Ischinger. 1995. Effects of salinity on establishment of *Populus fremontii* (cottonwood) and *Tamarix ramosissima* (saltcedar) in southwestern United States. *Great Basin Nat.* 55:58–65.

Shafroth, P. B., J. C. Stromberg and D. T. Patten. 2000. Woody riparian vegetation response to different alluvial water table regimes. *Western North Amer. Nat.* 60:66–76.

Sharma, V., R. D. MacNish and T. Maddock III. 1997. *Analysis of hydrologic data collected by the U.S. Bureau of Land Management 1987–1995 and recommendations for future monitoring programs.* University of Arizona. Dept. of Hydrology and Water Resources HWR 97-060.

Shaw, D. W. and D. M. Finch, eds. 1996. *Desired future conditions for southwestern riparian ecosystems: Bringing interests and concerns together.* USDA For. Serv. Gen. Tech. Rep. RM-GTR-272.

Sheppard, P. R., A. C. Comrie, G. D. Packin, K. Angersbach and M. K. Hughes. 2002. The climate of the US Southwest. *Climate Res.* 21:219–238.

Sher, A. A. and D. L. Marshall. 2003. Seedling competition between native *Populus deltoides* (Salicaceae) and exotic *Tamarix ramosissima* (Tamaricaceae) across water regimes and substrate types. *Amer. J. Bot.* 90:413–422.

Sher, A. A., D. L. Marshall, and S. A. Gilbert. 2000. Competition between native *Populus deltoides* and invasive *Tamarix ramosissima* and the implications for reestablishing flooding disturbance. *Conserv. Biol.* 14:1744–1754.

Sher, A. A., D. L. Marshall and J. P. Taylor. 2002. Establishment patterns of native *Populus* and *Salix* in the presence of invasive nonnative *Tamarix*. *Ecol. Appl.* 12:760–772.

Sheridan, T. 1995. *Arizona History*. University of Arizona Press.

Shields, F. D., Jr., S. S. Knight and C. M. Cooper. 1994. Effects of channel incision on base flow stream habitats and fishes. *Environ. Manag.* 18:43–57.

Shillingberg, W. B. 1999. *Tombstone, A.T.: A History of Early Mining, Milling, and Mayhem.* Arthur Clark Publishers.

Shure, D. J., P. D. Mooreside and S. M. Ogle. 1998. Rainfall effects on plant-herbivore processes in an upland oak forest. *Ecol.* 79:604–617.

Shuttleworth, W. J. 1988. Evaporation from an Amazonian rainforest. *Proc. Royal Soc. of London. Series B* 233:321–346.

——. 1991. Evaporation models in hydrology. Pages 93–120 in *Land Surface Evaporation.* T. J. Schmugge and J. André, eds. Springer-Verlag.

Sidle, R. C. and A. Sharma. 1996. Stream channel changes associated with mining and grazing in the Great Basin. *J. Environ. Qual.* 25:1111–1121.

Siegel, R. S. and J. H. Brock. 1990. Germination requirements of key southwestern woody riparian species. *Desert Plants* 10:3–8.

Simard, S. W., M. D. Jones and D. M. Durall. 2002. Carbon and nutrient fluxes within and between mycorrhizal plants. Pages 34–74 in *Mycorrhizal Ecology*. M. G. A. van der Heijden and I. R. Sanders, eds. Springer-Verlag.

Simberloff, D. S. and E. O. Wilson. 1969. Experimental zoogeography of islands: The colonization of empty islands. *Ecol.* 50:278–296.

Simonson, S. E., P. A. Opler, T. J. Stohlgren and G. W. Chong. 2001. Rapid assessment of butterfly diversity in a montane landscape. *Biodiversity and Conserv.* 10:1369–1386.

Simpson, B. B., J. L. Neff, and A. R. Moldenke. 1977. *Prosopis* flowers as a resource. Pages 84–107 in *Mesquite, Its Biology in Two Desert Ecosystems*. B. B. Simpson, ed. Dowden, Hutchinson, and Ross.

Simpson, G. G. 1964. Species diversity of North American recent mammals. *System. Zool.* 13:57–73.

Skagen, S. K. 1995. The importance of riparian corridors and oases to migrating birds. Final report to Arizona Game and Fish Dept. Heritage Fund and National Fish and Wildlife Foundation. National Biological Service. Ft. Collins, CO.

Skagen, S. K., J. F. Kelly, C. van Riper III, R. L. Hutto, D. M. Finch, D. J. Krueper and C. P. Melcher. 2005. Geography of spring landbird migration through riparian habitats in southwestern North America. *Condor* 107:212–227.

Skagen, S. K., C. P. Melcher, W. H. Howe and F. L. Knopf. 1998. Comparative use of riparian corridors and oases by migrating birds in southeast Arizona. *Conserv. Biol.* 12:896–909.

Skirvin, S. M., S. E. Drake, M. P. McClaran, S. E. Marsh and D. M. Meko. 2000. Climate change and land tenure: Potential impacts on vegetation and development in the San Pedro watershed, southeastern Arizona. 4th International Conference on Integrating GIS and Environmental Modeling (GIS/EM4): Problems, Prospects and Research Needs. Banff, Alberta, Canada, Sept. 2–8, 2000.

Smith, A. B., P. E. T. Dockens, A. A. Tudor, H. C. English and B. L. Allen. 2004. *Southwestern willow flycatcher 2003 survey and nest monitoring report*. Nongame and Endangered Wildlife Program Tech. Rep. 233. Arizona Game and Fish Dept. Phoenix.

Smith, A. B., C. E. Paradzick, A. A. Woodward, P. E. T. Dockens and T.D. McCarthey. 2002. *Southwestern willow flycatcher 2001 survey and nest monitoring report*. Nongame and Endangered Wildlife Program Tech. Rep. 191. Arizona Game and Fish Dept. Phoenix.

Smith, A. B., A. A. Woodward, P. E. T. Dockens, J. S. Martin and T. D. McCarthey. 2003. *Southwestern willow flycatcher 2002 survey and nest monitoring report*. Nongame and Endangered Wildlife Program Tech. Rep. 210. Arizona Game and Fish Dept. Phoenix.

Smith, D. M., N. A. Jackson, J. M. Roberts and C. K. Ong. 1999. Reverse flow of sap in tree roots and downward siphoning of water by *Grevillea robusta*. *Functional Ecol.* 13:256–264.

Smith, G. E. P. 1942. *Studies of Charleston Dam*. Manuscript. University of Arizona MS AZ563.

Smith, H. M., D. Chiszar, A. Chiszar, D. L. Auth, J. Auth, C. E. Bock, J. H. Bock, J. A. Rybak, R. L. Holland, K. Bonine and G. J. Watkins-Colwell. 1998. Slevin's bunch grass lizard (*Sceloporus slevini*) decimated on the Sonoita Plain, Arizona. *Herp. Rev.* 29:225–226.

Smith, J. D. 2004. The role of riparian shrubs in preventing floodplain unraveling along the Clark Fork of the Columbia River in the Deer Lodge Valley, Montana. *Amer. Geophysical Union Water Science and Application Series* 8:71–85.

Smith, S. D., D. A. Devitt, A. Sala, J. R. Cleverly and D. E. Busch. 1998. Water relations of riparian plants from warm desert regions. *Wetlands* 18:687–696.

Smith, S. D., A. B. Wellington, J. L. Nachlinger and C. A. Fox. 1991. Functional responses of riparian vegetation to streamflow diversion in the eastern Sierra Nevada. *Ecol. Appl.* 1:89–97.

Smith, S. E. and D. J. Read. 1997. *Mycorrhizal Symbiosis.* Academic Press.

Smith, W. 1986. *The effects of eastern north Pacific tropical cyclones on the southwestern United States.* NOAA Tech. Memorandum NWS WR-197.

Snow, T. K. and S. V. Castner. 1996. *Bat inventory of abandoned mines.* U.S. Bureau of Land Management. Tucson Resource Area.

Snow, T. K. and R. J. Olding. 2003. *Performance report, game stocking—beaver.* Arizona Fish and Game Dept.

Snyder, K. A. and D. G. Williams. 2000. Water sources used by riparian trees varies among stream types on the San Pedro River. *J. Ag. and For. Meterol.* 105:227–240.

——. 2003. Defoliation alters water uptake by deep and shallow roots of *Prosopis velutina* (velvet mesquite). *Functional Ecol.* 17:363–374.

Snyder, W. D. and G. C. Miller. 1992. Changes in riparian vegetation along the Colorado River and Rio Grande, Colorado. *Great Basin Nat.* 52:357–363.

Sogge, M. K. and R. M. Marshall. 2000. *A survey of current breeding habitats.* Pages 43–56 in Finch and Stoleson 2000.

Soil Survey Staff. 1999. *Soil taxonomy—a basic system of soil classification for making and interpreting soil surveys*, 2nd ed. USDA-NRCS Agricultural Handbook 436.

Solis, G., W. Brady and R. Lopez. 2000. Flora y vegetacion del Rio San Pedro, Mexico. Unpublished report.

Sonnichsen, C. L. 1974. *Colonel Greene and the Copper Skyrocket.* University of Arizona Press.

Southwest Cotton. 1931. *Maricopa County Municipal Water Conservation District No. One v. Southwest Cotton Company.* 1931. Ariz. 65, 4 p. 2d 369.

Soykan, C. U. and J. L. Sabo. In press. Spatio-temporal food web dynamics along a desert riparian-upland transition. *Ecography.*

Sperry, J. S. 1995. Limitations on stem water transport and their consequences. Pages 105–121 in *Plant Stems.* B. L. Gartner, ed. Academic Press.

Sperry, J. S., F. R. Adler, G. S. Campbell and J. P. Comstock. 1998. Limitation of plant water use by rhizosphere and xylem conductance: Results from a model. *Plant Cell and Environ.* 21:347–359.

Sponseller, R. A. 2007. Precipiation pulses and soil CO2 flux in a Sonoran Desert ecosystem. *Global Change Biology* 13:426–436.

Sponseller, R. A. and S. G. Fisher. 2006. Drainage size, stream intermittency, and ecosystem function in a Sonoran Desert landscape. *Ecosystems* 9:344–356.

Sprent, J. I. 1987. *The Ecology of the Nitrogen Cycle.* Cambridge University Press.

SRAG (Southwest Regional Assessment Group). 2000. Preparing for a changing climate: The potential consequences of climate variability and change. http://www.ispe.arizona.edu/research/swassess/pdf/complete.pdf

SRP. 2002. *Roosevelt Lake habitat conservation plan, Gila and Maricopa Counties, Arizona.* Submitted to USFWS, Arizona Ecological Services Office, Phoenix, Arizona pursuant to Section 10(a)(1)(B) of the Endangered Species Act.

Stamp, N. E. and R. D. Ohmart. 1979. Rodents of desert shrub and riparian woodland in the Sonoran Desert. *S.W. Nat.* 24:279–289.

Stanley, E. H., S. G. Fisher and N. B. Grimm. 1997. Ecosystem expansion and contraction in streams. *BioScience* 47:427–435.

Stebbins, R. C. 2003. *A Field Guide to Western Reptiles and Amphibians*, 3rd ed. Houghton Mifflin Co.

Steel, E. A., R. J. Naiman and S. D. West. 1999. Use of woody debris piles by birds and small mammals in a riparian corridor. *Northwest Science* 73:19–26.

Stefferud, J. A. and S. E. Stefferud. 2003. Systematic investigations of warm-water fish assemblages in San Pedro and Babocomari rivers, San Pedro Riparian National Conservation Area, Arizona, 1990–2003. Unpublished report to U.S. Bureau of Land Management, Hereford, AZ.

——. 2004. Aquatic and riparian surveys of selected stream courses on Sierra Vista and Nogales ranger districts, Coronado National Forest, Cochise and Santa Cruz counties, Arizona. Unpublished report to Coronado National Forest, Tucson.

Stefferud, S. E. and P. N. Reinthal. 2005. Fishes of Aravaipa Creek, Graham and Pinal counties, Arizona: Literature review and history of research and monitoring. Unpublished report to U.S. Bureau of Land Management, Safford, AZ.

Steiger, J. and A. M. Gurnell. 2002. Spatial hydrogeomorphological influences on sediment and nutrient deposition in riparian zones: Observations from the Garonne River, France. *Geomorph.* 49:1–23.

Steiner, F., J. Blair, L. McSherry, S. Guhathakurta, J. Marruffo and M. Holm. 2000. A watershed at a watershed: The potential for environmentally sensitive area protection in the upper San Pedro Drainage Basin (Mexico and USA). *Landscape and Urban Planning* 49:129–148.

Steinitz, C., H. Arias, S. Bassett, M. Flaxman, M. Goode et al. 2003. *Alternative Futures for Changing Landscapes, the Upper San Pedro Basin in Arizona and Sonora.* Island Press.

Stenlund, D. L. and I. D. Charvat. 1994. Vesicular arbuscular mycorrhizae in floating wetland mat communities dominated by *Typha*. *Mycorrhiza* 4:131–137.

Stevens, L. E. 1976. Insect production on native and introduced dominant plant species. Pages 146–154 in *An ecological survey of the riparian zone of the Colorado River between Lee's Ferry and the Grand Cliffs, Arizona.* S. W. Carothers and S. W. Aitchison, eds. National Park Service Colorado River Research Series Tech. Rep. No. 10, Grand Canyon.

——. 1985. Invertebrate herbivore community dynamics on *Tamarix chinensis* and *Salix exigua* in the Grand Canyon, Arizona. Master's thesis, Northern Arizona Univ.

Stevens, L. E., B. T. Brown, J. M. Simpson and R. R. Johnson. 1977. The importance of riparian habitat to migrating birds. Pages 156–164 in R. R. Johnson and Jones 1977.

Stevens, M. A., D. B. Simons and E. V. Richardson. 1975. Nonequilibrium river form. *Proc. Amer. Soc. of Civil Eng., J. Hydraulics Division* 101:557–566.

Stewart, I. T., D. R. Cayan and M. D. Dettinger. 2004. Changes in snowmelt runoff timing in western North America under a "business as usual" climate change scenario. *Climatic Change* 62:217–232.

Stohlgren, T. J., D. Binkley, G. W. Chong, M. A. Kalkhan, L. D. Schell, K. A. Bull, Y. Otsuki, G. Newman, M. Bashkin and Y. Son. 1999. Exotic plant species invade hot spots of native plant diversity. *Ecol. Monog.* 69:25–46.

Stohlgren, T. J., K. A. Bull, Y. Otsuki, C. A. Villa and M. Lee. 1998. Riparian zones as havens for exotic plant species in the central grasslands. *Plant Ecol.* 138:113–125.

Stoleson, S. W., J Agyagos, D. M. Finch, T. McCarthey, J. Uyehara and M. J. Whitfield. 2000. Research needs. Pages 119–127 in Finch and Stoleson 2000.

Stone, E. L. and P. J. Kalisz. 1991. On the maximum extent of tree roots. *For. Ecol. and Manag.* 46:59–102.

Stromberg, J. C. 1997. Growth and survivorship of Fremont cottonwood, Goodding willow, and salt cedar seedlings after large floods in central Arizona. *Great Basin Nat.* 57:198–208.

———. 1998a. Dynamics of Fremont cottonwood (*Populus fremontii*) and saltcedar (*Tamarix chinensis*) populations along the San Pedro River, Arizona. *J. Arid Environ.* 40:133–155.

———. 1998b. Functional equivalency of saltcedar (*Tamarix chinensis*) and Fremont cottonwood (*Populus fremontii*) along a free-flowing river. *Wetlands* 18:675–686.

———. 2007. Seasonal reversals of upland-riparian diversity gradients in the Sonoran Desert. *Diversity and Distrib.* 13:70–83.

Stromberg, J. C., K. J. Bagstad, J. M. Leenhouts, S. J. Lite and E. Makings. 2005. Effects of stream flow intermittency on riparian vegetation of a semiarid region river (San Pedro River, Arizona). *River Res. and Appl.* 21:925–938.

Stromberg, J. C., V. B. Beauchamp, M. D. Dixon, S. J. Lite and C. Paradzick. 2007. Importance of low-flow and high-flow characteristics to restoration of riparian vegetation along rivers in arid southwestern United States. *Freshwater Biol.* 52:651–679.

Stromberg, J. C., M. Briggs, C. Gourley, M. Scott, P. Shafroth and L. Stevens. 2004. Human alterations of riparian ecosystems. Pages 101–126 in M. B. Baker et al. 2004.

Stromberg, J. C., J. Fry and D. T. Patten. 1997. Marsh development after large floods in an alluvial, arid-land river. *Wetlands* 17:292–300.

Stromberg, J. C., S. J. Lite, R. Marler, C. Paradzick, P. B. Shafroth, D. Shorrock, J.White and M. White. 2007. Altered stream flow regimes and invasive plant species: The *Tamarix* case. *Global Ecol. and Biogeog.* 16:381–393.

Stromberg, J. C., S. J. Lite, T. J. Rychener, L. R. Levick, M. D. Dixon and J. M. Watts. 2006. Status of the riparian ecosystem in the Upper San Pedro River, Arizona: Application of an assessment model. *Environ Monitoring and Assess.* 115:145–173.

Stromberg, J. C. and D. T. Patten. 1990a. Riparian vegetation instream flow requirements: A case study from a diverted stream in the eastern Sierra Nevada, California, USA. *Environ. Manag.* 14:185–194.

———. 1990b. Variation in seed size of a southwestern riparian tree, Arizona walnut (*Juglans major*). *Amer. Midland Nat.* 124:269–277.

Stromberg, J. C., D. T. Patten and B. D. Richter. 1991. Flood flows and dynamics of Sonoran riparian forests. *Rivers* 2:221–235.

Stromberg, J. C., B. D. Richter, D. T. Patten and L. G. Wolden. 1993. Response of a Sonoran riparian forest to a 10-year return flood. *Great Basin Nat.* 53:118–130.

Stromberg, J. C., T. J. Rychener, and M. D. Dixon. Forthcoming. Return of fire to a free-flowing desert river: Effects on vegetation. *Restor. Ecol.*

Stromberg, J. C., R. Tiller and B. Richter. 1996. Effects of groundwater decline on riparian vegetation of semiarid regions: The San Pedro River, Arizona, USA. *Ecol. Appl.* 6:113–131.

Stromberg, J. C., J. A. Tress, S. D. Wilkins and S. Clark. 1992. Response of velvet mesquite to groundwater decline. *J. Arid Environ.* 23:45–58.

Strong, D. R., J. H. Lawton and R. Southwood. 1984. *Insects on Plants: Community Patterns and Mechanisms.* Harvard University Press.

Stutz, J. C., R. Copeman, C. A. Martin and J. B. Morton. 2000. Patterns of species composition and distribution of arbuscular mycorrhizal fungi in arid regions of southwestern North America and Namibia, Africa. *Can. J. Bot.* 78:237–245.

Stutz, J. C. and J. B. Morton. 1996. Successive pot cultures reveal high species richness of arbuscular endomycorrhizal fungi in arid ecosystems. *Can. J. Bot.* 74:1883–1889.

Swarth, H. S. 1905. Summer birds of the Papago Indian Reservation and of the Santa Rita Mountains, Arizona. *Condor* 7:22–28.

Swetnam, T. W. 1999. Fire-Southern Oscillation relations in the southwestern United States. *Sci.* 249:1017–1020.

Swetnam, T. W., C. H. Baisan and J. M. Kaib. 2001. Forest fire histories of the Sky Islands of La Frontera. Pages 95–119 in *Changing Plant Life of La Frontera*. G. L. Webster and C. J. Bahre, eds. University of New Mexico Press.

Swetnam, T. W. and J. L. Betancourt. 1990. El Nino-Southern Oscillation (ENSO) phenomena and forest fires in the southwestern United States. Pages 129–134 in *Proc. Pacific Climate (PACLIM) Workshop.* J. L. Betancourt and A. M. MacKay, eds. California Dept. of Water Resources, Interagency Ecological Studies Program Tech. Rep. 32.

——. 1998. Mesoscale disturbance and ecological response to decadal climatic variability in the American Southwest. *J. Climate* 11:3128–3147.

Szaro, R. C. 1980. Factors influencing bird populations in southwestern riparian forests. Pages 403–418 in *Workshop proceedings: Management of western forests and grasslands for nongame birds.* R. M. DeGraaf, ed. USDA For. Serv. Gen. Tech. Rep. INT-86.

——. 1989. Riparian scrubland and community types of Arizona and New Mexico. *Desert Plants* 9:1–138.

Szaro, R. C., S. C. Belfit, J. K. Aitkin, and J. N. Rinne. 1985. Impact of grazing on a riparian garter snake. Pages 359–363 in R. R. Johnson et al. 1985.

Szaro, R. C. and M. D. Jakle. 1985. Avian use of a desert riparian island and its adjacent scrub habitat. *Condor* 87:511–519.

Tabacchi, E., D. L. Correll, R. Hauer, G. Pinay, A. Planty-Tabacchi and R. C. Wissmar. 1998. Development, maintenance and role of riparian vegetation in the river landscape. *Freshwater Biol.* 40:497–516.

Tabacchi, E., L. Lambs, H. Guilloy, A. Planty-Tabacchi, E. Muller and H. Décamps.

2000. Impacts of riparian vegetation on hydrological processes. *Hydro. Proc.* 14:2959–2976.

Tabacchi, E. and A. M. Planty-Tabacchi. 2005. Exotic and native plant community distributions within complex riparian landscapes: A positive correlation. *Ecoscience* 12:412–423.

Tabacchi, E., A. Planty-Tabacchi, M. J. Salinas and H. Décamps. 1996. Landscape structure and diversity in riparian plant communities: A longitudinal comparative study. *Reg. Rivers Res. and Manag.* 12:367–390.

Taylor, J. N., W. R. Courtenay Jr. and J. A. McCann. 1984. Known impact of exotic fishes in the continental United States. Pages 322–373 in *Distribution, Biology, and Management of Exotic Fish.* W. R. Courtenay Jr. et al., eds. Johns Hopkins University Press.

Taylor, R. V. 2003. Factors influencing expansion of the breeding distribution of Bewick's wren into riparian forests of the Rio Grande in central New Mexico. *S.W. Nat.* 48:373–382.

Tellman, B. 2002. Human introduction of exotic species in the Sonoran region. Pages 25–46 in *Invasive Exotic Species in the Sonoran Region.* B. Tellman, ed. University of Arizona Press.

——. 2006. "Varmint" control in Cochise County over the years. Pages 287–294 in Gottffried et al. 2005.

Tellman, B., H. J. Cortner, M. G. Wallace, L. F. DeBano and R. H. Hamre, eds. 1993. *Riparian management: Common threads and shared interests.* USDA For. Serv. Gen. Tech. Rep. RM-226.

Tellman, B., R. Yarde and M. Wallace. 1997. *Arizona's Changing Rivers: How People Have Impacted Arizona's Rivers.* University of Arizona. Water Resources Research Center.

Tevis, J. H. 1954. *Arizona in the 50's.* University of New Mexico Press.

Tharme, R. E. 2003. A global perspective on environmental flow assessment: Emerging trends in the development and application of environmental flow methodologies for rivers. *River Res. and Appl.* 19:397–441.

Thomas, A. T. and I. D. Hodkinson. 1991. Nitrogen, water stress and the feeding efficiency of lepidopteran herbivores. *J. Applied Ecol.* 28:703–720.

Thomas, B. E. and D. R. Pool. 2006. *Trends in streamflow of the San Pedro River, southeastern Arizona, and regional trends in precipitation and streamflow in southeastern Arizona and southwestern New Mexico.* USGS Prof. Paper 1712.

Thompson, J. M., E. P. Bergersen, C. A. Carlson and L. R. Kaeding. 1991. Role of size, condition, and lipid content in the overwinter survival of age-0 Colorado squawfish. *Trans. Amer. Fish. Soc.* 120:346–353.

Thornber, J. J. 1910. The grazing ranges of Arizona. *Agricultural Experiment Station Bulletin* 65:245–360. University of Arizona.

Tibbets, T. M. and M. C. Molles. 2005. C:N:P stoichiometry of dominant riparian trees and arthropods along the Middle Rio Grande. *Freshwater Biol.* 50:1882–1894.

Tickner, D. P., P. G. Angold, A. M. Gurnell and J. O. Mountford. 2001. Riparian plant invasions: Hydrogeomorphical control and ecological impacts. *Prog. in Physical Geog.* 25:22–52.

Tiller, K. S. 1982. Charleston townsite revisited. *J. AZ. Hist.* 23:242–248.

Tiller, R. L. 2004. Seed ecology, productivity, and water source use by *Sporobolus wrightii*. PhD diss., Arizona State Univ.

Timmermann, A., J. Oberhuber, A. Bacher, M. Esch, M. Latif and E. Roeckner. 1999. Increased El Niño frequency in a climate model forced by future greenhouse warming. *Nature* 398:694–697.

Tisdall, J. M., S. E. Smith and P. Rengasamy. 1997. Aggregation of soil by fungal hyphae. *Australian J. Soil Res.* 35:55–60.

Titus, J. H., P. J. Titus, R. S. Nowak and S. D. Smith. 2002. Arbuscular mycorrhizae of Mojave Desert plants. *Western North Amer. Nat.* 62:327–334.

Toft, C. A. 1995. A 10-year demographic study of rabbitbrush (*Chrysothamnus nauseosus*): Growth, survival and water limitation. *Oecologia* 101:1–12.

Tollrian, R. and C. D. Harvell. 1999. *The Ecology and Evolution of Inducible Defenses.* Princeton University Press.

Tonn, W. M. and J. J. Magnuson. 1982. Patterns in the species composition and richness of fish assemblages in northern Wisconsin lakes. *Ecol.* 63:1149–1166.

Tooth, S. and G. C. Nanson. 2000. Equilibrium and nonequilibrium conditions in dryland rivers. *Physical Geog.* 21:183–211.

Towne, D. C. 2002. *Ambient groundwater quality of the Lower San Pedro River Basin: A 2000 baseline study.* Arizona Dept. of Environmental Quality.

Tracy, R. L. and G. E. Walsberg. 2002. Kangaroo rats revisited: Re-evaluating a classic case of desert survival. *Oecologia* 133:449–457.

Trappe, J. M. 1987. Phylogenetic and ecological aspects of mycotrophy in angiosperms from an evolutionary standpoint. Pages 2–25 in *Ecophysiology of VA Mycorrhizal Plants.* G. R. Safir, ed. CRC Press.

Trautman, M. B. 1981. *The Fishes of Ohio*, rev. ed. Ohio State University Press.

Trimble, S. W. and A. C. Mendel. 1995. The cow as a geomorphic agent: A critical review. *Geomorph.* 13:233–253.

Tuan, Y. 1962. Structure, climate, and basin landforms in Arizona and New Mexico. *Ann. Assoc. of Amer. Geog.* 52:51–68.

Turner, R. M. 1974. *Quantitative and historical evidence of vegetation changes along the Upper Gila River, Arizona.* USGS Prof. Paper 655H.

Turner, R. M., R. H. Webb, J. E. Bowers and J. R. Hastings. 2003. *The Changing Mile Revisited.* University of Arizona Press.

Tyree, M. T., K. J. Kolb, S. B. Rood and S. Patino. 1994. Vulnerability to drought-induced cavitation of riparian cottonwoods in Alberta—a possible factor in the decline of the ecosystem? *Tree Physiology* 14:455–466.

Tyus, H. M. 1990. Potamodromy and reproduction of Colorado squawfish in the Green River basin, Colorado and Utah. *Trans. Amer. Fish. Soc.* 119:1035–1047.

Tyus, H. M. and C. A. Karp. 1990. Spawning and movements of razorback sucker, *Xyrauchen texanus*, in the Green River basin of Colorado and Utah. *S.W. Nat.* 35:427–433.

Tyus, H. M. and J. F. Saunders III. 2000. Nonnative fish control and endangered fish recovery: Lessons from the Colorado River. *Fisheries* 25:17–24.

Udall Center. 1998. *Public input digest for the Upper San Pedro River Initiative.* Prepared for the Commission for Environmental Cooperation. University of Arizona. Udall Center for Studies in Public Policy.

Unitt, P. 1987. *Empidonax traillii extimus*: An endangered subspecies. *Western Birds* 18:137–162.

Unmack, P. J. and W. F. Fagan. 2004. Convergence of differentially invaded systems toward invader-dominance: Time-lagged invasions as a predictor in desert fish communities. *Biol. Invasions* 6:233–243.

Upper San Pedro Partnership. 2004. *2004 Water Management and Conservation Plan.* Sierra Vista, AZ.

USBLM. 1990. *San Pedro River ecosystem—an acquisition plan.* Prepared by U.S. Bureau of Land Management. Safford District Office, Arizona.

——. 1995. *Instream flow water rights request: The San Pedro River—Cascabel.* An analysis of water and water dependent resources in support of instream flow water rights. Report to Arizona Dept. of Water Resources and Arizona Game and Fish Dept.

——. 1998a. *Plan and NEPA decision record for re-establishment of North American beaver (Castor canadensis) into the San Pedro Riparian National Conservation Area.* U.S. Bureau of Land Management. Tucson.

——. 1998b. *The Upper San Pedro River Basin of the United States and Mexico: A resource dictionary and an overview of natural resource issues confronting decision-makers and natural resource managers.* U.S. Bureau of Land Management. Phoenix.

——. 2003. *Final Environmental Impact Statement, Dos Pobres/San Juan Project, Safford.* Vol. 2.

USBOR. 2002. *Lower Colorado River accounting system demonstration of technology calendar year 2001.* U.S. Bureau of Reclamation. Boulder City.

——. 2007. *Appraisal report: Augmentation alternative for the Sierra Vista sub-watershed.* U.S. Bureau of Reclamation.

USDA, NRCS. 2003a. *Soil survey of Cochise County, Arizona—Douglas-Tombstone part.* USDA, NRCS. Washington, D.C.

——. 2003b. G. W. Hurt, P. M. Whited and R. F. Pringle, eds. *Field indicators of hydric soils in the United States, Version 5.01.* USDA, NRCS in cooperation with National Technical Committee for Hydric Soils. Fort Worth, TX.

USDI. 2005. *Water management of the regional aquifer in the Sierra Vista Subwatershed, Arizona–2004.* U.S. Dept. of Interior, Report to Congress.

——. 2006. *Water management of the regional aquifer in the Sierra Vista Subwatershed, Arizona–2005.* U.S. Geological Survey, Report to Congress.

USDOD. 2003. *Ft. Huachuca consumer confidence report.* U.S. Dept. of Defense. Ft. Huachuca, AZ.

USEPA. 2005. Region 9: SuperFund, Apache Powder Company. EPA ID #AZD008399263.

USFWS. 1980. *Aquatic study. Special report on distribution and abundance of fishes of the lower Colorado River.* U.S. Fish and Wildlife Service. Phoenix.

——. 1984. *Sonoran topminnow (Gila and Yaqui) recovery plan.* U.S. Fish and Wildlife Service. Albuquerque.

——. 1985a. Endangered and threatened wildlife and plants; proposal to determine the loach minnow to be a threatened species and to determine its critical habitat. *Fed. Reg.* 50:25380–25387.

——. 1985b. Endangered and threatened wildlife and plants; proposal to determine the spikedace to be a threatened species and to determine its critical habitat. *Fed. Reg.* 50:25390–25398.

——. 1991a. *Colorado squawfish recovery plan.* U.S. Fish and Wildlife Service. Denver.

——. 1991b. *Loach Minnow (Tiaroga cobitis) recovery plan.* U.S. Fish and Wildlife Service. Phoenix.

———. 1991c. *Spikedace (Meda fulgida) recovery plan.* U.S. Fish and Wildlife Service. Phoenix.

———. 1993. *Desert pupfish (Cyprinodon macularius) recovery plan.* U.S. Fish and Wildlife Service. Albuquerque.

———. 1994a. Endangered and threatened wildlife and plants; animal candidate review for listing as endangered or threatened species; proposed rule. *Fed. Reg.* 59:58982–59028.

———. 1994b. Endangered and threatened wildlife and plants; notice of 90-day and 12-month findings on a petition to reclassify spikedace (*Meda fulgida*) and loach minnow (*Tiaroga cobitis*) from threatened to endangered. *Fed. Reg.* 59:35303–35304.

———. 1995. Final rule determining endangered status for the southwestern willow flycatcher. *Fed. Reg.* 60:10694.

———. 1996. *Biological opinion on operation of the modified Roosevelt Dam in Gila and Maricopa Counties, Arizona.* Fed. Reg. AESO 2-21-95-F-462. U. S. Fish and Wildlife Service. Albuquerque.

———. 1998. *Razorback sucker (Xyrauchen texanus) recovery plan.* U.S. Fish and Wildlife Service. Denver.

———. 2002a. *Colorado pikeminnow (Ptychocheilus lucius) recovery goals: Amendment and supplement to the Colorado squawfish recovery plan.* U.S. Fish and Wildlife Service, Mountain-Prairie Region 6. Denver.

———. 2002b. *Razorback sucker (Xyrauchen texanus) recovery goals: Amendment and supplement to the razorback sucker recovery plan.* U.S. Fish and Wildlife Service, Mountain-Prairie Region 6. Denver.

———. 2002c. *Final recovery plan, southwestern willow flycatcher (Empidonax traillii extimus).* U.S. Fish and Wildlife Service, Region 2.

———. 2005a. Endangered and threatened wildlife and plants; listing the Gila chub as endangered with critical habitat; final rule. *Fed. Reg.* 70:66664–66721.

———. 2005b. Final rule designating critical habitat for the southwestern willow flycatcher (*Empidonax traillii extimus*). *Fed. Reg.* 70:60866. Oct. 19, 2005.

———. 2006. Endangered and threatened wildlife and plants; 12-month finding on a petition to list a distinct population segment of the roundtail chub in the lower Colorado River basin and to list the headwater chub as endangered or threatened with critical habitat. *Fed. Reg.* 71:26007–26017.

———. 2007. Endangered and threatened wildlife and plants; designation of critical habitat for the spikedace (*Meda fulgida*) and the loach minnow (*Tiaroga cobitis*); final rule. *Fed. Reg.* 72:13355–13422.

USGS. 2005. *Water Quality.* USGS Water Science Glossary of Terms.

Utah Dept. of Natural Resources. 2004. *Range-wide conservation agreement for roundtail chub Gila robusta, bluehead sucker Catostomus discobolus, and flannelmouth sucker Catostomus latipinnis.* Division of Wildlife Resources. Salt Lake City.

Valenzuela, D. and G. Ceballos. 2000. Habitat selection, home range, and activity of the white nosed coati (*Nasua narica*) in a Mexican tropical dry forest. *J. Mammalogy* 81:810–819.

Valett, H. M. 1993. Surface-hyporheic interactions in a Sonoran Desert stream—hydrologic exchange and diel periodicity. *Hydrobiologia* 259:133–144.

Valett, H. M., S. G. Fisher, N. B. Grimm and P. Camill. 1994. Vertical hydrologic exchange and ecological stability of a desert stream ecosystem. *Ecol.* 75:548–560.

Van Auken, O. W. 2000. Shrub invasions of North American semiarid grasslands. *Annual Rev. of Ecol. and Systematics* 31:197–215.

van der Heijden, E. W. 2001. Differential benefits of arbuscular mycorrhizal and ectomycorrhizal infection of *Salix repens*. *Mycorrhiza* 10:185–193.

van der Heijden, E. W., T. Boller, A. Wiemken and I. R. Sanders. 1998. Different arbuscular mycorrhizal fungal species are potential determinants of plant community structure. *Ecol.* 79:2082–2091.

van der Heijden, E. W., J. N. Klironomos, M. Ursic, P. Moutogis, R. Stretwolf-Engel et al. 1998. Mycorrrhizal fungal diversity determines plant biodiversity, ecosystem variability and productivity. *Nature* 396:69–72.

van der Heijden, E. W. and T. W. Kuyper. 2001. Does the origin of mycorrhizal fungus or mycorrhizal plant influence effectiveness of mycorrhizal symbiosis? *Plant and Soil* 230:161–174.

van der Heijden, E. W. and M. Vosatka. 1999. Mycorrhizal associations of *Salix repens* L. communities in succession of dune ecosystems. II. Mycorrhizal dynamics and interactions of enctomycorrhizal and arbuscular mycorrhizal fungi. *Can. J. Bot.* 77:1833–1841.

Van Horne, B. 1983. Density as a misleading indicator of habitat quality. *J. Wildllife Manag.* 47:893–901.

Van Hylckama, T. E. A. 1980. *Weather and evapotranspiration studies in a saltcedar thicket, Arizona.* USGS Prof. Paper 491-F.

Vander Wall, S. B. 1990. *Food Hoarding in Animals.* University of Chicago Press.

——. 1997. Dispersal of singleleaf pinon pine (*Pinus monophylla*) by seed-caching rodents. *J. Mammalogy* 78:181–191.

Vandersande, M. W., E. P. Glenn and J. L. Walworth. 2001. Tolerance of five riparian plants from the lower Colorado River to salinity, drought and inundation. *J. Arid Environ.* 49:147–159.

Vanicek, C. D. and R. H. Kramer. 1969. Life history of the Colorado squawfish, *Ptychocheilus lucius*, and the Colorado chub, *Gila robusta*, in the Green River in Dinosaur National Monument, 1964–1966. *Trans. Amer. Fish. Soc.* 98:193–208.

Vannote, R. L., G. W. Minshall, K. W. Cummins, J. R. Sidell and C. E. Cushing. 1980. The river continuum concept. *Can. J. Fish. and Aquatic Sci.* 37:130–137.

Varady, R. G., ed. 1991. *Preserving Arizona's Environmental Heritage.* Arizona Town Hall. Phoenix.

Varady, R. G., M. A. Moote, and R. Merideth. 2000. Water management options for the upper San Pedro basin: Assessing the social and institutional landscape. *Nat. Res. J.* 40:223–235.

Varela-Romero, A. 1992. *Prospeccion de peces nativos en la cuenca del Rio San Pedro, Sonora, Mexico.* Centro Ecológico de Sonora. Hermosillo, Sonora, Mexico.

Varela-Romero, A., C. Galindo-Duarte, E. Saucedo-Monarque, L. S. Anderson, P. L. Warren, S. E. Stefferud, J. A. Stefferud, S. Rutman, T. Tibbitts and J. Malusa. 1992. Re-discovery of *Gila intermedia* and *Gila purpurea* in northern Sonora, Mexico. *Proc. Desert Fishes Council* 22:33–34.

Velasco, A. L. 1993. *Summary report of the San Pedro River mesohabitat mapping survey conducted June–July 1993.* U.S. Fish and Wildlife Service, Ecological Services.

Velez, C. E., L. L. Leslie and S. A. Bonar. 2005. Impact of predation by nonnative fishes on native fishes in the Verde River, Arizona. *Proc. Desert Fishes Council* 35:89.

Verner, J., M. L. Morrison and C. J. Ralph. 1986. *Wildlife 2000: Modeling Habitat Relationships of Terrestrial Vertebrates.* University of Wisconsin Press.

Villnow, J. 1986. Groundwater management under conflicting criteria: A "manual" approach. Master's thesis, University of Arizona.

Vionnet, L. B. and T. Maddock. 1992. *Modeling of groundwater flow and surface water/groundwater interactions in the San Pedro River Basin*—Part I—*Cananea, Mexico to Fairbank, Arizona.* University of Arizona. Dept. of Hydrology and Water Resources HWR 92-010.

Virginia, R. A. 1986. Soil development under legume tree canopies. *For. Ecol. and Manag.* 16:69–79.

Virginia, R. A. and W. M. Jarrell. 1983. Soil properties in a mesquite-dominated Sonoran Desert, California, USA ecosystem. *Soil Sci. Soc. of Amer. J.* 47:138–144.

Vitt, L. J. and R. D. Ohmart. 1978. Herpetofauna of the lower Colorado River: Davis Dam to the Mexican border. *Proc. Western Found. of Vertebrate Zool.* 2:35–72.

Voeltz, J. B. 2002. *Roundtail chub (Gila robusta) status survey of the lower Colorado River basin.* Arizona Game and Fish Dept. Tech. Rep. 186:1–221.

Voeltz, J. B. and R. H. Bettaso. 2003. *2003 status of the Gila topminnow and desert pupfish in Arizona.* Arizona Game and Fish Dept. Tech. Rep. 226:1–124.

Voss, R. S. and L. H. Emmons. 1996. Mammalian diversity in neotropical lowland rainforests: A preliminary assessment. *Bull. Amer. Museum of Nat. Hist.* 230:1–115.

Wagoner, J. J. 1962. Overstocking of the ranges in southern Arizona during the 1870s and 1880s. *Arizoniana* 2:23–27.

Wakeman, T. H. 1997. Engineering leadership in public policy resolution. *J. Manag. in Eng.* 13:57–60.

Walker, L. R. and S. Smith. 1997. Impacts of invasive plants on community and ecosystem properties. Pages 69–86 in *Assessment and Management of Plant Invasions.* J. O. Luken and J. W. Thieret, eds. Springer-Verlag.

Walters, C. J. 1986. *Adaptive Management of Renewable Resources.* MacMillan.

——. 1997. Challenges in adaptive management of riparian and coastal ecosystems. *Conservation Ecology* 1:art1(online) http://www.ecologyandsociety.org/vol1/iss2/art1/

Walther, G., E. Post, P. Convey, A. Menzel, C. Parmesan et al. 2002. Ecological responses to recent climate change. *Nature* 416:389–395.

Walther-Hellwig, K. and R. Frankl. 2000. Foraging habitats and foraging distances of bumblebees, *Bombus* spp. (Hym., Apidae) in an agricultural landscape. *J. Applied Ecol.* 124:299–306.

Walvoord, M. A., F. Phillips, D. Stonestrom, D. Evans, P. Hartsough, B. Newman and R. Striegl. 2003. A reservoir of nitrate beneath desert soils. *Sci.* 302:1021–1024.

Wang, G. L. and D. Schimel. 2003. Climate change, climate modes, and climate impacts. *Ann. Rev. of Environ. and Res.* 28:1–28.

Ward, J. V. 1989. The four-dimensional nature of lotic ecosystems. *J. North Amer. Benth. Soc.* 8:2–8.

Ward, J. V. and K. Tockner. 2001. Biodiversity: Towards a unifying theme for river ecology. *Freshwater Biol.* 46:807–819.

Ward, J. V., K. Tockner, D. B. Arscott and C. Claret. 2002. Riverine landscape diversity. *Freshwater Biol.* 47:517–539.

Wardel, D. A. 2002. *Communities and Ecosystems: Linking the Aboveground and Belowground Components.* Princeton University Press.

Warner, N. J., M. F. Allen and J. MacMahon. 1987. Dispersal agents of vesicular-arbuscular mycorrhizal fungi in a disturbed arid ecosystem. *Mycologia* 79:721–730.

Warren, M. S. 1985. The influence of shade on butterfly numbers in woodland rides, with special reference to the wood white *Leptidea sinapis*. *Biol. Conserv.* 33:147–164.

Warren, P. L. and C. R. Schwalbe. 1985. Herpetofauna in riparian habitats along the Colorado River in Grand Canyon. Pages 347–354 in R. R. Johnson et al. 1985.

Waters, M. R. 1985. Late Quaternary alluvial stratigraphy of White Water Draw Arizona: Implications for regional correlation of fluvial deposits in the American Southwest. *Geol.* 13:705–708.

Waters, M. R. and C. V. Haynes. 2001. Late Quaternary arroyo formation and climate change in the American Southwest. *Geol.* 29:399–402.

Wauer, R. H. 1977. Significance of Rio Grande riparian systems upon the avifauna. Pages 165–177 in R. R. Johnson and Jones 1977.

Webb, R. H. and J. L. Betancourt. 1992. *Climatic variability and flood frequency of the Santa Cruz River, Pima County, Arizona*. USGS Water-Supply Paper 2379.

Webb, R. H. and S. A. Leake. 2006. Ground-water surface-water interactions and long-term change in riverine riparian vegetation in the southwestern United States. *J. Hydr.* 320:302–323.

Webb, R. H., S. A. Leake and R. M. Turner. 2007. *The Ribbon of Green: Change in Riparian Vegetation in the Southwestern United States*. University of Arizona Press.

Weber, T. P., B. J. Ens and A. I. Houston. 1998. Optimal avian migration: A dynamic model of fuel stores and site use. *Evol. Ecol.* 12:377–401.

Weedman, D. A., A. L. Girmendonk and K. L. Young. 1996. *Status review of Gila chub, Gila intermedia, in the United States and Mexico*. Arizona Game and Fish Dept. Tech. Rep. 91:1–120.

Weeks, E. P., H. L. Weaver, G. S. Campbell and B. D. Tanner. 1987. *Water use by saltcedar and by replacement vegetation in the Pecos River floodplain between Acme and Artesia, New Mexico*. USGS Prof. Paper 491-G.

Weiss, S. B. and D. D. Murphy. 1990. Thermal microenvironments and the restoration of rare butterfly habitat. Pages 50–60 in *Environmental Restoration*. J. J. Berger, ed. Island Press.

Weiss, S. J., E. O. Otis and O. E. Maughan. 1998. Spawning ecology of flannelmouth sucker, *Catostomus latipinnis* (Catostomidae), in two small tributaries of the lower Colorado River. *Environ. Biol. of Fishes* 52:419–433.

Welter, J. W., S. G. Fisher and N. B. Grimm. 2005. Nitrogen transport and retention in an arid land watershed: Influence of storm characteristics on terrestrial-aquatic linkages. *Biogeochem.* 76:421–440.

Weltzin, J. F., S. Archer and R. K. Heitschmidt. 1997. Small-mammal regulation of vegetation structure in a temperate savanna. *Ecol.* 78:751–763.

Weltzin, J. F. and G. R. McPherson. 1997. Spatial and temporal soil moisture partitioning by trees and grasses in a temperate savanna, Arizona, USA. *Oecologia* 112:156–164.

Wentz, F. J., L. Ricciardulli, K. Hilburn and C. Mears. 2007. How much more rain will global warming bring? *Sci.* 317:233–235.

Werner, E. E. and M. A. McPeek. 1994. Direct and indirect effects of predators on two anuran species along an environmental gradient. *Ecol.* 75:1368–1382.

West, H. M. 1996. Influence of arbuscular mycorrhizal infection on competition between *Holcus lanatus* and *Dactylis glomerata*. *J. Ecol.* 84:429–438.

Westerhoff, P. and D. Anning. 2000. Concentrations and characteristics of organic carbon in surface water in Arizona: Influence of urbanization. *J. Hydr.* 236:202–222.

White, T. C. R. 1984. The abundance of invertebrate herbivores in relation to the availability of nitrogen in stressed food plants. *Oecologia* 63:90–105.

Whitson, T. D. and L. C. Burrill, eds. 1992. *Weeds of the West.* Western Soc. of Weed Science.

Whittaker, R. H. 1972. Evolution and measurement of species diversity. *Taxon* 22:213–251.

——. 1975. *Communities and Ecosystems.* Macmillan.

Whittier, J. 2004. Groundwater flow model of the Lower San Pedro River Basin for the conservation of riparian habitats. Master's thesis, University of Arizona.

Whittier, J. D. and T. Maddock III. 2004. *A lower San Pedro River basin groundwater flow model.* University of Arizona. Dept. of Hydrology and Water Res. Tech. Rep. Series HWR 04-050.

Widmer, A. M., C. J. Carveth, S. A. Bonar and J. R. Simms. 2006. Upper temperature tolerance of loach minnow under acute, chronic, and fluctuating thermal regimes. *Trans. Amer. Fish. Soc.* 135:755–762.

Wiens, J. A. 1973. Patterns and processes in grassland bird communities. *Ecol. Monog.* 43:237–270.

Wilding, L. P. 2000. Introduction: General characteristics of soil orders and global distribution. Pages 175–183 in *Handbook of Soil Science.* M. E. Summer, ed. CRC Press.

Williams, D. G., J. P. Brunel, S. M. Schaeffer and K. A. Snyder. 1998. Biotic controls over the functioning of desert riparian ecosystems. Pages 43–48 in *Proceedings from the Special Symposium on Hydrology.* E. F. Wood, A. G. Chehbouni, D. C. Goodrich, D. J. Seo and J. R. Zimmerman, eds. Amer. Meteor. Society. Boston.

Williams, J. E., M. P. Dombeck and C. A. Wood. 2003. *From Conquest to Conservation: Our Public Lands Legacy.* Island Press.

Williams, K. S. 1993. Use of terrestrial arthropods to evaluate restored riparian woodlands. *Restor. Ecol.* 2:107–116.

——. 1997. Terrestrial arthropods as ecological indicators of habitat restoration in southwestern North America. Pages 238–258 in *Restoration Ecology and Sustainable Development.* K. M. Urbanska, N. R. Webb and P. J. Edwards, eds. Cambridge University Press.

——. 2000. *Assessing success of restoration attempts: What can terrestrial arthropods tell us?* USGS Open-File Report 00-62.237-244.

Williams, N., R. Minckley and F. Silviera. 2001. Variation in native bee faunas and its implications for detecting community changes. *Conserv. Ecol.* 5:7.

Willmer, P. G. 1982. Microclimate and the environmental physiology of insects. *Advances in Insect Physiology* 16:1–57.

——. 1983. Thermal constraints on activity patterns in nectar-feeding insects. *Ecol. Entomology* 8:455–469.

Wilson, E. O. 1987. The little things that run the world (the importance of conservation of invertebrates). *Conserv. Biol.* 1:344–346.

Wilson, J. B., J. B. Steel, W. M. King and H. Gitay. 1999. The effect of spatial scale on evenness. *J. Veg. Sci.* 10:463–468.

Wilson, J. P. 1995. *Islands in the Desert: A History of the Uplands of Southeastern Arizona.* University of New Mexico Press.

Wilson, K., A. Goldstein, E. Falge, M. Aubinet, D. Baldocchi, P. Berbigier, C. Bernhofer, R. Ceulemans, H. Dolman and C. Field. 2002. Energy balance closure at FLUXNET sites. *Ag. and For. Meterol.* 113:223–243.

Wimp, G. M., G. D. Martinsen, K. D. Floate, R. K. Bangert and T. G. Whitham. 2005. Plant genetic determinants of arthropod community structure and diversity. *Evolution* 59:61–69.

Winograd, I. J. and F. N. Robertson. 1982. Deep oxygenated ground water: Anomaly or common occurrence? *Science, New Series* 216:1227–1230.

Witham, J. H. and E. L. Smith. 1979. Desert bighorn movements in a southwestern Arizona mountain complex. *Desert Bighorn Council Trans.* 1979:20–24.

Wohl, E. and P. Pearthree. 1991. Debris flows as geomorphic agents in the Huachuca Mountains of southeastern Arizona. *Geomorphology* 4:273–292.

Wolden, L. G., J. C. Stromberg and D. T. Patten. 1994. Flora and vegetation of the Hassayampa River Preserve. *J. AZ-Nev. Acad. Sci.* 28:76–111.

Wolman, M. G. and R. Gerson. 1978. Relative scales of time and effectiveness of climate in watershed geomorphology. *Earth Surface Processes* 3:189–208.

Wolman, M. G. and L. Leopold. 1957. *River flood plains: Some observations on their formation.* USGS Prof. Paper 282-C.

Wolman, M. G. and J. P. Miller. 1960. Magnitude and frequency of forces in geomorphic processes. *J. Geol.* 68:54–74.

Wood, D. J., S. G. Fisher and N. B. Grimm. 1990. Pools in desert streams: Limnology and response to disturbance. *J. AZ-Nev. Acad. Sci.* 26:171–182.

Wood, M. L. 1997. *Historical channel changes along the lower San Pedro River, southeastern Arizona.* Arizona Geological Survey Open-File Report 97-21.

Woolsey, N. 1987. Furbearing animals of the San Pedro Preserve. Unpublished report for U.S. Bureau of Land Management, San Pedro Project Office.

Wright, H. S. and P. Martin. 1967. *Pleistocene Extinctions: The Search for a Cause.* Yale University Press.

Wright, J. W. 1993. Evolution of the lizards of the genus Cnemidophorus. Pages 27–81 in *Biology of Whiptail Lizards (Genus Cnemidophorus).* J. W. Wright and L. J. Vitt, eds. Oklahoma Museum of Natural History.

Wright, J. W. and C. H. Lowe. 1965. The rediscovery of *Cnemidophorus arizonae* Van Denburgh. *J. AZ. Acad. Sci.* 3:164–168.

——. 1968. Weeds, polyploids, parthenogenesis, and the geographical and ecological distribution of all-female species of *Cnemidophorus. Copeia* 1968:128–138.

Wright, S. F., M. Franke-Snyder, J. Morton and A. Upadhyaya. 1996. Time-course study and partial characterization of a protein on hyphae of arbuscular mycorrhizal fungi during active colonization of roots. *Plant and Soil* 181:193–203.

Xiong, S., M. E. Johansson, F. M. R. Hughes, A. Hayes, K. S. Richards and C. Nilsson. 2003. Interactive effects of soil moisture, vegetation canopy, plant litter and seed addition on plant diversity in a wetland community. *J. Ecol.* 91:976–986.

Yard, H. K., C. van Riper III, B. T. Brown and M. J. Kearsley. 2004. Diets of insectivorous birds along the Colorado River in Grand Canyon, Arizona. *Condor* 106:106–115.

Yatskievych, G. A. and C. E. Jenkins. 1981. Fall vegetation and zonation of Hooker Cienega, Graham County, Arizona. *J. AZ-Nev. Acad. Sci.* 16:7–11.

Yepez, E. A., D. G. Williams, R. L. Scott and G. Lin. 2003. Partitioning overstory and understory evapotranspiration in a semiarid savanna woodland from the isotopic composition of water vapor. *Ag. and For. Meteor.* 119:43–68.

Yong, W. and D. M. Finch. 2002. *Stopover ecology of landbirds migrating along the Middle Rio Grande in spring and fall.* USDA For. Serv. Gen. Tech. Rep. RMRS-GTR-99.

Yong, W., D. M. Finch, F. R. Moore and J. F. Kelly. 1998. Stopover ecology and habitat use of migratory Wilson's Warblers. *Auk* 115:829–842.

Young, J. A. and C. D. Clements. 2003. Seed germination of willow species from a desert riparian ecosystem. *J. Range Manag.* 56:496–500.

Young, K. L. and M. A. Lopez. 1995. *Fall fish count summary: 1988–1994.* Arizona Game and Fish Dept. Tech. Rep. 81:1–118.

Yuncevich, G. M. 1993. The San Pedro National Conservation Area. Pages 369–372 in Tellman et al. 1993.

Zavaleta, E. S. and J. L. Royval. 2002. Climate change and the susceptibility of U.S. ecosystems to biological invasions: Two cases of expected range expansion. Pages 277–342 in *Wildlife Responses to Climate Change: U.S. Case Studies.* S. H. Schneider and T. L. Root, eds. Island Press.

Zimmerman, J. C., L. E. DeWald and P. G. Rowlands. 1999. Vegetation diversity in an interconnected ephemeral riparian system of north-central Arizona, USA. *Biol. Conserv.* 90:217–228.

Zimmerman, R. L. 1969. *Plant ecology of an arid basin, Tres Alamos–Redington area.* USGS Prof. Paper 485-D.

Index

A

C

G

H

U

V

W

X

About the Editors

Juliet Stromberg is an associate professor in the School of Life Sciences at Arizona State University. She and her graduate students have been studying the ecology of desert rivers for over two decades to understand the processes that shape riparian plant populations, communities, and landscapes. She has authored numerous publications on riparian plant ecology in scientific journals and has served on editorial boards of the journals *Wetlands*, *Restoration Ecology*, and *Ecological Applications*. She has served on various riparian-related boards and commissions including the Southwestern Willow Flycatcher Recovery Team and Arizona Riparian Council. She lives with her husband and their three dogs (plus foster puppies) on a four-acre parcel in south Phoenix that harbors a small orchard, desert shrubs, and riparian mesquite bosque amidst the urban matrix.

Barbara Tellman is a retired senior research specialist from the Water Resources Research Center, the University of Arizona. She is the editor of *Invasive Exotic Species in the Sonoran Region* (University of Arizona Press) and has authored publications on Arizona water policy and law, environmental history of Arizona rivers, and invasive species. She mentors environmental studies students at Prescott College's Tucson campus. She has served on boards and commissions at the local and state levels, dealing with riparian protection, constructed wetlands, wastewater treatment, and water supply. She and her two dogs live in an historic Tucson home near the dry Santa Cruz River, while her two children and three grandchildren live in far less arid areas.

List of Contributors and Their Affiliations

Kenneth J. Bagstad (Arizona State University, School of Life Sciences)

Kathryn J. Baird (University of Arizona, Dept. of Hydrology and Water Resources)

Vanessa B. Beauchamp (Towson University, Dept. of Biological Sciences)

Julio L. Betancourt (U.S. Geological Survey, Desert Laboratory)

L. Arriana Brand (University of Arizona, Dept. of Hydrology and Water Resources)

Paul Brooks (University of Arizona, Dept. of Hydrology and Water Resources)

Anne Browning-Aiken (University of Arizona, Udall Center for Studies in Public Policy)

David J. Cerasale (Cornell University, Department of Ecology and Evolutionary Biology)

Robert W. Clarkson (U.S. Bureau of Reclamation)

Mark D. Dixon (University of South Dakota, Dept. of Biology)

Alexander K. Fremier (University of California Davis, Department of Environmental Design)

Hector Galbraith (University of Colorado Boulder, Institute of Arctic Alpine Research)

Robert Glennon (University of Arizona, College of Law)

David C. Goodrich (U.S. Dept. of Agriculture-Agricultural Research Service)

Douglas M. Green (Arizona State University-Polytechnic, Dept. of Applied Biological Sciences)

Nancy B. Grimm (Arizona State University, School of Life Sciences)

Jeanmarie A. Haney (The Nature Conservancy in Arizona)

Laura E. Hannon (Northern Arizona University, Center for Environmental Sciences and Education)

Tamara K. Harms (Arizona State University, School of Life Sciences)

David B. Harris (The Nature Conservancy)

Richard Hereford (U.S. Geological Survey, retired)

Katherine K. Hirschboeck (University of Arizona, Laboratory of Tree-Ring Research)

Gary Huckleberry (Washington State University, Dept. of Anthropology)

Kathyrn M. Jacobson (Grinnell College, Department of Biology)

Jeffery Johnson (Arizona State University, School of Life Sciences)

Gabrielle Katz (Appalachian State University, Dept. of Geography and Planning)

Linda J. Kennedy (National Audubon Society, Appleton-Whittell Research Ranch)

David J. Krueper (U. S. Fish and Wildlife Service)

Eric W. Larsen (University of California Davis, Landscape Architecture Program)

David Bruce Lewis (Pennsylvania State University, Dept. of Crop and Soil Sciences)

Sharon J. Lite (Arizona State University, School of Life Sciences)

Kathleen Lohse (University of Arizona, School of Natural Resources)

Robert Mac Nish (U.S. Geological Survey, retired)

Thomas Maddock III (University of Arizona, Dept. Hydrology and Water Resources)

Elizabeth Makings (Arizona State University, School of Life Sciences)

Paul C. Marsh (Marsh and Associates)

Charles E. Paradzick (Salt River Project, Environmental Services Department)

Phil Pearthree (Arizona Geological Survey)

Jeff T. Price (California State University Chico, Geological and Environmental Sciences)

Terry D. Rich (U. S. Fish and Wildlife Service)

Brantlee Spakes Richter (North Carolina State University, Dept. Plant Pathology)

Holly Richter (The Nature Conservancy in Arizona)

Leslie Ries (University of Maryland, Dept. Biology)

Philip C. Rosen (University of Arizona, School of Natural Resources)

John L. Sabo (Arizona State University, School of Life Sciences)

John D. Schade (St. Olaf College, Biology Dept.)

Russell L. Scott (U.S. Dept. of Agriculture-Agricultural Research Service)

Candan U. Soykan (Arizona State University, School of Life Sciences)

Jerome A. Stefferud (U.S. Forest Service, retired)

Sally E. Stefferud (U. S. Fish and Wildlife Service, retired)

Juliet C. Stromberg (Arizona State University, School of Life Sciences)

Jean C. Stutz (Arizona State University-Polytechnic, Dept. of Applied Biological Sciences)

Barbara Tellman (University of Arizona, Water Resources Research Center, retired)

Ronald L. Tiller (The Nature Conservancy)

Robert Varady (University of Arizona, Udall Center for Studies in Public Policy)

David G. Williams (University of Wyoming, Dept. of Renewable Resources)

Kathy S. Williams (San Diego State University, Department of Biology)